Guide to Electronic Measurements and Laboratory Practice

PRENTICE-HALL ELECTRICAL ENGINEERING SERIES

Guide to Electronic Measurements and Laboratory Practice

STANLEY WOLF

Dept. of Electrical Engineering
University of California, Santa Barbara

Prentice-Hall, Inc., Englewood Cliffs, New Jersey

Library of Congress Cataloging in Publication Data

WOLF, STANLEY.
Guide to electronic measurements and laboratory practice.

Includes bibliographical references.
1. Electronic measurements. 2. Electronic instruments. I. Title.
TK7878.W64 621.381′043 72–13669
ISBN 0–13–369587–5

10

Printed in the United States of America

Prentice-Hall International, Inc., *London*
Prentice-Hall of Australia, Pty. Ltd., *Sydney*
Prentice-Hall of Canada, Ltd., *Toronto*
Prentice-Hall of India Private Limited, *New Delhi*
Prentice-Hall of Japan, Inc., *Tokyo*

to **CARROL ANN**

Contents

Preface

This is a practical book designed to provide non-experts in electronics with a guide to the use of electronic measuring instruments. Its aim is to teach these readers how to become proficient *users* of such equipment. In this regard, it sets out to explain how to select instruments for various measurement applications, how to evaluate their capabilities, how to connect them together, and how to operate them properly. In addition, descriptions of the terminology, apparatus, and measurement techniques which are unique to the electrical laboratory environment are provided. In sum, the book is meant to serve as a self-contained vehicle for carrying the reader through most electronic measurement tasks.

The presentation is kept at a basic level, so that the book can be used without a prior background in electronics. At this level we can afford the luxury of developing the material in ample detail. Furthermore, many important subjects that tend to be overlooked in more advanced texts can be covered. For example, we place considerable emphasis on the discussion of concepts that are usually regarded by instrumentation experts as self-evident (e.g. grounds and grounding, electrical safety, input and output impedance, loading, shielding against unwanted external signals, ground loops, and impedance matching). Such concepts often remain as puzzles to the beginner unless they are explicitly explained.

Several chapters are allotted to a description of the most common components and quantities encountered in electrical laboratory work. These chapters present practical information dealing with the construction, appearance, and uses of such items as resistors, capacitors, inductors, transformers, relays, batteries, power supplies, cables, switches, connectors, fuses, transducers, and amplifiers.

The remaining chapters are concerned with the use and characteristics of electronic measuring equipment. Each instrument is introduced by explaining the principles that underlie its operation. Then the specifications (i.e. capabilities and limitations) of commercially available models are listed. Next, a general description of how to actually operate and connect the instrument to test circuits is presented. Finally, a list is provided of the most common mistakes committed when using the instrument.

An especially comprehensive treatment is devoted to the operation and capabilities of the oscilloscope. Since the oscilloscope is such a complex and highly versatile measuring tool, a thorough introduction to its use is a necessity.

The book should be especially useful as a text for two types of courses:

First, the introductory course in electrical engineering given to most students of engineering and technology. In this role, the book is meant to serve as a supplemental text. It will supply information on many of the practical aspects of electrical engineering and serve as a manual for the laboratory portion of the course. Because the primary textbooks used in such courses must cover such a wide range of topics, they cannot contain extensive material on either of these areas. However, since this course may be the only one in which an engineering student is taught how to use electronic measuring instruments, it is important that a source of background information on electronic instrument practice be available to him. This will help him gain mastery of a skill which he is likely to employ regularly throughout his career.

Second, the course taught to students majoring in the sciences (physics, chemistry, biology, geology, medicine, etc.) on how to use electronic instruments for measurements. Such courses are becoming more popular as the number of applications of electronic measuring techniques continue to multiply.

In addition, the book is designed to serve as a reference work outside the classroom, and many technical workers who have completed their formal training will find it valuable for refreshing their backgrounds in electronic measurement.

Although many solved example problems are included in the text, no attempt is made to develop an accompanying set of experiments. Most laboratory programs are already based on a set of experiments tailored to fit the existing courses. (What these laboratory programs usually do need is a source of reference material to back up the experimental procedures and instructions.) For instructors who are seeking a set of such experiments, I recommend Cooper and Tait, *A Laboratory Manual To Accompany Electronic Instrumentation and Measurement Techniques* (Prentice-Hall, 1970).

In addition, a set of exercises is provided at the end of each chapter in this text. These exercises can be used by the instructor as required by the course content.

I wish to thank the faculty and staff of the Electrical Engineering Department of the University of California, Santa Barbara, for their encouragement during the time this text was being developed. In particular, the author is deeply grateful to Prof. Joseph Sayovitz, Earl Hall, Donald Zak, and Steven Cowen for their many detailed comments and suggestions. The cooperation of Prof. John Baldwin in allowing portions of the text to be class tested in an introductory electrical engineering course is also appreciated. Finally, I thank my wife Carrol Ann for her continual support and encouragement in this project and Joan Stanton for her excellent typing of the manuscript.

Stanley Wolf
Santa Barbara, California

Guide to Electronic Measurements and Laboratory Practice

1

Language of Electrical Measurements

Electrical measurements and instruments are described with the help of various symbols, conventions, and terms, many of which are unique to electrical science. One should be acquainted with the most common of these terms and symbols before studying the details of electrical instrument operation and measurement techniques. Familiarity with the "language" paves a solid path for continuing study in any subject.

In keeping with this principle, our discussion will begin with a chapter that introduces some of the most general concepts associated with electrical measurements. These particular concepts are part of the vocabulary used to describe all phases of electrical measurement work. Because of the basic nature of these concepts, some readers may already be acquainted with them and should feel free to bypass familiar material. However, if questions arise during later study, the information in this chapter can be used as a reference.

Charge, Voltage, and Current

The concepts of electric charge, electric current, and voltage will be introduced first. Electrical measurements almost always involve determining one or more of these quantities.

ELECTRIC CHARGE

Electrical phenomena arise from the nature of the particles that constitute matter. For example, atoms are composed largely of electrically charged particles. The nucleus of an atom is a central core consisting of protons (which possess a positive charge) and neutrons (which are electrically neutral). Each nucleus is surrounded by a swarm of electrons. Electrons are particles which possess a negative electric charge and a mass that is about 1/2000 of the mass of a proton or neutron. Despite its much smaller mass, the electron's charge is equal in magnitude but opposite in polarity to the charge of a proton. Therefore, an electrically neutral atom must contain an equal number of electrons and protons. A solid body (composed typically of 10^{23} atoms per cubic centimeter) will also be electrically neutral if the number of protons and the number of electrons contained in it are kept equal.

If one or more electrons are removed from an atom, it is no longer neutral, that is, its positive and negative charges are no longer equal. There are now fewer electrons than protons and the atom has a net positive charge. If electrons are removed from many neutral atoms of a substance and are then removed from the boundaries of the body itself, the entire body has a net positive charge. Likewise, if extra electrons are somehow injected into a body of electrically neutral matter, the body acquires a net negative charge.

The unit used to describe an amount of charge is the *coulomb* (abbreviated C). One coulomb is equivalent to the total electric charge possessed by 6.2×10^{18} electrons; therefore, one electron has a charge of 1.6×10^{-19} C.[1]

A body exhibiting a net charge will experience a force when placed in the neighborhood of other charged bodies. The magnitude of such an electrostatic force between two charged bodies is found from Coulomb's law

$$F = \frac{kQ_1Q_2}{d^2} \tag{1-1}$$

where Q_1 is the charge, in coulombs, on one body, and Q_2 is the charge on the other. F is the force in *newtons*,[2] d is the distance, in meters, separating the charged bodies, $k = (4\pi\epsilon_o)^{-1}$ is a constant of value 9×10^9 newton-meters^2/coulomb, and ϵ_o is the permittivity of free space.

If charges possess *like* polarities (i.e., both positive or both negative), the force between them is *repulsive*. If the charges are of opposite polarity, the force is *attractive* (Fig. 1-1). The attractive force which binds the nucleus and electrons of an atom together, arises in part from their opposite electrical charges.

The force experienced by a charged body which is placed near other charged bodies is represented by the strength of the *electric field* at that point. Thus, electric fields surrounding charged bodies can be represented by electric force lines.

[1] The value of the charge possessed by a single electron was discovered about a century after the coulomb unit was established. This is why the relationship between a coulomb and the charge of one electron is not a simple number.

[2] A *newton* is the force required to accelerate a mass of one kilogram at a rate of one meter per second each second. The force of one newton is equal to the force of 0.2248 pounds.

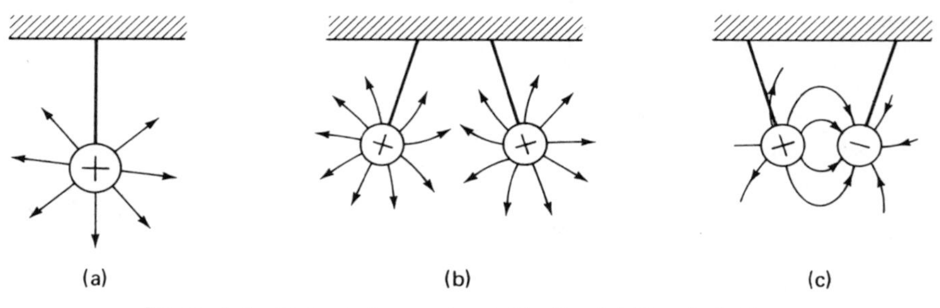

Figure 1-1. Forces between charged bodies (a) Electric field lines around charged body (b) Like charges repel (c) Opposite charges attract

EXAMPLE 1-1

If 9.3×10^{14} electrons are removed from body A and are transferred to body B, what is the net charge on each body? If the two bodies are separated by 10 cm, what is the force of attraction between them?

Solution: (a) One coulomb equals the charge of 6.2×10^{18} electrons. Therefore the charge of 9.3×10^{14} electrons is

$$Q = \frac{9.3 \times 10^{14}}{6.2 \times 10^{18}} = 1.5 \times 10^{-4}\ \text{C} = 0.00015\ \text{C}$$

(b) Since $d = 10$ cm $= 0.1$ m, by using Eq. (1-1), we find

$$F = \frac{kQ_1Q_2}{d^2} = \frac{(9 \times 10^9) \times (1.5 \times 10^{-4})^2}{(0.1)^2} = 2.02 \times 10^4 \text{ newtons}$$

From this example we see that it does not require very much electric charge to exert a rather large force.

VOLTAGE

The concept of *voltage* is related to the concepts of potential energy and work. That is, when electric charges are moved against the force of an electric field, work must be done to move them. This work involves an expenditure of energy. Since the law of conservation of energy says that energy cannot be created or destroyed, the energy used to move charges against an electric field must be converted to another form. This conversion is similar to the energy conversion involved in lifting a weight against the force of gravity. The energy expended in lifting a weight from the floor to a table top is stored by the weight in its location on the table top. The stored energy is called *potential energy* because it has the potential to be released and reconverted to the energy (kinetic) associated with a moving mass. This would occur if the weight were dropped from the table (Fig. 1-2(a)).

If an electric charge is infinitely far away from other electric charges, it

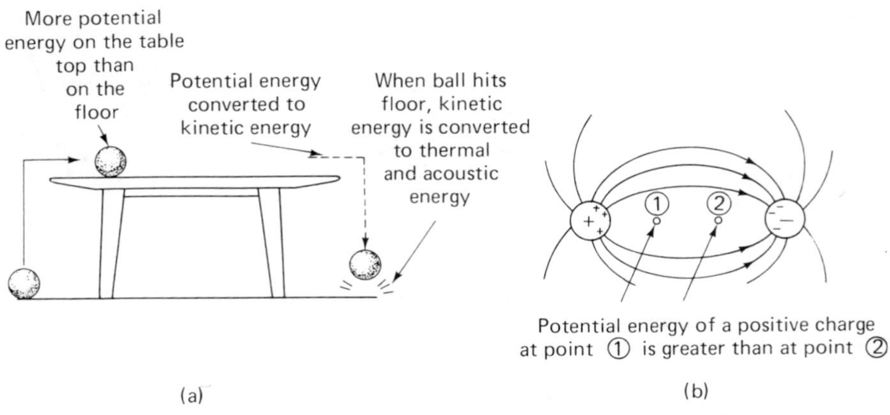

Figure 1-2. Potential difference in gravitational and electric fields (a) Potential energy of a ball on a table (b) Potential energy of a charge in an electric field

will not feel any force of repulsion or attraction due to them. At that point, the *electrostatic potential* of the charge is defined to be zero. If the charge is then brought closer to other electric charges, its electrostatic potential (and potential energy) will change. That is, if the charge is moved closer to a charge of the same polarity, it must be moved against the force of the electric field, and this will increase its *potential energy*. (If the charge is moved closer to charges of opposite polarity, it is moved *with* the force of the electric field, and thus will lose potential energy.) The *electrostatic potential* of any point in space is thus defined as the energy, per unit charge, that would be required to bring the charge to that point from a point of zero electrostatic potential. If a charged body is moved from one point in an electrical system to another, the two points which locate the positions of a charged particle before and after being moved can be characterized by the *difference of* (electrostatic) *potential* between them. This *potential difference* is also called *voltage*, and it indicates how much energy would be acquired or lost (per unit of charge) by a particle as it was moved within the electric field (Fig. 1-2(b)).

Two points in a system (e.g., points 1 and 2) are said to differ by a potential of one *volt* if one *joule*[3] of energy is required to move one coulomb of charge from the one point to the other. This is written mathematically as

$$\text{Potential difference}_{1\text{-}2} = \text{volts} = \frac{\text{joules}}{\text{coulomb}} \tag{1-2}$$

The unit of potential difference is the *volt*.

[3]One *joule* is the work done when a force of one newton must be used to move an object a distance of one meter.

Note once again that it is the difference of potential between *two* points that is being measured by the value of the voltage. However, in many practical systems, one particular potential level is chosen as the reference level and assigned an arbitrary value of zero. Then the potential of all other points of the system are compared to this level. In such systems we can speak of the voltage value of single points in the system, because the zero reference is already implied in each such voltage value.

The planet Earth is the most commonly used zero reference[4] (and is also called *ground*). This means that the potential of the Earth at any point to which an electric circuit is connected is considered to be zero. However, in circuits not connected to the Earth (such as airplanes, automobiles, and ships), another surface or point (such as the airplane fuselage) may be assigned a zero potential level for convenience.

Positive electric charges flow from points of higher potential to points of lower potential, just as water flows downhill because that is the direction toward a lower gravitational potential. This means that if the charges in an electrical conductor are not in motion (i.e., there is no current flowing), all the points in the conductor are at the same potential. When this condition exists, the surface of the conductor is said to be an *equipotential surface.*

ELECTRIC CURRENT

Electric current is defined as the number of charges moving past a given point in a circuit in one second. This definition is written mathematically for a steady current as

$$i = \frac{q}{t} \tag{1-3}$$

where i is the current and q is the net charge which moved past the point in t seconds. The unit of current is the ampere (A), and one ampere indicates that one coulomb of charge is transported past a point in one second. (Since the charge on a single electron is about 1.6×10^{-19} coulombs, a current of one ampere corresponds to a flow of about 6×10^{18} electrons per second.) Smaller currents are more conveniently described by using the milliampere (mA) or the microampere (μA). These measures correspond to 10^{-3} A and 10^{-6} A, respectively.

The moving charges which make up current can take such forms as the motion of electrons in a vacuum or solid or the motion of ions in liquids

[4]We defined the point at which an electric charge is infinitely far from other charges as a point of zero electrical potential (i.e., it does not feel any repulsive or attractive electric forces). Since the Earth is electrically neutral and so very large, any man-made charge will not appreciably affect this neutrality. Hence, for all practical purposes, the Earth can also be defined as having an electrical potential of zero.

or gases. Most of the currents found in electric circuits involve the motion of electrons in solids or vacuums. However, in devices such as batteries or in certain transducers, the current may also involve the motion of positive and negative ions. Nevertheless, in this section we will describe only the phenomenon of current flow in a solid conductor since it is the type of current most frequently encountered in measurement circuits.

Electrical conductors contain essentially free electrons which can move about quite easily within the boundaries of the conductor. When an electric field is applied to the conductor, these electrons move in response to the applied electric field. If the conductor is a wire (as shown in Fig. 1-3) and the electric field is applied in the direction shown, the motion of the free electrons in the wire will be from left to right. The current resulting from the charge motion is said to flow from right to left. (The reason the electron motion and the direction of current flow are in opposite directions will be explained in the next section.) The total number of electrons which move past some cross-sectional area of the wire per unit of time yields the magnitude of the current.

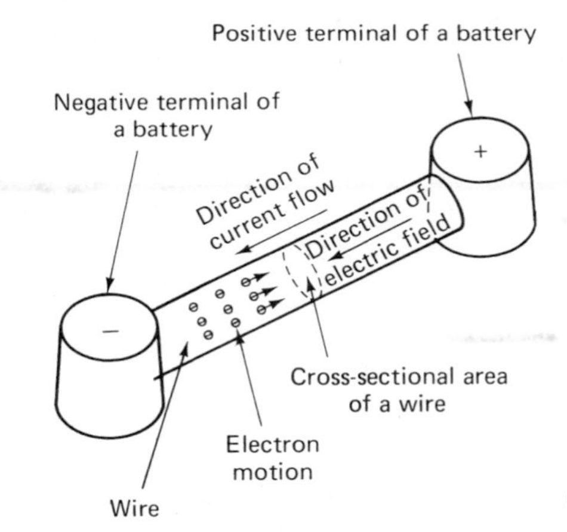

Figure 1-3. Current flow in a wire

When a voltage is applied across a conductor, a current flows almost instantaneously throughout the conductor. The rapidity with which current appears in the entire conductor is due to the velocity with which the electric field propagates within the conductor. (This velocity is effectively the speed of light—approximately 3×10^8 m/s)

The electric field acts on each electron in the conductor. The electrostatic force on the electrons due to the field causes them to acquire a component of average velocity in the direction of the field. Although this component has a rather small value (on the order of 0.001 cm/s), the magnitude of the current can still be quite large because there are about 10^{23} free electrons per cubic centimeter of conductor material.

Conventions for Describing Electrical Quantities

The values of voltage and current which exist at points throughout a circuit are often the quantities we seek to measure. In some applications, however, we need to know more than just the magnitudes of these quantities in order to describe them completely. For example, it may be necessary to know the directions of the currents and the polarities of the measured voltages. There are a number of commonly agreed upon conventions and rules used to describe these directions and polarities.

The rule which fixes the direction of *current flow* with respect to the electron motion in a conductor was formulated after a decision by Benjamin Franklin in 1752. He defined positive charge as flowing out of the *positive terminal* of a voltage source and into the negative terminal. Unfortunately, about 150 years later, it was discovered that *negative charge* moves in the opposite direction, i.e., electrons flow from the *negative terminal* of a battery into the positive terminal. This is exactly opposite to the direction of current flow as defined by Benjamin Franklin. The current described by the electron motion was named *electron current*, and the original (Franklin) concept of current was named *conventional current*. Thus, two conventions are used to describe the direction of current flow in a conductor.

Because of widespread use, it was decided to continue to use the conventional current to describe the direction of current flow. As a result, the current referred to in electric circuits is still most often the conventional current described by Franklin. We will also use the conventional current description. This explains why the current in the wire of Fig. 1-3 is said to move in one direction, while the electron motion is in the opposite direction.

In electric circuits there are usually many branches and loops in which currents are flowing. Each branch or loop may be assigned a direction which defines the direction of a positive conventional current in it. If the conventional current flowing in a branch or loop happens to flow in the same direction as the assigned one, the current is said to be a *positive current*. If the current flows in the direction opposite to the assigned direction, it is given a negative sign (Fig. 1-4). Thus, once all the parts of the circuit are assigned current directions and the sign of the current flowing in a branch is known, its direction of flow is described unambiguously.

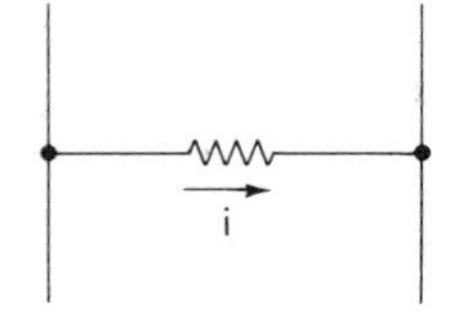

Figure 1-4. A circuit branch is assigned a positive current direction as indicated by the arrow in the picture. If the conventional current in this branch moves in the same direction the current is given a positive sign. If the current flows in the opposite direction, it is given a negative sign.

The elements of a circuit are often marked with plus and minus signs at each terminal. These signs designate the voltage polarities of the terminals. For *passive*[5] two-terminal elements, the plus terminal should generally correspond to the terminal into which a positive current is flowing. For *active*[6] circuit elements, the plus terminal should be the one *from* which a positive current is flowing (as per Franklin's decision). Then, if a positive current flows through any element from a positive to a negative terminal, the voltage across these terminals is called a *voltage drop* (Fig. 1-5(a)). If a positive current flows from the negative to the positive terminal, this is spoken of as a *voltage rise* (Fig. 1-5(b)).

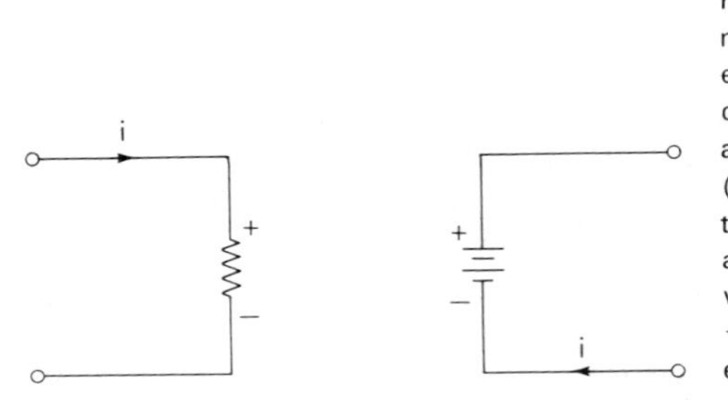

Figure 1-5. (a) If a positive valued current flows into the + terminal of an element, this indicates that a voltage drop exists across the element. That is, if a + charge falls from + to − in a circuit, it is associated with positive released energy. (b) If a positive valued current flows into the − terminal of an element, this indicates a voltage *rise* across the element. In other words, when a + charge rises from − to +, it is associated with positive generated energy.

When describing voltages, currents, and other variables with letter symbols, we use both capital and lowercase letters (i.e., v and V are both used for voltage). When a capital letter is used, it means that the quantity being described is a constant-valued quantity (such as a dc value, an rms value, or the amplitude of a periodic waveform[7]). Lowercase letters, on the other hand, denote instantaneous quantities of time-varying waveforms. As an example, the expression

$$v = A \sin \omega t$$

indicates that v is the instantaneous value of the sinusoidal waveform at time t. A is the amplitude of this waveform; a capital letter is used because the amplitude of the sine wave is a constant quantity. (Note that capital letters are usually used to describe the parameters of passive circuit elements such as R, L, and C. This indicates that these parameters are considered to be constant with time for the sake of analysis.) Table 1-1 is a summary of this notational convention.

[5]Passive elements are circuit elements which are capable only of dissipating or storing energy, but not producing energy. Resistors, capacitors, and inductors are passive elements.

[6]Active elements are circuit elements which serve as energy sources in electric circuits. Examples of active elements include batteries, generators, and transistors.

[7]The terms *rms* and *waveform* are defined later in this chapter.

Table 1-1 Use of Capital and Small Letters as Symbols of Electrical Quantities

Type of Letter Used	Connotation
Capital letters (i.e., V, I, P)	1. dc value. 2. Rms value of ac quantity. 3. Amplitude of ac waveform.
Lowercase letters (i.e., v, i, p)	Instantaneous value of a quantity that varies with time.

Electrical Units

To be able to speak quantitatively about any group of quantities, we must devise a set of *units* which describes a fixed amount of each quantity. The International System of Units (SI), which includes the units used to describe electrical quantities, will be the unit system used in this book. Table 1-2 gives those SI units most commonly used in connection with electrical measurements. A more complete list of electrical units is given in Appendix A.

The SI system was formerly known as the meter-kilogram-second-ampere (MKSA) system because these four quantities are used to define all the other units used by the system. Prior to 1960, when the SI units were adopted as the standard, other systems were also acceptable for use. Therefore, they still may be found in some older publications. These other systems include the CGS (centimeter-gram-second) and the English Gravitational (foot-

Table 1-2 SI Electrical Units

Quantity	Unit	Abbreviation
Length	meter	m
Mass	kilogram	kg
Time	second	s
Current	ampere	A
Temperature	degree Kelvin	°K
Voltage	volt	V
Resistance	ohm	Ω
Capacitance	farad	F
Inductance	henry	H
Energy	joule	J
Power	watt	W
Frequency	hertz	Hz
Charge	coulomb	C
Force	newton	N
Magnetic flux	weber	Wb
Magnetic flux density	webers/meter^2	Wb/m^2

pound-second) systems. Appendix A also contains conversion factors for both CGS and the English units. If these other systems are encountered elsewhere, the conversion factors may be used to convert the units into SI units.

Circuit Models and Ideal Circuit Elements

A circuit model of an electrical device or circuit is an ideal mathematical model that acts approximately like the actual circuit. Circuit models are developed from a set of ideal circuit elements which are connected together to yield the representation of the device being modeled. In some of the discussions that will be undertaken later in the book, the functioning of instruments is described with the help of their circuit models. This approach is used because the circuit model is sometimes easier to understand than the device itself. The circuit model of an instrument only includes information about the essential electrical characteristics. Therefore, many of the details which involve the mechanical or structural aspects of the actual instrument (but which are not necessary for an understanding of its electrical operation) can be ignored if we examine its circuit model alone. In addition, the circuit model usually satisfies a set of relatively simple mathematical relations, and these can be solved to yield quantitative answers as to how the instrument will function.

There are only five ideal electrical elements which are used to develop all linear circuit models. Three of these are *passive* elements (the resistor, the capacitor, and the inductor), and the other two are *active* elements (the voltage source and the current source). This section briefly introduces these ideal elements and the mathematical relations they follow.

Our interest in these ideal elements involves important practical considerations as well as theoretical ones. Many circuit elements used in constructing actual circuits are devices which behave very much like their ideal counterparts. By learning about the behavior of the ideal elements, we also learn a great deal about the actual elements. In later chapters we will discuss how closely such real circuit elements correspond to their ideal models.

In this section we will also discuss some additional symbols which specify various circuit conditions but which are not considered to be circuit elements themselves.

PASSIVE CIRCUIT ELEMENTS

The three *passive* elements used in the development of ideal circuit models are the resistor, the capacitor, and the inductor. They are called passive elements because they can only store or dissipate energy and cannot act as energy sources.

The *ideal resistor* is an element that *dissipates* energy. When a current passes through a resistor, the dissipated energy appears as heat, and this energy cannot be returned to the electrical system in which the resistor is connected. The mathematical equation governing the relation between the current and voltage in a resistor is called *Ohm's law*:

$$v = Ri \tag{1-4}$$

where R is the resistance value of the resistor. This law says that the voltage across the resistor is always proportional to the current flowing through it. The circuit symbol of the resistor is shown in Fig. 1-6(a). Actual resistors are built which closely follow the behavior predicted by Ohm's law, that is, they exhibit a linear current-voltage relationship. Resistors are constructed with wires or strips of special materials which possess the specific values of resistance required. A more detailed discussion on the construction of various resistor types is undertaken in Chapter 5.

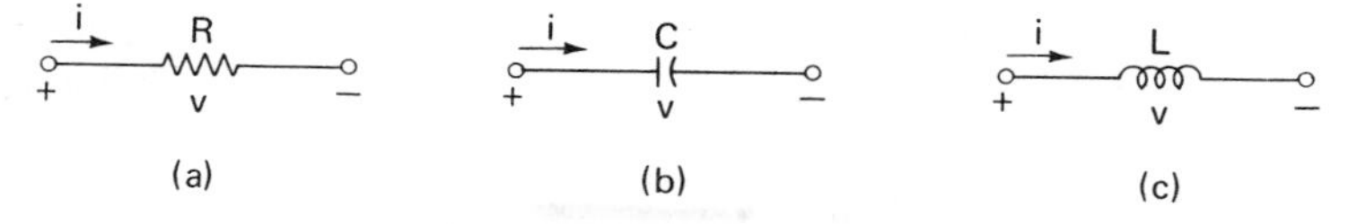

Figure 1-6. Symbols for the ideal passive circuit elements (a) Resistor (b) Capacitor (c) Inductor

The *ideal capacitor* is an element which stores energy in the form of an electric field. It can also be visualized as an element that stores energy in the form of separated charges. The ratio of the amount of charge (q) stored in a capacitor to the voltage (v) across the capacitor is called its *capacitance*, C:

$$C = \frac{q}{v} \tag{1-5}$$

In ideal capacitors this ratio remains constant as q and v change. Most common capacitor structures consist of two parallel metal plates separated by an insulating space.

If the voltage across a capacitor changes with time, a current flows onto and out from its plates. The instantaneous value of this current depends on how quickly the voltage is changing with time (dv/dt) and on the value of the capacitance of the structure.

$$i = C\frac{dv}{dt} \tag{1-6}$$

When a capacitor is charged to a voltage, v, the energy it is storing (w) is calculated from

$$w = \tfrac{1}{2}Cv^2 \tag{1-7}$$

If the ideal capacitor is discharged and v goes to zero, all the energy it was storing is returned to the circuit by the current flowing out from the capacitor. The circuit symbol of the capacitor is shown in Fig. 1-6(b). Many actual capacitors come quite close to behaving like their ideal models.

The *ideal inductor* is an element that stores energy in the form of a magnetic field. Since a moving charge always creates a surrounding magnetic field, an inductor that carries current stores energy in this magnetic field. If the source providing the current that flows in an inductor is shut off, the energy stored in the magnetic field is returned to the circuit. The returning energy imparts motion to the charge-carriers in the conductor from which the inductor is constructed. Thus a transient current briefly continues to flow in an inductor even after its energy source is removed. This tendency[8] of an inductor to continue supporting a current flow is called *inductance*, L.

If the current in an inductor changes with time, a voltage appears across the inductor. The ratio of the voltage across an inductor to the rate at which current is changing in the inductor determines the value of its inductance:

$$L = \frac{v}{di/dt} \tag{1-8}$$

The circuit symbol of the inductor is shown in Fig. 1-6(c). The energy stored by an inductor depends on the inductance and on the magnitude of the current flowing in the inductor:

$$w = \tfrac{1}{2}Li^2 \tag{1-9}$$

The electrical characteristics of actual inductors do not compare as well with the ideal model as do the characteristics of the other two passive elements. This is partly due to the fact that real inductors always contain some resistance, a property not found in ideal inductors. The resistance exists because the changing current in the inductor must flow in a real wire, and all real wires contain resistance. (Actual inductors are constructed by winding a wire in the shape of coil. Sometimes the coil is wound on a *core* of magnetic material to increase the inductance value of the element.)

IDEAL ACTIVE ELEMENTS

The two ideal active elements are the *voltage source* and the *current source*. Unlike the passive elements, active elements are sources of electrical energy. Furthermore, ideal active sources contain no resistive, capacitive, or inductive effects.

[8]The same tendency also makes the inductor resist currents that try to start flowing in it. In this case the inductor extracts and stores energy (as a magnetic field) from the source which moves the charge through it.

Actual energy sources such as batteries, generators, solar cells, and transistors are all approximated by one of the two ideal active elements. (Sometimes additional passive elements are also connected to the ideal active element to improve the approximation between an actual energy source and its circuit model.)

The chief characteristic of an *ideal voltage source* is that it supplies a voltage output which is independent of its required current output. In other words, an ideal voltage source can continue to supply the same voltage output regardless of the amount of current drawn from it. This characteristic of the ideal voltage source implies that its impedance (i.e., resistance) is zero. In fact, if the voltage output is reduced to zero, the voltage source acts like a short circuit.

The circuit symbol of the voltage source is given in Fig. 1-7(a). The plus sign indicates the positive voltage terminal of the source and the minus sign indicates the negative terminal. If the voltage source is a dc voltage source, the symbol in Fig. 1-7(b) is sometimes used instead of that in Fig. 1-7(a).

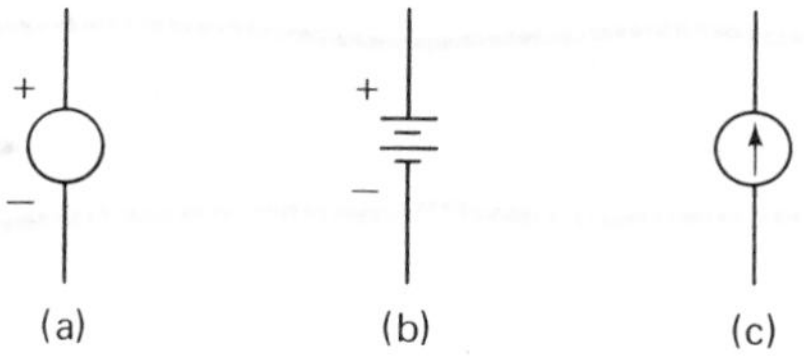

+ − (a) + − (b) (c)

Figure 1-7. Symbols for the ideal active elements (a) Voltage source (b) Alternate symbol for dc voltage source (c) Current source

As an example of an actual voltage source, consider the lead storage battery used in automobiles. For current outputs below about 50 A, the automobile battery closely approximates an ideal voltage source, that is, from zero to about 50 A, the voltage output of the automobile battery remains quite constant. If the current output is increased beyond this limit, the voltage output of the battery drops off. (At that point, the ideal voltage source is no longer as good an approximation to the battery.) This decrease in voltage output can be observed if an automobile engine is started while the headlights are on. The excessive current drawn by the engine starter-motor causes the voltage output to drop below 12 V (as evidenced by the dimming of the headlights).

The *ideal current source* produces a current output which is independent of the voltage required of it. Hence, an ideal current source continues to supply the same current regardless of the voltage drop across its terminals. If an actual energy source provides a relatively constant current over a specific voltage range, the ideal current source provides a better approximation of its operation than does the voltage source. Examples of actual elements that serve as current sources are the pentode vacuum tube and the transistor. Both of these elements can put out a relatively constant current while the voltage across them changes over a limited range.

The circuit symbol of the ideal current source is shown in Fig. 1-7(c).

The direction of the arrow indicates the direction of the current flow from the source. Since v can change while i remains constant, the impedance of the current source is *infinite*.

ADDITIONAL IDEAL-CIRCUIT SYMBOLS

Some additional circuit symbols should be mentioned at this point to complete the discussion on ideal elements. These symbols represent circuit conditions in ideal-circuit models (and actual circuits), but are not themselves considered to be circuit elements.

The short-circuit symbol of Fig. 1-8(a) signifies the existence of a current path across which no voltage drop exists. All the current that can be supplied will flow through a *short circuit*. However, a real "short" always exhibits at least a small voltage drop.

The *open circuit* of Fig. 1-8(b) indicates a condition of infinite resistance to current flow. There may be a voltage across an open circuit, but no current can flow. The *ideal switch* of Fig. 1-8(c) is a device which changes a portion of a circuit from an open to a short circuit and vice versa. However, other than the function of connecting or disconnecting portions of a circuit, this device has no other electrical characteristics. The *ideal diode* of Fig. 1-8(d) is a circuit element that allows current to flow through itself in one direction only. When current flows in this preferred direction, the diode appears as a short circuit. If current attempts to flow in the opposite direction, the diode appears as an open circuit. The short-circuit direction of the diode is the direction in which its arrowhead points. (Some actual devices have characteristics that approximate those of ideal diodes, but the comparison is not quite exact.) The *ideal-meter* movement of Fig. 1-8(e) is a device which can monitor current or voltage in a circuit without disturbing their values. Real meter movements are not completely able to match this ideal.

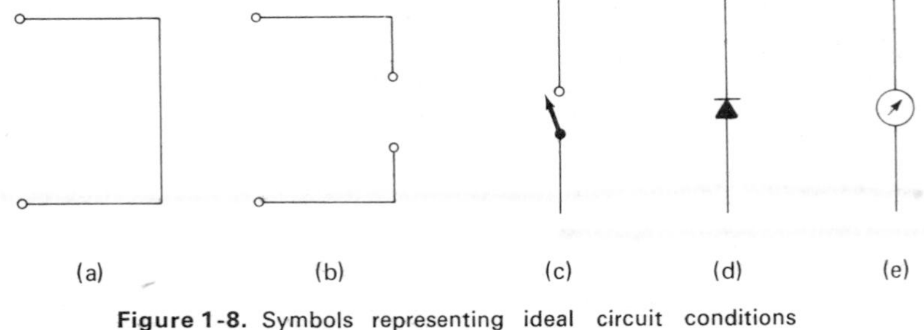

Figure 1-8. Symbols representing ideal circuit conditions (a) Short circuit (b) Open circuit (c) Switch (d) Diode (e) Meter

Electrical Diagrams

Standard diagrams are used to record the configurations of electrical components that make up an electric circuit or instrument. Although actual sketches of the components, wires, switches, etc., could be made to represent the appearance of a circuit on the workbench, such representations are usually not very satisfactory (e.g., Fig. 1-11(a)). The time required to draw sketches is relatively long, and the resultant illustrations are not standard (and this can lead to confusion or ambiguity of interpretation).

The standard diagrams used to represent electrical circuits overcome these limitations by using a form of electrical shorthand. That is, a set of standard, easy to draw symbols, along with rules for their use, are associated with each type of diagram. Utilizing such shorthand diagrams, one can quickly and accurately set down all the necessary information about a circuit. As long as the reader of the diagram knows the meanings of the symbols, he can clearly and rapidly understand the conveyed information.

This section provides a brief description of the most common types of electrical diagrams and the applications where each may be used most effectively. It also discusses the manner in which each type of diagram is read.

The most commonly utilized types of diagrams are the following:

1. Circuit or schematic diagrams.
2. Equivalent circuit diagrams.
3. Block diagrams.

Before we discuss their various characteristics, let us first define the different sections of an electrical circuit and some approximations used in constructing electrical diagrams.

NOMENCLATURE OF ELECTRICAL CIRCUITS

Each element of a circuit possesses two or more *terminals* which serve to connect it to other elements or to act as points at which measurements can be made. A *node* is the point or junction where three or more circuit element terminals are connected (Fig. 1-9(a)). If the junction joins only two elements, the point is called a *secondary node*. (Secondary nodes are rarely defined or used in describing circuits. They are mentioned here only to keep the other circuit definitions consistent.)

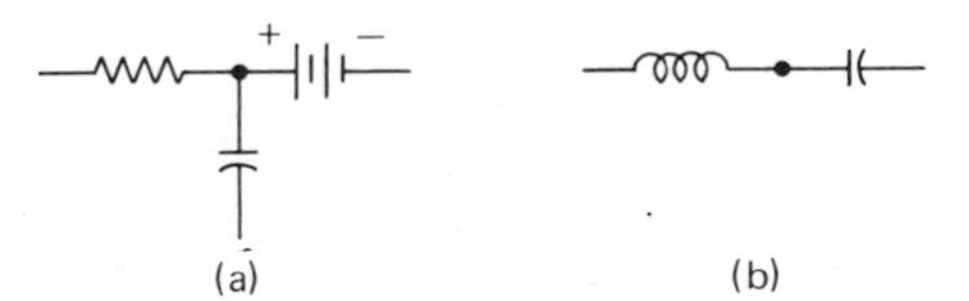

Figure 1-9. (a) Node (b) Secondary node

A *branch* of a circuit consists of a single circuit element (Fig. 1-10(a)) or of a series connection of elements between two *nodes* (Fig. 1-10(b)). (Note that one or more secondary nodes may be found along a branch.) A *loop* is a closed path in which current can flow (Fig. 1-10(c)).

An *electric circuit* is formally defined as a connection of circuit elements

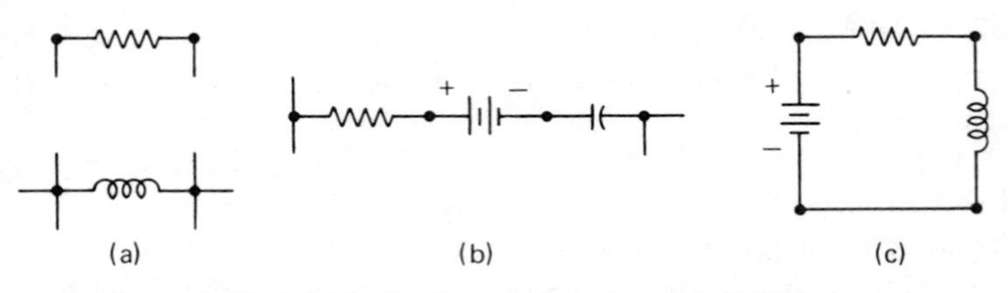

Figure 1-10. (a) Single element branches (b) Multielement branch (c) Loop

that forms at least one closed loop in which current is able to flow. Thus the system of a battery, switch, lamp, resistor, and connecting wires of Fig. 1-11(a) would fit this description if the switch were closed.

In all circuit diagrams, an approximation is used which implies that wires and connections have no role in influencing circuit behavior except to transport current. In other words, wires and connections are treated as if they are perfect conductors and play no other part in the circuit's behavior. In reality, wires and connections do possess some resistance, but the value is usually much lower than the resistance of the other circuit elements. Therefore this resistance can often be neglected without introducing significant error. (We will see in Chapter 5 how much resistance actual wires do possess.)

SCHEMATIC DIAGRAMS OF ELECTRICAL CIRCUITS

Figure 1-11(b) is a *circuit diagram* or *schematic* of the circuit shown in Fig. 1-11(a). We can see that the only information included in a schematic

Figure 1-11. (a) Actual circuit (b) Circuit or schematic diagram (c) Equivalent circuit diagram

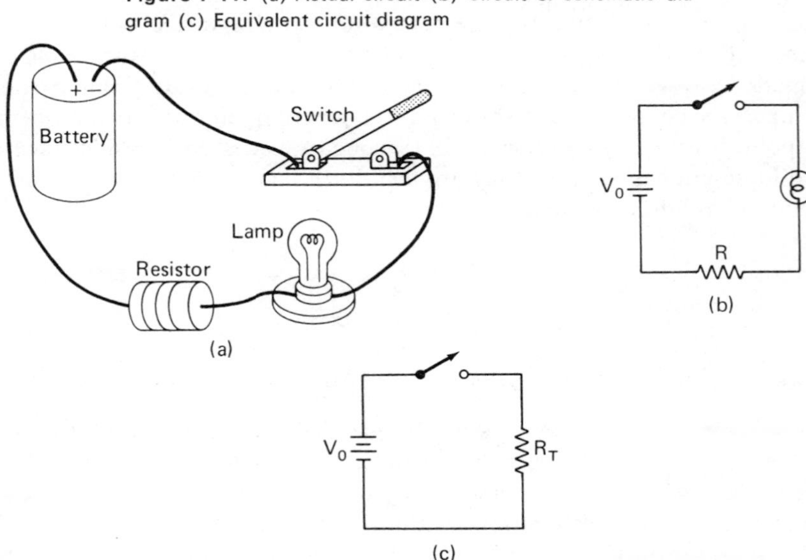

pertains to the electrical operation of the circuit or instrument. Excluded are all such nonelectrical items as the external cases or packaging of the components, mechanical supports, dials, etc.

Schematics can be used to build a replica of actual circuits or to help locate faults and malfunctions in existing circuits. In complex circuits, one can use the schematic diagram as a guide in tracing a signal through the circuit. Without such a "map," it ordinarily takes much longer to follow complicated connections and combinations of elements in the search for a source of trouble. Circuit schematics can also be read by the user of an instrument as a means of gaining a greater understanding of the instrument's operation.

In a schematic, each element of the real circuit is represented by a standard pictorial symbol. The value and type of each circuit component are also included in the diagram (if this information is applicable). Therefore, one of the keys to understanding a circuit schematic is to know the meaning of each of the standard symbols used in them. Figure 1-12 shows most of the symbols used in schematics and the components to which they correspond.

The symbols shown in a schematic are usually drawn in a fashion that makes it convenient to think about the circuit components in a functional manner. But the schematic does not actually show or contain specific information as to where the elements might be located in an actual layout. For example, the separate resistors and capacitors used in amplifier circuits are shown in a schematic so that it is easy to see the circuit role they play. When one looks at an actual layout, this functional coordination is not as evident, i.e., these elements might instead be located at a spot where there is sufficient room to put them. Thus, in tracing an actual layout with the help of a schematic, one must follow the wire connections and be able to recognize circuit components as he comes to them.

When reading a schematic, we usually start at the upper left corner of the diagram. This part of the diagram ordinarily contains the input point to the device being described. From this input point we read the diagram from left to right. At each block of elements, it is best to stop and form a mental image of how the electrical signal or quantity is altered by the group of elements. The altered signal is then used as the input to the next block. By continuing this thought process until the output terminals of the device are reached, we can get a good grasp of the electrical operation of the instrument represented by the diagram.

EQUIVALENT-CIRCUIT DIAGRAMS

The equivalent-circuit diagram of Fig. 1-11(c) is closely related to the idea of the circuit model. The circuit model of a real circuit is a mathematical model which approximates the actual behavior of the real circuit. An equivalent-circuit diagram can be constructed from a schematic by replacing the pictorial symbol of each component by an *equivalent-circuit model.*

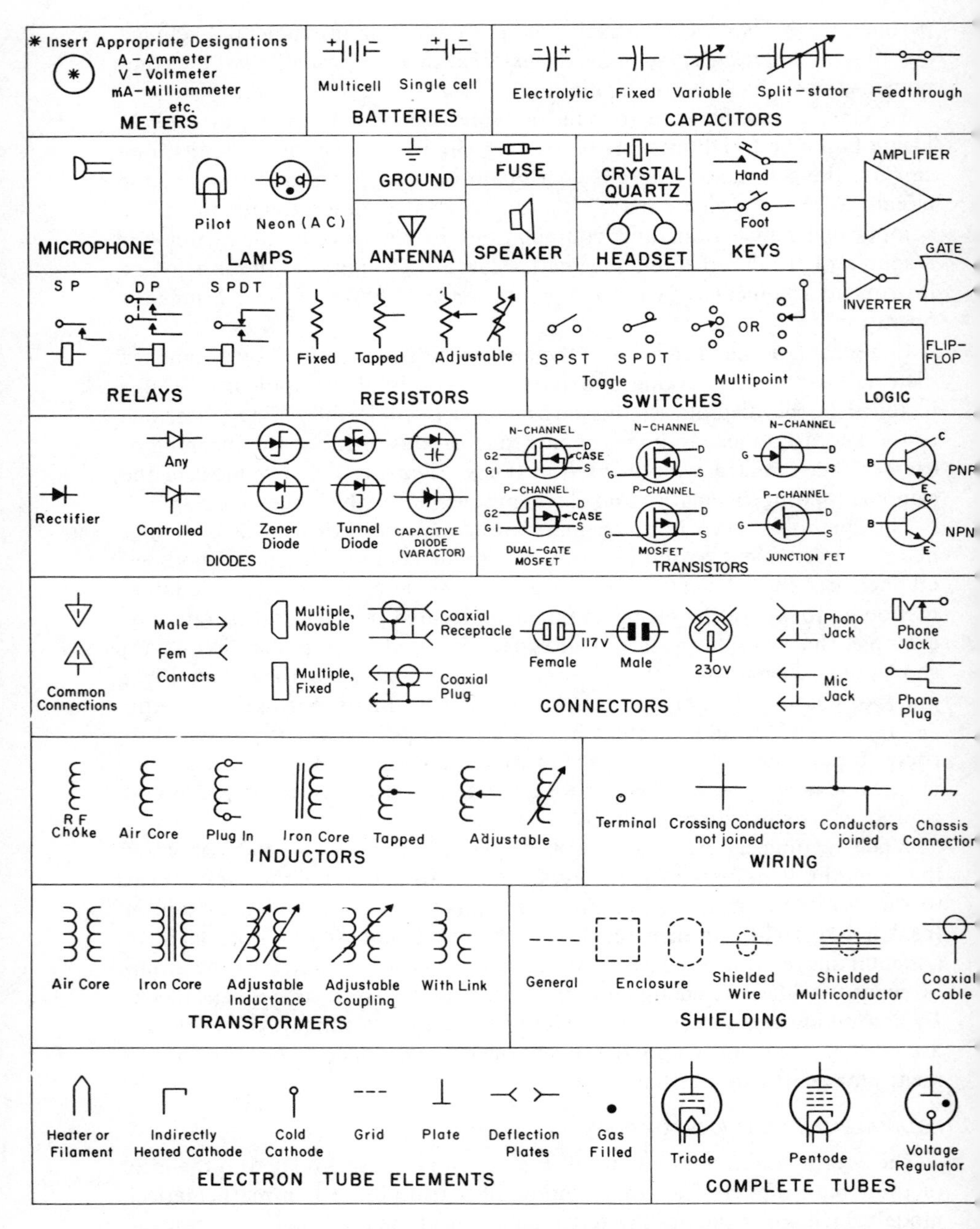

Figure 1-12. Schematic symbols used in circuit diagrams (Courtesy of American Radio Relay League)

Since circuit models are developed from the five ideal-circuit elements and the extra symbols that designate ideal-circuit conditions, equivalent-circuit diagrams are also constructed by using the symbols for these ideal elements. Figure 1-13 shows a few schematic symbols of actual elements along with some possible equivalent-circuit models.

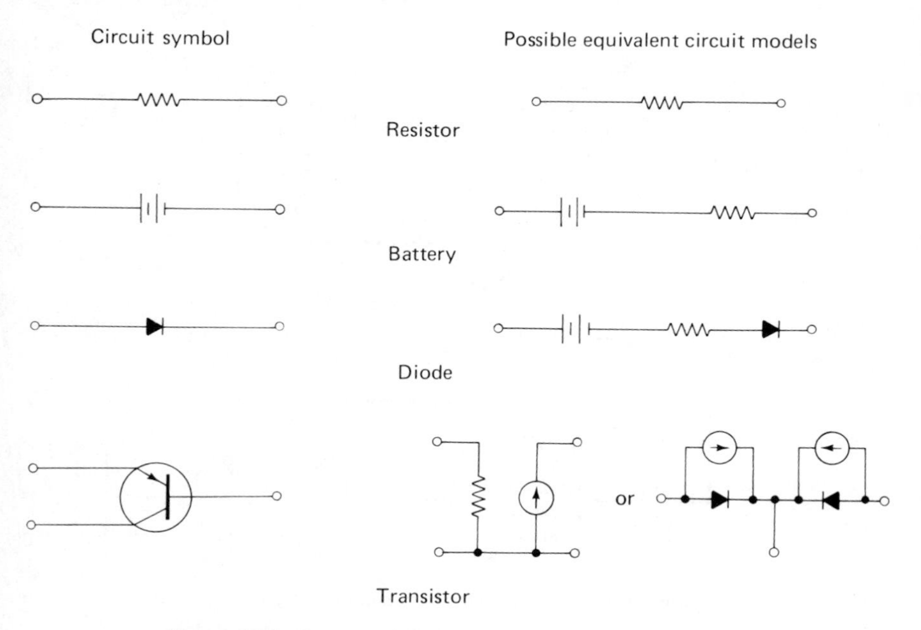

Figure 1-13. Some equivalent circuit models used in equivalent circuit diagrams

Since each component in a circuit model is an ideal element, we know that these components must obey specific mathematical relations. Therefore, the equations governing an actual circuit's operation can be set down from its equivalent-circuit model. Constructing an equivalent-circuit diagram from a circuit schematic is one way to develop the equivalent-circuit model. In this way, an equivalent-circuit diagram can be used to help analyze an actual circuit's behavior. If the equivalent-circuit models (as described by the diagram) provide a close approximation to each of the actual device characteristics, the equations developed with the help of the diagram can predict the behavior of the circuit quite accurately.

BLOCK DIAGRAMS

Block diagrams are often used to help describe the overall operation of rather complex instruments or measurement systems. The major components or subassemblies of the system are represented as *blocks*, and the interrelationships between them are easily seen. Figure 1-14 is a block diagram of a stereophonic record player. The diagram is read from left to right. In the diagram of Fig. 1-14, we start at the stereo phonograph cartridge block and follow the arrows. (Note that the power supply furnishes power to all the other amplifier blocks.)

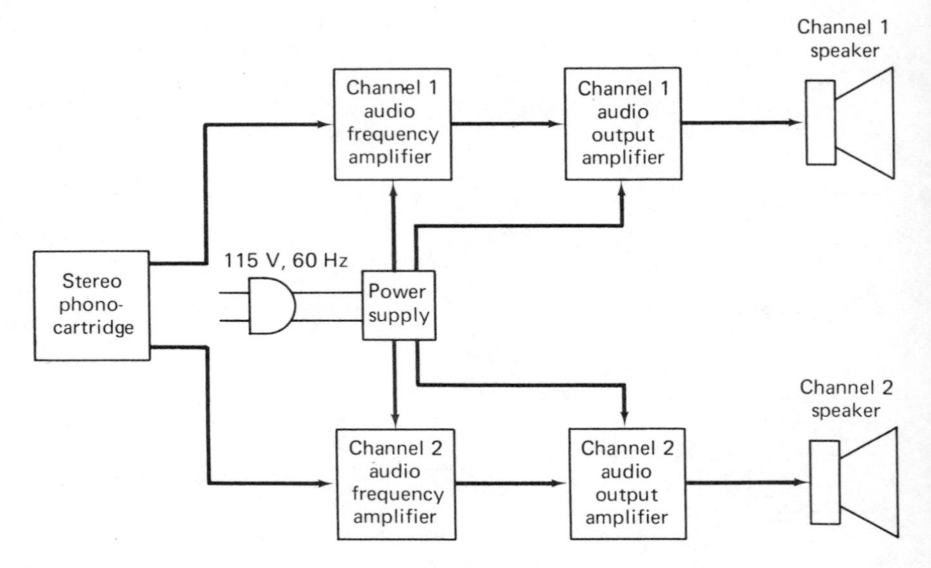

Figure 1-14. Block diagram of stereo record player

We can see that such a block diagram allows us to trace the path of a signal through the system. It also gives a concise and overall view of the operation and functioning of the system. However, no information is given on detailed component connections or wiring.

Sine Waves, Frequency, and Phase

The instantaneous values of electrical signals can be graphed as they vary with time. Such a graph of the signal is called its *waveform*. Signal waveforms are analyzed and measured in many electrical applications.

Generally speaking, if the value of a signal waveform remains constant with time, the signal is referred to as a *direct-current* (*dc*) *signal*. If a signal is time-varying and has positive and negative instantaneous values, the waveform is known as an *alternating-current* (*ac*) *waveform*. If the variation is continuously repeated (regardless of the shape of the repetition), the waveform is called a *periodic waveform*.

The most common periodic waveform encountered in electrical systems is the sinusoid. Figure 1-15 shows an example of a sinusoid. The mathematical expression for this waveform is

$$v = V_o \sin \omega t \tag{1-10}$$

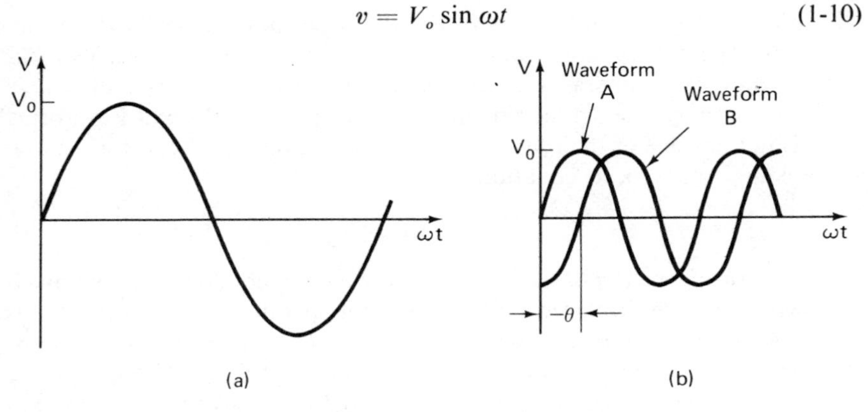

Figure 1-15. (a) Sinewave (b) Waveform

V_o is called the amplitude of the sine wave and denotes the maximum value of the sine function. The frequency, f, of the sine wave (and of other periodic waveforms) is defined as the number of cycles traversed in one second. Accordingly, the frequency is measured in cycles per second, or hertz (Hz). The time duration (in seconds) of one cycle is called the *period*, T, of the waveform. The frequency and period of the waveform are related by the expression

$$f = \frac{1}{T} \tag{1-11}$$

In addition, one cycle of a waveform is defined as spanning 2π radians. Thus, if 2π is multiplied by the frequency, we obtain the angular frequency (ω) of the sine wave:

$$\omega = 2\pi f = \frac{2\pi}{\mathrm{T}} \tag{1-12}$$

The units of ω are *radians per second.*

In electrical work, there are several frequency ranges which are so commonly used that they are given special names. The three most widely used frequency ranges which we will encounter in this text are the following:

1. 5 Hz to 400 Hz (low frequencies).
2. 20 Hz to 20,000 Hz (audio frequencies or af).
3. 500 kHz to 100 MHz (radio frequencies or rf).

Although the numbers given in these frequency ranges are only approximate, they provide a reasonable indication of the frequencies under discussion when the abbreviations above are used.

In Fig. 1-15(b) two sine waves of equal frequencies are depicted on one time axis. The equations for both of these waveforms cannot be identical because they each possess a different instantaneous value from the other at any given time. The manner in which the equations of the two waveforms differ is in the value of their *phase angles.* If we define waveform A as having a phase angle of zero, its equation is written as

$$v = V_o \sin \omega t \tag{1-10}$$

Then the waveform B will have a phase angle θ which indicates how much the waveforms are displaced from each other in time. If waveform B has a zero value (for a positive slope) that occurs *later* in time than the zero value (for a positive slope) of waveform A, then waveform B is said to *lag* waveform A, and vice versa. For example, in Fig. 1-15(b) we can say that waveform B lags waveform A by θ degrees. The equation of waveform B is then written as

$$v = V_o \sin (\omega t - \theta) \tag{1-10a}$$

where the minus sign of θ indicates that the waveform of Eq. 1-10(a) lags the waveform of Eq. 1-10 by an angle θ.

Average and RMS Values

If the signals applied to a circuit are exclusively dc signals, it is relatively easy to calculate such quantities as the number of amperes flowing in the circuit or the energy dissipated by the components of the circuit over a period of time. In addition, one measurement of the dc waveform at any time will reveal all that must be known about the quantity it represents. However, the magnitudes of electrical quantities usually vary with time rather than keep constant values. If a signal is time-varying, its waveform is no longer as simple as the dc waveform shown in Fig. 1-16(a). Instead the waveform has some time-varying shape. The variation may be periodic (as in Fig. 1-16(b) and 1-16(c)) or it may be a more random variation.

When waveforms possess time-varying shapes, it is no longer sufficient to measure the value of the quantity they represent at only one instant of time. It is not possible from one measurement to determine all that must be known about the signal. However, if the shape of a time-varying waveform can be determined, it is possible to calculate some characteristic values of the waveform shape (such as its average value). These values can be used to compare the effectiveness of various waveforms with other waveforms, and

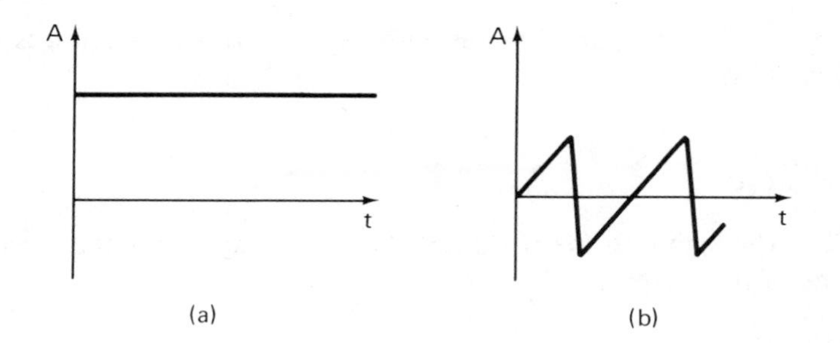

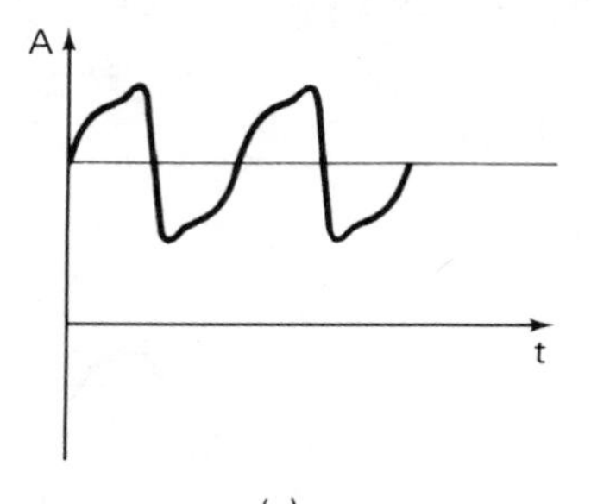

Figure 1-16. Signal waveforms (a) dc waveform (b) Time-varying periodic waveform (c) Time-varying periodic waveform superimposed on a dc level

they can also be used to predict the effects that a particular signal waveform will have on the circuit to which it is applied.

The two most commonly used characteristic values of time-varying waveforms are their average and their root-mean-square (rms) values. We will see how and why both of these values are determined.

AVERAGE VALUE

The meaning of the *average value* of a waveform can best be understood if we use a current waveform as an example. The average value of a time-varying current waveform over the period, T, is the value that a dc current would have to have if it delivered an equal amount of charge in the same period, T. Mathematically, the average value of any periodic waveform is found by dividing the area under the curve of the waveform in one period, T, by the time of the period. This can be written as

$$A_{av} = \frac{\text{area under the curve}}{\text{length of the period (seconds)}} \tag{1-13}$$

where A_{av} is the average value of the waveform.

For the readers who are familiar with calculus, this expression is written more generally as

$$A_{av} = \frac{1}{T}\int_0^T f(t)\,dt \tag{1-14}$$

where T is the length of the period of the curve and $f(t)$ is the equation of the shape of the waveform.[9]

EXAMPLE 1-2

Find the average value of the curves given in Fig. 1-17.

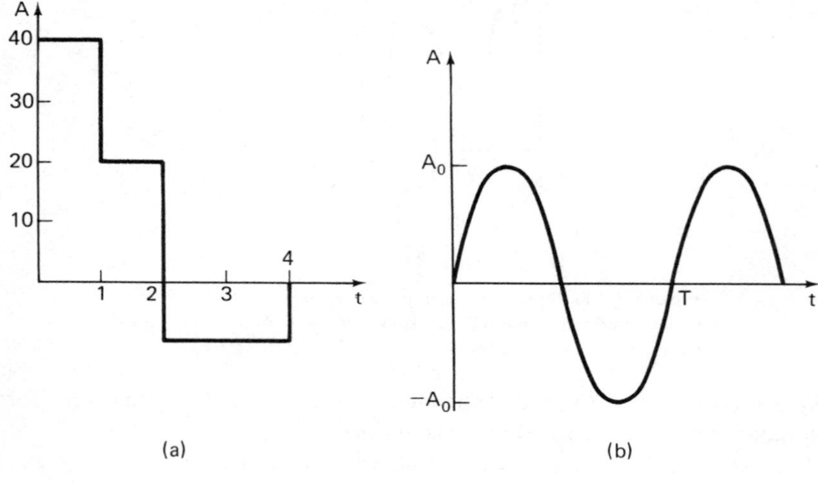

Figure 1-17

Solution: The average value of the curve of Fig. 1-17(a) is found from Eq. (1-13)

(a) $$A_{av} = \frac{(40 \times 1) + (20 \times 1) - (10 \times 2)}{4} = \frac{40}{4} = 10$$

(Note that the area under the curve from $t = 2$ to $t = 4$ is a negative area.)

(b) In this case the curve is a sinusoid and the area under the curve is found using Eq. (1-14).

$$A_{av} = \frac{1}{T}\int_0^T A_o \sin\frac{2\pi t}{T}\,dt = \frac{A_o}{T}\int_0^T \sin\frac{2\pi}{T}t\,dt$$

[9]Note that it is not absolutely necessary to know calculus to continue using this text; but the problem of calculating average values is an example of one area where calculus is almost indispensable in order to obtain analytical results. For those who are not familiar with calculus, the average values of several common waveforms encountered while making measurements are presented in Fig. 1-18.

$$= -\frac{A_o}{2\pi}\left[\cos\frac{2\pi t}{T}\right]_0^T$$

$$= -\frac{A_o}{2\pi T}[(1) - (+1)] = 0$$

$$A_{av} = 0$$

This important result shows that the average value of a purely sinusoidal waveform is zero!

ROOT-MEAN-SQUARE (rms) VALUES

The second common characteristic value of a time-varying waveform is its *root-mean-square* (*rms*) value. In fact, the rms value is used more often than the average value to describe electrical signal waveforms. The major reason for this is that the average value of symmetrical[10] periodic waveforms is zero. (Example 1-2 showed that this was indeed the case for a purely sinusoidal waveform.) A value of zero certainly does not provide much useful information about the properties of a signal. In contrast, the rms value of a waveform does not suffer from this limitation.

The rms value of a waveform refers to its *power delivering* capability. In connection with this interpretation, the rms value is sometimes called the *effective value.* This name is used because the rms value is equal to the value of a dc waveform which would deliver the same power if it replaced the time-varying waveform.

To determine the rms value of a waveform, we first square the magnitude of the waveform at each instant. (This makes the value of the magnitude positive even when the original waveform has negative values). Then the average (or mean) value of the squared magnitudes is found. Finally, the square root of this average value is taken to get the result. Because of the sequence of calculations that is followed, the result is given the name root-mean-square. The situation which led to an average value of zero for some waveforms is avoided because the squaring process makes the entire quantity positive before the mean is taken.

Mathematically, the rms value of a waveform is written as

$$A_{rms} = \sqrt{\langle f(t)^2 \rangle} \tag{1-15}$$

where the symbol $\langle\ \ \rangle$ means that the average of the quantity within the brackets is taken. For a given waveform, $f(t)$, the rms value is found by using the expression

$$A_{rms} = \sqrt{\frac{1}{T}\int_0^T [f(t)]^2\, dt} \tag{1-16}$$

where T is the length of one period of the waveform (in seconds).

[10] By *symmetrical*, we mean in this context that a periodic waveform has equal positive and negative areas.

EXAMPLE 1-3

Find the rms value of the sinusoidal waveform of Fig. 1-17(b).

Solution:

$$A_{\text{rms}} = \sqrt{\frac{1}{T}\int_0^T [f(t)]^2\,dt} = \sqrt{\frac{1}{T}\int_0^T A_o^2\left(\sin\frac{2\pi}{T}t\right)^2 dt}$$

$$= \sqrt{\frac{A_o^2}{T}\left[\frac{t}{2} - \frac{T\sin\left(\frac{4\pi}{T}t\right)}{8\pi}\right]_0^T} = \frac{A_o}{\sqrt{2}}$$

Therefore the rms value of sinusoidal waveforms is

$$A_{\text{rms}} = \frac{A_o}{\sqrt{2}} = 0.707A_o$$

When referring to sinusoidal signals, it is most common to describe them in terms of their rms values. For example, the 115-V, 60-Hz voltage which is delivered by electric power companies to domestic consumers is really a sinusoidal waveform whose amplitude is about 163 V and whose rms value is therefore 115 V.

Figure 1-18 shows six time-varying waveforms which are commonly encountered in electrical measurement work. The average and rms values of each waveform are given in relation to their amplitudes.

Problems

1. How many coulombs are represented by the following numbers of electrons?
   a) 62.4×10^{18}
   b) 1.24×10^{18}
   c) 8.66×10^{15}
2. Calculate the force (in pounds) that exists between a positive charge of 0.2 coulombs and a negative charge of 0.6 coulombs, 20 cm apart.
3. A copper penny has a mass of 3.1 grams and contains about 2.9×10^{22} atoms. Let us assume that one electron is removed from each atom and these electrons are all removed from the penny to a distance such that the force of attraction between the positively charged penny and the removed group of electrons is 1 newton. How far apart must the penny and the group of electrons be in such a case?
4. If 81 joules of energy are required to move 3 coulombs of charge from infinity to some point in space designated as point A, a) what is the potential difference between infinity and point A, b) if 24 additional joules are required to move the three coulombs from point A to point B, what is the potential difference between point A and point B?
5. If the potential difference between two points in an electric circuit is 42 V, how much work is required to move 14 coulombs of charge from one point to the other?

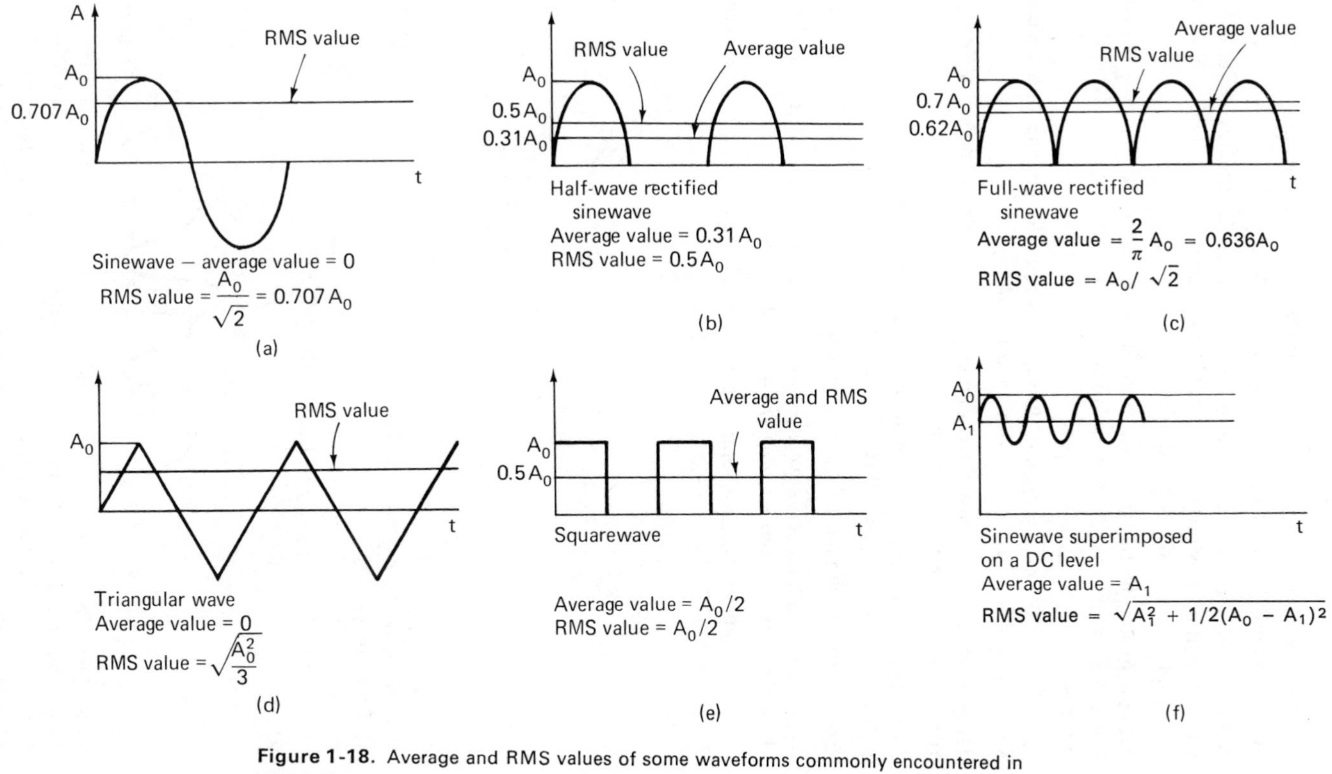

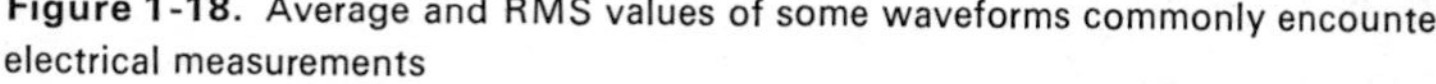

Figure 1-18. Average and RMS values of some waveforms commonly encountered in electrical measurements

6. List three different possible locations at which the electrostatic potential of a point can be defined as having a value of zero.

7. A body attached to earth ground by an electrical conductor is at zero electrical potential. A charge of 25 coulombs is added to the body. What is the potential of the body now? Explain.

8. What is meant by the term "free electron" in reference to electrons in material bodies? Explain why some materials are conductors of electricity while others are not.

9. If 21.847 $\times$ 10^{18} electrons pass through a wire in 70 seconds, find the current that was flowing in the wire during that time.

10. How many electrons pass through a conductor in one minute if a steady current of 20 mA flows during that time period?

11. If the current flowing in a conductor is 2 mA, how much time is required for 0.0046 coulombs to pass through the conductor?

12. Explain the difference between "conventional current" and electron current.

13. What is the difference between *passive* and *active* circuit elements? What is the difference between a *voltage source* and a *current source*?

14. Name the electric components represented by the circuit diagram symbols shown in Fig. P1-1.

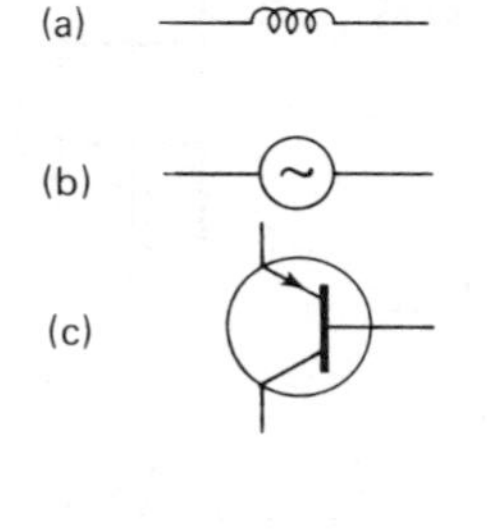

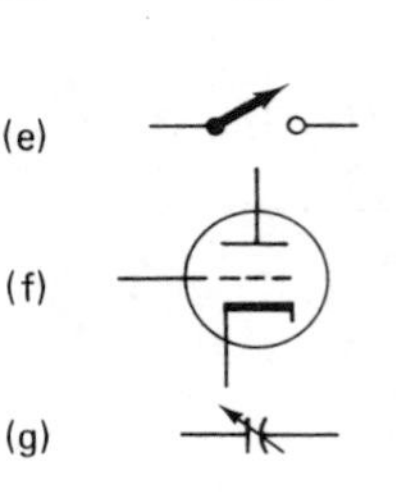

Figure P1-1

15. For the waveform shown in Fig. P1-2:
    a) Find the period T.
    b) How many cycles of the waveform are shown?
    c) What is the frequency of the waveform?
    d) What is its amplitude?

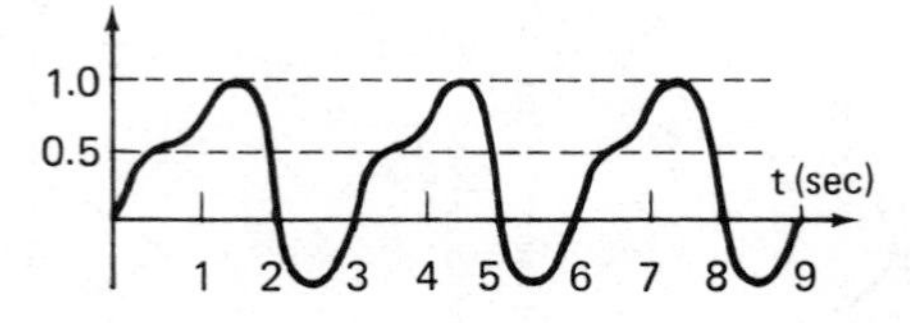

Figure P1-2

16. Draw two sine waves and label one of them *current* and the other *voltage*. Draw the voltage waveform with an amplitude of twice the current waveform but with the same frequency. Draw the waveforms on the same set of axes and have the voltage waveform lead the current waveform by 90°.

17. Repeat Prob. 16, but draw the current waveform leading the voltage waveform by 30°.

18. Convert the following numbers of degrees to radians:
    a) 45°
    b) 60°
    c) 270°

19. Find the angular velocity of the waveforms which have the following frequencies:
    a) 50 Hz
    b) 600 Hz
    c) 0.03 MHz

20. Find the amplitudes and frequencies of the following waveforms:
    a) 20 sin 377t
    b) −7.6 sin 43.6t
    c) 0.001 sin 942t

21. Find the average value of the current waveform shown in Fig. P1-3.

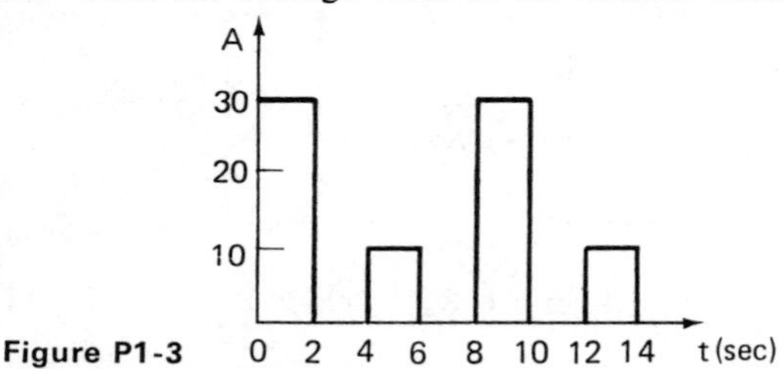

Figure P1-3

22. Find the average value of the waveform shown in Fig. P1-4. Find the average value of this waveform over the first half cycle. Find the rms value of the entire waveform and the rms value of the waveform over the first half of one cycle of the waveform.

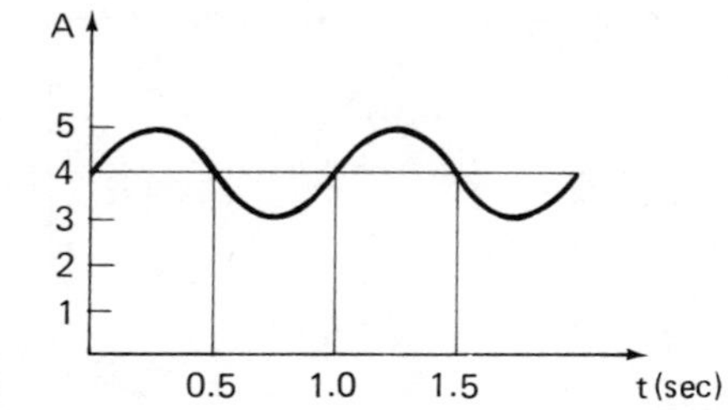

Figure P1-4

23. A light bulb is used in 40 V dc system. What rms value of ac voltage is necessary to have the bulb light as brightly as it does when powered by the dc system?

24. If two or more resistors are connected as shown in Fig. P1-5, they are said to be connected in *series*. The total series connection has a resistance given by

$$R_T = R_1 + R_2 + \cdots + R_n$$

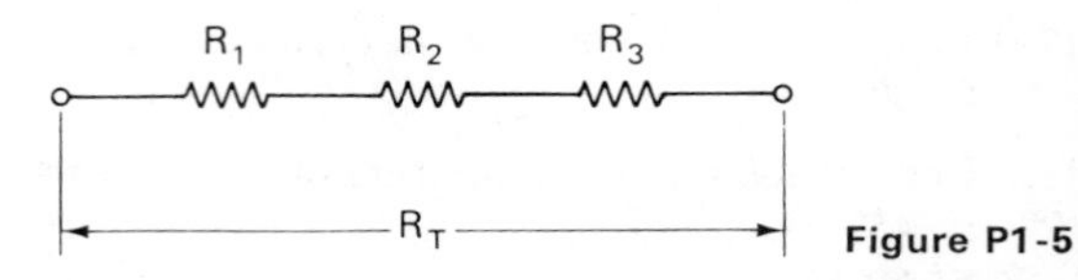

Figure P1-5

If the values of the resistors in Fig. P1-5 are $R_1 = 53\ \Omega$, $R_2 = 270\ \Omega$ and $R_3 = 14\ \Omega$, what is R_T?

25. If two or more resistors are connected as shown in Fig. P1-6, the connection is called a *parallel* connection. The resistance of the total parallel connection is given by

$$\frac{1}{R_T} = \frac{1}{R_1} + \frac{1}{R_2} + \cdots + \frac{1}{R_n}$$

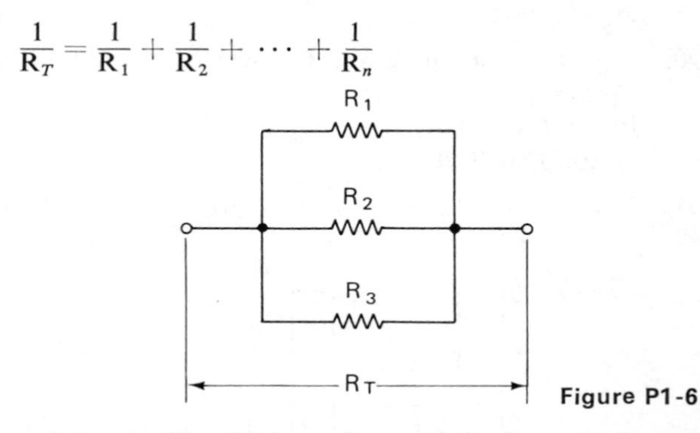

Figure P1-6

If the values of the resistors in Fig. P1-6 are $R_1 = 50\ \Omega$, $R_2 = 70\ \Omega$ and $R_3 = 25\ \Omega$, what is the value of R_T?

26. Find R_T for the connection of resistors shown in Fig. P1-7.

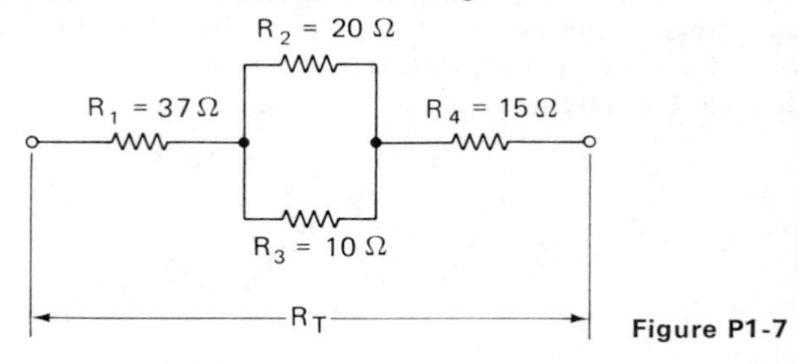

Figure P1-7

References

1. HALLIDAY, D., and RESNICK, R., *Physics.* New York, John Wiley & Sons, Inc., 1962.

2. BOYLESTAD, R., *Introductory Circuit Analysis.* Columbus, Ohio, Charles E. Merrill, Co., 1968.

3. DEFRANCE, J. J., *Electrical Fundamentals*, 3rd ed. Englewood Cliffs, N. J., Prentice-Hall, Inc., 1969.

4. SMITH, R. J., *Circuits, Devices and Systems*. New York, John Wiley & Sons, Inc., 1966.

5. FITZGERALD A. E., HIGGENBOTHAM, D. E., and GRABEL, A., *Basic Electrical Engineering*, 3rd ed. New York, McGraw-Hill Book Company, 1967.

2

Experimental Data and Errors

Measurements play an important role in substantiating the laws of science. They are also essential for studying, developing, and monitoring many devices and processes. However, the process of measurement itself involves many steps before it yields a useful set of information. To study the methods which will produce effective measurement results, let us consider the measurement process as a sequence of five operations. These operations can be listed as follows:

1. Designing of an efficient measurement setup. This step includes a proper choice of available equipment and a correct interconnection of the separate components and instruments.
2. Intelligent operation of the measurement apparatus.
3. Recording of the data in a manner that is clear and complete. The recorded information should provide an unambiguous record for future interpretation.
4. Estimating the accuracy of the measurement and magnitudes of possible attendant errors.
5. Preparation of a report which describes the measurement and its results for those who may be interested in using them.

All five of these items must be successfully completed before a measurement is truly useful.

The first two items of this list are the legitimate concerns of the remainder of our book. The latter three will be discussed to some extent in this chapter. The material presented here will only be a short introduction to these topics;

however, for many measurement applications, this level of presentation should be adequate. If some situations are encountered where more details are required, the reader can consult other references dealing more extensively with the subject. Some of these are listed at the end of the chapter.

Measurement Recording and Reporting

The *original data sheet* is a most important document. Mistakes can be made in transferring information, and therefore copies cannot have the validity of an original. If disputes arise, the original data sheet is the basis from which they are resolved (even in courts of law). Thus it is an excellent practice to carefully label, record, and annotate data as they are taken. A short statement at the head of the data sheet should explain the purpose of the test and list the variables to be measured. Items such as the date, wiring diagrams used, equipment serial numbers and models, and unusual instrument behavior should all be included. The measurement data themselves should be neatly tabulated and properly identified. (All this should emphasize the fact that jotting down data on scrap papers and trusting the memory to record data are not acceptable procedures for recording data. Such bad habits will certainly lead to the eventual loss of valuable pieces of information.) In general, the record of the experiment on the data sheet should be complete enough to specify exactly what was done and, if need be, to provide a guide for duplicating the work at a later date.

The report presented at the end of a measurement should also be carefully prepared. Its objective is to explain what was done and how it was accomplished. It should give the results that were obtained, as well as an explanation of their significance. In addition to containing all pertinent information and conclusions, the report must be clearly written with proper attention to spelling and grammatical structure. To aid in organizing the report and avoid omitting important information, an outline and rough draft should always be used. The rough draft can later be polished to produce a concise and readable document.

The form of the report should consist of three sections:

1. Abstract of results and conclusions.
2. Essential details of the procedure, analysis, data, and error estimates.
3. Supporting information, calculations, and references.

In industrial and scientific practice, the abstract is likely to be read by higher-level managers and other users who are scanning reports for possible information contained in the report body. The details, on the other hand, will usually be read by those needing specific information contained in the report or by others wanting to duplicate the measurement in some form. The latter groups will be interested in the details on the data sheets, the analysis of the level of accuracy, and the calculations and results that support the conclusions and recommendations. For these readers, the references from which source material and information were obtained should also be provided.

The results and conclusions of the report form its most important parts. The measurement was made to determine certain information and to answer some specific questions. The results indicate how well these goals were met. Since the results or supporting data and calculations are often presented graphically, it is important that such information be clearly exhibited. Toward this end, a short set of guidelines on how to prepare graphs is listed below.

GUIDELINES FOR GRAPHING DATA

1. Make graphs neat. Be sure they include a graph title, clearly labeled axes, and the date.
2. Draw graphs with sharp pencils, rulers, and French curves.
3. Plot data while taking measurements, even if graphs are to be replotted later for neatness. This often helps one spot unusual difficulties or faults of the measurement.
4. Choose the independent variable as the abscissa (horizontal axis) and the dependent variable as ordinate (vertical axis).
5. Data points are usually indicated by circles about the actual data points or by a bar intersecting the data point and extending over the possible ranges of error. If more than one variable is plotted on the same graph, use other symbols in addition to the circle (such as small squares and triangles), with each curve having its own symbol. Include a legend identifying each symbol.
6. Choose proper graph paper to best display the variation of the quantity. The types available include linear, semilog (one axis linear and one axis semilogarithmic), log-log (both axes have a logarithmic variation), and polar.

Precision and Accuracy

In measurement analysis the terms *accuracy* and *precision* are often misunderstood and used incorrectly. Although they are taken to have the same meaning in everyday speech, there is a distinction between their definitions when they are used in descriptions of experimental measurements.

The *accuracy* of a measurement specifies the difference between the measured and the true value of a quantity. The deviation from the *true value* is the indication of how accurately a reading has been made. *Precision*, on the other hand, specifies the repeatability of a set of readings, each made independently with the same instrument. An estimate of precision is determined by the deviation of a reading from the *mean* (average) *value*. To illustrate the difference between accuracy and precision more graphically, consider an instrument which has a defect in its operation. The instrument may be giving a result which is highly repeatable from measurement to measurement, yet far from the true value. The data obtained from this instrument would be highly precise but quite *inaccurate*. It may also occur that two instruments produce readings which are equally precise, but which differ in accuracy because of differences in the instrument design. Such examples emphasize that *precision does not guarantee accuracy*, although accuracy requires precision.

The concept of *accuracy*, when applied to instruments that display a reading by the use of a scale and pointer, usually refers to their full-scale reading (unless otherwise specified). When a meter is said to be accurate to 1 percent, this means that a reading taken anywhere along one of its scales will not be in error by more than 1 percent of the full-scale value.

EXAMPLE 2-1

A voltmeter is specified as being accurate to 1 percent of its full-scale reading. If the 100-V scale is used to measure voltages of (a) 80 V and (b) 12 V, how accurate will the readings be (assuming all other errors besides the meter reading error are negligible)?

Solution: Since the meter is accurate to within 1 percent of its full-scale value, any reading taken will be accurate to $(1\%) \times (100 \text{ V}) = 1$ V. Thus the error of the 80-V reading will be 80 ± 1 V. The possible percent error is

$$\text{percent error} = \frac{\text{true value} - \text{measured value}}{\text{true value}} \times 100\%$$

$$= \frac{(80 - 79)}{(80)} \times 100\% \cong 1.25\%$$

The error of the meter while the 12-V measurement is made can still be ± 1 V. Then the possible percent error is

$$\text{Percent error} = \frac{(12 - 11)}{(12)} \times 100\% \cong 8\%$$

Example 2-1 shows that the use of a small segment of a meter's scale to make a reading can result in a larger measurement error than if a greater segment of the scale is utilized. In Example 2-1, a meter with a smaller full-scale voltage than 100 V could have been used to reduce the error of the 12-V measurement.

Recently, some manufacturers of indicating instruments have been changing the method of specifying scale errors as a result of design improvements in the meters. Scale errors are sometimes being stated as a *percentage of a reading* rather than as a percentage of the full-scale value. Instruments designed to satisfy this type of accuracy rating will eliminate some of the errors possible in the older style instruments.

Computational Aids

POWERS OF TEN AND THEIR ABBREVIATIONS

Both very large and very small numbers are often used in measuring and expressing electrical quantities. It is usually more convenient and more precise to express these numbers in terms of their powers of ten rather than to write the whole numbers. By utilizing this procedure, we can clearly state the exact number of significant digits of a quantity. In addition, we can avoid the use of many zeros when dealing with both large and small numbers. The following example shows how to express numbers by using the powers-of-ten notation.

EXAMPLE 2-2

$$1,390,000 = 1.39 \times 10^6$$
$$0.000032 = 3.2 \times 10^{-5}$$

Certain standard prefixes and symbols are used to denote particular multipliers. They are shown in Table 2-1. Examples of how these abbreviations are used in connection with electrical units is shown below:

Table 2-1 Powers of Ten

Multiplier	Prefix	Abbreviation
10^{12}	tera	T
10^{9}	giga	G
10^{6}	mega	M
10^{3}	kilo	k
10^{2}	hecto	h
10	deka	da
10^{-1}	deci	d
10^{-2}	centi	c
10^{-3}	milli	m
10^{-6}	micro	μ
10^{-9}	nano	n
10^{-12}	pico	p
10^{-15}	femto	f
10^{-18}	atto	a

EXAMPLE 2-3

$$10,000 \text{ ohms} = 10 \text{ kilohms} = 10 \text{ k}\Omega$$
$$1.0 \times 10^{-6} \text{ farads} = 1 \text{ microfarad} = 1\ \mu\text{F}$$
$$\tfrac{1}{1000} \text{ ampere} = 1 \text{ milliampere} = 1 \text{ mA}$$

SIGNIFICANT FIGURES

When making calculations using data from measurements, care must be taken to manipulate the data numbers so that additional uncertainty does not creep into the final results. This entails an awareness of the role of *significant figures* in measurement data, as well as knowing how accurate the experimental results may be. A *significant figure* is a digit which is believed to be closer to the true value of the quantity being measured than any other digit. The number of significant figures expresses the precision of a measurement. (Note that zeros in a number which are not preceded by any other digits are not significant figures, e.g., 0.00021 contains only two significant figures.) When measurements are taken, only the number of digits that are meaningful should be set down. The last digit should represent the point of uncertainty. For example, the number 243.1 indicates that this particular measurement value lies somewhere between 243.2 and 243.0. Hence, there are four significant or meaningful digits in this number. The overall rule for making calculations is that the final answer cannot be any more accurate than the least accurate of its components.

As a classic example of this rule, consider the grade point average of students who are graded according to a four-point grading system (i.e., $A = 4$, $B = 3$, $C = 2$, $D = 1$, and $F = 0$). The number of significant digits in each individual grade received by the students is only one. Thus, a grade point average of two significant figures is not absolutely meaningful. For example, a student with a 2.1 average could have consistently received grades that were just below the standard required for *B* grades. However, since only one significant figure is used in each individual grade, the 2.1 grade point average erroneously makes him appear as being very close to a *C* student. To make a grade point average containing two significant figures truly indicative of the student's performance, two significant figures would have to be given in each individual grade.

The significant figures of a number can be expressed in different ways. Examples of three different ways are shown in the next section (along with a discussion of their various merits). All three methods are used in the literature, depending on the number and the method by which it was determined.

METHODS OF EXPRESSING THE SIGNIFICANT FIGURES OF A NUMBER

1. *Use of the entire number* (e.g., 52,400): If this method is used, it indicates that the last digit of the number is significant, and the possible uncertainty (in this case) is ± 1. The implication is that the actual value of the example number lies between 52,399 and 52,401. Another implication is that this number contains five significant digits. Note that this method is not suitable for portraying large numbers which are approximated by placing one

or more zeros between their last significant digit and the decimal point. Thus, it would not be a correct representation if 4 was the last significant digit of the number 52,400.

2. *Use of powers of ten* (e.g., 5.24×10^4): This method implies that the example number has three significant digits and that the digit 4 is the last significant digit. It can be used to overcome the restrictions of method 1. The value of the example number is known to within $\pm 0.01 \times 10^4 = \pm 100$.
3. *Range-of-error method* (e.g., $52{,}400 \pm 50$): This method implies that the example number has three significant digits. It also implies that the value of the number is known to a certainty between 52,350 and 52,450. This way of expressing the uncertainty may be the most satisfactory method if the error can be accurately estimated.

CALCULATIONS INVOLVING SIGNIFICANT FIGURES

The following guidelines should be used when making calculations from measurement data.

1. *Rounding Off*—If there are more digits in a number or a calculated answer than are known with certainty, discard those digits. If the last digit of those dropped is 5 or more, add one unit to the last retained digit. If the last dropped digit is 4 or less, do not change the last retained digit.

EXAMPLE 2-4

532.2~~57~~ ⟶ 532.3

0.041~~42~~ ⟶ 0.041

2. *Addition or Subtraction*—No digit should be kept in the result whose position with respect to the decimal point is to the right of the last significant digit of any number entering the calculation.

EXAMPLE 2-5

(a)
$$\begin{array}{r} 3.7 \\ +26.142 \\ \hline 29.8 \end{array}$$

(b)
$$\begin{array}{r} 15.32 \\ -3.1 \\ \hline 12.2 \end{array}$$

(c)
$$\begin{array}{r} 7.33 \\ +6.2 \\ +18.251 \\ \hline 31.8 \end{array}$$

3. *Multiplication or Division*—Retain in each number (especially the answer) only the percentage of uncertainty which exists in the number containing the fewest significant digits. In other words, reject digits until another rejection would add a larger percentage of uncertainty than the largest one that already exists.

EXAMPLE 2-6

Multiply 8.6 and 212.2. The number 8.6 has an uncertainty of ± 0.1 (which is about 1 percent of 8.6). Thus we can drop the last significant figure of 212.2 and still keep its uncertainty less than 1 percent. Then $8.6 \times 212. = 1823.2$. Now since the uncertainty in the number 8.6 was about 1 percent, the final answer should reflect this uncertainty too. The appropriate answer could then be written as either 1.8×10^3 or as $(1.82 \pm 0.02) \times 10^3$.

4. Calculations Involving Figures Which Contain a Range of Possible Errors— When manipulating figures which contain a range of possible errors, always choose the worst case possible to calculate the error.

EXAMPLE 2-7

$$\begin{array}{rr} \text{Add the figures:} & 12.34 \pm 0.04 \\ & \underline{8.62 \pm 0.02} \\ \textit{Ans:} & 20.96 \pm 0.06 \end{array}$$

In this case the worst situation would occur if both errors were as high or as low as could be. Note, however, that the percentage of the overall error after addition is still about equal to the errors present in the numbers which are being summed.

EXAMPLE 2-8

$$\begin{array}{rr} \text{Multiply the figures:} & 8.62 \pm 0.02 \\ & \underline{12.34 \pm 0.04} \end{array}$$

Solution: Do two multiplications, one with the original figures, and the other with the worst-case numbers (i.e., in the second multiplication, both numbers should be as far from the stated values in the plus limit as possible).

$$(8.62) \times (12.34) = 106.3728$$

$$(8.64) \times (12.38) = 106.9612$$

Then subtract these results. Use the difference as the possible range of errors. In this case, the answer is $106.37 \pm .59$.

EXAMPLE 2-9

$$\begin{array}{rr} \text{Subtract the numbers:} & 12.34 \pm (0.04) \\ & \underline{-8.62 \pm (0.02)} \end{array}$$

Solution: Again, as in addition, choose the worst case (i.e., $+0.04$ and $+0.02$, respectively). *Ans.* 3.72 ± 0.06. Note that in this case (subtraction) the possible errors can add, while the overall result decreases. This means that the percentage of the overall error of the result is much larger than the error contained in either number of the original figures. Guideline 5 follows as a consequence of this fact.

5. Avoid Designing Experiments that Involve Subtraction of one Result from Another. —The error of a subtracted result (as a percentage of its value) can grow to be quite large. Example 2-9 has shown this quite clearly.

Errors in Measurement

Errors are present in every experiment. They are inherent in the act of measurement itself. Since perfect accuracy is not attainable, a description of each measurement should include an attempt to evaluate the magnitudes and sources of its errors. From this point of view, an awareness of errors and their classification into general groups is a first step toward reducing them. If an experiment is well designed and carefully performed, the errors can often be reduced to a level where their effects are smaller than some acceptable maximum. Figure 2-1 classifies the main categories of errors and describes some causes and methods of correcting them.

Sometimes a specific reading taken during a measurement is rather far from the mean value. If faulty functioning of the measurement instruments is suspected as the cause of such unusual data, the value can be rejected. However, even such data should be retained on the data sheet (although they should be labeled as suspect data). Nevertheless, even when all the items involved in a measurement setup appear to be operating properly, unusual data may still be observed. We can use a guide to help decide when it is permissible to reject some suspect data. This guide is derived by using the methods of statistical evaluation of errors which will be covered in the following sections. It is stated as follows: Individual measurement readings taken when all the instruments of a measurement system appear to be operating properly may be rejected when their deviation from the average value is four times larger than the *probable error* of one observation (the procedure for calculating the probable error is described in the following section). It can be shown that such a random error will not occur in more than one out of one hundred observations, and the probability that some unusual external influence was at play is thus very high. For example, if the probable error in measuring a specific voltage of 5.21 V is ± 0.21 V, and a measurement yields a value of 6.2 V, this piece of datum can probably be rejected. However, when an unusually large error does occur as described above, this event may be a signal that some systematic error is being committed. An attempt to locate the cause of the error should be undertaken. Keeping rejected data in the data sheet can be of assistance in finding the extent and cause of error.

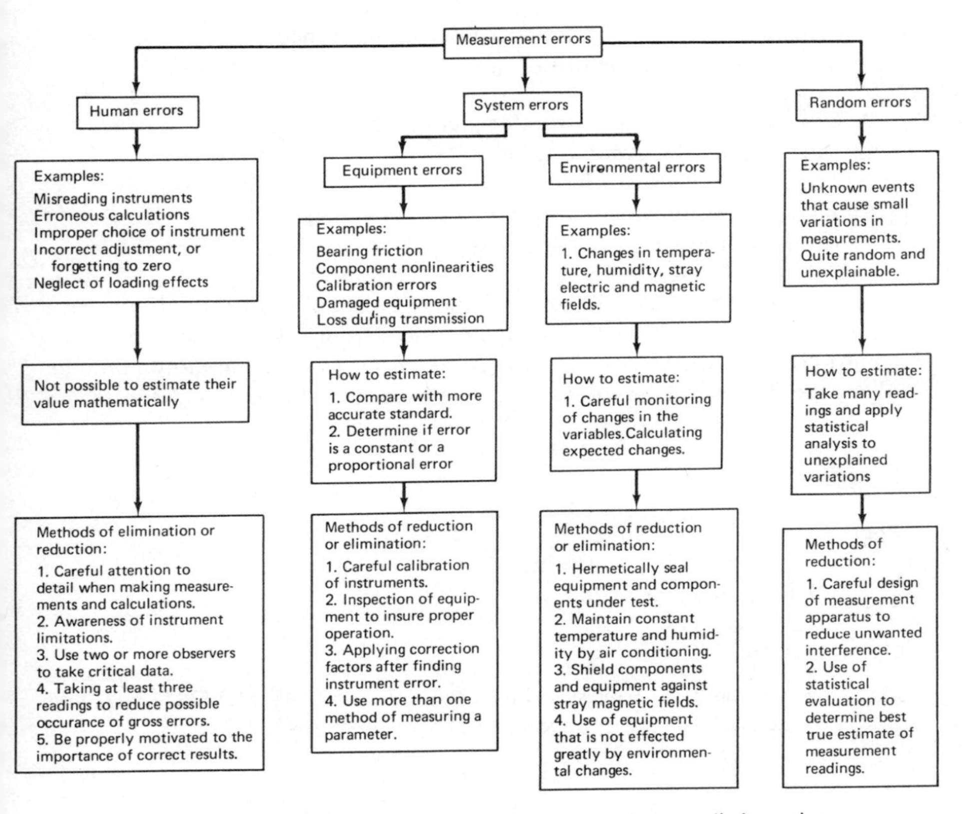

Figure 2-1. Measurement errors: how to estimate, reduce, or eliminate them

Statistical Evaluation of Measurement Data and Errors

Statistical methods can be very helpful in allowing one to determine the most probable value of a quantity from a limited group of data. That is, given an experiment and the resulting data, we can tell which value is most likely to occur. Furthermore, the probable error of one observation and the extent of uncertainty in the best answer obtained can also be determined. However, a statistical evaluation cannot improve the accuracy of a measurement. The laws of probability utilized by statistics operate only on random errors and not on system errors. Thus the system errors must be small compared to the random errors if the results of the statistical evaluation are to be meaningful. For example, if a zero adjustment is incorrect, a statistical treatment will not remove this error. But a statistical analysis of two different measurement methods may demonstrate the discrepancy. In this way, the measurement of precision can lead to a detection of inaccuracy. We will now show how the following quantities can be calculated using statistics;

1. Average or mean value of a set of measurements.
2. Deviation from the average value.
3. Average value of the deviations.
4. Standard deviation (related to the concept of rms).
5. Probability of error size in one observation.

1. Average or Mean Value—The most likely value of a measured quantity is found from the arithmetic average or mean (both words have the same definition) value of the set of readings taken. Of course, the more readings taken, the better will be the results. The average value is calculated from:

$$a_{av} = \frac{a_1 + a_2 + \cdots + a_n}{n} \tag{2-1}$$

where: a_{av} = average value,
$a_1, a_2, a_3, \cdots$ = the value of each reading
n = number of readings.

EXAMPLE 2-10

If the measurement of a resistor yields the values 52.3 Ω, 51.7 Ω, 53.4 Ω, and 53.1 Ω, what would be the average value of these numbers?

$$R_{av} = \frac{52.3 + 51.7 + 53.4 + 53.1}{4} = 52.6\ \Omega$$

2. Deviation from the Average Value—This number indicates the departure of each measurement from the average value. The value of the deviation may be either positive or negative.

EXAMPLE 2-11

For the average value calculated above, the deviation of the first number (52.3) is

$$d_1 = (52.3) - (52.6) = -0.3\ \Omega$$

3. Average Value of the Deviations—This value will yield the precision of the measurement. If there is a large *average deviation*, it is an indication that the data taken varied widely and the measurement was not very precise. The average value of the deviations is found by taking the absolute magnitudes (i.e., disregarding any minus signs) of the deviations and computing their mean. The average of the deviations of the resistance values found in Example 2-10 is:

$$D = \frac{|d_1| + |d_2| + |d_3| + |d_4|}{4} = \frac{0.3 + 0.9 + 0.8 + 0.6}{4}$$

$$= \frac{2.6}{4} = 0.65 \text{ or } 0.7\ \Omega \tag{2-2}$$

4. Standard Deviation and Variance—The average deviation of a set of measurements is only one of the methods of determining the dispersion of a set of readings. However, the average deviation is not mathematically as convenient for manipulating statistical properties as the *standard deviation* (also known as the root-mean-square or rms deviation). Although the difference between the average and the standard deviation cannot be completely appreciated at our level of presentation, the fact remains that the standard deviation is a much more useful statistical quantity. As such, it is used almost exclusively in expressing dispersions of data. The standard deviation is found from the formula:

$$\sigma = \sqrt{\frac{d_1^2 + d_2^2 + d_3^2 + \cdots + d_n^2}{n - 1}} \tag{2-3}$$

where: σ = standard deviation,
$d_1, d_2, d_3, \ldots$ = deviation from the average value, and
$n - 1$ = one less than the number of measurements taken.

The variance (V) is the value of the standard deviation (σ) squared.

$$V = \sigma^2 \tag{2-4}$$

5. Probable Size of Error and Gaussian Distribution—If a *random* set of errors about some average value is examined, we find that their frequency of occurrence relative to their size is described by a curve (Fig. 2-2) known as a Gaussian curve (or bell-shaped curve). Gauss was the first to discover the

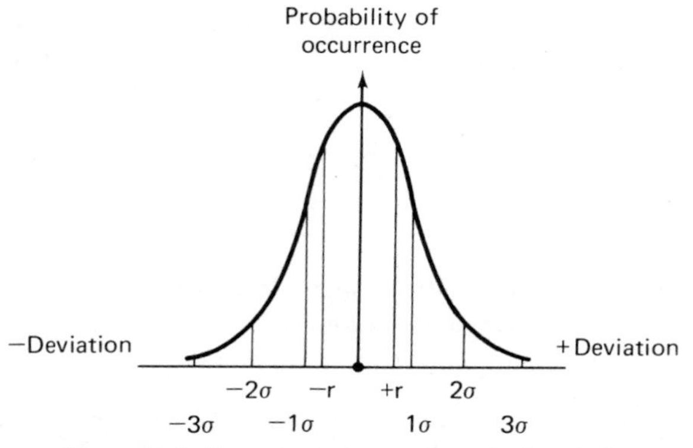

Figure 2-2. Error size in terms of standard deviations

relationship expressed by this curve. It shows that the occurrence of small random deviations from the mean value are much more probable than large deviations. In fact, it shows that large deviations are extremely unlikely. The curve also indicates that random errors are equally likely to be positive or negative. If we use the standard deviation as a measure of error, we can use the curve to determine what the probability of an error greater than a certain σ value will be for each observation. Table 2-2 shows the probability of an error occurring greater than a specific σ value for each observation.

Table 2-2

Error (±) (Standard Deviations)	Probability of Error Being Greater Than Given $+\sigma$ or $-\sigma$ Value in One Observation
0.675	0.250
1.0	0.159
2.0	0.023
3.0	0.0015

EXAMPLE 2-12

One resistor is chosen from a group whose average resistance is known to be 20.8 Ω The standard deviation of their resistance values is 1 Ω. What is the probability that a resistor chosen at random will have a resistance greater than 23 Ω?

Solution: Using Table 2-2 we see that a 23-Ω resistor would have a deviation whose value is about equal to two standard deviations. Hence, the probability of choosing such a resistor at random would be less than 2.3 percent.

6. Probable Error—From Table 2-2 we can also calculate the probable error that will occur if only one measurement is taken. Since a random error can be either positive or negative, an error greater than $|0.675\sigma|$ is probable in 50 percent of the observations. Hence, the probable error of one measurement is

$$r = \pm 0.675\sigma \tag{2-5}$$

EXAMPLE 2-13

Given the following set of current measurements taken from an ammeter, find their (a) average value, (b) average deviation, (c) standard deviation, and (d) probable error.

Data: 153 mA, 162 mA, 157 mA, 161 mA, 155 mA

Solution:

(a) Average value: $I_{av} = \frac{I_1 + I_2 + I_3 + I_4 + I_5}{5} = \frac{785}{5} = 157\text{mA}.$

(b) Average deviation: $D = \frac{|d_1| + |d_2| + |d_3| + |d_4| + |d_5}{5}$

$= \frac{4 + 5 + 0 + 4 + 2}{5} = 3\text{mA}.$

(c) Standard deviation: $\sigma = \sqrt{\frac{(d_1^2 + d_2^2 + d_3^2 + d_4^2 + d_5^2)}{4}} = \sqrt{15} \cong 3.8\text{mA}.$

(d) Probable error: $r = 0.675\sigma \cong 2.6\text{mA}.$

The Decibel

Occasionally, when measuring a quantity such as power, it may not be as important to know the absolute value of the quantity as it is to know its value relative to some other quantity. For instance, one might want to know the ratio of the power developed in one part of a circuit to the power in another part. Also, it is often easier to measure this ratio rather than the absolute value itself. For example, one prime quantity of interest in an amplifier is the *power gain* (defined as the ratio of the power at its output to the power at its input, P_{out}/P_{in}). A logarithmic scale can be used to describe such ratios quite conveniently and the one that is commonly employed to do it is called the *decibel* scale ("bel" after Alexander Graham Bell.) The ratio in decibels of two power values P_2 and P_1 is defined as

$$G = 10 \log_{10}\left(\frac{P_2}{P_1}\right) \text{ decibels} \tag{2-6}$$

and the result is expressed as the difference in decibels (dB). For example, when there is a difference of 1 dB in power, $P_2/P_1 = 1.26$. A 3-dB difference corresponds to $P_2/P_1 = 2$. A 10-dB difference indicates that the power ratio is also 10. Figure 2-3 is a chart which shows how a ratio of powers corresponds to positive decibel (dB) values (power *gain*). If the power of P_2 is smaller than that of P_1, there is a power *loss* and the dB value is negative.

The reasons for the convenience of the decibel scale are twofold. The first is seen if we consider two or more circuits that are connected together (cascaded). Assume that the power ratios of P_o/P_i are known for each circuit and are expressed as ratios rather than in decibels. Then these power ratios must be multiplied to get the power gain (or loss) of the entire assembly (Fig. 2-4).

$$\frac{P_{out}}{P_{in}} = \frac{P_1}{P_{in}} \times \frac{P_2}{P_1} \times \frac{P_{out}}{P_2}$$

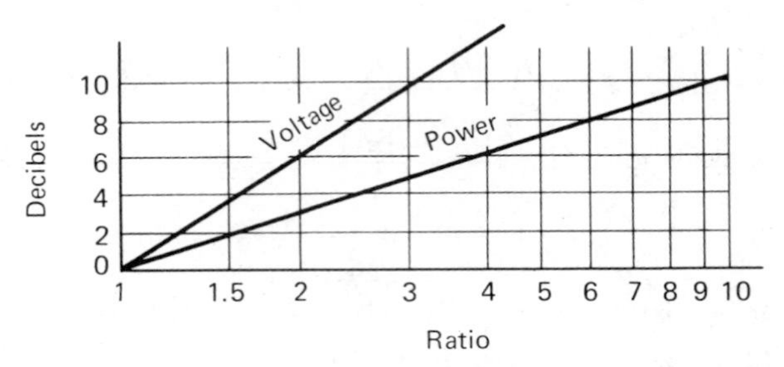

Figure 2-3. Chart showing ratio of decibels to power and voltage

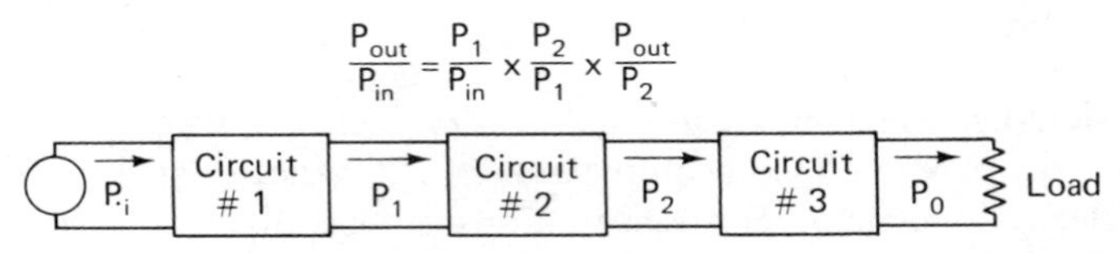

Figure 2-4. Cascaded circuits

On the other hand, if the gains are expressed in decibels, they can simply be added together to yield the same result (also in decibels).

The second advantage comes from being able to graphically display a change in the value of the ratio as it goes from a very small to a very large quantity. For example, the ratio of P_{out} to P_{in} of an amplifier may vary from much less than one to many thousand as the frequency of its input signal changes. To describe such a change on a graph, a logarithmic scale is more useful and compact than a linear scale.

In addition to being used to describe power ratios, the decibel has also come to be used in describing voltage and current ratios or gains. Voltage and current are more frequently measured than power, and their ratios can also be expressed logarithmically. Strictly speaking, the dB scale for expressing the ratios of voltages is valid only if the input and load impedances of the circuit being measured are equal ($R_i = R_L$). (See Chapter 3 for a discussion of input and output impedances.) In this case, from the definition of gain in decibels:

$$G = 10 \log_{10}\left(\frac{P_o}{P_i}\right) = 10 \log_{10}\left(\frac{V_o^2}{R_L} \times \frac{R_i}{V_i^2}\right) = 20 \log_{10}\left(\frac{V_o}{V_i}\right) \qquad (2\text{-}7)$$

However, common usage has changed the correct meaning as defined above to the definition that $G_v = 20 \log_{10} (V_o/V_i)$ regardless of whether R_i and R_L are equal. Under such circumstances, it is not always possible to convert the voltage gain to a power gain. G_v is called the *voltage gain* in decibels.

When a cerain power level is chosen as a standard reference, the ratio of a measured power to this level can also be described in decibels. For example, in telephone circuit use, the milliwatt (1×10^{-3} watt) is used as the standard power reference level. Thus a 10-mW signal is referred to as +10 dBm (i.e., +10 dB referred to the milliwatt standard). Since 600 Ω is the standard impedence level of telephone circuits, voltage can also be expressed in decibels. Many voltmeters have calibrated scales in decibels for use in such applications. A reading of 0 dBm on such a scale corresponds to a voltage of 0.775 V (this number is derived from the voltage existing at 1mW and 600Ω).

Problems

1. Describe in your own words the difference between the terms *accuracy* and *precision* as they are used in reference to experimental measurements.
2. What are the three general classes of measurement errors?
3. List five different specific errors that frequently occur in the process of making measurements.
4. A voltmeter whose accuracy is guaranteed to within 2% of its full-scale reading is used on its 0–50 V scale. The voltage measured by the meter is 34 V. Calculate the possible percentage error of the reading.
5. A 0–50 mA ammeter has an accuracy of 0.5%. Between what limits may the actual current be when the meter indicates 10 mA?
6. Make the following conversions:
   a) 0.035 A to milliamperes and microamperes,
   b) 0.02 mV to microvolts and volts,
   c) 250,000 Ω to megohms,
   d) 370 μV to millivolts and volts.
7. State the number of significant digits in each of the following numbers:
   a) 6,437
   b) 0.32
   c) 43.02
   d) 0.00017
   e) 32.3×10^6
   f) 350,000
8. Four resistors are connected in series. The values of the resistors are 32.6 Ω, 5.32 Ω, 63.221 Ω, and 0.43 Ω, with an uncertainty of one unit in the last digit of each value. Calculate the resistance of the total connection. (See Prob. 1-23 if you need to know the formula for calculating the resistance of a series connection.)
9. Subtract 536 ± 4 from 692 ± 4, and express the uncertainty which exists in the answer as a percentage of the answer.
10. Ten measurements of current in a circuit branch yield values of 50.2, 50.6, 49.7, 51.1, 50.3, 49.9, 50.4, 50.6, 50.3, and 50.0 mA. Assume that only random errors were present in the measurement system. Calculate a) the average value, b) the standard deviation of the readings, c) the probable error of the readings.
11. The following voltage values are listed on a data sheet as the values obtained from measuring a certain voltage: 21.35, 21.84, 21.56, 19.07, 21.63, 21.29 V.

By examining the numbers calculate a) the average value, b) the probable error. If only random errors are present, how does one treat the 19.07 value?

12. Calculate the decibel power gain for the following input and output powers:
    a) $p_o = 100$ mW, $p_i = 5$ mW
    b) $p_o = 50\ \mu$W, $p_i = 10\ \mu$W

13. Find the magnitude power gain corresponding to a decibel power gain of +50 dB.

14. Find the magnitude power gain corresponding to a decibel power gain of −12 dB.

15. The input power to a circuit is 10,000 W at a voltage of 1000 V. The power output is 500 W, with the output impedance being 20 Ω.
    a) Find the power gain in decibels.
    b) Find the voltage gain in decibels (use the relation $P = V^2/R$ to find V_0).
    c) Explain why the results of (a) and (b) agree or disagree.

16. An amplifier which is rated at 20 W output is connected to a speaker whose impedance is 10 Ω.
    a) If the power gain of the amplifier is +30 dB, what is the input power required to obtain the full output from the amplifier?
    b) If the voltage gain of the amplifier is 40 dB, what is the required input voltage if the amplifier is to produce its rated output?

References

1. STOUT, M. B., *Basic Electrical Measurements*, 2nd ed., Chap. 2. Englewood Cliffs, N.J., Prentice-Hall, Inc., 1960.

2. KINNARD, I. F., *Applied Electrical Measurements*, Chaps. 2 and 3. New York, John Wiley & Sons, Inc., 1956.

3. TUVE, G. L., and DOMHOLDT, L. C., *Engineering Experimentation*, Chap. 2. New York, McGraw-Hill Book Company, 1966.

4. COOPER, W. D., *Electronic Instrumentation and Measurement Techniques*, Chaps. 1 and 2. Englewood Cliffs, N.J., Prentice-Hall, Inc., 1970.

5. WEDLOCK, B. D., and ROBERGE, J. K., *Electronic Components and Measurements*, Chaps. 1, 2, and 3. Englewood Cliffs, N.J., Prentice-Hall, Inc., 1969.

6. BARTHOLOMEW, D., *Electrical Measurements and Instrumentation*, Chaps. 1 and 2. Boston, Allyn and Bacon, Inc., 1963.

3

Electrical Laboratory Practice

In addition to electrical measuring instruments and the body of terms and symbols used to describe them, there are a number of laboratory techniques and associated concepts that are uniquely related to electrical measurements. Sometimes these techniques involve the direct application of a physical principle (such as impedance matching); at other times they are more in the realm of an art (as the laying out of circuits or the methods used to eliminate "ground loops"). Usually a student is exposed to such laboratory techniques indirectly and learns about them through a kind of osmosis. Often, this type of learning is not completely effective. There is still a need for reinforcing the student's intuitive knowledge by a physical understanding of the processes that underlie the techniques. The objective of this chapter is to provide some of this reinforcement by examining these techniques from a physical point of view.

The techniques to which we refer include grounding practices, electrical safety and shocks, the laying out of circuits, and the elimination of external interference signals from measurement systems. Additional concepts of a similar nature include impedance matching, input impedance, and the loading effects of measuring devices. We will present a discussion on the introductory aspects of each of these subjects.

Even an introductory discussion will often put us into a position of using concepts that have not yet been defined in the text. Unfortunately, this is an unavoidable problem. If the chapter were placed toward the end of the book after all the necessary concepts had been developed, there would be a greater likelihood of it being overlooked by the reader. Since the information contained in this chapter is quite essential for making proper electrical measurements, and since such information is often hard to find in other texts, it is important that the reader be aware of its presence in this book. Thus, in the interests of maximum exposure, the chapter appears near the beginning. Once the reader has finished the remainder of the book, he should be able to reread this chapter and understand it with greater clarity.

There are some topics related to laboratory techniques that will be presented in more detail later in the text, rather than in this chapter. Usually such topics logically belong to the body of knowledge presented at that point and will thus assume greater relevance by being presented there. For example, the resistance of wire leads and contacts is discussed in the chapter on *Resistors*. Likewise, the principles related to the loading effects of meters and oscilloscopes are more fully developed in the chapters that deal specifically with these instruments.

Safety

When working in the electrical laboratory or when using electrical equipment, observing proper safety precautions is as important as making accurate measurements. Potentially lethal hazards exist in the electrical laboratory environment, and failure to follow careful safety procedures can make you or a fellow worker the victim of a serious accident. The best way to avoid accidents is to recognize their causes and carefully adhere to well-established safety procedures. A full awareness of the dangers and the possible consequences of accidents helps develop the proper motivation for following such procedures.

The most common and serious hazard of the electrical laboratory is electric shock. Other hazards which should also be recognized include dangerous chemicals, moving machinery, and soldering irons.

ELECTRIC SHOCK

When electric current is passed through the human body, the effect that it causes is called *electric shock*. Electric shock can accidentally occur due to poor equipment design, electrical faults, human error, or a combination of unfortunate circumstances. The lethal aspect of electric shock is a function of the amount of current which is forced through the human conducting path. It is not necessarily dependent on the value of the applied voltage. A shock from 100 V may turn out to be as deadly as a shock from 1000 V.

The severity of an electric shock varies somewhat with the age, sex, and physical condition of the victim. But in general, the level of current required to kill any human being is remarkably small. For this reason, extreme care must always be exercised in order to prevent electric shock from occuring.

The threshold of perception of current in most humans is about 1 mA. The sensation due to this current level takes the form of an unpleasant tingling or heating at the point of contact. Currents above 1 mA but below 5 mA are felt more strongly, but usually do not produce severe pain. However, 1–5 mA current levels can still be dangerous because of the startling reaction they may cause. For example, a shock from such currents might lead one to jump back against a hot oven or a moving piece of machinery or to fall off a ladder, thereby causing an injury. (Note that 5 mA is the maximum current allowed to leak from home appliance circuits to their cases and still be able to pass the Underwriter's Laboratory specifications.)

At levels above 10 mA, current begins to cause involuntary muscular contractions. Due to these spasms, the victim loses the ability to control his muscles. Even though the pain is severe, he is unable to release his grip on the electrical conductor being held. For this reason, such a current level is called "can't let go" current. If it is sustained, "can't let go" current can lead to fatigue, collapse, and even death.

If the current level flowing in the body exceeds 100 mA, it begins to interfere with the coordinated motion of the heart. This *fibrillation* prevents the heart from pumping blood, and death will occur in minutes unless the fibrillation is stopped. Above 300 mA, the heart's muscular contractions are so severe that fibrillation is prevented. If the shock is halted quickly enough, the heart will probably resume a normal rhythm. In such cases, breathing may have stopped and artificial respiration may have to be applied. If proper first aid is provided, the shock may not be fatal, even though severe burns may have resulted. (In fact, a method of administering large current pulses to fibrillating hearts is used to restore them to their normal rhythm.)

From this discussion we see that the most fatal current range lies approximately between 100 and 300 mA. (Note that 100 mA is about one tenth of the current flowing in a 100-W lamp.) Figure 3-1 summarizes the effects of dangerous currents levels on the human body.

The voltage required for a fatal current level to flow in the human conducting path can vary. Its value depends upon the skin resistance at the point of contact. Wet skin may have a resistance as low as 1 kΩ, while dry skin may have as much as 500 kΩ resistance. (Once the current passes into the body, the resistance is much less—due to the conductivity of body fluids.) Thus a 100-V potential applied to wet skin can be fatal. In fact, even 50 V under certain circumstances can be as deadly as 5000 V. Furthermore, skin resistance falls rapidly as current passes through the point of contact because the current breaks down the protective, dry, outer-skin layer. This makes it important to break the contact with the live conductor as soon as possible. Since the voltage at the point of contact usually remains constant, and since the resistance decreases, the current can soon rise to a lethal level.

The best method for protecting oneself from the hazard of shock when using electrical equipment is to rely on proper grounding of the equipment employed. The details of how and why to properly ground equipment are

Effects of 60 Hz Electric Shock* (current) on an average human through the body trunk

Current intensity — 1 second contact	Effect
1 milliampere	Threshold of perception
5 milliamperes	Accepted as maximum harmless current intensity
10–20 milliamperes	"Let-go" current before sustained muscular contraction
50 milliamperes	Pain. Possible fainting, exhaustion, mechanical injury, heart and respiratory functions continue.
100–300 milliamperes	Ventricular fibrillation will start but respiratory center remains intact
6 amperes	Sustained myocardial contraction followed by normal heart rhythm. Temporary respiratory paralysis. Burns if current density is high.

Figure 3-1. Effects of various current levels on the human body (Courtesy of Hewlett-Packard Co.)

given in a later section entitled *Grounding for Safety*. In addition to good grounding techniques, one should avoid handling equipment which has exposed wires or conductors. Always try to shut off power when touching any circuits. Furthermore, always wear shoes to further insulate yourself from ground. Avoid coming in contact with such grounds as metal plumbing while handling the wires or instruments. If "hot" equipment must be repaired, use only one hand, keeping the other far away from any part of the circuit. Do not wear metal, rings, bracelets, or wrist watches when working with electrical systems.

FIRST AID FOR ELECTRIC SHOCK

The first step in aiding a victim of electric shock is to try and shut off the power to the conductor with which he is in contact. If the attempt is not successful and the victim is still receiving a shock, break the contact of the victim and the source of electricity without endangering yourself. Do this by using an insulator [such as a piece of wood (dry), rope, cloth, or leather] to pull or separate the victim and the live conductor. Do not touch the victim with bare hands as long as he is electrified. (Even momentary contact with the victim can be fatal if the current level is sufficiently high.) The contact must be broken quickly because skin resistance falls rapidly with time and a fatal current of 100–300 mA can be reached if the shock is allowed to continue long enough.

If breathing has stopped and the individual is unconscious, start giving artificial respiration immediately. Do not stop until a medical authority pronounces the victim beyond further help. This may take up to eight hours. Symptoms of rigor mortis and the lack of a detectable pulse should be disregarded, as these are sometimes results of the shock. They are not necessarily proof that the victim has expired.

OTHER HAZARDS OF THE ELECTRICAL LABORATORY

When using power tools such as drills or saws, care must also be taken to prevent serious injury. Power tools should not be operated unless instructions on how to operate them have been received. In addition, loose clothing or long hair which could become caught in moving machinery should not be worn when such equipment is being operated. Finally, always wear goggles or safety glasses when drilling or cutting with power tools.

The soldering iron is another instrument which can cause accidents if used carelessly. Unattended hot irons can burn unsuspecting workers and may set fire to the surroundings. To prevent soldering-iron accidents, always replace the iron into its holder when not soldering. Also, make sure to turn off soldering irons after use.

When using cleaning solvents (such as trichlorethylene) or corrosive chemicals (such as acids in semiconductor laboratories), care must be exercised in their use and disposal. Well ventilated fume hoods must be used when working with these chemicals to dispose of the corrosive or poisonous chemical fumes. Gloves, special clothing, and goggles should be used to protect against chemical splattering and contamination. When the corrosive chemicals are dumped into sinks, a large volume of water should be allowed to flow after them to dilute their harmful properties. In case of acid spills, flush the skin with copious amounts of water.

SAFETY RULES

1. Never work alone. Be sure there are others in the laboratory to summon and provide aid in case of accidents.
2. Use only instruments and power tools provided with three-wire power cords. (See Section on *Grounding for Safety*.)
3. Always shut off power before handling wiring.
4. Check all power cords for sign of damage. Replace or repair damaged cords and leads.
5. Always wear shoes. Keep shoes dry. Avoid standing on metal or damp concrete. (All these precautions prevent you from becoming a low-impedance path to ground.) Do not wear metal, rings, etc.
6. Never handle electrical instruments when your skin is wet (the moisture decreases your skin's resistance and allows a greater current to flow through you).

7. Hot soldering irons should not be left unwatched. Keep hot soldering irons in holders when not soldering.
8. Never wear loose clothing around machinery. Always wear safety goggles when using chemicals or power tools.
9. Always connect a cable or lead to the point of high potential as the last step. That is, do not connect the lead to the "hot" side of a circuit first, or you will end up holding a "hot" connector in your hand.

Circuit Layout and Assembly

When circuits are to be physically constructed from circuit diagrams, some guidelines should be followed to ensure that the resulting assembly will function in the prescribed manner. First, of course, one must ascertain that the circuit diagram being used as a guide is actually a schematic (and not some other type of electrical diagram such as an equivalent-circuit diagram). Next, in the layout of the circuit, the positions of the components should correspond as closely as possible to the locations on the schematic. Neatness is important because, if malfunctions occur, they can be traced far more easily on an orderly layout. The connections of a circuit may be soldered if the circuit is to be semipermanent or permanent, but tight and clean solderless connections can also be satisfactory (especially if the connections are only temporary). However, if the connections are loose, flimsy, or dirty, and if such obstacles as wire insulation are preventing good contacts from being made, connections can become a source of frustration and loss of time.

The leads (wires) connecting the circuit components should be kept as short as possible to help provide a neat circuit layout. In addition, shorter leads pick up less noise and exhibit less capacitance. This results in less interference in the desired circuit signals (particularly when measuring or amplifying small currents and voltages).

Grounds

IMPORTANCE OF GROUNDS

The concepts of *ground* and *grounding* are basic and integral concepts utilized in the design of electrical measurement systems. For proper operation of such systems, these concepts should be well understood. However, grounds are often not clearly defined during a student's training. As a result he may end up working with measurement systems that are not properly grounded. If such situations lead to erroneous measurements, a consequent confusion may also develop as to why the error exists. To keep such problems from arising, a discussion concerning some of the basic principles of grounds is presented in the following sections.

Some readers may be wondering at this point why a thorough understanding of grounding is such an important matter? The following reasons may help to answer this question:

1. Measurement circuits and instruments often rely on the implementation of various grounding principles for proper operation. Thus, by realizing the purposes of grounding, one acquires a more thorough comprehension of the entire measurement system.
2. A greater awareness of electrical safety considerations in the laboratory (and in the home) is developed from an appreciation of grounding methods.
3. Unwanted "noise" and interference that arises in measurement systems can often be reduced or eliminated by the application of proper grounding and shielding techniques.

GROUNDING

Since all measurements of potential difference (voltage) are relative, the voltage level of any point in a circuit must always be compared to some reference level. This means that there must be a voltage level at one point which is defined as the reference voltage. Usually this reference level is assigned a voltage value of zero and is known as the *circuit ground* or *common point* of the system.

To provide one common and convenient reference potential for the majority of measurements, the potential of the Earth (the planet) was chosen as zero. The soil of the Earth contains water and electrolytes which conduct currents quite easily. If voltage differences exist between two points on its surface, currents can flow and equalize these potentials. When a conductor or a circuit is connected at some point to the Earth by a low-impedance electrical connection, that point will be at essentially the same potential as the Earth (zero). The conductor or circuit is then said to be *earth grounded, earthed,* or *grounded.*

Now note that the *circuit ground* mentioned at the outset of the discussion may be an earth ground, or it may simply be a point in the circuit to which all other voltages are referred without being connected to earth ground. For example, a flashlight may operate from a 6-V battery. The *circuit ground* of its system is not connected to earth and may or may not be at zero potential with respect to earth. However, the positive terminal of the battery is always at 6 V relative to the flashlight circuit ground. Other examples of such non-earth grounded circuits include the automobile and the airplane. For the electrical systems of these machines, the *circuit ground* may be the metal body of the automobile or the fuselage of the airplane. In such cases, the chassis takes the place of the earth in serving as a zero potential level. When the chassis acts as the zero reference potential, the system is said to be *chassis grounded.* Note that the chassis voltage may be many volts above that of earth ground and yet provide a zero reference level for its own internal circuits. When a circuit is connected to a chassis which is deliberately disconnected from earth ground, the circuit is said to be *floating*. (This condition is often specifically created when a power supply or oscilloscope is used for making certain types of voltage measurements.) A potential can exist between the chassis and earth grounds, and if a conducting path is connected between them, a current will flow. The conducting path could be a human being; thus *floating* equipment must be handled as if it were at some higher potential to avoid electric shock.

This discussion emphasizes the fact that the term *ground* can have distinctly different meanings, all related closely enough to occasionally cause confusion. A *circuit ground* is the most general definition because it can be an *earth ground*, a *chassis ground*, or neither (it may just be a convenient point in the circuit to which other voltages can be referred). One must clearly determine which meaning is the relevant one each time the term is encountered. The symbol ⏚ is often used interchangeably to refer to all types of grounds. Sometimes, however, if there is no earth ground connection, the symbol ⊥ is used to denote a floating chassis ground.

Connections to earth ground are constructed by burying or driving conductors into the earth. Such connections are effective if they provide a very-low-resistance path to earth ground. The best and most common method used to achieve a low-resistance path to ground is to drive a rod into the earth. A driven ground-rod can provide a path to earth whose resistance is less than 5 ohms. To improve this connection even more, metal plates may be buried near the end of the rod and a salt solution pumped around them. Another way of making the ground connection, which is not quite as effective, is to tie the ground wire to underground metallic pipes, metal building frameworks, or other metallic objects which are in good contact with the earth. This usually yields resistance paths to ground of 5 to 25 ohms. Figure 3-2 shows some methods used in making earth ground connections.

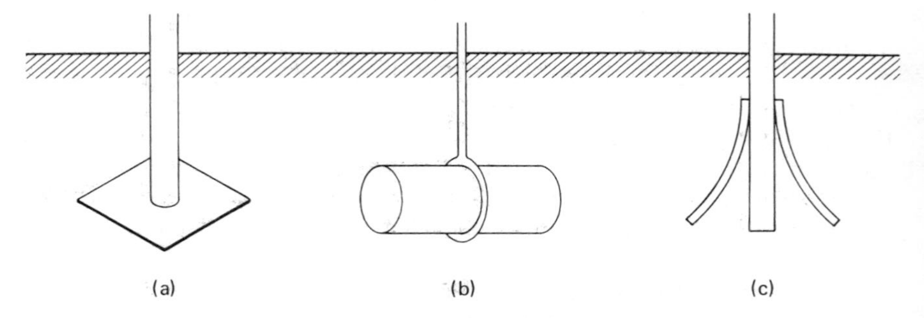

Figure 3-2. Methods for making ground connections (a) Rod and buried metal plate (b) Metal pipe (c) Rods

In the common three-wire wall outlet from which electric power is obtained in the home, there are two wires which are connected to ground (Fig. 3-3). Wire 2 is connected to ground and carries the return current from the load back to ground. It is called the *neutral* and the color used for it is white. Wire 3 is a non-current-carrying wire also connected to ground. Its color is designated green and it is called the *ground wire*. (Wire 3, as we will see in the section on *Grounding for Safety*, is used as a protective measure.) Wire 1 is not connected to ground but is connected to the terminal of higher potential of the ac source; its color designation is black (for death).

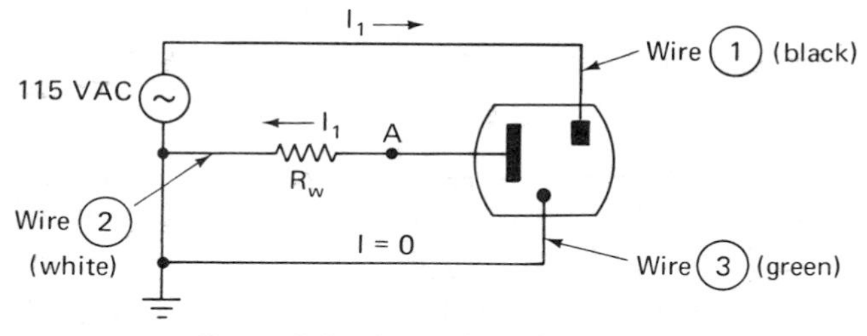

Figure 3-3. Three-wire wall outlet

A current flows in wires 1 and 2 when an appliance or instrument is connected to the outlet. Since all wire contains some resistance per foot of length, point A on wire 2 will not be at zero potential, even though it is connected to ground. The potential of the wire at point A will instead be

$$V_A = R_W I_1 \tag{3-1}$$

where R_W is the total resistance of the path from point A to ground. (The sum of the resistance of the wire and the resistance of the contact to ground determines R_W.) This means that, although wire 2 is connected to ground, it is not really a true point of zero reference.

A common example of a failure to observe proper grounding techniques involves the use of measuring equipment connected to earth ground through the third (ground) wire of the three-wire power cord [e.g., vacuum-tube voltmeters (VTVMs) or oscilloscopes which have one input terminal connected to the chassis (and the chassis is connected to earth ground)]. For example, consider the voltage measurement being attempted in Fig. 3-4. Since one of the input terminals of the VTVM is grounded, an attempt to measure the nongrounded voltage between points A and B of the circuit results in short-circuiting point B to ground. This short circuit effectively eliminates the remainder of the elements from the circuit. Thus, the voltage value measured by the VTVM is erroneous because the circuit has been drastically altered by the connection of the VTVM.

A method that allows the VTVM to be used for the measurement of nongrounded voltages is to *float* the VTVM. One way of doing this is to use a

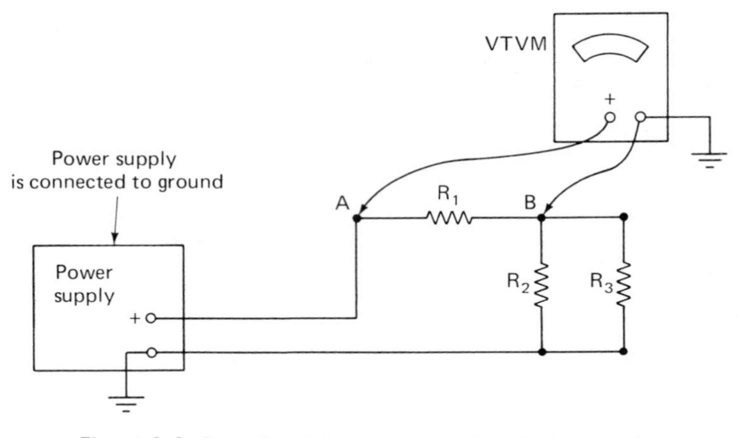

Figure 3-4. Example of incorrect grounding technique. By connecting the VTVM into the circuit as shown, point B becomes grounded.

three-to-two-wire plug adaptor on the three-wire plug of the VTVM power cord. This adaptor disconnects the third wire of the three-wire cord from earth ground and thereby floats the chassis of the VTVM. The danger that arises from using this method of floating electrical equipment, is that the chassis now assumes the potential of the point in the circuit to which the negative terminal of the instrument is connected. Thus an electrical shock hazard exists. If someone were to simultaneously touch both the chassis of the floating equipment and ground, he would form a conducting path to ground, and current could flow through him. Thus extreme care must be used if this method is employed to float electrical equipment.

EQUIPMENT GROUNDING FOR SAFETY

Ultimately, the most important reason for grounding electrical equipment is to provide additional protection against electrical shocks. Electrical instruments and household appliances are built so that their equipment cases (also called the *chassis*) are electrically isolated from the wires that carry power to their circuits. The isolation is provided by the insulation of the wires and the chassis is thereby prevented from becoming electrically "hot."

If the chassis of the equipment somehow comes into contact with an exposed part of one of the current-carrying wires (possibly due to wear or damage of the wire insulation), it will attempt to assume the same potential above the earth ground as the wire with which it is in contact. If there is no good electrical connection from the chassis to earth ground, the chassis will remain at the higher potential. If the user of this piece of equipment is unaware that the case is hot, and touches it while he is simultaneously in contact with a good ground connection (e.g., a water pipe, wet floor, etc.), current will flow through him to ground. If the current is high enough, the resulting shock may be fatal. Such an accident can occur if the appliance or equipment utilizes a two-wire power cord. In such cases, both wires carry current when the equipment is in normal operation. There is no wire available for grounding the chassis in case of accidental electrical contact. (Fig. 3-5(b)).

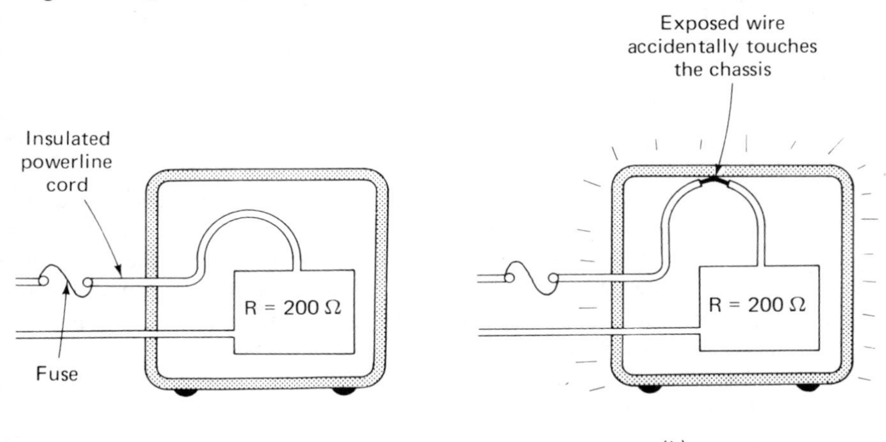

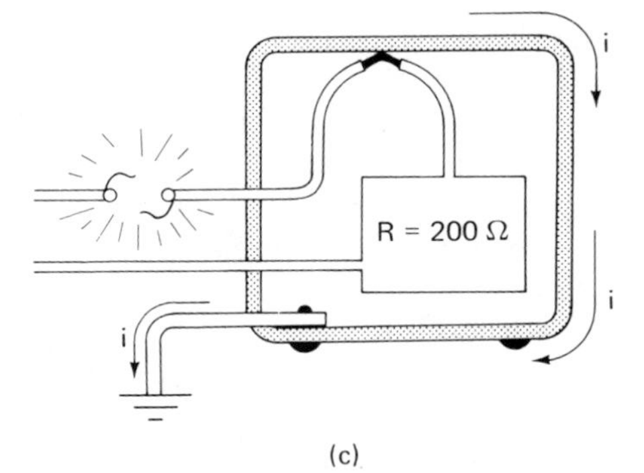

Figure 3-5. Grounding of equipment for safety

On the other hand, if there is a good connection from the chassis to ground and an exposed wire touches the chassis, the current can flow directly to ground through a very-low-resistance path. This low-resistance path usually offers less impedance to current flow than the conducting path through the appliance. A large current flow in the circuit is the result. Such a large current surge should cause the circuit fuse to burn out or the circuit breaker to open.

Since this action causes a break in the circuit, power will be cut off and the equipment will no longer be at the dangerous potential.[1]

When appliances equipped with a three-wire power cord are used, such a ground connection is provided. The third wire of the cord (see Fig. 3-5(c)) is attached to the chassis of the equipment and then to ground through the ground-wire connection in the wall-plug outlet. This makes such appliances much safer than those equipped with a two-wire cord. (Nevertheless, one can still ground the chassis of a piece of equipment even if a two-cord wire is being used. A ground connection to the chassis can be made by connecting a metal portion of the chassis to a drain pipe, or other earth ground, with an additional wire.)

Unfortunately, the three-wire cord is not a foolproof safety device. Damage may occur to the ground conductors in the three-wire cord without coming to the attention of the user. If the ground wire is severed, the equipment again becomes ungrounded. In addition, the wall-outlet-to-ground connection may be inadequate (due to a high-resistance path to ground or a break in its path). This may again leave a supposedly grounded piece of equipment ungrounded. Therefore the ground wire and building ground system should be visually inspected or electrically checked from time to time to ensure that an effective ground connection exists. Inadequate building grounds are also a source of instability and drift in sensitive measuring instruments connected to the inadequate grounding system.

Circuit Protection Devices

Large current surges from overloads or short circuits sometimes accidentally occur in electric circuits. Such current surges can lead to component destruction, electric shocks, or fires if not stopped in time. To guard electrical systems against damage from such unexpected overloads, certain protective devices are used. The most common of these are the fuse and the circuit breaker. They function by rapidly interrupting the flow of current in a circuit if it exceeds a specified value.

A fuse is basically a length of fine wire that is designed to heat up and melt if its maximum current rating is exceeded (Fig. 3-6). It is placed in series with the circuit it is meant to protect. (Circuits should be installed with fuses on the hot side of the input line unless both sides are fused. If the fuse were placed in series with the low side of the line, the electrified circuit would remain at the potential of the hot wire, even if the fuse burned out. Therefore

[1] In addition to the danger from a *dead short* as described above, current leakage from the hot wires to the chassis can occur even with no break in the wire insulation. In fact, some home appliances suffer from such current leakage due to faulty design or sloppy manufacture. Without a ground connection to the chassis, a shock hazard may again develop. The presence of a ground connection reduces the potential of the chassis to a safer level, even if the leakage current is not large enough to cause a fuse to blow.

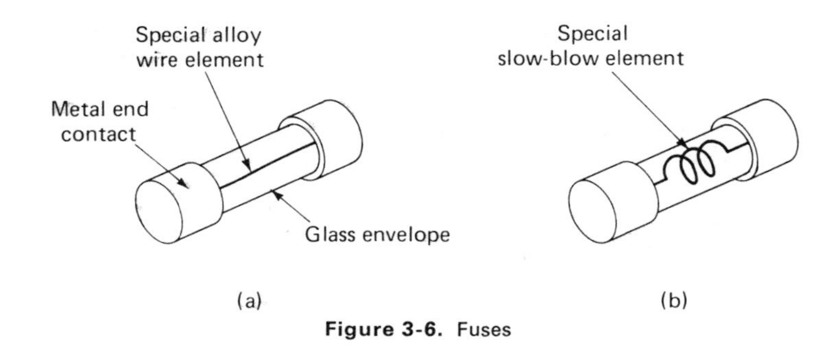

Figure 3-6. Fuses

a shock hazard would still exist.) By melting when the current flowing in the circuit exceeds its capability, the fuse destroys a portion of the conducting path. This halts the current flowing in the rest of the circuit (Fig. 3-7). Usually this break must occur very rapidly so that damage to wiring, circuit components, or the power source is prevented. In such ordinary applications, so-called *fast fuses* are used. Some circuits, however, are designed to produce or withstand short, high current pulses without damage. Such circuits still need to be protected against current surges which are too large or too long in duration. In these cases, delayed-action or *slow-blow* fuses may be used as the protective element. A slow-blow fuse resists melting if its current rating is exceeded for a short period of time. However, if the overload is too large or persists too long, the fuse also eventually melts and opens.

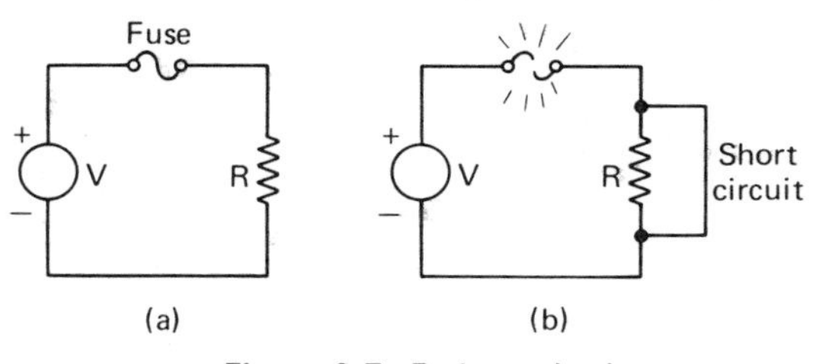

Figure 3-7. Fusing a circuit

The blowing of a fuse is an indication that there is a malfunction within the circuit the fuse is guarding. Before a blown fuse is replaced with a new one, the fault should be located and repaired.

The *circuit breaker* is a protective device which also opens a circuit if an overload is applied to it. However, unlike the fuse, the breaker is not destroyed by the overload.

Breakers generally consist of a switch which is held closed by a catch. To open the circuit, the catch is released. Two common breaker-release mechanisms are the electromagnet and the bimetallic strip (Fig. 3-8). When the current exceeds the critical value in the electromagnet coil, its magnetic field draws in the metallic bar, and the breaker catch is released. In the bimetallic strip

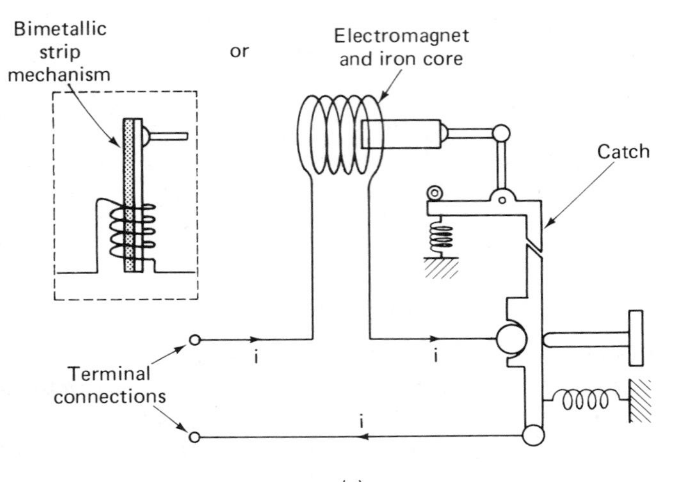

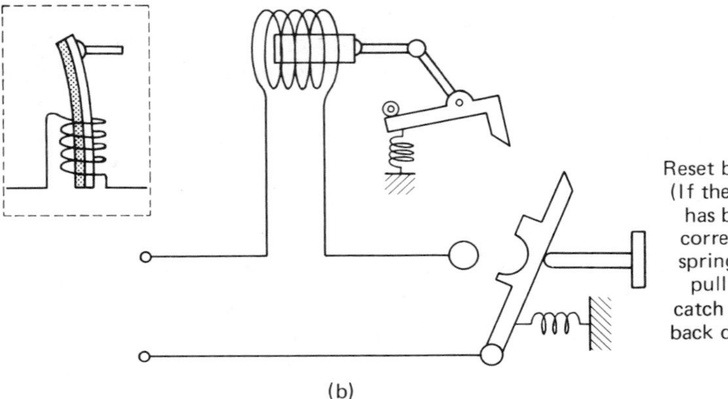

Figure 3-8. Simplified sketch of two of the mechanisms used in circuit breakers (a) Bimetallic and electromagnetic breaker release mechanisms are shown in their "closed" positions (b) If excess current passes through the mechanism: 1) heat from current bends backward the bimetallic strip causing the catch to release or 2) current passing through coil draws in iron rod, thereby releasing the catch.

type of mechanism, the current passing through the breaker heats the strip and causes it to bend. If the strip is heated by an excessive current, it bends backward so far that it causes the catch to spring open. In the thermal-magnetic breaker, both mechanisms are used. Normal overloads cause the bimetallic strip to release the breaker catch, whereas short circuits cause the electromagnet to activate the release mechanism (Fig. 3-9). When the cause of the excess current has been located and repaired, the circuit breaker can be reset to its conducting position by a switch or pushbutton. Because of the switch, circuit breakers can also be used as on-off switches.

Current limiters are another type of circuit protection device. In some instruments which contain elements that are extremely sensitive to currents (such as moving-coil meter movements), a current limiter in the form of a variable shunt is installed along with ordinary fuses. If the current in the circuit ever gets high enough to damage the sensitive element, the current limiter will come into play and divert most of the current from the path of the sensitive element.

Figure 3-9 Cutaway view of a circuit breaker (Courtesy of Westinghouse Electric Corp.)

Input Impedance, Output Impedance, and Loading

The concepts of *input impedance*, *output impedance*, and *loading* are all commonly used in the description of electrical instruments. The terms are interrelated and are often indications of how effectively a measuring instrument can perform its specified function. These terms can best be explained if we first define the concept of impedance.

Impedance, in general terms, is the ratio of the voltage to the current, and it is denoted by the letter Z. The units of impedance are ohms (Ω). In dc circuits, the impedance is equal to the ratio of the dc voltage to the dc current.

Because resistors are the only effective elements in dc circuits, the impedance is just equal to the resistance of the part of the circuit in which V and I are determined.

$$Z_{dc} = \frac{V}{I} = R \tag{3-2}$$

In ac circuits, impedance is defined as the ratio of the rms voltage to the rms current in the part of the circuit being considered.

$$Z_{ac} = \frac{V_{rms}}{I_{rms}} \tag{3-3}$$

However, in ac circuits, the impedance is no longer strictly resistive. Since capacitors and inductors also contribute to the impedance of ac circuits, the impedance contains a reactive as well as a resistive component.

If we have an electrical instrument and if we make a connection to its input or output terminals, the instrument will exhibit some characteristic impedance as seen from these terminals. For the sake of analysis, we can always replace the instrument by this impedance (along with an appropriate voltage source, if the instrument contains active as well as passive elements). If the instrument is a measuring instrument (such as a meter, oscilloscope, etc.), the ratio of the voltage across the input terminals to the current flowing into them is known as the *input impedance* of the instrument.

$$Z_{in} = \frac{V_{in}}{I_{in}} \tag{3-4}$$

This input impedance can be measured if we connect a voltage source across the input terminals and measure the current that flows through the instrument at a particular voltage setting (Fig. 3-10). Note that this ratio (and consequently the input impedance) may be so high in some instruments that it would be very hard to actually measure. If dc signals are used

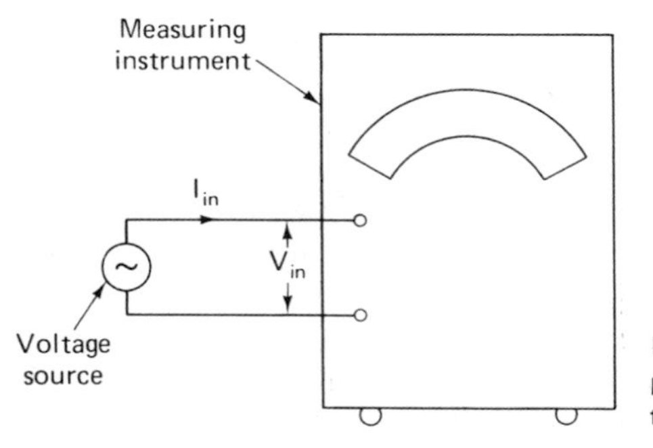

Figure 3-10. Determining the input impedance of a measuring instrument with the help of a voltage source

to excite the instrument, V_{in} and I_{in} are dc quantities. If the input signals to an instrument are ac quantities, V_{in} and I_{in} refer to the rms values of the quantities.

The *output impedance* of a device is defined as

$$Z_{out} = \frac{V_{out}}{I_{out}}$$

In most cases we will be interested in the output impedance of devices or instruments which contain active elements and thus serve as signal sources in measurement systems. (Instruments and devices such as power supplies, oscillators, batteries, amplifiers, and active transducers all fit into this category.) For these sources, V_{out} is the voltage appearing across the open-circuited output terminals of the device. I_{out} is the calculated current that would flow if the output terminals were connected by a short circuit. However, the output impedance of sources is not measured by actually short-circuiting the output terminals. (An attempt to measure the value of I_{out} in this manner could result in burning out the source.) As an example of how the output impedance of some devices can be measured, see the discussion in Chapter 8 on the measurement of the internal resistance of a battery. (The internal resistance of a battery corresponds to its output impedance.) Now let us see how the concepts of input and output impedance are related to the concept of *loading*.

Instruments that are used to measure voltage are placed *across* (in parallel) the element or circuit being measured. Ideally, a measuring instrument should not disturb or change the values of the current and voltage in the circuit under test. In the case of voltage-measuring devices, the instruments should draw no current when connected to the two points across which the voltage is being measured. This condition would be satisified if the voltage measuring device appeared to the test circuit as an open circuit. The input impedance of the voltage measuring device describes how it actually appears to the test circuit. Since an open circuit is equivalent to an infinite impedance, the value of the input impedance of a voltage-measuring instrument determines how closely it approaches the open-circuit ideal. However, because voltage-measuring instruments are not ideal, they do draw some current from the circuit being measured. The effect of drawing current is known as *loading*.[2]

If a voltage-measuring device does not have a high input impedance and consequently draws a significant percentage of the current flowing in the test circuit, the measuring device is said to be *loading down* the test circuit. The

[2]Loading is also occasionally mentioned in connection with ammeters. However, since an ammeter is connected in *series* with the branch, its resistance reduces the original value of current in the branch, rather than drawing it away into another branch. Thus, for the sake of consistency, we will not refer to the disturbance of a circuit by an ammeter as loading, since it does not satisfy our definition of drawing current.

greater the percentage of the current drawn from the circuit under test, the more a voltage-measuring device disturbs the circuit it is monitoring. Thus, the higher the value of input impedance that a voltage measuring instrument possesses the more accurate a voltage measurement it can make. Additional details on the loading effects of voltmeters and oscilloscopes are given in the chapters dealing specifically with these instruments.

If a signal source (e.g., a generator or an oscillator) is providing energy to a circuit, the term loading is also used to describe the fact that current is being drawn from the signal source. As the impedance connected across the output terminals of such sources is made smaller, the current output of the source will increase. Hence oscillators or generators are said to be *loaded down* when a low impedance is connected across their output terminals. Additional information relating to the effects of loading on the operation of oscillators is presented in Chapter 13.

Interference Signals and Shielding

Many electrical measurements involve the detection and measurement of low-level signals.[3] Usually these low-level signals are amplified by a part of the measuring system so that they can be more easily displayed or otherwise utilized.

Unfortunately, the electrical laboratory environment usually contains many sources of electric and magnetic energy that can induce extra, unwanted signals in the wires that carry low-level signals. These unwanted signals are then amplified along with the signal being measured. Under some circumstances, the magnitudes of such induced signals can become so large that they distort or obscure the signal of interest and lead to inaccurate or meaningless measurement results. Therefore, as a part of making accurate measurements, the sources of interfering signals need to be identified. In addition, if the sources create significant interference effects, steps must be taken to eliminate or minimize the interference. This usually means that wires carrying low-level signals must be shielded against the influences of such external interfering sources.[4]

In typical measurement systems, low-level signal wires could be the wires that transfer signals from the circuit under test to the inputs of the amplifying or measuring instruments. In addition, any other portions or components of measuring instruments that are sensitive to external sources of interference must also be protected from their influence.

[3] In this context low-level means that the magnitude of a signal is on the order of microvolts or millivolts.

[4] Ordinarily, it is not as important to shield wires carrying large signals against external interference signals. Interfering signals are usually in the millivolt range and won't significantly affect the values of the larger signals.

SOURCES OF INTERFERENCE SIGNALS

Interference signals can be classified according to the physical phenomena that are responsible for their generation or transmission. The five major types of interference signals are therefore known as:

1. Capacitive interference
2. Inductive interference
3. Electromagnetic interference
4. Resistively coupled interference
5. Ground-loop interference

Capacitive and inductive interference signals are caused by the so-called *near fields* which exist in the space close to any charged or current-carrying conductor. Electromagnetic interference signals arise from the *far* or *radiation fields* that are created at somewhat farther distances from such conductors. Ground-loop interference can arise from the effects of both near fields and far fields, as well as from other electrical effects. Table 3-1 summarizes the effects of each of these types of interference. The characteristics of interference signals and the necessary shielding precautions that must be taken against them are a function of whether the source of the interference is a near field or a radiation field. Therefore, the distinctions between the two types of fields need to be discussed in more detail.

When a conductor possesses a net electric charge, an electric field exists in the space around the conductor. If the conductor carries a current, the region surrounding the conductor contains a magnetic field.[5] If the charge or current varies with time, the associated fields at any point in the surrounding space will also vary, that is, they will shrink and grow in magnitude in proportion to the strength of their source. In the regions nearby the conductor, these fields will pulsate in synchronization with the time variation of the current or charge. Such synchronous fields are known as the *near* (or *induction*) fields.

If an electric charge is placed at some distance from the conductor, it can feel the effects of any changes in the current flow or charge in the conductor. The effects are felt as variations in the strengths of the electric and magnetic fields at the point where the charge is located. However, the effects of any

[5] A conductor with a low resistance can carry a large current (and hence have a large magnetic field) and have only a small electric field between it and ground. On the other hand, a conductor can also have a very large amount of charge (and therefore a large surrounding electrical field) without supporting any current flow. Then no magnetic field will be present. Finally, the conductor can have both a large current flowing through it and a large potential difference (signifying a large charge difference) between it and other conductors. Then both a large electric field and a large magnetic field will exist around the conductor.

changes in the charge or current flow are carried away from the conductor by the fields at the speed of light. Therefore, at points far from the conductor, the effects of such changes are not felt until some time after the actions have taken place. In fact, once the generated fields themselves are at appreciable distances from the conductor, they are also affected by this time lag. Due to the time lag, a portion of such distant fields can no longer be recalled to the conductor quickly enough when the sources of the fields decrease in strength. Instead, this portion of such fields continues to move away from the conductor as radiated energy. Once such energy has broken away from the conductor's fields in this manner, it can move independently throughout space and influence other conductors far from the location of the originating conductor. In fact, since the field is periodic, it can induce an alternating signal of the same frequency as the original source in any suitable conductor it encounters. Such fields are called the *far* or *radiation* fields.

In order for electric or magnetic fields to exist in space without being associated with the charge or current in a conductor, they must assume the form of waves. These waves must simultaneously contain electric and magnetic field components. As a result, the waves are given the name *electromagnetic* (*EM*) *radiation.* Radio waves are one common form of such EM radiation. The signals induced in the antennas of radio receivers are an example of the way EM waves influence conductors far from their point of origin.

To designate the distance from the conductor at which EM radiation begins to be produced, we used the term "an appreciable distance." The magnitude of this distance (let us call it D) depends on the frequency of the time-varying source. If the frequency of the changes in the current or charge on the conductor is low, D is quite large. As the frequency of these changes gets higher, D gets progressively smaller.

A numerical value for D can be roughly calculated from the expression

$$D = \frac{1}{2}\frac{c}{f} \tag{3-5}$$

where c is the speed of light (in m/s) and f is the frequency of the current or charge variation (in Hz).

EXAMPLE 3-1

For conductors carrying signals that vary at (a) 60 Hz, (b) 100 kHz, and (c) 100 MHz, find the value of D.

Solution The speed of light, c, is 3×10^8 m/s.

(a) For $f = 60$ Hz

$$D = \frac{1}{2}\frac{(3 \times 10^8)}{60} = 2.5 \times 10^6 \text{ m}$$

(b) For $f = 100$ kHz

$$D = \frac{1}{2}\frac{(3 \times 10^8)}{(1 \times 10^5)} = 1.5 \times 10^3 \text{ m}$$

(c) For $f = 100$ MHz

$$D = \frac{1}{2}\frac{(3 \times 10^8)}{(1 \times 10^8)} = 1.5 \text{ m}$$

If a conductor is known to be a source of interference, then at distances smaller than D we can say that any interference signals will be caused primarily by its *near fields*. For distances larger than D, the interference signals will be predominantly due to *radiation fields*.

For low-frequency sources, the primary cause of interference signals will therefore be the near fields. For high-frequency sources, the interference can be due to either near fields or radiation fields, depending upon how close the source is located to the wire in which it induces a signal.

CAPACITIVE INTERFERENCE

When two conductors at different voltages are placed near one another, a charge difference, and an electric field, exist between them. Both of these effects arise in association with the potential difference. Since a charge difference and voltage difference exist between the two conductors, a capacitive effect also exists between them. Such capacitive effects can give rise to interference signals in both conductors. This interference is therefore called *capacitive interference* (or *capacitive pickup*.)

In measurement systems, one of two such conductors is usually considered to be the *source* of interference and the other the *object* of the interference. Low-level-signal-carrying wires which are connected to the inputs of instruments that possess high input impedances are most likely to be *objects* of capacitive interference. The *sources* of interference in the laboratory are any wires in which the potential varies periodically. The wires of unused wall outlets or fluorescent bulb fixtures are prime examples of such sources. They carry a potential that varies from +163 V to −163 V (peak value) at 60 Hz. This potential variation continues even though no current flows in the wires.[6] (As long as the outlets or fixtures remain unconnected, virtually no current flows in their wires.) Other typical sources of capacitive interference are wires from other equipment being run from the power line, from oscillators, and from signal generators.

[6] All interference signals caused by the voltage or current variations in the 60-Hz power-line wires are called *hum*. They get this name from the fact that, if the interference is amplified and fed to a loudspeaker, the resulting sound is a low hum. The sound is that of a 60-Hz audio tone.

As an example of how capacitive effects induce interference signals in a low-level-signal wire, consider wire A shown in Fig. 3-11(a). Let us assume the wire is connected to ground by a resistor of value $2R$ at each of its ends. As long as there is no current flowing in the wire, it remains at zero potential (the same potential as the ground point to which it is connected). Let us then place another wire (wire B of Fig. 3-11(b)) near wire A. The voltage of wire B varies from $+V_0$ to $-V_0$ volts at some specific frequency. When the voltage of wire B equals $+V_0$, electrons are drawn up to wire A from the ground plane due to the capacitive effect between wires A and B. The charges must move through the resistors to get to wire A, and moving charges constitute a current flow. This current, i, flowing through the resistors will cause the wire to acquire some nonzero voltage according to

$$-v = -iR_1 \tag{3-6}$$

Likewise, when wire B is at $-V_0$ volts, electrons are repelled from wire A and will move away from it. A current thereby flows through the resistors again, this time in the opposite direction The potential of wire A also reverses and becomes equal to $+V_1$ when wire B is at voltage $-V_0$.

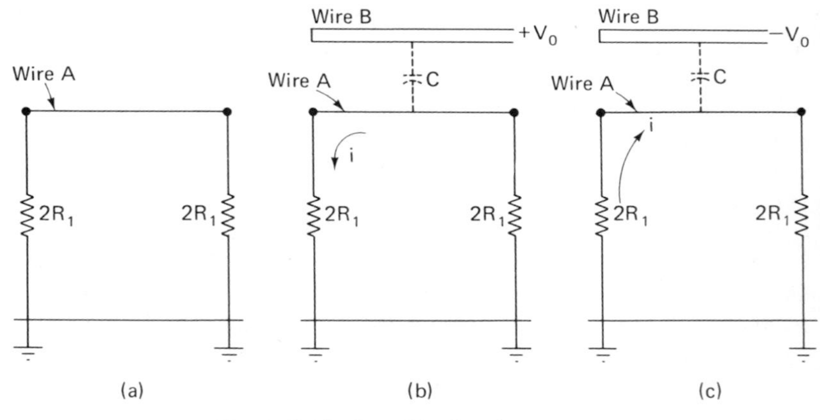

Figure 3-11. Capacitive interference

Thus, as the voltage of wire B varies from $+V_0$ to $-V_0$, the voltage induced in wire A will vary from $-V_1$ to $+V_1$. The magnitude of V_1 will depend upon the capacitance that exists between wires A and B, the value of the resistors connecting wire A to the ground, the value of V_0, and the frequency of the variation of the voltage V_0. This information can be utilized to develop a circuit model of the events occuring in the system (Fig. 3-12). In the model, wire B is connected to a voltage source which varies at a frequency of ω radians per second. A capacitive effect exists between wires A and B, and this is represented by a capacitor of value C.

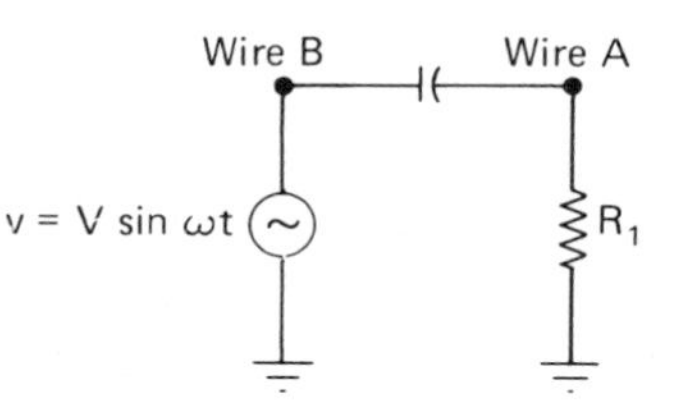

Figure 3-12. Model for calculating the magnitude of capacitive interference between two wires

If we know the frequency of V_0 and its magnitude, the value of R_1, and the magnitude of the interference signal induced in wire A, we can find C.

EXAMPLE 3-2

A test lead is connected to an instrument which has a 2-MΩ input impedance and to a circuit which has a 2-MΩ output impedance (Fig. 3-13). If a 1-mV rms signal is induced in the lead from nearby 60-Hz power wires, what is the capacitance that must exist between the lead and the power wires?

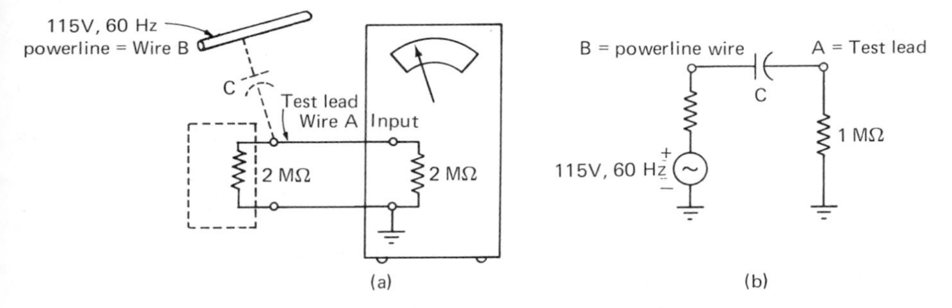

Figure 3-13. (a) Instrument with 2MΩ input impedance connected to test circuit having an output impedance of 2MΩ (b) Equivalent circuit model of (a).

Solution: Using the model shown in Fig. 3-12, we see that the rms voltage of point A will be 1 mV. R_1 of the model is 1 MΩ for this case. Then the current flowing in the circuit can be found from

$$I_{\text{rms}} = \frac{V_{A_{\text{rms}}}}{R_1} = \frac{1 \text{ mV}}{1 \text{ M}\Omega} = \frac{0.001\text{V}}{10^6\Omega} = 10^{-9}\text{A}$$

The capacitance between conductors A and B is found from

$$V_{B_{\text{rms}}} = I_{\text{rms}} Z = I_{\text{rms}}\left(R^2 + \frac{1}{\omega^2 C^2}\right)^{1/2} \qquad (3\text{-}7)$$

Since in this case $1/\omega C \gg R$,[7] we can neglect R in Eq. (3-7) and write

$$V_{B_{\text{rms}}} \approx \frac{I_{\text{rms}}}{\omega C} \qquad (3\text{-}8)$$

[7]Actually in this step we are assuming that the value of C is so small that $\omega C = (377C)$, is such a tiny number that $1/\omega C \gg 10^6$. The final answer bears out the validity of this assumption.

or

$$C = \frac{I_{\text{rms}}}{\omega V_{B_{\text{rms}}}} = \frac{10^{-9}}{2\pi(60) \times (120)} = 2 \times 10^{-14}\text{F} = 0.02 \text{ pF}$$

Example 3-2 indicates how small the capacitance can be to induce a 1-mV interference signal in such a high impedance circuit. This example is especially relevant to instrument practice because the input impedances of oscilloscopes, electronic voltmeters, and most amplifiers are 1 MΩ or more. If such instruments are used to measure low-level signals in high-impedance circuits, the test leads are very much like the test lead examined in Example 3-2. We can see that such leads will be very prone to picking up capacitive interference from any nearby conductors possessing a time-varying potential. In addition to the 60-Hz signal, Eq. (3-8) shows that, as the frequency of the source of the interference increases, the induced current due to this capacitive pickup also becomes larger. Therefore conductors carrying high-frequency signals are also liable to be sources of capacitive pickup.

In summary, we see that unwanted capacitive pickup is increased as:

1. The input impedance of the instrument increases.
2. The capacitance between the interfering source and test lead increases.
3. The voltage between the interfering source and test lead increases.
4. The frequency of the voltage in the interfering source increases.

To prevent wires which carry low-level voltage signals from being influenced by capacitive effects, several courses of action are possible. The most common countermeasure is to surround the low-level signal wire with an electrostatic shield. The shield in actual wires usually consists of a braided metal sleeve which surrounds the test lead (Fig. 3-14). This type of shielding is effective because external electric fields cannot penetrate an enclosure surrounded by an electrical conductor. However, the metal shield surrounding the test signal cable must also be connected to ground through a low-impedance path so that it will remain at a potential very close to zero. Then no appreciable potential and no interfering signal will arise between the signal-carrying test lead and the shield itself. Since the test lead is enclosed by

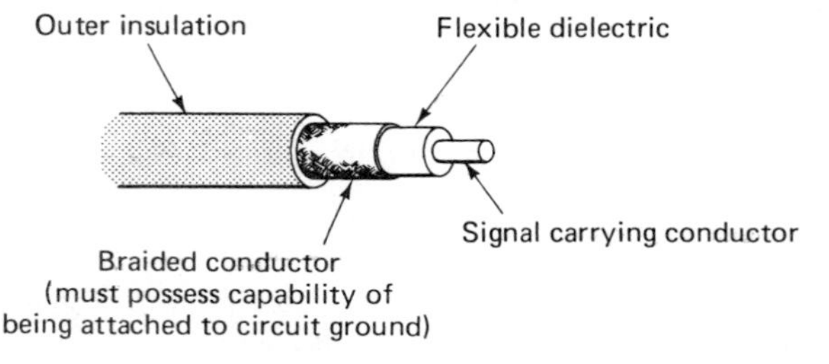

Figure 3-14. Sketch of shielded cable

the electrostatic shield, it will be protected from being influenced by any external varying voltages. In practice, shielded cables are about 90 percent effective in reducing capacitive interference due to external fields.

Another more effective but also more complex method of reducing capacitive pickup is to use a differential amplifier with a balanced pair of twisted leads to transfer the signal from the test circuit to the measuring instrument. The reason this method is so effective can best be explained along with the discussion on differential amplifiers in Chapter 16.

INDUCTIVE INTERFERENCE AND SHIELDING

Capacitive interference involves electrostatic fields that exist between conductors at different potentials. Inductive interference stems from magnetic fields that are associated with current-carrying conductors. As mentioned earlier, magnetic fields can exist in the regions nearby current-carrying conductors, or they can exist as part of electromagnetic (EM) waves in regions of space far from any conductors. In this section we will consider the effects of near-field magnetic fields.

A current can be induced to flow in a closed-loop conductor if the loop is in motion within a magnetic field. By the same token, a current in a loop can also be induced if the magnetic field cutting the loop changes with time. The magnitude of the induced current will depend upon the strength of the magnetic field, the frequency of its variation, and the area of the loop. (The loop area is a factor because the larger the loop, the more magnetic flux is encompassed by it.) In addition, the current depends upon the resistance of the loop. If the resistance of the current path forming the loop is high, the magnetic fields usually encountered in the laboratory environment will induce only insignificantly small currents. For this reason most test circuits connected to measuring instruments which have high input impedances are rarely bothered by significant inductive interference signals. However, test leads carrying low-level signals which are being fed to instruments that have low input impedances can be affected by inductive interference. For example, instruments which use current amplifiers have low input impedances and are subject to picking-up inductive interference in their low-level signal-carrying leads. (Current amplifiers are used in such applications as monitoring signals from strain gauges and in certain biomedical measurements.)

There are also other instruments which are subject to being adversely affected by external magnetic fields in other ways. These instruments include electromechanical meter movements (especially the iron vane and electrodynamometer movements), cathode ray tubes (CRTs), and relays. The meter movements detect current with the help of rather weak magnetic field effects, and these magnetic effects can be disturbed by additional magnetic fields in their vicinity. Unwanted changes in the magnetic fields of the movements can cause inaccurate meter readings. (The basic meters to be discussed in

Chapter 4 are sensitive to such magnetic interference.) The electron beam in the CRT can be deflected by magnetic fields, and such deflections can lead to misleading shifts in the position of the beam. Finally, since relays are switches activated by magnetic fields, any additional magnetic fields may cause sensitive relays to be activated at other than their proper magnetic field strengths.

The most common sources of magnetic-field interference in the laboratory are the inductors and transformers found in electrical instruments. (Transformers are used in the power supplies that are a part of most electronic equipment. Inductors are used in wide variety of circuits.) In addition, any wires carrying large currents are surrounded by magnetic fields. In some cases such fields are strong enough to cause interference effects in nearby circuits.

Shields designed to prevent low-frequency magnetic fields from entering an enclosure are made of ferromagnetic conductors (Fig. 3-15(a)). This type of material provides a preferred path for the flux lines of the magnetic field. Thus the magnetic field concentrates itself in the ferromagnetic material, rather than penetrating strongly into the region surrounded by the shield. When such material surrounds a test lead, the lead is protected from the largest part of the magnetic field. If the material is used to build an enclosure around an element which is a source of magnetic fields, it prevents the fields from spreading into the nearby space and disturbing sensitive circuits which may be in the vicinity (Fig. 3-15(b)).

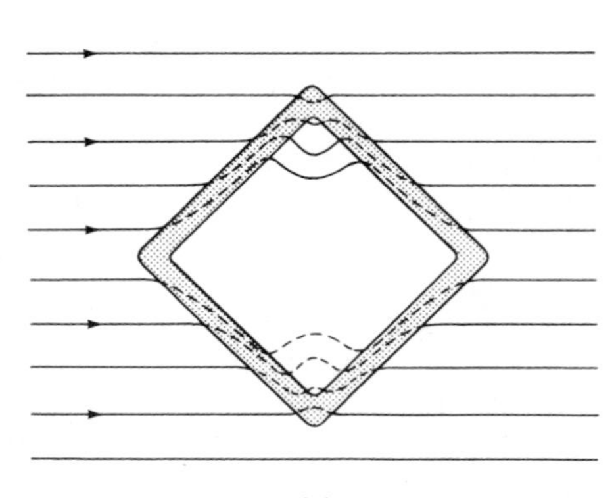

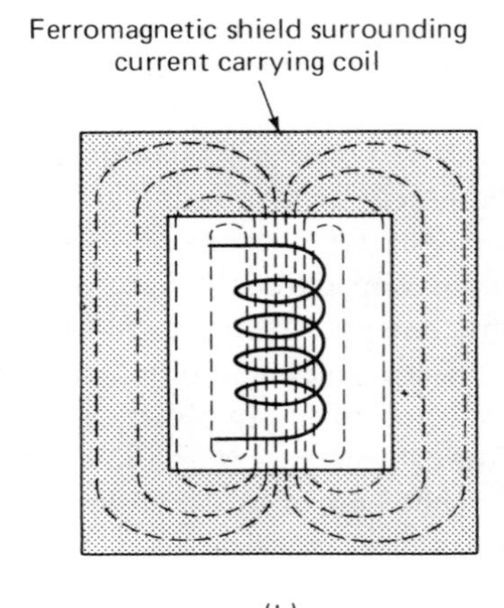

Figure 3-15. Shielding against magnetic effects (a) Ferromagnetic shield which protects device from external, low frequency magnetic fields (b) Ferromagnetic shield which surrounds a coil carrying a low frequency current weakens the magnetic field of the coil outside the space of the shield

If the source generates a high-frequency magnetic field, a copper housing is a better shield than a ferromagnetic housing. At high frequencies, the shielding against magnetic fields involves the generation of counter magnetic fields by eddy currents induced in the surrounding electrical conductor.

In sum, the following steps are taken to eliminate most inductive interference effects. Any obvious sources of large magnetic fields, such as coils or transformers, are enclosed in a ferromagnetic or copper housing (depending on the frequency of the source). Also, large-current-carrying wires (which could be sources of magnetic fields) are usually twisted together with wires that carry their return currents. The magnetic fields of the oppositely moving currents then cancel each other. (Vacuum-tube heater wires are often twisted together for this reason.) Finally, any instruments or components susceptible to external magnetic field interference are enclosed in soft-iron or other suitable shields. (To minimize weight, the CRT may be painted with special ferromagnetic materials to provide a similar shielding effect.)

ELECTROMAGNETIC INTERFERENCE AND SHIELDING

At high frequencies, a part of the energy associated with the fluctuating current or charge in a conductor is radiated away from it in the form of electromagnetic radiation. This phenomenon is specifically used to generate radio waves for communication and radar applications. However, it has become common parlance to refer to any EM waves that have frequencies comparable to radio or radar waves as *radio-frequency* or *rf* waves—whether they are actually signals from radio or radar transmitters or not. Besides radio waves, there are many other sources (both man-made and natural) which produce rf signals. However, in measurement systems, *all* types of rf signals are considered to be sources of unwanted EM interference. As a result, sensitive circuits must be protected from all rf signals no matter what their source might be.

Man-made sources of rf signals include gas discharges in fluorescent lights and X-ray tubes; arcing in electric motors, generators, switches, and relays; and high-frequency oscillations in pulse circuits, discharge circuits, and oscillators. (Of course, radio transmitters can also be sources of rf interference.) Natural sources of rf radiation include thunderstorms and other electrical atmospheric phenomena.

Since EM waves have both electric and magnetic field components, interference effects comparable to those generated by capacitive and inductive pickup can be caused by EM radiation. If the circuits of a measuring system are susceptible to one of these kinds of pickup more than the other, then the EM waves will predominantly produce that type of interference in the circuit. Furthermore, both inductive and capacitive interference effects increase as the frequency of the disturbing effect increases. Since EM waves are usually high-frequency signals, this makes them even more likely to induce interference. For these reasons, a nearby source of EM radiation is likely to cause interference in the widest variety of circuits. Fortunately, it is also possible to effectively shield against EM radiation.

EM shields make use of the fact that EM radiation must simultaneously contain electric and magnetic fields in order to propagate independently through space. If we eliminate either the electric or the magnetic component of an EM wave, the other component is also halted. Therefore, since a shield designed to eliminate electrostatic fields can be made quite easily, this type of shield is also used to eliminate the electric-field component of an EM wave. Without the electric component, the magnetic field cannot continue to propagate and is thereby extinguished too. Thus, an enclosure consisting of a good electric conductor connected to ground through a low-impedance path will be an effective shield against EM interference.

If a known source of EM radiation is in the vicinity of the measurement setup, it is a wise policy to turn off the source while making measurements or to enclose it in a housing made of a good conductor which is tied to ground. Such a grounded conductor will stop the EM radiation from propagating away from the source and causing interference in circuits that are close enough to be affected by the emitted radiation.

RESISTIVELY COUPLED INTERFERENCE

Interference can also be caused by signals which originate in electrical equipment that is somehow connected in the same circuit as the measuring instrument. Since the connection takes place through electrical conductors (i.e., wires), such interference is called *resistively coupled interference*. The most common conducting path which introduces such resistively coupled interference into the measurement system is the ac power line. Any electrical equipment which generates signals during its operation and which is connected in the same power-line circuit as the measuring instruments is likely to cause unwanted signals in the measurement system. Motors, temperature-controlled furnaces, and other such equipment are some of the more common sources of such interference signals. If sensitive measurements must be made, these sources must be eliminated from any ac power-line circuits to which the measurement system is connected.

GROUND-LOOP INTERFERENCE

Ground loops and ground-loop interference are often a nemesis in low-level and high-frequency measurement systems. Because they are elusive and little understood entities, ground loops are often blamed for unexplained interference signals.

Although their causes and methods of elimination become quite complex at high frequencies, at low frequencies ground-loop effects can be more easily understood. Hence, we will undertake a discussion that will explore the origins of ground-loop currents and their effects in low-frequency circuits. The objective of the discussion will be to indicate how to eliminate ground-loop interference in such measurement systems.

The countermeasures against ground-loop interference which will be introduced by our discussion will also be partially effective in eliminating ground loops at higher frequencies. However, because the origins of ground loops at higher frequencies involve more complex effects, it is also necessary to employ more sophisticated preventive measures to completely eliminate them. A full exploration of such measures would require an entire text and is actually not necessary for the type of measurements which we will be discussing. The reader interested in more information on the subject is referred to a book by R. Morrison entitled *Grounding and Shielding Techniques* (John Wiley & Sons, Inc., 1970).

Ground loops are closed electrical paths in which the sections of the path consist of the ground wires of a system and the ground plane. Ground loops are created whenever two or more terminals of a ground conductor of an electrical system are connected to the ground plane at different points (Fig. 3-16). Since the ground wires of most systems and the ground plane are usually low-impedance conducting paths, ground loops as a whole are conducting paths of low impedance. Thus, if even small voltage differences exist between any points along the loop, large currents will flow in them.

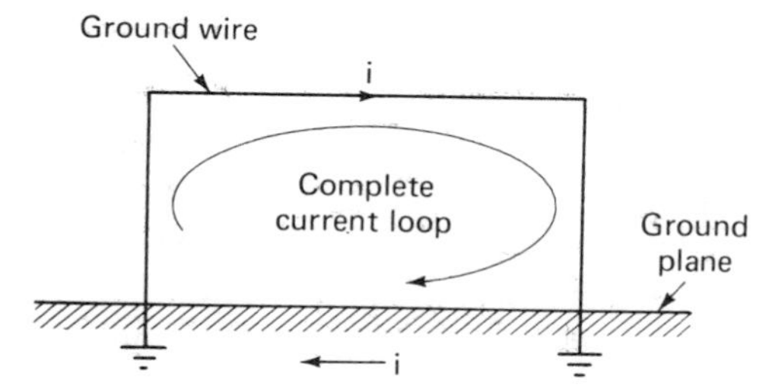

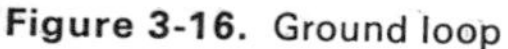
Figure 3-16. Ground loop

The two principal causes of current flow in ground loops are:

1. Differences in potential between the points of the ground plane to which the ground terminals are connected.
2. Inductive pickup—due to stray magnetic and rf fields.

Differences in potential can and often do exist between two points of a grounding system because the earth and the conductors of a grounding system which are tied to earth are not *perfect conductors*. Since the grounding system of a building carries return currents back to ground, the fact that resistance exists in the wires of the system means that different points along the grounding system can be at different potentials. For example, Fig. 3-17 shows a possible grounding system of a building. We see that wall outlets

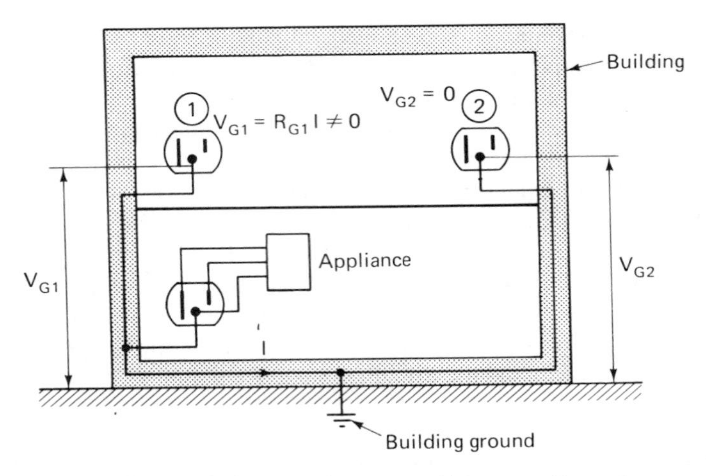

Figure 3-17. Building ground system—which causes V_{G1} and V_{G2} to be at different levels. If a conductor is connected between the ground points of plug 1 and 2 a current will flow in the resultant ground loop.

① and ② are connected to the earth ground of the building by ground wires. These ground wires possess a certain amount of resistance per foot of length. Therefore, if a current is flowing in the ground wire, the potential at the point where the ground wire is connected to the wall outlet will not be zero. In the situation shown in Fig. 3-17, one of the grounding wires is carrying a current. The current may be due to a faulty instrument or appliance which is connected to the ground circuit at some other point. Because of this current, the potential of the ground wire at wall outlet ① will be at some non-zero value ($V_{G1} = R_{G1} I \neq 0$). Meanwhile the potential of the ground wire at wall outlet ② will be equal to zero ($V_{G2} = 0$), since no current is flowing in its circuit. If a connection were established between sockets ① and ②, a ground loop would be created. Furthermore, a potential difference would exist between two points in the loop. Thus a current flow in the ground loop would also be present. If the grounding system of a building is poor (that is, if the conducting path to ground has a high resistance), ground-loop currents will be even more pronounced.

Inductive pickup by a ground loop occurs the same way as inductive pickup in any closed conducting loop. Since a ground loop has a relatively large area and very low impedance, inductive pickup from stray magnetic or electromagnetic fields in the measurement environment can occur quite easily. Orthodox shielding methods are not as effective against inductive pickup in ground loops because the shielding conductors and the ground plane themselves form a part of the loop.

These resulting ground-loop currents from potential differences or inductive pickup produce interference in a system by causing inductive pickup in nearby low-level-signal-carrying wires themselves. They also change the apparent values of signals being measured by shifting the potential level of the ground point to which the signals are being referenced.

The key to preventing or eliminating ground-loop interference relies on applying special grounding techniques to the entire measurement system of interest. These techniques are based on the fact that current cannot flow in any path unless a complete loop exists. Therefore, to eliminate ground-loop interference, we must avoid establishing any complete ground-loop paths and break up any such loops which already exist. The best way to ensure that no ground-loop paths exist is to design the measuring system so that only one point of it is ever connected to ground. In that way there can be no closed loop through the ground plane because there is only one link to ground in the entire system.

Figure 3-18 shows how this principle is applied to a simple system made up of two measuring instruments powered from the 115-VAC, 60-Hz power line. The chassis of both pieces of equipment are connected to ground by the third wire of the three-wire power cord. The grounded input terminal of each instrument is also tied (connected) to the chassis. When the two instruments are connected together to form a measurement system, a ground loop like the one shown in Fig. 3-18 is created. To break the ground loop of this system, a 2-to-3-wire adapter is used on the wall plugs of one of the instruments. This adapter breaks the third-wire ground connection at that point. As a result, the system is connected to ground at only one point (i.e., through

Figure 3-18. Test setup for avoiding ground loop (Courtesy of Hewlett-Packard Co.)

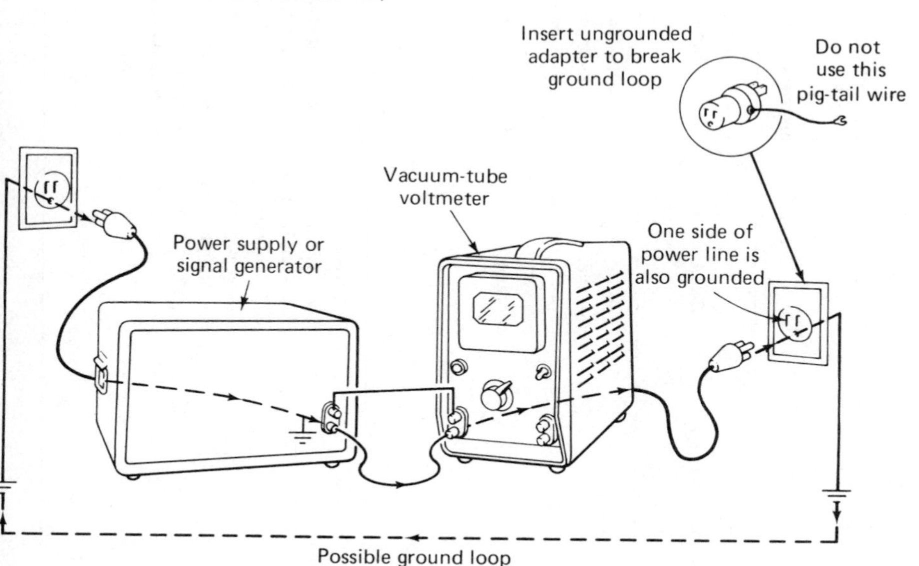

Table 3-1 Summary of the Characteristics of External Interference Signals

Type of Interference	Source	Susceptible Elements	Conditions Leading to Large Interfering Signals	Countermeasures
Capacitive	Time-varying potential difference between two non-connected conductors.	Small-signal-carrying wires or test leads in high-impedance circuits (oscilloscope or VTVM test leads; amplifier input leads).	1. High impedance between test lead and ground. 2. Long lead lengths. 3. Increases with increasing frequency or potential difference.	1. Enclose small-signal-carrying wires with an electrostatic shield that is connected to ground, i.e., shielded or coaxial cables. 2. Use a differential input with twisted pair input leads.
Inductive	Time-varying currents in a conductor (coils are particularly strong sources of time-varying inductive fields).	Wires or test leads carrying low-level signals in low-impedance circuits. Iron vane and electrodynamometer meters Cathode-ray tubes Relays	1. Closed loop in which current can flow is necessary for inductive pickup. 2. Increases with increasing loop area and frequency of source.	1. Surround obvious sources of magnetic fields with a ferromagnetic material. 2. Shield low-level-carrying wires with ferromagnetic shields. 3. Keep large-current-carrying wires and small-signal-carrying wires as far apart in a circuit as possible. 4. Keep current-carrying conductors perpendicular to one another when possible. 5. Use differential inputs and twisted pair leads.

Radio Frequency	1. *Man-made*—radio transmitters; arcing in motors and contacts; gas discharges, etc. 2. *Natural*—thunder storms and other ionospheric phenomena.	Small-signal-carrying wires in all circuits.	1. Nearby sources of rf are producing large amounts of interference signals.	1. Enclose obvious sources of rf in electrical shield. 2. Shield low-level-carrying signals with a well-conducting enclosure tied to ground with a low-impedance path.
Ground Loops	1. Potential differences existing between the two points in the ground plane to which the system is connected. 2. Inductive coupling of the ground loop and neighboring magnetic fields.	Ground-wire connections in any system of instruments.	1. Multiple connections to the ground plane in the measurement system. 2. High-frequency signal sources in vicinity of measurement system.	1. Connect all ground points in the circuit to only one point which is connected to earth ground.
Resistive Coupling	1. Ac power line	1. Measuring instruments connected in any way to power line.	1. Noise producing instruments in same power circuit.	1. Avoid using an ac power line which is also connected to possible noise sources.

the ground wire of the other plug). Since there is no longer a closed loop in which to flow, currents which cause ground-loop interference are eliminated.

In systems of greater complexity there may be more than two separate connections to ground. In such cases, several ground loops can be established. Some or all of these loops may cause significant, unwanted interference. Since cable shields are often tied to ground, they are subject to being the causes of ground loops. Therefore, in more complex systems it is advisable to connect all the shields together and tie them to ground at the same point where the measurement system is grounded (Fig. 3-19). In general, if more than one point in a system must be tied to ground, these ground connections should all be made at the point where the input signal is grounded.

Cables, Connectors, and Switches

CABLES

A large proportion of electrical signals are transmitted through solid electrical conductors. Most such signal-carrying conductors are in the form of wires or cables. (A wire is a single conductor; a cable is a configuration of two or more wire conductors.) The type of cable used depends on the specific application.

Usually the best electrical conductor is most suitable as a carrier of electrical signals. In other words, the better the electrical conductor, the lower the resistance losses which occur during the transmission of electrical signals. Therefore, the conductors in wires and cables are usually made of copper. (In some cases aluminum is replacing copper, due to the high cost of copper.)

Wires and cables are also usually surrounded by some type of electrical insulator. The insulation prevents the current they carry from leaking away to any conducting materials with which the cable makes contact. The materials chosen to make this insulation have a high insulation resistance, high mechanical strength, and durability. Furthermore, they are designed to be able to operate over a fairly wide temperature range and to withstand oil and corrosive chemicals without deterioration. The most popular insulating materials are PVC (polyvinyl chloride), Teflon, polyethylene, and rubber. PVC and polyethylene are the most widely used materials of this group. Teflon is very inert, but quite costly, and it is therefore limited to use in cables which must undergo extreme environmental conditions.

The most common types of wires and cables (Fig. 3-20) include the following:

1. *Hookup Wire*—Generally consists of a multistranded single conductor surrounded by PVC or polyethylene. Used for connecting elements in ordinary low-frequency circuits.

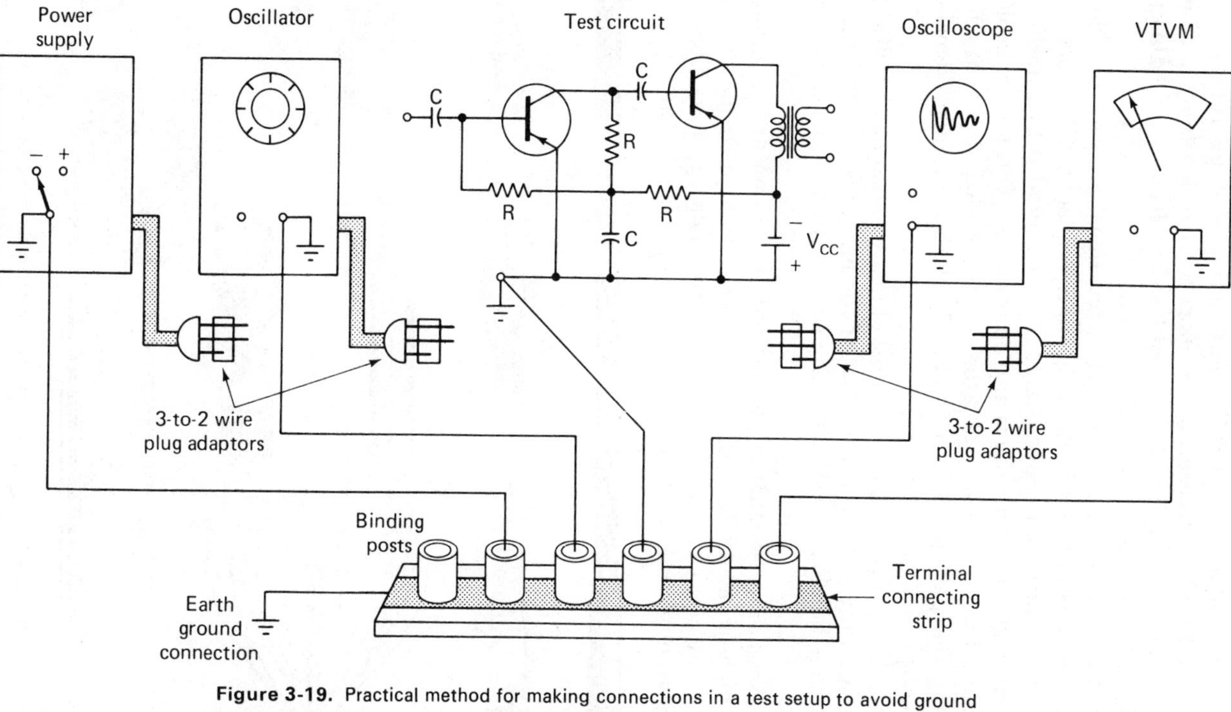

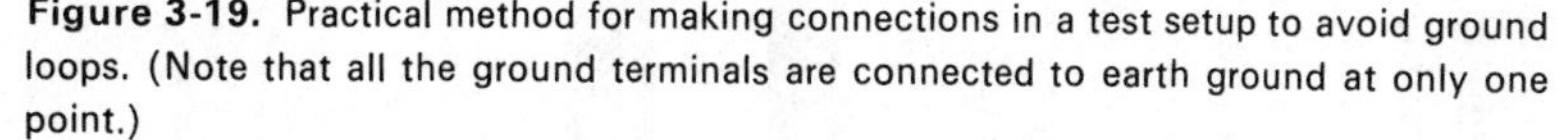
Figure 3-19. Practical method for making connections in a test setup to avoid ground loops. (Note that all the ground terminals are connected to earth ground at only one point.)

2. *Test-Prod Wire*—Very flexible wire surrounded by rubber insulation. Used in test leads of measuring instruments. High flexibility is desirable so that wire does not break with repeated bending. Rubber insulation provides high insulation-resistance as well as flexibility.

3. *Shielded Cable*—Consists of an inner signal-carrying conductor and braided metal sheath surrounding the inner conductor. The inner conductor and surrounding sheath are separated from one another by a flexible insulating layer. Outer layer of wire is also an insulator. This type of cable is used as cable for carrying low-level signals. The braided sheath is effective in reducing the pickup of interference signals by the inner signal-carrying conductor.

4. *Multiple-Conductor Cables*—Consist of many conductors bundled together in one sheath. Can have any number of conductors as well as different types in the same bundle. In Fig. 3-20 there are several multiple-conductor cables shown.

5. *Coaxial Cable*—Similar to the shielded cable in construction, but used to carry high-frequency and pulse-type signals. At high frequencies, ordinary single-conductor cable would radiate too much energy away from the cable during transmission. Coaxial cables eliminate this problem.

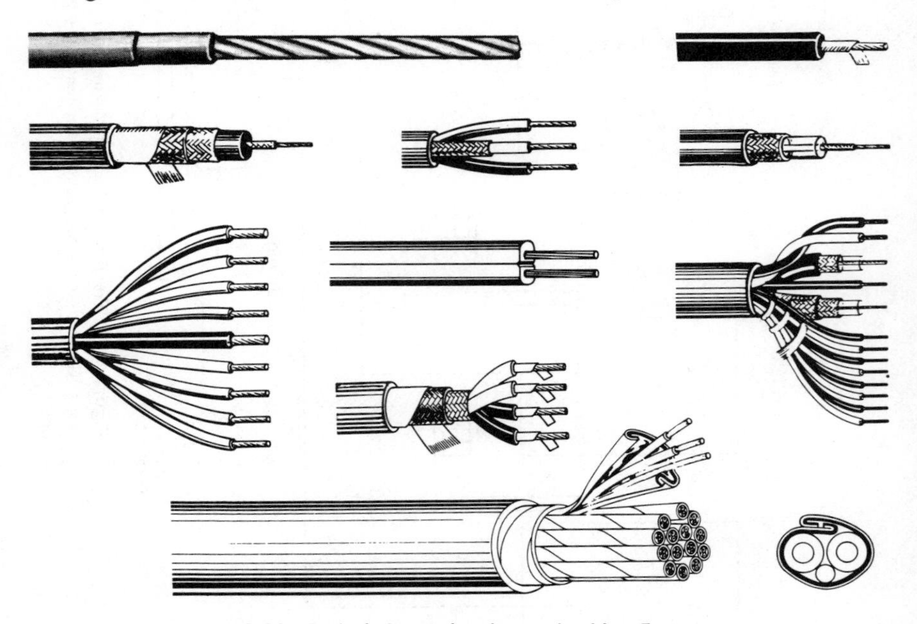

Figure 3-20. Typical electronic wires and cables. From top to bottom: hookup wire, test prod wire, shielded cable, special audio and sound cable, coaxial cable, multiple conductor cable, twin-lead cord, shielded multiple conductor cable, TV Eye cable, multipurpose cable with shielded pairs. (Courtesy of Belden Manufacturing Corp.)

CONNECTIONS AND CONNECTORS

As described in the previous section, the components of electric circuits and instruments are usually interconnected by wires and cables. At the points where the wires themselves are joined, suitable means must be provided for making satisfactory electrical connections. A connection is deemed satisfactory when it furnishes a path that does not alter the characteristics of signals that are transmitted through it. Thus one general requirement which must be met by connections is that they introduce as little resistance into the electrical path as possible. The methods used to join wires or cables together can be classified according to whether the connections are to be *permanent*, *semipermanent*, or *separable*.

Permanent connections are usually made by *soldering*, *welding*, or *crimping*. Soldering is probably the most common of these methods. In soldering, two metal surfaces are united when a solder junction is formed between them. *Solder* itself is a metal or metal composition which melts at a fairly low temperature ($\approx$400°C) and "wets" the surfaces to which it is applied. Upon cooling, solder forms a low-resistance, permanent connection between these surfaces. Soldered connections can be made rapidly between wires, and multiple solder connections sometimes can be made simultaneously. This means that solder connections lend themselves to mass soldering and automated techniques. It should also be mentioned that solder connections are not strictly permanent, that is, they can actually be disconnected and remade a limited number of times.

The tools and equipment necessary for making solder connections are a soldering iron, solder, tinned wires, and terminals. The quality of the connections depends to some extent on the skill of the operator.

The second type of permanent connection, the *weld*, is the strongest and most permanent type of electrical connection. In welding, a direct contact is made by heating and fusing the metals of the cables being joined. A very strong connection results. However, welding requires special equipment and is suitable only for solid-wire and single-lead connections. Since other methods yield suitable connections which can be fabricated more easily, welded connections are limited in their use to special applications (i.e., for connections that must be able to withstand high temperatures).

Crimping is the third method of permanently joining wires. In crimping, two metals are pressed together with a special *crimping tool*. The high pressure forces the metals into intimate contact with one another and forms a low-resistance contact through deformation. The reliability of well-made, crimped connections is high. Crimping is the most common method used to join solderless wire terminals to wires (Fig. 3-21).

Semipermanent connections between elements are usually made by screwing on various types of terminals to *binding posts* or *terminal blocks*. Some various screw-on *terminals* (also called lugs) are shown in Fig. 3-22. Such terminals are attached to wires by crimping or soldering. The wire with the

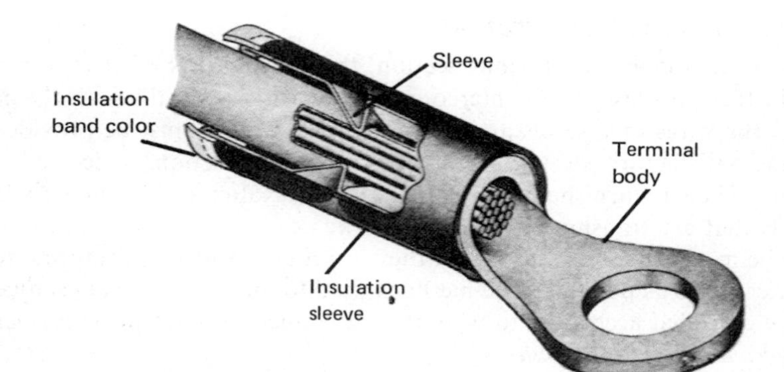

Figure 3-21. Joining a terminal and a wire (Courtesy of American Pamcor, Inc.)

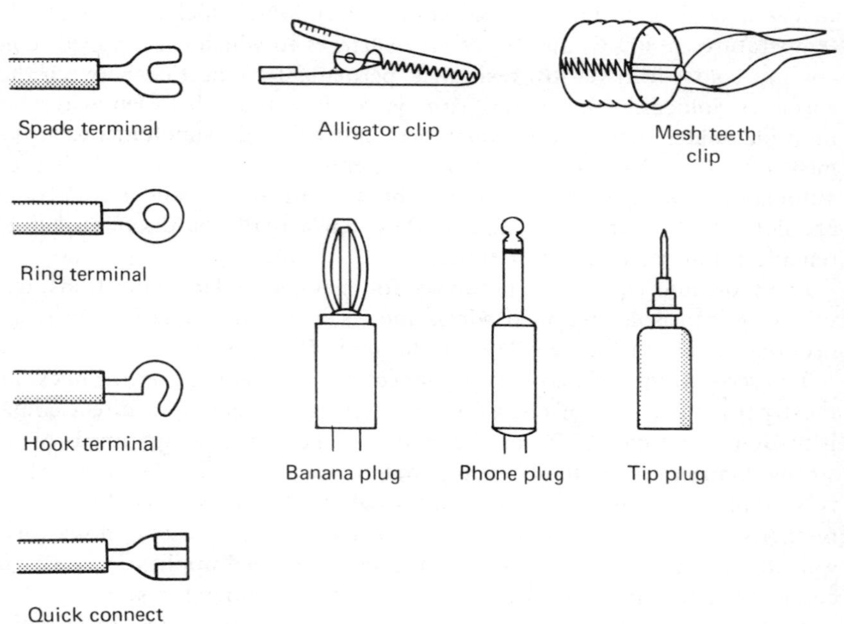

Figure 3-22

attached terminal can then be connected to a binding post or terminal block with a screw clamp.

The typical *binding post* is shown in Fig. 3-23. The nuts of binding posts can be made of metal or insulator material. Such posts are designed to accomodate bare wire, banana and phone plugs, as well as spade and hook wire terminals (Fig. 3-22).

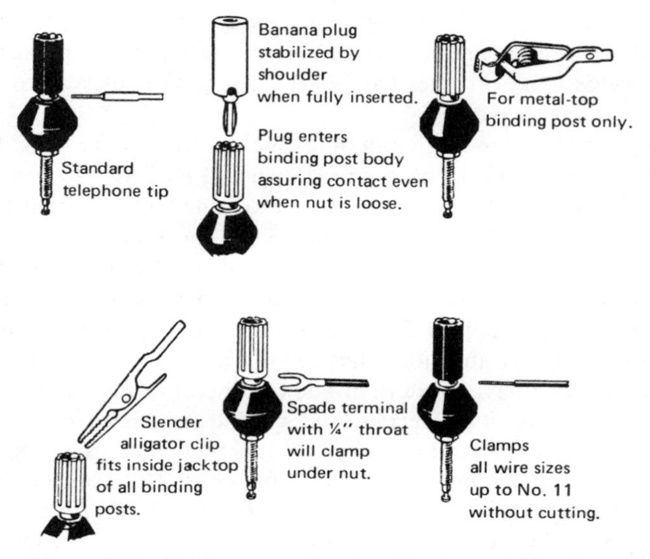

Figure 3-23. Making connections to a binding post (Courtesy of General Radio Corp.)

Two types of terminal blocks are shown in Fig. 3-24. The *insulating-barrier type* has insulating materials between each connecting strip to isolate it from the neighboring strips. Terminal lugs can be attached to the terminal block by screwing them into place. The lug type of terminal strip is designed to accept solder connections. Terminal block connections are used most often in low-voltage, low-power applications that need infrequently detachable connections.

Whenever circuits or instruments are designed to be readily joined or connected to other electrical components, some form of *separable connection* is required. The class of components used to provide separable connections are called *connectors*. Connectors usually have two mating halves. One of the halves are called *plugs*, *pins*, or *male ends*. The corresponding halves are

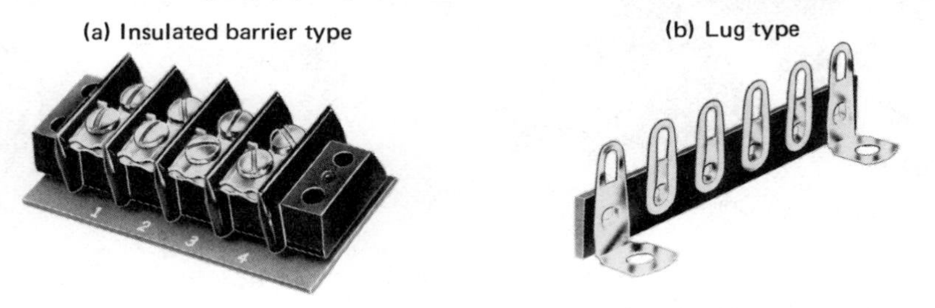

Figure 3-24. Terminal blocks (Courtesy of Cinch, Div. of TRW Inc.)

known as *receptacles, jacks, sockets,* or *female ends.* Usually the receptacle is mounted on the more permanent member of the equipment being connected (i.e., the chassis, box, or other fixed part.) The plug is usually connected to the cable or movable part.

The simplest plugs are single-pin plugs such as the banana plug, the phone plug or the tip plug. These plugs are shown in Fig. 3-22. (In some applications double-tip banana plugs are also available.) Other simple plug-like terminations are the alligator clip and the claw (or meshed teeth) type of clip. These last two clips are ordinarily found on the probes of test leads, instrument accessories, and on cords used for making quick connections.[8] As will be noted in Chapter 5, alligator and claw-type tips are not always suitable for making connections that require a very low resistance.

For cables which have more than one conductor, a multipin connector is used. Since each pin is connected to a specific conductor of the cable, all conductors can be correctly connected together each time the multipin connector is mated. To ensure that multipin connectors can be mated only in the proper way, the pins can be aligned in special patterns. This built-in method of providing correct orientation of the connectors is called *polarization.*

There are many different types of multipin connectors. In the home, the common power cord is a multipin connector which has either a two- or three-pin plug and socket. Figure 3-25 shows some other examples of the many types of multipin connectors which are available. The circular connectors shown are usually used for connecting two cables together. The rectangular connectors are more often employed for making connections between a cable

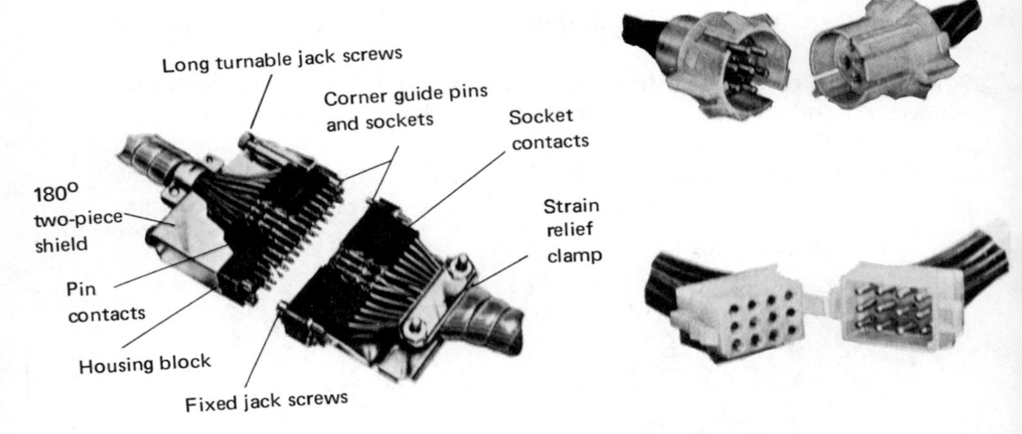

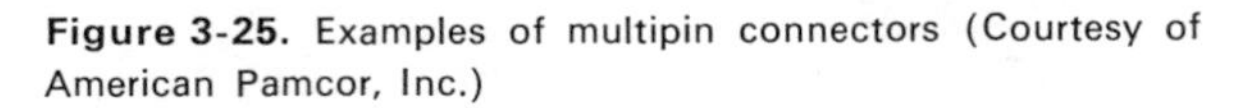
Figure 3-25. Examples of multipin connectors (Courtesy of American Pamcor, Inc.)

[8]Quick-connect cords are often called *patch cords.*

and chassis. The contacts of the multipin connectors are usually mounted in an insert in such a way that they can be self-aligning. The insert is then fitted in a connector housing. An insulator such as phenolic or melamine is used to form the insert. The contacts of connectors are made of nickel or gold-plated brass or bronze, and they have a spring-like action incorporated into their design. The spring tension provides for low-resistance contacts and a sure connection. The housings of the two mating halves are most often joined by screw threads or a so-called bayonet-type design. A few other types use jackscrews or latches.

For coaxial cables, special connectors must be utilized. These connectors are usually designed so that their impedances are matched to the coaxial cables for which they form the connection. This allows them to furnish a low-distortion path for the signals they carry. The most common type of "coax" connectors are called BNC types. These have a *bayonet type* of connector (Fig. 3-26). Their threaded equivalents are called TNC connectors. Some other widely used coax connector classifications are the N, HN, C and UHF types.

SWITCHES

A switch is a device for turning on, turning off, or directing electric current. The most common types of switches used in electrical instruments and measurement systems are the following:

1. Toggle switch
2. Pushbutton switch
3. Rotary switch
4. Slide switch
5. Snap switch
6. Mercury switch

We will examine the operation and some applications of each of these types. First, however, let us define some terms used when describing the construction of switches. The arm or part of the switch that is moved to open or close a circuit is called the *switch pole.* If a switch has only one pole, it is known as a *single-pole switch.* If it has two poles it is a *double-pole switch.* Switches can also have three, four, or any other number of poles (e.g., triple-pole, four-pole, multipole, etc.).

If each contact alternately opens and closes only one circuit, the switch is a single-throw type. On the other hand, if the contact is double-acting (i.e., if it breaks one circuit while simultaneously closing another), the switch is referred to as a *double-throw* type.

A switch therefore can be a single-pole, single-throw (SPST), single-pole, double-throw (SPDT), double-pole, single-throw (DPST), double-pole, double-throw (DPDT), or any other combination of multipole and either

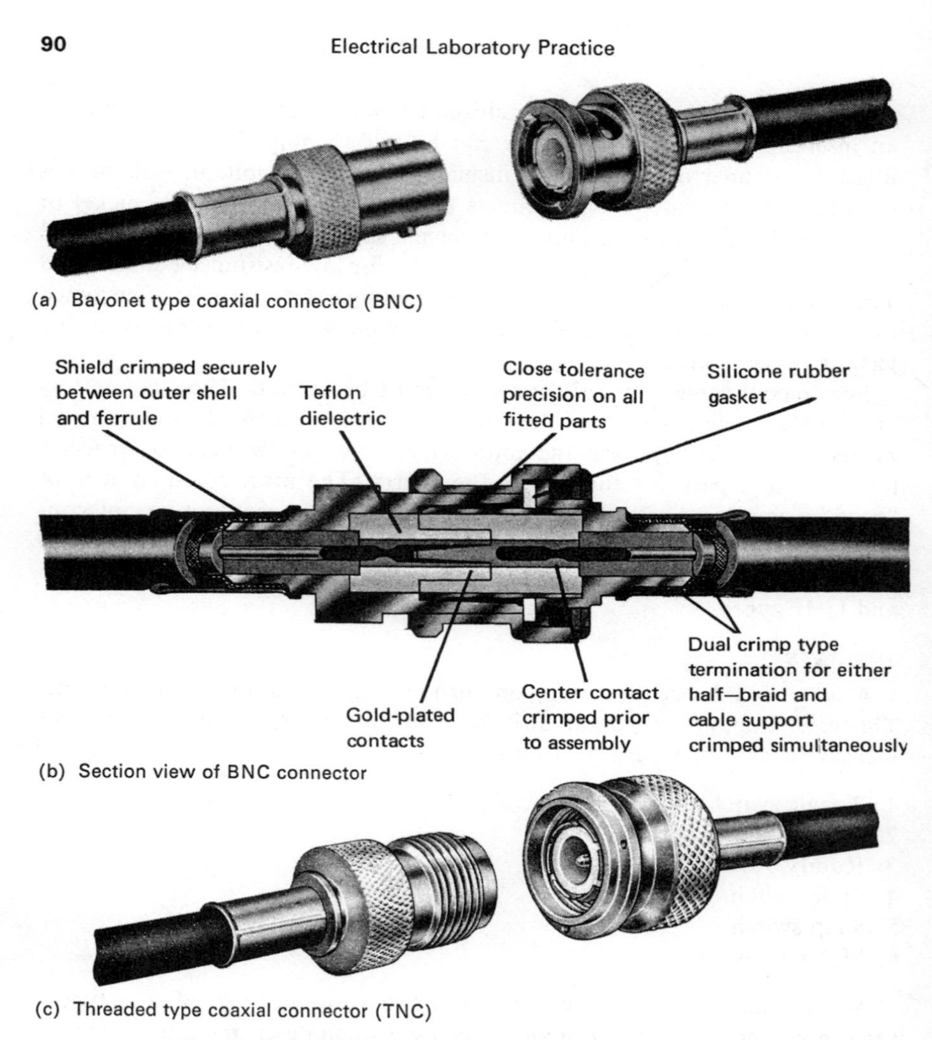

(a) Bayonet type coaxial connector (BNC)

(b) Section view of BNC connector

(c) Threaded type coaxial connector (TNC)

Figure 3-26. Examples of coaxial connectors (Courtesy of American Pamcor, Inc.)

single- or double-throw types. Figure 3-27 shows the various symbols for these switch types.

Relays are switches that are operated by magnetic action, and they employ the same type of notation to describe their contacts. Relays will be described in more detail in Chapter 7.

The *toggle switch* is a switch in which a projecting arm or knob, moving through a small arc, causes the contacts of a circuit to open or close suddenly. The fact that the contact is made and broken suddenly reduces arcing and yields a sure contact. For these reasons toggle switches are used in a wide range of switching applications. Lighting switches in most houses and the

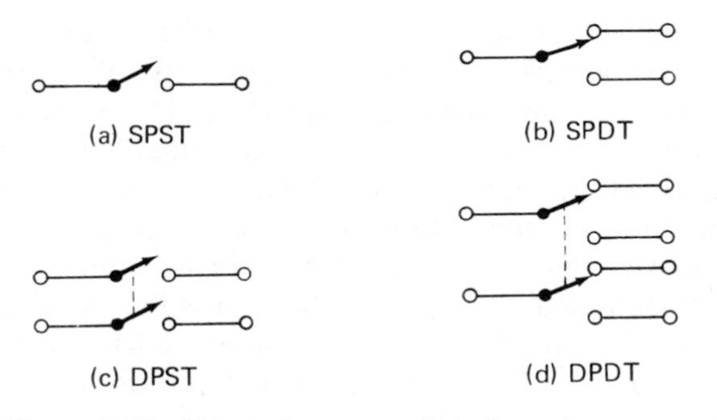

Figure 3-27. Circuit diagram symbols for various types of switches

on-off switch on many electrical tools and industrial instruments are toggle switches. The reset switch on circuit breakers is also usually a toggle switch. Figure 3-28 shows how a simple SPST toggle switch operates. The flexible activating arm allows the switch to snap quickly between *on* and *off*. More complex toggle switches can have more than two positions and may open and close more than one circuit branch simultaneously.

The *pushbutton switch* is designed to open or close a circuit when depressed, and to return to a normal position when released. In some pushbutton switches the contacts remain open or closed after the pushbutton has returned to its normal position (alternate-action type). In the alternate action types, the pushbutton must be depressed twice to return to the original position. In other types, the contact is closed or open only as long as the pushbutton is depressed (momentary type). Pushbutton switches are especially useful in limited-space applications. They are also easy to activate quickly. Some common uses of pushbutton switches include the selector switches on automobile radios, the dimmer switch on automobile headlights, the doorbell switch, and safety switches on motors.

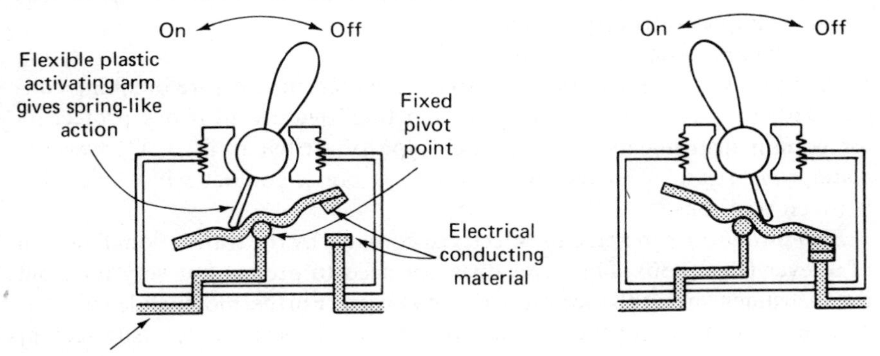

Figure 3-28. Operation of a SPST toggle switch

The *rotary switch* is a switch that makes or breaks circuits as it is rotated between positions (Fig. 3-29). A contact attached to a shaft is turned by means of a knob connected to the other end of the shaft. The contact moves along a fixed *insulated wafer* which has strips of conducting material placed along its circumference. As the shaft is rotated from one position to the next, the rotating contact makes a connection to these conducting strips. This closes and opens desired circuits. In some rotary switches, more than one wafer and contact are connected to the shaft. This allows the switch to be used as a multipole switch. A spring-loaded ball bearing aligns itself with indentations in part of the switch, thereby locating the switch positions (Fig. 3-29(b) and (c)).

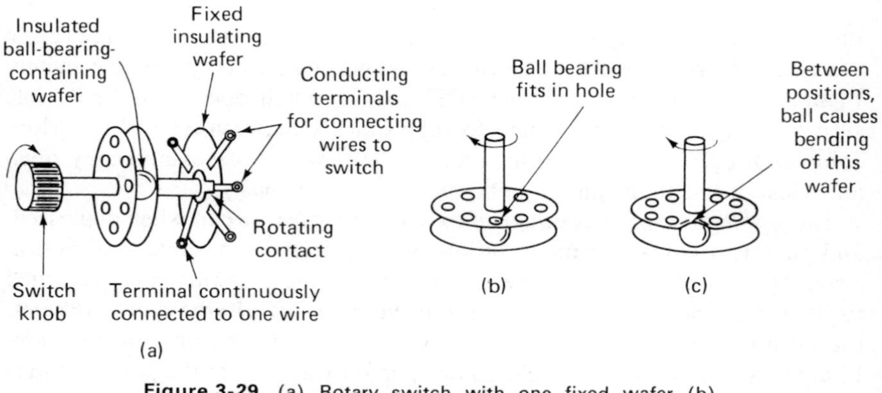

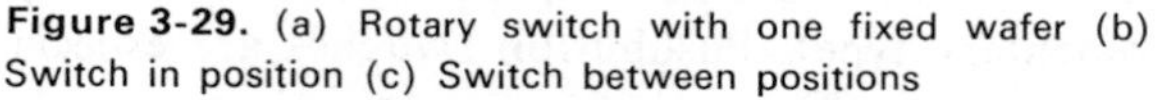
Figure 3-29. (a) Rotary switch with one fixed wafer (b) Switch in position (c) Switch between positions

The rotary switch is used when one switch must be able to be set to many different positions. The channel-selecting dial on a television, the switches on decade resistors, and the function switches on meters and oscilloscopes are all examples of rotary switches.

If the rotary switch is a *shorting-type* switch, this indicates that the rotating contact always makes a connection to the next terminal before breaking contact with the previous one. Such a "shorting" feature provides protection for certain instruments in which this type of switch is used. Conversely, *nonshorting types* of rotary switches always break contact when switching between positions.

Slide switches open and close electrical contacts by the translational motion of a lever (Fig. 3-30). This lever does not need to project out very far from the instrument or appliance on which it is used. Furthermore, slide switches can be marked so that the various positions of the switch are easily seen at a glance. Finally, slide switches are very simple, and this makes them at-

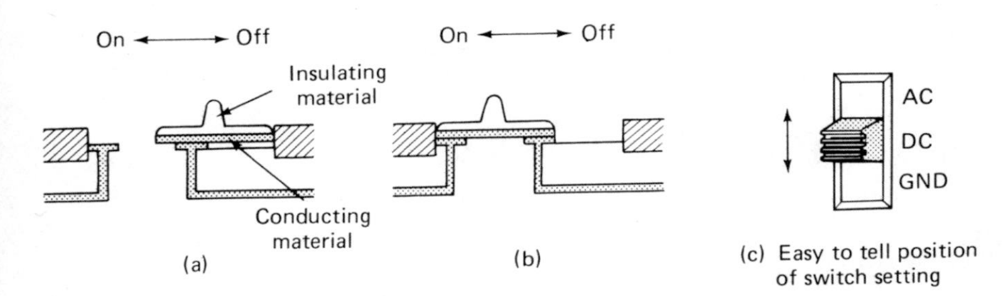

Figure 3-30. Slide switch

tractive for use in inexpensive, low-voltage devices. However, the contacts made by slide switches are not as reliable as toggle or pushbutton contacts. Furthermore, they are subject to arcing, and this is why they are limited to low-voltage applications. Flashlight switches, on-off switches of electric razors, automobile headlight switches, and some control switches on electrical instruments are examples of slide switches.

Snap-action switches are designed to be activated by machines rather than by humans. They are similar to pushbutton switches, except that the pushbutton is often depressed by a pivoted lever (Fig. 3-31). The two most common levers are the straight and the rolling lever.

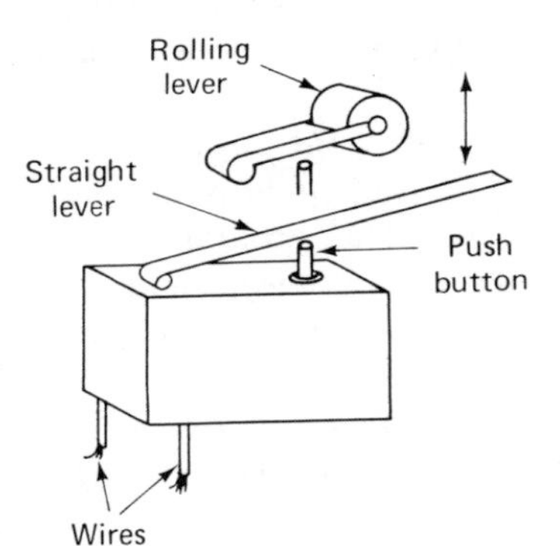

Figure 3-31. Snap-action switch

The *mercury switch* is a position-sensing switch (Fig. 3-32). If the switch is tilted, the contacts remain open because the pool of mercury in the switch remains at the bottom. If the switch is upright, the mercury closes the gap between the two contacts, placing the switch in the *on* position. Because the mercury switch has no moving parts, it is a silent switch and is less subject to wear. Common uses of mercury switches include the furnace switch in household thermostats, automobile trunk-lid light switches, and silent lighting switches.

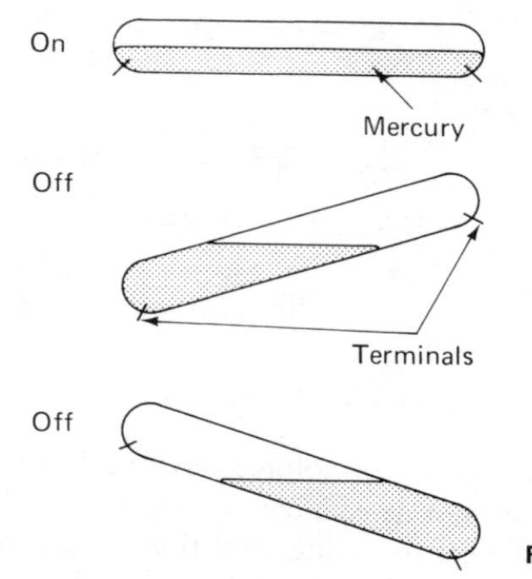

Figure 3-32. Mercury switch

Power Transfer and Impedance Matching

If we have a system in which power is being expended, the part of the system to which power should be delivered is called the *load*. For example, a man on a bicycle may be the system of interest. The man is the power supply, and (on a level path) the mass and friction of the bicycle and the wind resistance comprise the load. However, the man dissipates energy internally (in his heart, muscles, etc.), and thus all the energy he expends cannot be transferred to the load.

In an electrical system, the power source might be a generator, a battery, or an amplifier. The load could be an electric lamp, loudspeaker, or meter. The generator, battery, and amplifier also have internal losses and hence cannot deliver all the power they generate to a connected load. These internal losses become important when we consider the type of load that should be connected to the source in order to transfer a desired amount of power.

There are usually two desired conditions under which power is drawn from a source for delivery to a load. In the first we want maximum *efficiency*; in the second, maximum *amount* of power transferred.

If we are interested in delivering the maximum percentage of the power generated by a source to the load, we want the source to be operating at the *highest efficiency* possible. That is, the ratio of the power generated by the source to the power put into the source should be maximized. But when a generator or other source is operated at maximum efficiency, the source is not transferring the maximum *amount* of power it can generate to the load. At maximum efficiency, the source is run at far below its maximum output capability. This is the price that must be paid for having as little energy dissipated in the source as possible.

For a generator supplying many megawatts of power to a city, a condition of maximum operating efficiency is a necessity. If a sizable fraction of the generator output were dissipated in the generator itself, the resulting heat would quickly melt the entire generating system.

To achieve high efficiency, the output impedance of the generator is kept to only a fraction of an ohm. The loads to which power is supplied have impedances which are larger than the generator output impedances by several orders of magnitude.

In many measurement systems, however, the signal source being measured is very weak and generates only a tiny amount of power. In such cases, it is necessary that the maximum *amount* of power that the source can generate be delivered to the measuring instruments. For example, a transducer may be the source and it may generate only a small electrical signal in response to a change in a nonelectrical quantity. We would therefore want as much of the power of this signal to be transferred to the load (i.e., an amplifier or meter) as possible.

When a condition of maximum power transfer from the source to the load is achieved, we say that the load is *matched* to the power source. The procedure used to achieve this condition is called *impedance matching*. To achieve such impedance matching in an electrical system, the impedance of the *load* is usually changed to match the impedance of the source.

Physical principles specify that at the condition of *maximum power transfer*, 50 percent of the power being produced by the source is delivered to the load, while 50 percent is being dissipated internally in the source. Of course, the efficiency of this system is only 50 percent, but more power is being delivered by the source than at a point of higher efficiency.

For an electrical system containing only resistance, a condition of maximum power transfer exists if the resistance of the load R_L equals the internal resistance of the source R_G.

$$R_L = R_G \tag{3-9}$$

A specific example of impedance matching is the connection of a vacuum-tube amplifier and a loudspeaker. Usually the loudspeaker (load) has a low resistance and the amplifier (source) has a high output resistance. In order to transfer the maximum power from the amplifier to the speaker, a transformer is used. The transformer makes the loudspeaker appear to have a higher resistance in order to match the amplifier's output resistance.

EXAMPLE 3-3

Find the value of R_L for maximum power transfer in the circuit shown in Fig. 3-33.

Solution: The equivalent resistance of the known resistors is

$$R = 6 + \frac{10 + 5}{(10)(5)} = 6 + 3\frac{1}{3} = 9\frac{1}{3}\,\Omega$$

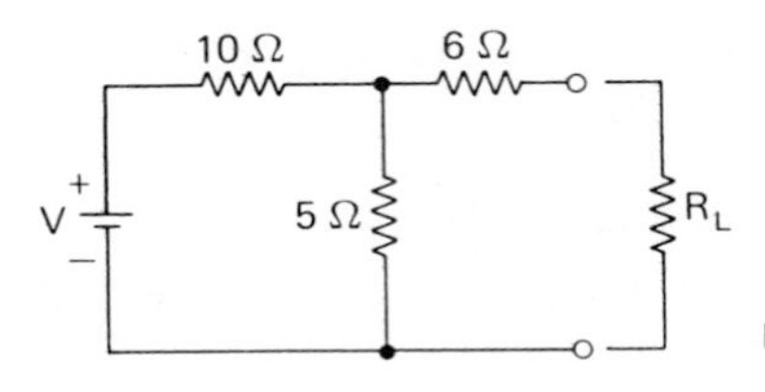

Figure 3-33. Circuit for Ex. 3-3

Then the $R_L = R$ for maximum power to R_L and

$$R_L = 9\frac{1}{3}\ \Omega$$

When applied to ac circuits, the maximum-power-transfer theorem states that the impedance of the load should be equal to the complex conjugate of the equivalent impedance of the source, or

$$Z_L = Z_G^* \tag{3-10}$$

For further discussion of complex numbers and complex impedance, see, for example, R. Boylestad's *Introductory Circuit Analysis* (Columbus, Ohio: Charles E. Merril Co., 1968).

Temperature Effects on Component Operation and Measurements

Of all the nonelectrical parameters that affect a component's operation, temperature can have the most significant influence. Low temperatures may change the electrical characteristics of a device to the extent that it will not operate properly (or not operate at all). Such adverse effects due to low temperatures are common in batteries and semiconductor devices. At high temperatures, devices may also fail to operate properly or may be entirely destroyed by the excess heat. Sometimes the heat is generated by the device itself and must be dissipated or kept below a specific rate of production. In general, the operating life of electrical components is longer if the temperatures at which they operate are kept low. Therefore, to ensure reliable system operation, careful attention must be paid to thermal design and methods of heat dissipation.

In the various sections on circuit element construction and operation, relevant temperature effects are discussed in more detail. In this section we will discuss general temperature ratings and methods used for heat dissipation.

Manufacturers of temperature-sensitive devices (such as resistors, semiconductor devices, and batteries) usually supply proper operating temperature figures and related power dissipation ratings. Most such devices have upper and lower temperature limits beyond which proper operation cannot be guaranteed. Where device operation generates heat, the observance of maximum power dissipation levels will prevent such elements from heating themselves to destructive temperatures. Since transistor operation will not be discussed at length in the rest of this text, let us at least mention the transistor here as an example of how temperature effects in an element are estimated, modeled, and controlled.

The operation of a transistor develops heat internally, and this heat must be conducted away to prevent the transistor from destroying itself. In equilibrium, such heat is carried away to the packaging and then to the surrounding environment by heat conduction and radiation. Since it is most efficiently removed by conduction, a good heat-conducting path should be provided. The rate at which heat is conducted away depends primarily upon the surface area of the conduction boundary.

A *heat sink* reduces the level of the thermal resistance between a component and its ambient environment. In actual use, it is usually a device of a high heat-transfer capability such as a metal structure with fins. Sometimes forced-air or even forced-liquid cooling is used to carry away heat more quickly (and thus increase its heat-transfer capabilities). Figure 3-34 shows some pictures of typical heat sinks.

The elements are mounted on these heat sinks so that they will be in close thermal contact with them. In addition, a special *thermal-joint* or *heat-sink* compound is applied at the contact surface so that the thermal conduction path to the heat sink is even more efficient.

(a) (b)

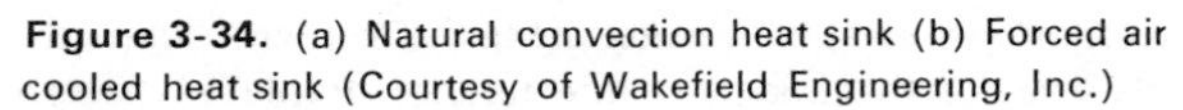

Figure 3-34. (a) Natural convection heat sink (b) Forced air cooled heat sink (Courtesy of Wakefield Engineering, Inc.)

Problems

1. List several factors that can influence the severity of an electric shock in a human being.

2. If the resistance of dry skin is 100 kΩ and accidental contact is made with a conductor which is at a voltage of 120 V, how much current can flow through the human conducting path? If contact with this same conductor is maintained for 10 seconds and the skin resistance at the point of contact falls to 1 kΩ, what current can then flow through the body to ground?
3. Describe how the third wire of the three wire power cord acts as a protective mechanism against electric shock hazards.
4. Explain why the following acts lead to hazardous safety conditions when working with electrical equipment:
   a) wearing metal rings or bracelets,
   b) being barefoot,
   c) working on a damp or concrete floor,
   d) touching pipes or other grounded conductors while working on electrical equipment,
   e) working on electrical equipment with sweaty hands.
5. If the resistance of the path from the wall outlet end of the neutral wire to ground is 20 Ω, calculate the voltage of that point (relative to ground) if a 100 W bulb is connected across the outlet.
6. The two-wire 120 V supply line from the electric power company to the consumer has a resistance of 0.07 Ω. If a short circuit occurs at the consumer's residence, find the current that flows through the short circuit.
7. Describe (with the help of sketches) the path taken by the currents flowing in circuits whose grounds are the following ones: a) earth ground, b) electric circuit in an airplane, c) electric circuit in an instrument whose circuit is floating.
8. Describe the difference between fuses and circuit breakers. Where might each type of device find its best use?
9. Explain why it is desirable for the input impedance of a voltmeter to have a very high value? What are the effects on the test circuit if the input impedance of the voltmeter is low?
10. Define the word "loading" and explain in what context it is used when referring to a) loading by an oscilloscope, b) loading by an amplifier, c) loading of an oscillator.
11. Name and briefly describe 3 types of electrical interference.
12. Explain why it is necessary to shield coils and transformers in some applications.
13. List three common electronic measuring instruments that are likely to be significantly affected by capacitive interference. Explain what properties these instruments share that make them sensitive to this type of interference.
14. If an electronic voltmeter has an input impedance of 11 MΩ and the capacitance between its probe leads and the surrounding 60 Hz ac powerlines is 0.5 pF, find the rms voltage that will be indicated by the meter if the probes are not connected across the test circuit (i.e. they are open-circuited).

References

1. Bair, E. J., *Introduction to Chemical Instrumentation*, Chap. 8. New York, McGraw-Hill Book Company, 1962.
2. Tektronix, *Biophysical Measurements*. Beaverton, Ore.: Tektronix, Inc., 1971.
3. Ficchi, R., *Electrical Interference*. New York, Hayden Book Company, Inc., 1964.
4. Morrison, R., *Grounding and Shielding Techniques in Instrumentation*. New York, John Wiley & Sons, Inc., 1970.
5. Shiers, G., *Design and Construction of Electronic Equipment*, Chaps. 6, 7 and 9. Englewood Cliffs, N.J., Prentice-Hall, Inc., 1966.

4

Basic dc and ac Meters

There are many different methods and instruments used for measuring current and voltage. Voltage measurements are made with such varied devices as electromechanical voltmeters, vacuum-tube voltmeters, digital voltmeters, oscilloscopes, and potentiometers. Current-measuring methods use instruments called *ammeters*. Some ammeters operate by actually sensing current, while others determine the current indirectly from an associated variable such as voltage.

An ammeter is always connected in *series* with a circuit branch and measures the current flowing in it. An ideal ammeter would be capable of performing the measurement without changing or disturbing the current in the branch. (Such a disturbance-free measurement would be possible if the meter appeared as a short circuit to the current flow.) However, real ammeters always possess some internal resistance, and the current in the branch may change due to the insertion of the meter.

Conversely, a voltmeter is connected in *parallel* with the elements being measured. It measures the potential difference (voltage) between the points across which it is connected. Like the ideal ammeter, the ideal voltmeter should not change the current and the voltage in the test circuit. Such an ideal voltage measurement can be achieved only if the voltmeter does not draw any current from the test circuit. (It should appear as an *open circuit* between the two points to which it is connected.) However, most actual voltmeters operate by drawing a small but finite current and thereby also disturb the test circuit to some degree. We will discuss the extent of the measurement errors caused by the nonideal aspects of real meters.

The simplest instruments commonly used to measure the electrical quantities of voltage and current in the laboratory are the electromechanical direct-current (dc) and the alternating-current (ac) meters. These meters will be discussed in this chapter, while the oscilloscope and other types of voltage- and current-measuring instruments will be studied in later chapters. Table 4-2 at the end of this chapter summarizes the characteristics of the various types of meters.

At the heart of each electromechanical meter there is some type of current- or voltage-detecting device. We will begin our discussion by examining the most common of these detecting mechanisms and then see how they are applied to constructing actual ammeters and voltmeters.

Electromechanical Meter Movements

D'ARSONVAL GALVANOMETER MOVEMENT

The most common sensing mechanism used in basic dc ammeters and voltmeters is a current-sensing device. This mechanism was developed by D'Arsonval in 1881 and is called the *D'Arsonval* or *permanent-magnet–moving-coil movement*. It is also used in ohmmeters, rectifier ac meters, and impedance bridges. Its wide applicability arises because of its extreme sensitivity and accuracy. Currents of less than 1 μA can be detected by commercially available movements. (Certain special laboratory instruments which use D'Arsonval movements can measure currents as tiny as 1.0×10^{-13} A.) The movement[1] detects current by using the force arising from the interaction of a magnetic field and the current flowing through the field. The force is used to generate a mechanical displacement, which is measured on a calibrated scale.

Charges moving perpendicular to the flux of a magnetic field are acted on by a force which is perpendicular to both the flux and the direction of motion of the charges. Since current flowing in a wire is due to a motion of charges, these charges will be subject to the magnetic force if the wire is oriented properly in a magnetic field. The force is transmitted by the charges to the atoms of the wire, and the wire itself is caused to move. As an example, let us place such a wire in a field oriented as shown in Fig. 4-1. If the current flows upward in this wire, the force will cause the wire to move to the right. If we bend the wire into the shape of a rectangular coil and suspend it into the same magnetic field, the resulting force on the wire will now tend to rotate the coil as shown in Fig. 4-2.

[1] The word *movement* is used to dénote the sensing devices in electromechanical meters because movements display the electrical quantity being measured by moving a pointer along a calibrated scale. Thus, the words "mover" or "movement" describe their action faithfully.

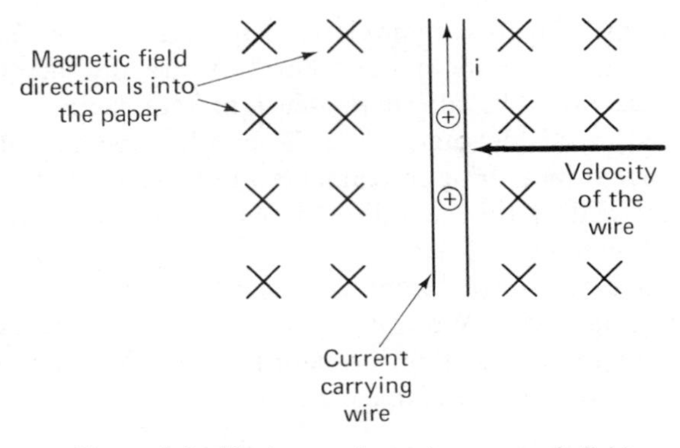

Figure 4-1. Moving conductor in a magnetic field

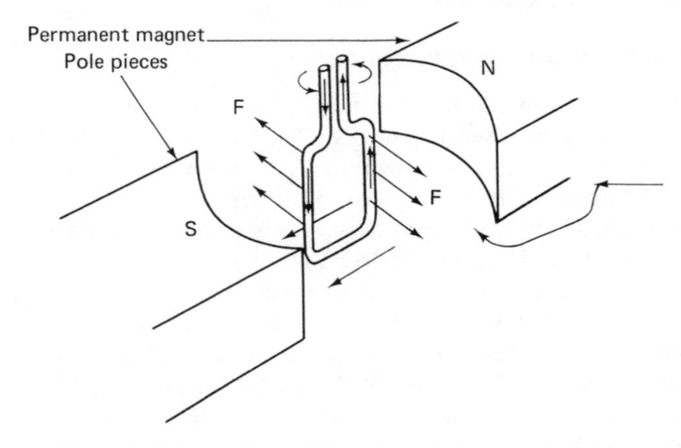

Figure 4-2. Loop of current carrying wire in magnetic field

The movement that D'Arsonval patented is based on this principle and is shown in Fig. 4-3. A wire coil is attached to a shaft which pivots on two jewel bearings. The coil can rotate in a space between a cylindrical soft-iron core and two magnetic pole pieces. The pole pieces create the magnetic field, and the iron core restricts the field to the air gap between it and the pole pieces. If a current is applied to the suspended coil, the resulting force will cause the coil to rotate. This rotation is opposed by two fine springs which supply a torque (rotational force) that opposes the magnetic torque. The spring strengths are calibrated so that a known current causes a rotation through a specific angle. (The springs also provide the electrical connections to the coil.) The lightweight pointer displays the extent of the rotation on a calibrated scale.

The deflection of the pointer is directly proportional to the current flowing in the coil, provided the magnetic field is uniform and spring tension is

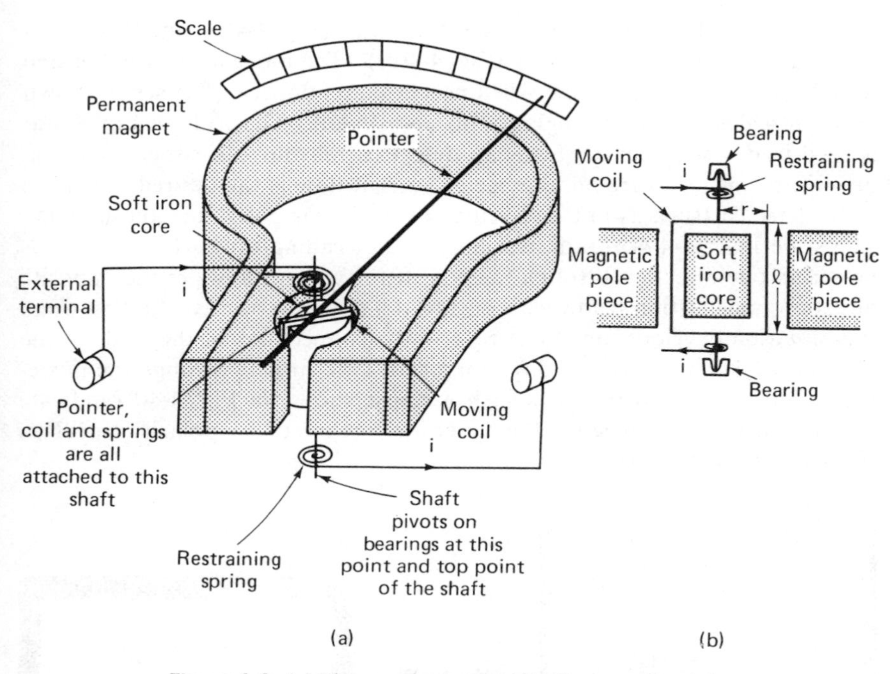

Figure 4-3. (a) D'Arsonval movement (b) Cross-sectional view of the moving coil and magnet of the D'Arsonval movement

linear. Then the scale of the meter is also linear. The accuracy of D'Arsonval movements used in common laboratory meters is about 1 percent of the full-scale reading.

The torque τ_D (force times radial distance) developed for a given current, i, determines the sensitivity of the movement. The greater the torque for a given current, the smaller the current which can be detected. This torque depends on the number of turns (N), the length (l) of the conductor perpendicular to the magnetic field, and the strength (B) of the magnetic field. The mathematical expression for the torque is given by[2]

$$\tau_D = f \cdot r = B(2Nl)i \cdot r = 2NBlri \tag{4-1}$$

Because a greater number of turns in the coil also increases the overall length of the wire, this increases the resistance of the movement. Thus, this method of increasing the coil sensitivity also makes it less of an ideal movement.

Two types of scales are generally employed with the D'Arsonval move-

[2]The factor of 2 arises because there are two vertical wire sections in each turn of the coil in the field. Typical values of $B = 0.15 - 0.5$ Wb/m^2, and $N = 20$–100 turns.

ment: those with a zero at the center of the scale (Fig. 4-4(a)) and those with a zero at the left end of the scale (Fig. 4-4(b)). (The movements are adjusted to indicate zero on each scale when no current is flowing. The screw shown on both scales is used for making this adjustment.) The scale in Fig. 4-4(a) is used in dc instruments which can detect current flow in either direction, or in instruments where an absence of current flow is the desired condition to be detected (such as in the Wheatstone bridge or potentiometer circuits). The scale in Fig. 4-4(b) indicates an upscale reading only when current is passed in one direction through the coil. If the current flows in the opposite direction, a deflection below zero occurs. To obtain a positive reading when this situation develops, one must reverse the connections of the leads to the movement. This reverses the direction of the current flow through the movement. Most meters with scales such as those shown in Fig. 4-4(b) indicate the proper way of connecting the meter into the circuit by polarity markings on the meter terminals.

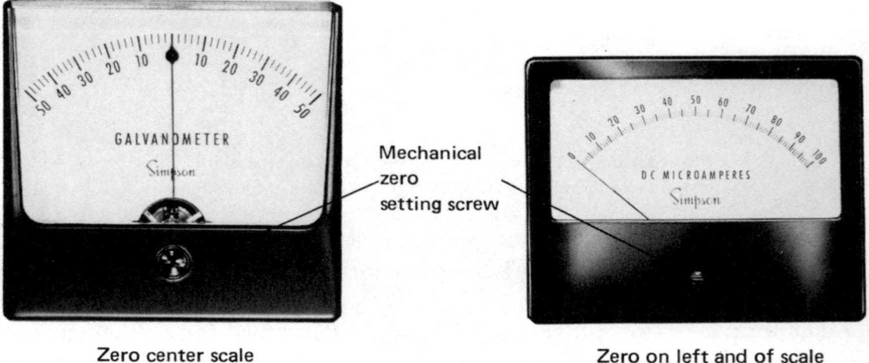

Figure 4-4. D'Arsonval movement scales (Courtesy of Simpson Electric Co.)

MOVING IRON-VANE MOVEMENT

Moving iron-vane movements are mostly used in ac meters. Like the D'Arsonval movement, they are also current-detecting devices. They operate on the principle that a magnetic field induces magnetism in iron and that magnetic poles of the same polarity will repel each other. A coil carrying the current being measured is used to create a magnetic field. Inside the coil there are two pieces of iron, one fixed and one movable. The field within the coil magnetizes both irons with the same polarity, and they repel one another because both north poles and south poles are adjacent (Fig. 4-5). The strength of the repulsion is proportional to the strength of the coil's

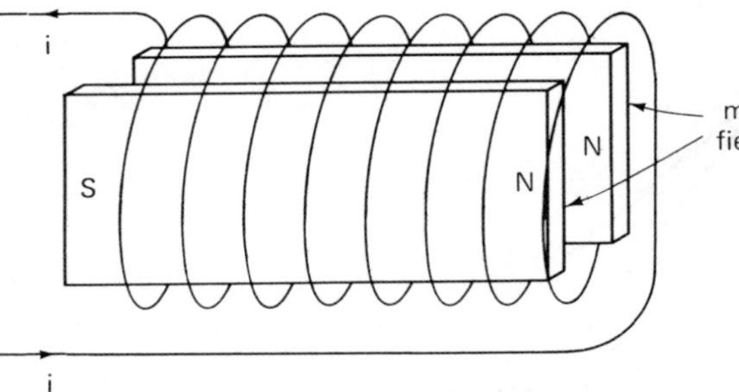

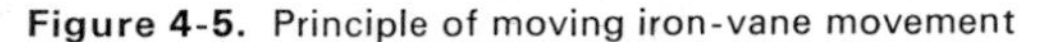

Figure 4-5. Principle of moving iron-vane movement

magnetic field. If ac current is applied to the coil, the magnetic field will change direction along with the direction of current flow. The irons' magnetization will also switch, but since all pole positions reverse, the net effect is still one of repulsion. The movable iron piece is mounted on a shaft which has restraining springs as in the D'Arsonval movement. Since only it can move, the force of repulsion is made to deflect a pointer (connected to the shaft) which indicates the magnitude of current passing in the coil.

Because the iron-vane movement does not distinguish polarity, it can be used to detect ac currents. If the frequency of the current being measured is greater than a few hertz, the pointer is not able to follow the changes because of its inertia. Instead, the deflection of the pointer takes a position proportional to the rms value of the ac current. Generally, the movement is not used to measure dc quantities. However, if it is used for this purpose, a correction in torque must be applied to account for residual magnetism remaining in the iron. Without the correction, the dc reading is not an accurate indication of the current flowing in the movement.

For ac measurements, the accuracy of the movement is limited by the fact that the magnetization of the iron vanes is nonlinear. In addition, hysteresis effects and eddy currents also arise in the vanes. (See Chapter 7 for more details on hysteresis and eddy currents.) Together, these effects cause the readings of the movement to depart from a true rms reading, particularly at higher frequencies. Furthermore, the flux density of the coil's magnetic field is quite small. As a result of this low flux density, the movement has a rather low current sensitivity (i.e., it does not respond well to very small currents).

Many different configurations of coils and iron vanes are used in meter construction. The most sensitive configuration is the radial iron-vane movement shown in Fig. 4-6(a). Others include the simple attraction (rather than repulsion) movement (Fig. 4-6(b)) and the concentric iron-vane (Fig. 4-17). None of the iron-vane movements have current-carrying parts in their moving elements; this makes them very rugged and resistant to damage (even under severe overload conditions)

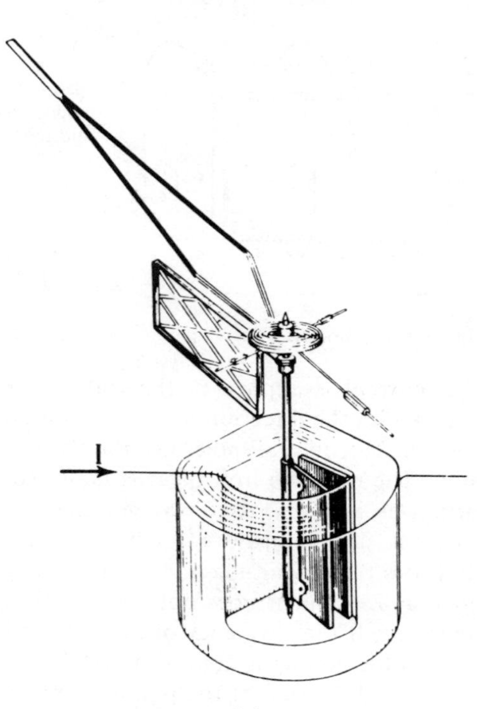

(a)

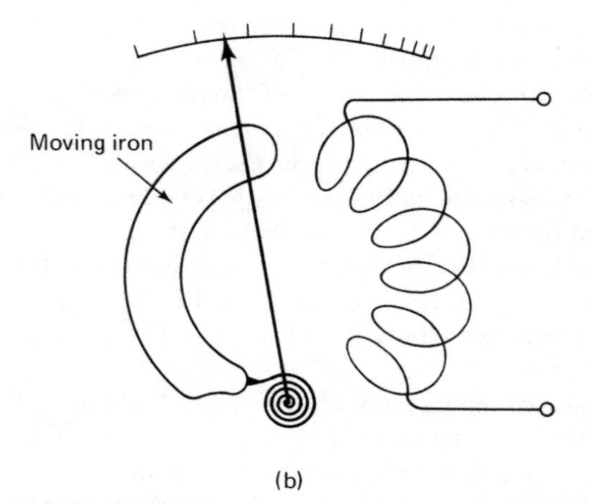

(b)

Figure 4-6. Moving iron-vane movements (a) Radial iron-vane movement (b) Simple attraction movement (Courtesy of Weston Instrument, Inc.)

ELECTRODYNAMOMETER MOVEMENT

The electrodynamometer movement is used in the construction of highly accurate ac voltmeters and ammeters and in wattmeters and power-factor meters. Like the D'Arsonval and iron-vane movements, it also operates as a current-sensitive device. Extremely high accuracies can be obtained using this movement because it uses no magnetic materials (and magnetic materials possess nonlinear properties).

In contrast to the D'Arsonval movement, which uses a permanent magnet to provide a magnetic field, the electrodynamometer creates a magnetic field from the current being measured. This current passes through two fixed coils and establishes the magnetic field which interacts with the current in the moving coil. The force on the moving coil due to the magnetic fields of the fixed coils causes the moving coil to rotate (Fig. 4-7). The moving coil is attached to a pointer that moves along a scale marked to indicate the value of the quantity being measured. The entire movement assembly is mounted in an iron-lined case to shield it from any external stray magnetic fields. Because the current being measured determines both the strength of the magnetic field and the moving coil's interaction with the field, the resulting deflection of the pointer is proportional to i^2. In ac use, the pointer takes up a position proportional to the *average* of current squared. The scale can be calibrated to read the square root of this quantity (rms). Note that the scale of the electrodynamometer movement shown in Fig. 4-7 is marked in the manner of a typical dynamometer meter.

The electrodynamometer movement produces an extremely accurate read-

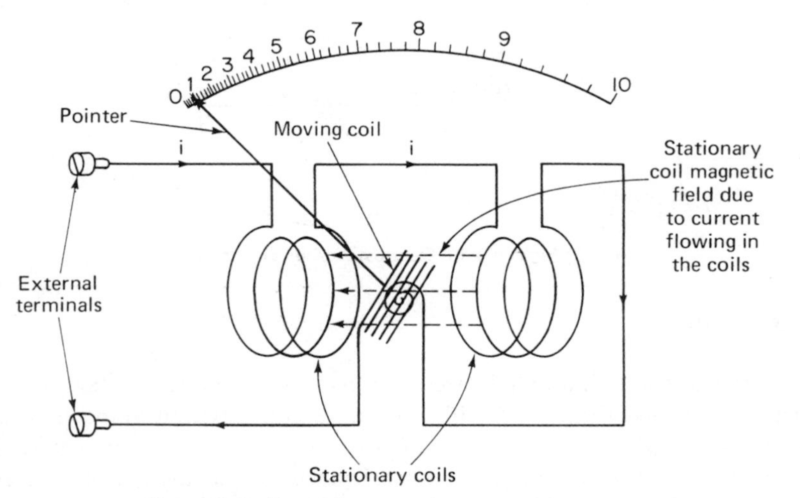

Figure 4-7. Electrodynamometer movement

ing but is limited by its power requirement. The magnetic field of the stationary coils produced by a small current is much weaker than the permanent field of the D'Arsonval movement. (Typical comparison is 6×10^{-3} Wb/m^2 vs 0.2 Wb/m^2.) Thus the sensitivity of the electrodynamometer movement is comparatively poor. When it is used as a voltmeter, the sensitivity is 10–30 ohms/volt which, as we will see, is very low.

ELECTROSTATIC MOVEMENT

Unlike the other movements discussed, the electrostatic movement senses voltage rather than current. However, because the smallest full-scale voltages it can detect are on the order of 100 V, the electrostatic movement is not very commonly used. Its operation depends on the force of attraction between oppositely charged bodies. The movement (shown in Fig. 4-8) consists of two plates of metal (one stationary and one movable) that can be connected across a potential difference. The movable plate is attached to a shaft equipped with restraining springs. The voltage induces charges of opposite polarity to appear on the plates (as in a capacitor), and the resulting attraction between the plates causes them to try and align. The pointer on the movable plate stops when the attraction is balanced by the opposing torque of the springs. The scale is marked to display the voltage applied. The movement can measure dc or ac voltages over a large range of frequencies, and its response to ac quantities is a true rms reading.

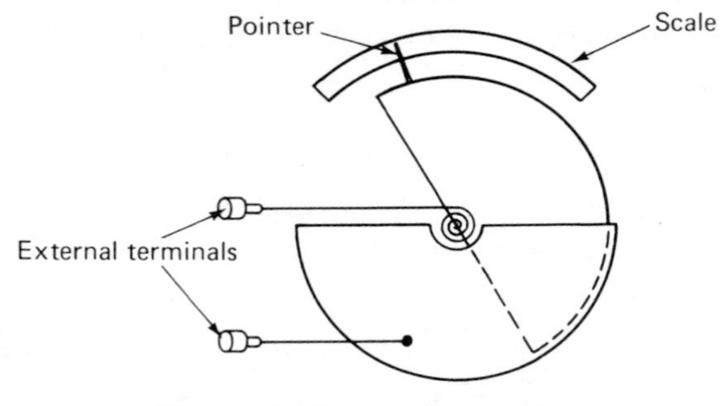

Figure 4-8. Electrostatic movement

dc Ammeters

Laboratory and industrial electromechanical dc ammeters are used to measure currents from 1 μA (10^{-6} A) to several hundred amperes. Figure 4-9 shows a photograph of the interior of a typical dc ammeter. The D'Arsonval movement is used in most dc ammeters as the current detector. Typical laboratory bench meters of this type have accuracies of about 1 percent of their full-scale readings due to inaccuracies of the meter move-

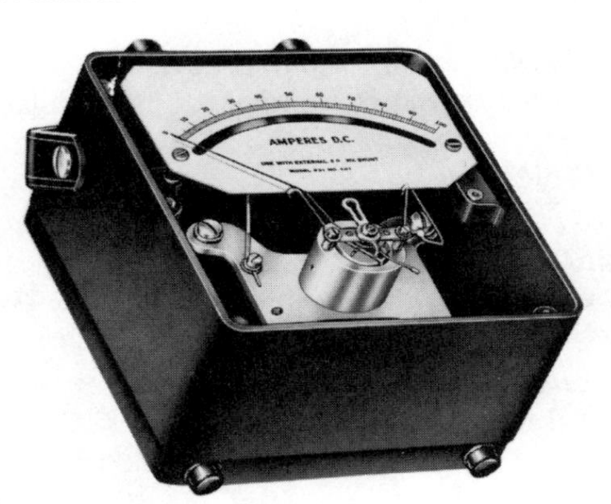

Figure 4-9. Interior view of an ammeter, with the magnet inside the moving coil (Courtesy of Weston Instruments, Inc.)

ment. In addition to this error, the resistance of the meter coil introduces a departure from the ideal ammeter behavior described in the introduction of this chapter. The *model* usually used to describe a real ammeter in equivalent-circuit use is a resistance R_m (equal in value to the resistance of the meter coil and leads) in series with an ideal ammeter (which is assumed to have no resistance) (Fig. 4-10).

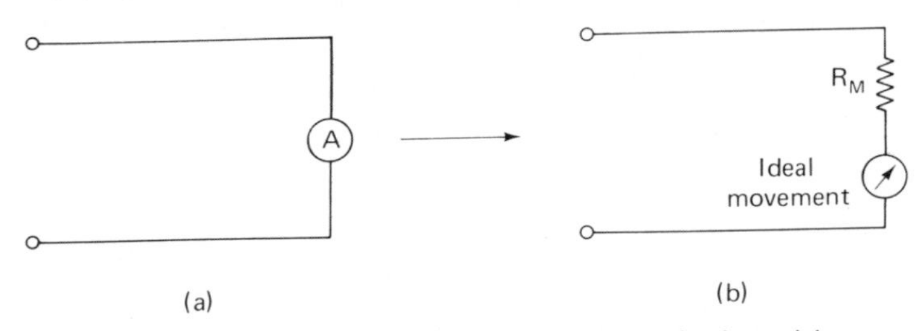

Figure 4-10. Ammeter symbol and equivalent circuit model (a) Circuit symbol (b) Equivalent circuit model

By using this model, we can calculate the error caused by introducing an ammeter into a circuit, or we can specify the maximum allowable resistance that will cause the ammeter to have a negligible effect. This effect is similar to the effect of voltmeter loading because the additional meter resistance causes less total current to flow in the circuit branch being measured. Table 4-1 shows the internal resistances of typical D'Arsonval movements.

Table 4-1 Internal Resistances of Typical D'Arsonval Movements

50 μA	1000–5000 Ω
500 μA	100–1000 Ω
1 mA	30–120 Ω
10 mA	1–4 Ω

The following example indicates how the error due to the extra resistance of an ammeter in a circuit can be calculated.

EXAMPLE 4-1

We are given a 1-mA ammeter which has an internal resistance of 50 Ω and we wish to measure the current flowing in a branch which contains a 1000-Ω resistor. Calculate (a) the error introduced by the extra resistance of the ammeter on the circuit and (b) the ammeter reading if one volt is applied across the branch?

Solution: (a) Without the ammeter in the circuit, 1 V applied across 1000 Ω will yield a current

$$I = \frac{V}{R_1} = \frac{1}{1000} = 0.001 \text{ A} = 1.0 \text{ mA}$$

When the ammeter is inserted in series with this resistance (Fig. 4-11), the total resistance of the branch is 1050 Ω. Thus, 1 V applied across this resistance will produce a current of

$$I = \frac{V}{R_1 + R_M} = \frac{1}{1050} = 0.95 \text{ mA}$$

The error in the reading caused by R_M of the ammeter is

$$\text{Error} = \frac{(1.0 \times 10^{-3}) - (0.95 \times 10^{-3})}{1.0 \times 10^{-3}} \times (100\%) = 5\%$$

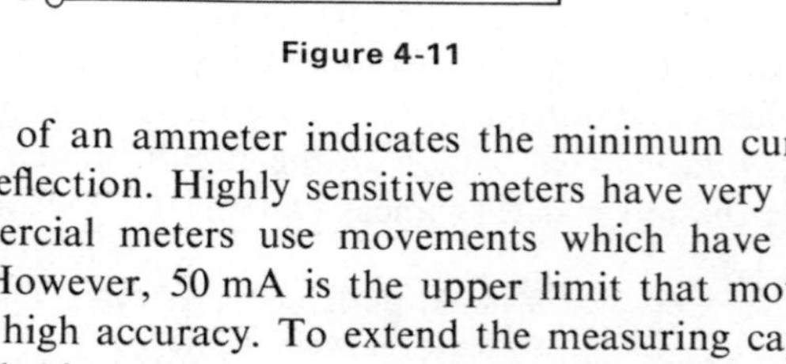

Figure 4-11

The sensitivity of an ammeter indicates the minimum current necessary for a full-scale deflection. Highly sensitive meters have very small full-scale readings. Commercial meters use movements which have sensitivities as small as 1 μA. However, 50 mA is the upper limit that movement springs can handle with high accuracy. To extend the measuring capabilities of dc ammeters beyond this upper bound, shunts must be used.

A *shunt* is a low-resistance path connected in *parallel* to the meter movement. Figure 4-12(a) shows an ammeter with a shunt. The shunt allows a specific fraction of the current flowing in the circuit branch to bypass the meter movement. If we know exactly how the current is divided, the fraction of the current flowing in the movement can indicate the total current flowing in the branch in which the meter is connected.

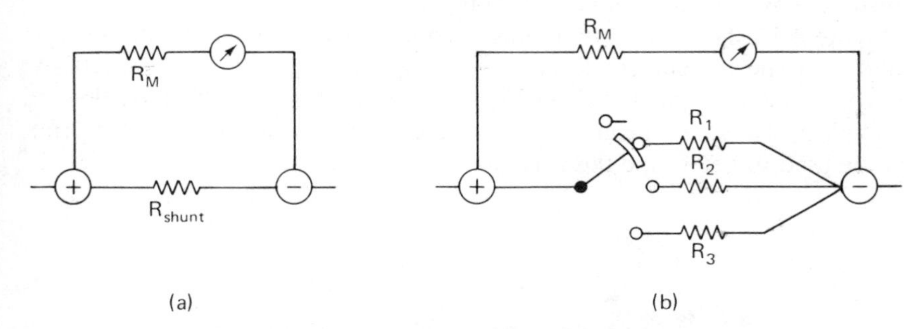

Figure 4-12. (a) Ammeter with shunt (b) Multirange ammeter

EXAMPLE 4-2

Given a 1-mA meter movement with an internal (coil) resistance of 50 Ω. If we want to convert the movement into an ammeter capable of measuring up to 150 mA, what will be the required shunt resistance?

Solution: If the movement can handle a maximum of 1 mA, the shunt will have to carry the remainder of the current. Thus, for a full-scale deflection

$$\begin{aligned} I_{\text{shunt}} &= I_{\text{total}} - I_{\text{movement}} \\ &= 150 - 1 \\ &= 149 \text{ mA} \end{aligned}$$

Since the voltage drops across the shunt and the movement are equal (by virtue of their being connected in parallel), then

$$\begin{aligned} V_{\text{shunt}} &= V_{\text{movement}} \\ I_{\text{shunt}} R_{\text{shunt}} &= I_M R_M \\ R_{\text{shunt}} &= \frac{I_M R_M}{I_{\text{shunt}}} = \frac{(0.001)(50)}{0.149} \\ R_{\text{shunt}} &= 0.32\ \Omega \end{aligned}$$

Many ammeters are multirange instruments. Some of these use several external terminals (binding posts) as a means of changing ranges; others use a rotary switch. If a rotary switch is used to change ranges (as shown in Fig. 4-12(b)), the switch pole must make contact with the adjacent shunt resistance before breaking contact with the shunt resistance in the former branch. By using such a shorting-type rotary switch, insurance that the movement will not be accidentally subjected to the full current in the branch is provided. For meters that can measure up to 50 A, the shunts are usually mounted within the instrument. Higher-range ammeters use external, high-current shunts made of special materials which maintain stability (constant resistance) over a wide temperature range.

Figure 4-13 shows two commonly used external ammeter shunts. Each is rated at a specific current level and voltage drop. For example, a 100-A, 50-mV shunt is designed to drop 50 mV across itself when 100 A are flowing through it. Thus any meter which indicates 50 mV at a full-scale deflection can be used to determine the current in this shunt.

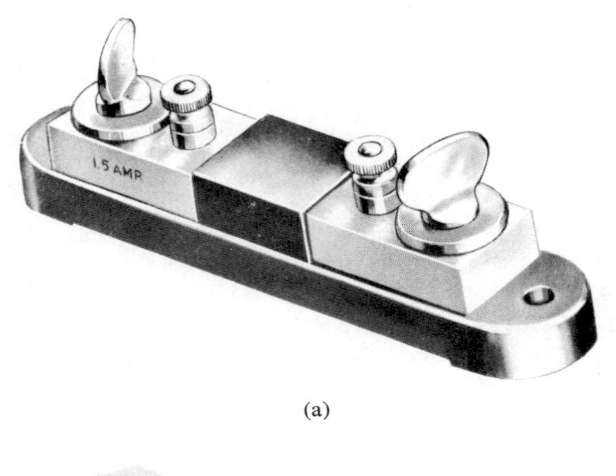

(a)

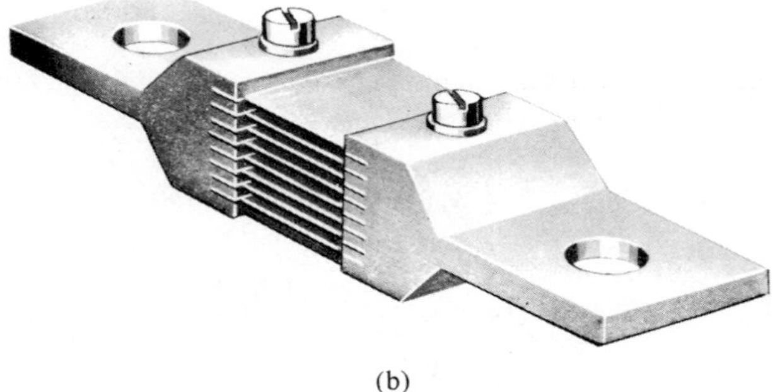

(b)

Figure 4-13. Ammeter shunts (a) External shunt for portable ammeters (1-200A) (b) External shunt for switchboard ammeter (up to 8000A) (Courtesy of Weston Instruments, Inc.)

The current is fed through the shunt by heavy current-carrying terminals. These heavy terminals are used to keep the contact resistance of this connection as small as possible. The voltage drop across the shunt is measured by the meter movement which is connected to the two inner "potential" terminals. By measuring voltage drop across the potential terminals, the effect of any contact resistance on the measured value is eliminated.

The two heavy copper blocks which make up the ends of the shunt are welded to sheets of resistive material as shown in Fig. 4-13(b). The resistive material is specially chosen to keep a constant resistance value, even with changes in its temperature. Precision external shunts are built in ranges from 0.1 to 2000 A with accuracies of 0.1 percent.

dc Voltmeters

Most dc voltmeters also use D'Arsonval movements. The D'Arsonval movement itself can be considered to be a voltmeter if we note that the current flowing in it, multiplied by its internal resistance, causes a certain voltage drop. For example, a 1-mA full-scale, 50-Ω movement has a 50-mV drop across it when 1 mA is flowing in the movement. If the scale reads volts rather than amperes, the movement is acting as a 50-mV voltmeter. To increase the voltage that can be measured by such a meter, an additional resistance is added in series with the meter resistance. The extra resistance (called a *multiplier*) limits the current flowing in the meter circuit (Fig. 4-14(a)).

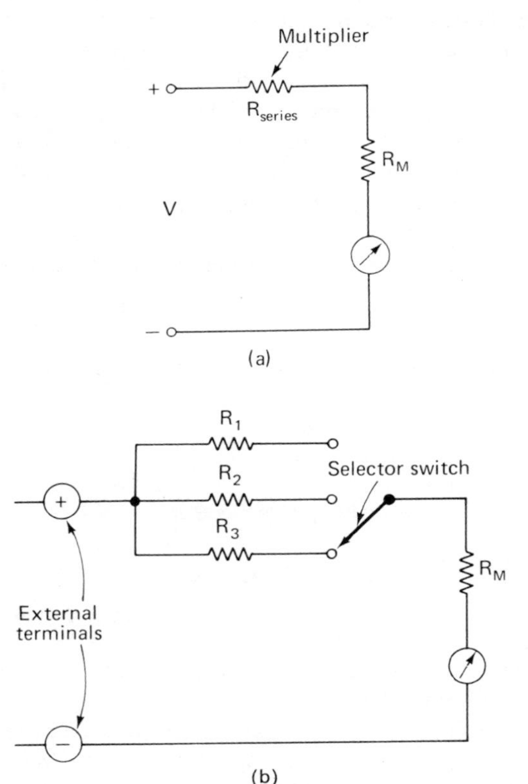

Figure 4-14. (a) Basic dc voltmeter (b) Multirange voltmeter

EXAMPLE 4-3

If we want to use a 1 mA, 50-Ω movement as a 10-V full-scale voltmeter, how much resistance must we add in series with the movement?

Solution: At full scale, 1 mA flows in the movement. If the meter is to measure 10 V, the total resistance required of the meter is

$$R_{\text{total}} = \frac{V}{I} = \frac{10\text{ V}}{0.001\text{ A}} = 10{,}000\,\Omega$$

Since the movement resistance is 50 Ω, the added series resistance must be

$$R_{\text{series}} = R_{\text{total}} - R_{\text{movement}}$$

or

$$R_{\text{series}} = 9950\,\Omega$$

To construct a multirange voltmeter, we can use a switch that connects various values of resistance in series with the meter movement (Fig. 4-14(b)). In order to get an upscale deflection, the leads must be connected to the voltmeter with the same polarity as the meter terminal markings. Typical laboratory dc voltmeters have accuracies of ±1 percent of full scale.

A voltmeter's *sensitivity* can be specified by the voltage required for a full-scale deflection. But another more widely used sensitivity criterion is the ohms/volt rating. For each voltage range, the total resistance exhibited by the voltmeter, R_T, divided by the full-scale voltage, yields a quotient, S. This quotient is a constant of the voltmeter, and it is called the *ohms/volt* rating. The easiest method for calculating S is to find the reciprocal of the *current* sensitivity of the movement being used in the voltmeter.

EXAMPLE 4-4

What is the ohms/volt rating of a voltmeter with (a) a 1-mA movement? (b) a 50-μA movement?

Solution:

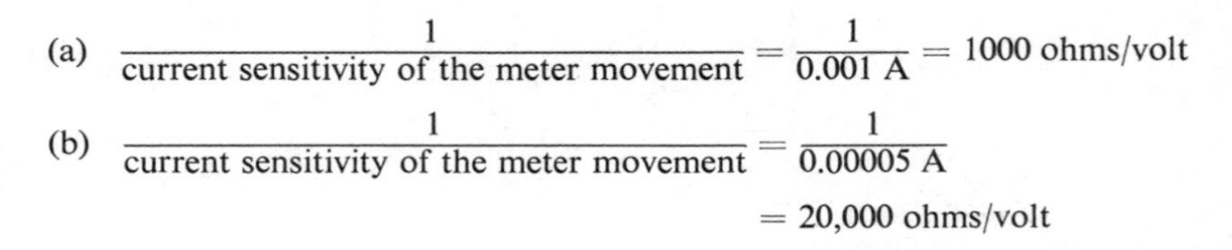

(a) $$\frac{1}{\text{current sensitivity of the meter movement}} = \frac{1}{0.001\text{ A}} = 1000\text{ ohms/volt}$$

(b) $$\frac{1}{\text{current sensitivity of the meter movement}} = \frac{1}{0.00005\text{ A}} = 20{,}000\text{ ohms/volt}$$

The ohms/volt rating is essentially an indication of how well an actual voltmeter approaches the behavior of an ideal voltmeter. An ideal voltmeter would have an infinite ohms/volt ratio and would appear as an infinite resistance (or open circuit) to the circuit it was measuring. Typical basic dc laboratory voltmeters have a 20,000 ohms/volt rating.

Because the voltmeter is not ideal, it draws some current from the circuit it is measuring. If a low-sensitivity (small ohm/volt rating) meter is used to measure the voltage across a high resistance, the meter will actually act like a shunt and will reduce the equivalent resistance of the branch. A highly unreliable reading will result. Such a circuit disturbance caused by current being drawn by a voltmeter is called a *loading effect*. Example 4-5 demonstrates a classic situation in which the choice of a voltmeter makes a significant difference in the accuracy of the measured result.

EXAMPLE 4-5

We want to measure the voltage across the 10-kΩ resistor of the circuit shown in Fig. 4-15. We have two voltmeters with which to do it. Voltmeter A has a sensitivity of 1000 Ω/V and voltmeter B has a sensitivity of 20,000 Ω/V. Both use their 50-V scales. Calculate (a) what each meter will read and (b) what is the error from the true reading?

Solution: The true reading should be

$$V_T = \frac{V_s}{R_1 + R_2} R_2 = 100 \times \left(\frac{10\ \text{k}\Omega}{30\ \text{k}\Omega}\right) = 33.3\ \text{V}$$

Now voltmeter A has a 1000 Ω/V sensitivity. It has an equivalent R of 50,000 Ω when using its 50-V scale. The total resistance from point 1 to point 2 with the voltmeter connected in the circuit is found from

$$\frac{1}{R_{12}} = \frac{1}{10\ \text{k}\Omega} + \frac{1}{50\ \text{k}\Omega} \quad \text{or} \quad R_{12} \cong 8300\Omega$$

The total resistance, R_T, of the circuit is $R_T = R_1 + R_{12} = 28{,}300\ \Omega$. Then the voltage between points 1 and 2 of Fig. 4-15, as indicated by the voltmeter, is

$$V_{12} = V_s \frac{R_{12}}{R_T} = 100 \times \frac{8300\ \Omega}{28{,}300\ \Omega} = 29.0\ \text{V}$$

This is an error of

$$\text{Error} = \frac{33.3 - 29.0}{33.3} \times 100\% = 13\%$$

Voltmeter B has a 20,000-Ω/V sensitivity, and thus its equivalent $R = 50 \times 20{,}000 = 1\ \text{M}\Omega$. Then

$$\frac{1}{R_{12}} = \frac{1}{10\ \text{k}\Omega} + \frac{1}{1\ \text{M}\Omega} \quad \text{or} \quad R_{12} = 9900\ \Omega$$

Therefore $R_T = 29{,}900\ \Omega$, and the voltage indicated by the voltmeter is

$$V_{12} = 100 \times \frac{9900\ \Omega}{29{,}900\ \Omega} = 33.1\ V$$

The error of this reading is

$$\text{Error} = \frac{33.3 - 33.1}{33.3} \times 100\% = 0.6\% \text{ low}$$

Figure 4-15

The example shows that the meter with the highest ohms/volt rating will yield the most reliable reading in terms of the possible loading error. We can use the same type of calculation to determine how sensitive a voltmeter must be if we want to reduce the loading error to some maximum percentage of the true reading. We also note that the loading error which may occur when measuring voltages in high-resistance circuits can often be far greater than the error due to other inherrent meter inaccuracies. In some such cases, accurate readings can be obtained only with electronic-type voltmeters which have input resistances of 10 MΩ or more. As a rule of thumb, to reduce the loading error of the voltmeter reading to less than one percent, the resistance of the voltmeter should be at least 100 times as large as the resistance of the path across which the voltage is being measured.

ac Ammeters and Voltmeters

Ac meters measure electrical quantities of current and voltage that change in amplitude and direction periodically with time (Fig. 4-16). However, for

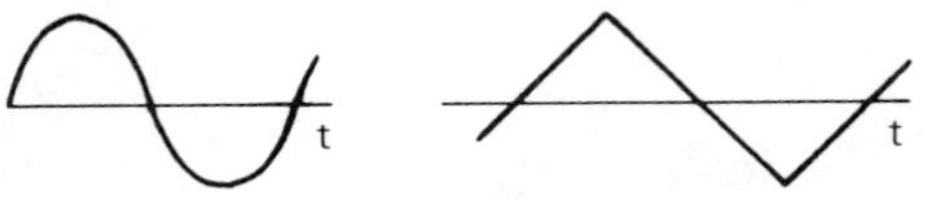

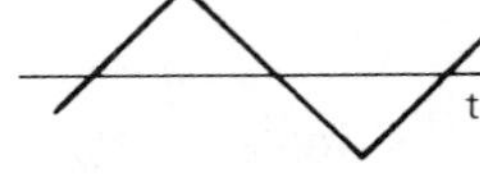

(a) (b)

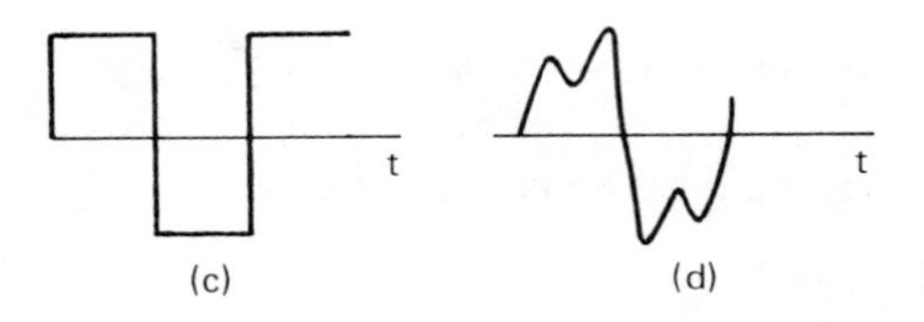

(c) (d)

Figure 4-16. Examples of ac waveforms

frequencies greater than a few cycles per second (Hz), meter movements cannot follow the rapid variations because of inertia and damping. Instead, the pointers of the meter movements take up a position in which the average torque is balanced by the torque of the shaft springs. Those meter mechanisms that respond in a linear fashion to applied ac variables have deflections proportional to the average value of the alternating quantities. The movements that react to the value of the quantity squared (called *square-law response*) have deflections proportional to the average of the *square* of the value of the quantity being measured. Since most ac measurements require the rms (effective) value of the variable rather than the average value, most ac meter scales are calibrated to read rms values,[3] regardless of whether their movement response is linear or square-law.

Ac meters generally use one of two possible methods for measuring ac quantities. For frequencies below several hundred hertz and for quantities whose amplitudes are not too small, electromechanical meter movements that respond directly to ac excitation can be used. For higher frequencies, the ac quantities are first converted to dc and then are applied to a D'Arsonval movement. If one uses ac meters for measurements above their specified maximum frequency limit, the accuracy of the meters is (often drastically) decreased.

ELECTROMECHANICAL ac METERS

The iron-vane, electrodynamometer, and electrostatic movements can all be used to build electromechanical ac voltmeters and ammeters. These movements can all respond directly to ac quantities because the magnetic field flux changes at exactly the same instant as the current flowing in the coils.

The iron-vane meters are the simplest type of ac meter and thus can be cheaply and ruggedly manufactured. They are very popular for use in industrial and commercial ac measuring applications. Typical simple laboratory ac meters are also of moving iron-vane construction (the most common being the concentric-vane type shown in Fig. 4-17). The usual commercial units are rated for accuracy only between 25 and 125 Hz. However, this makes them attractive for use at 60 Hz, the common power line frequency. (Special instruments can be made to extend the upper frequency limit to 2500 Hz.) Because iron-vane meters are inexpensive but somewhat less accurate than the electrodynamometer meters, they are used

[3]To see one important reason why the rms value gives more information about an ac signal than the average value, consider sinusoids, the most common ac signal. As shown in Chapter 1, the average value of any sinusoid is zero. If an ac meter were to indicate the average value of a sinusoid, we would not be able to distinguish between sinusoidal voltages of various amplitudes. However, the rms value of a sinusoid is not zero, but 0.707 of the amplitude of a sinusoid. Thus an rms reading yields much more information about a sinusoid than the average value reading. See Chapter 1 for more details on rms and average values.

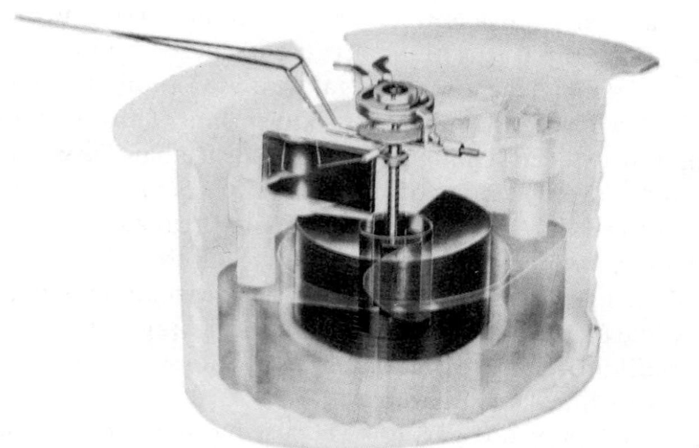

Figure 4-17. Phantom photograph of a concentric-vane moving-iron instrument. The figure shows details of the indicator with its counterweight, control spring, and damping vane. The moving vane may be seen as distinguished from the fixed vane in its brass retainer and is indicated by the lightly shaded area. (Courtesy of Weston Instruments, Inc.)

when cost rather than high accuracy is the prime consideration. Accuracies vary from 2 percent for simple lab meters, up to 0.5 percent of full scale. Ranges between 10 mA and 50 A for the ammeter and 1 to 750 V for the voltmeter are available.

The electrodynamometer movement responds to the square of the applied current and gives a true rms reading. Ac meters built using this movement can be extremely accurate (especially at power-line frequencies of 60 Hz) but are relatively expensive. They are also limited by their minimum power requirement for activation (1 to 3 watts). Their upper frequency limit is 200 Hz. Above 200 Hz, the inductance of the coils of the movement begins to introduce significant errors. However, for measuring ac signals with frequencies lower than 200 Hz, they are the most accurate instruments available. Meters with ranges between 1 and 50 A and 1 to 300 V, and with accuracies of up to 0.1 percent are manufactured. Figure 4-18 is a cutaway photograph of the electrodynamometer movement.

The electrostatic movement also responds to the square of the applied voltage; hence, its reading is likewise a true rms value. The frequency response is limited only by the ac current-carrying capacity. The movement can be used as a voltmeter to measure voltages between 100 V and 100 kV. It can be accurate to 0.5–1.0 percent and can measure ac voltages up to 50 MHz in frequency. Practically speaking, electrostatic meters are usually limited in use to special measurements such as high voltages (around 1000 V) and in certain ac circuits where other instruments would give erroneous readings.

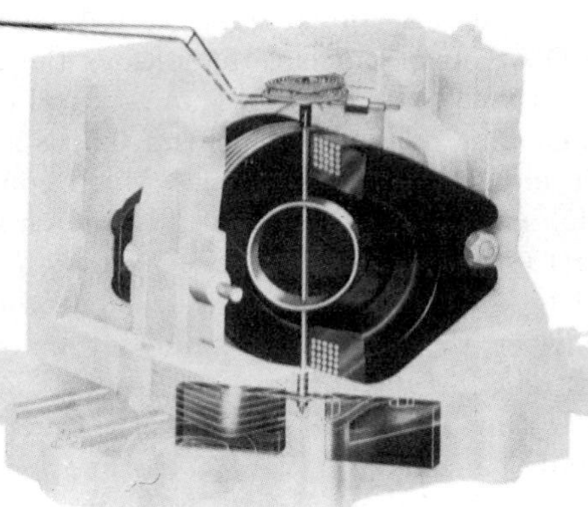

Figure 4-18. Phantom photograph of an electrodynamometer movement, showing the arrangement of fixed and movable coils. The rigidly constructed mechanism is surrounded by a laminated magnetic shield to minimize the effect of external fields on the meter indication. (Courtesy of Weston Instruments, Inc.)

RECTIFIER-TYPE ac METERS

We noted that electromechanical movements which respond directly to ac excitation (iron-vane, dynamometer) are limited in their accuracy to frequencies below a few hundred hertz. Furthermore, they are not sensitive enough to indicate ac signals whose power is less than about 1 W. To allow ac measurements of higher frequency or lower power signals, rectifier, thermocouple, and electronic instruments are used. Signals of frequencies greater than 500 MHz and magnitudes less than 1 mA can thus be measured.

Rectifier ac meters use a two-step process in measuring ac quantities (Fig. 4-19). A rectifier first converts the ac quantity to a varying unidirec-

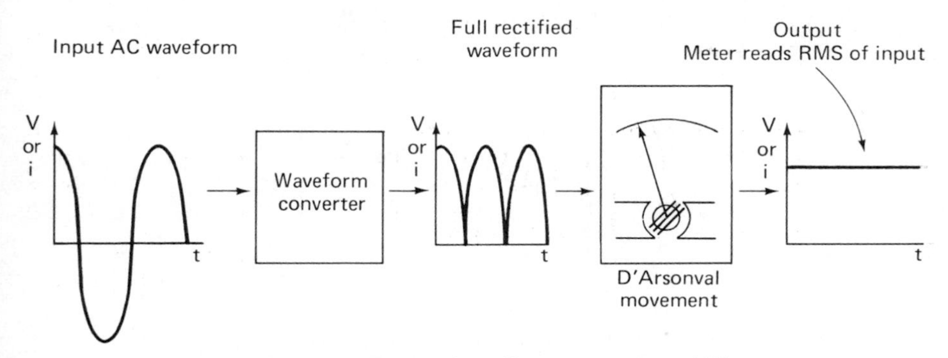

Figure 4-19. Block diagram of rectifier type ac meter which uses a full-wave rectifier for the waveform converter.

tional quantity. This unidirectional quantity is then fed to a D'Arsonval movement. The D'Arsonval movement indicates the average value of the quantity which is applied to it. (Naturally, the scale of the meter is calibrated to display the rms value of the ac waveform that is applied to the meter.)

A rectifier is a circuit containing diodes which convert ac quantities to unidirectional (dc) quantities. For example, the circuit shown in Fig. 4-20(a) shows the simplest type of rectifier. It contains only a single diode. If an ac sinusoidal signal is applied to this diode, current will be conducted only during the time when the ac current is flowing in the positive direction. During the times that the current has a negative value, the diode prevents the current from flowing. As a result, the output of a rectifier is a unidirectional quantity. Since only half of the initial waveform appears at its output, this type of rectifier is called a *half-wave rectifier*. Figure 4-20(b) shows the circuit diagram of one type of *full-wave rectifier*. In this type of rectifier, current is conducted in one direction through the rectifier no matter what the direction of current in the input waveform. Since a current appears at the output of the rectifier regardless of the direction of the current of the input waveform, the output is called *full-wave rectified waveform*. A third type of rectifier whose output is equal to the amplitude (or peak) value of the input waveform is called a *peak rectifier*.

As we noted, the waveform converter feeds a varying dc current to a D'Arsonval movement. Now the D'Arsonval movement senses the *average* value of an applied varying voltage. Since rms is the value usually required of an ac measurement, the scale must be designed by the manufacturer to read rms rather than average values. This process requires two necessary steps. First, an assumption is made that most of the ac signals to be measured

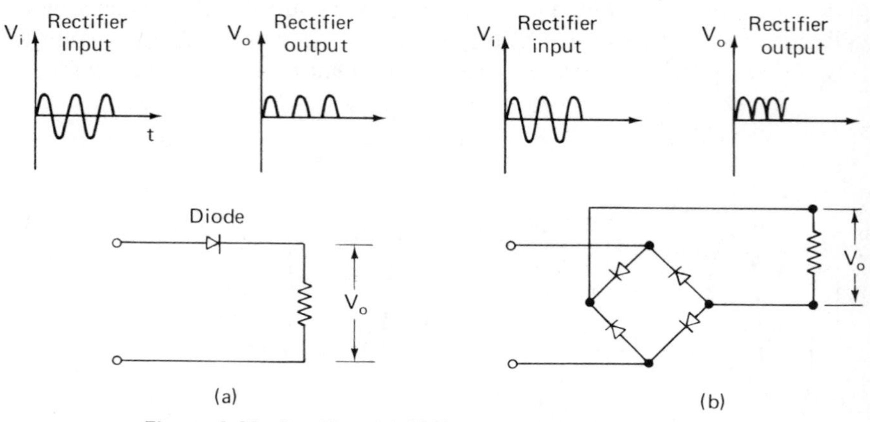

Figure 4-20. Rectifiers (a) Half-wave rectifier (b) Full-wave rectifier

will be pure sine waves. Second, the characteristics of the rectifier circuit are used to calculate the ratio between the average and rms values of the rectified wave. For example, if the waveform converter generates a full-wave rectified waveform to activate the D'Arsonval meter, we can calculate that the ratio of the rms value of a *sinusoid* to the average value of a *full-rectified* sinusoid is 1.11. (See Chapter 1.) Then the rms scale of the meter will actually be indicating 1.11 times the actual average current flowing in the D'Arsonval movement. For a *half-wave rectified* sinusoid, this ratio is 2.22.

The first assumption is not always valid, and this can lead to errors in the readings of rectifier meters. If the ac signal is not a pure sine wave (it could be a square wave, Fig. 4-16(a), triangular wave, Fig. 4-16(b), or many sinusoids of different frequencies added together), the meter will still be reading an average value of the rectified waveform. However, the ratio between this average value and the rms value of the rectified signal may no longer be the same as the ratio applicable for sinusoidal signals. A correction factor based on the shape of the ac signal and the type of rectifier used in the meter must be applied to the readings to eliminate the error.

EXAMPLE 4-6

A full-wave rectifier type of ammeter displays a current reading of 4.44 mA rms when measuring the triangular-wave current shown in Fig. 4-21(a).
(a) What is the true rms value of the waveform?
(b) What is the error of the meter?

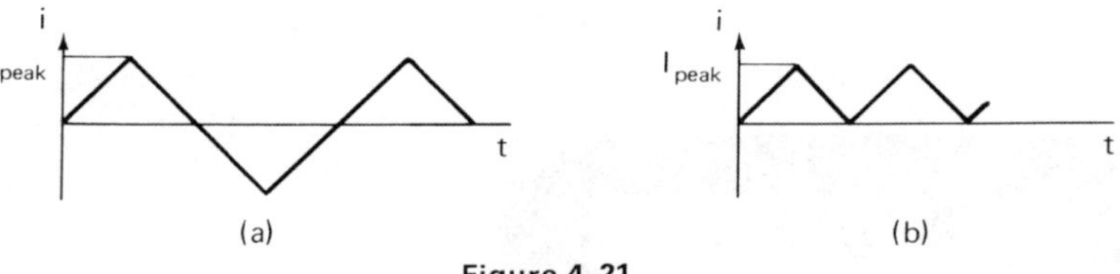

Figure 4-21

Solution: The instrument rectifier produces the resulting wave shown in Fig. 4-21(b). The meter displays 1.11 times the average value of the rectified wave. Therefore, the average value of the rectified wave is $I_{av} = 4.44 \text{ mA}/1.11 = 4 \text{ mA}$, and the peak value for a triangular wave is $I_{peak} = 2I_{av} = 8$ mA. The true rms value of the original wave is found in this case from[4]

$$I_{rms} = \sqrt{\frac{I_{peak}^2}{3}} = \sqrt{\frac{(0.008)^2}{3}} \tag{4-4}$$

$$I_{rms} = 4.6 \text{ mA}$$

[4]Figure 1-18 has the formula for the rms value of this triangular waveform and other common ac waveforms.

Hence the meter error is

$$\text{Meter error} = \frac{(\text{true value}) - (\text{indicated value})}{(\text{true value})} \times 100\%$$

$$= \frac{4.60 - 4.44}{4.6} \cong 4\% \text{ low}$$

In addition to waveform errors, rectifier instruments also suffer from variations in their readings with frequency (even though their upper limits are much higher than those of electromechanical ac meters). If accurate readings need to be made by using rectifier instruments, the manufacturer's specification of the upper frequency limit should be noted. (A rule of thumb which can be applied to rectifier ac meters is that their error can increase by about 0.5 percent for each 1-kHz increase in frequency above their maximum specified frequency.)

THE VOLTMETER-OHMMETER-MILLIAMMETER (VOM)

The VOM is a very useful and versatile laboratory instrument (Fig. 4-22). It is a meter which is capable of measuring dc and ac voltages, dc currents, and resistance. The various circuits required for measuring each of these quantities are all built into the meter design. For dc measurements, the VOM incorporates a D'Arsonval movement. For ac measurements, a rectifier ac

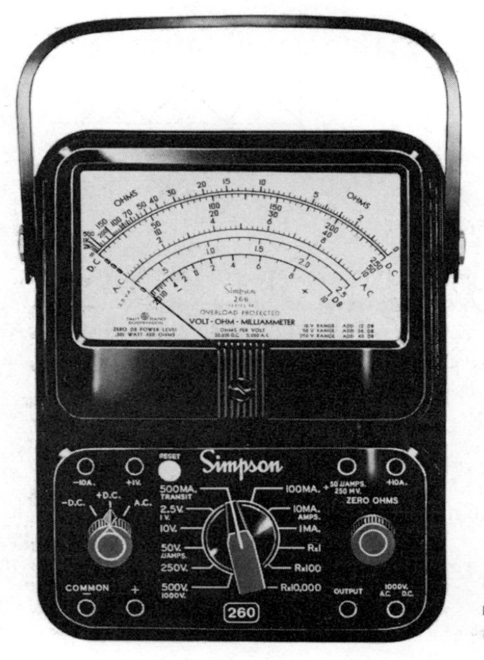

Figure 4-22. Typical Volt-Ohm-Milliameter (VOM) (Simpson Model 260-6P) (Courtesy of Simpson Electric Co.)

meter arrangement is used (because the ac rectifier meter also makes use of the D'Arsonval movement.) Resistance is measured by means of an ohmmeter circuit. The ohmmeter circuit applies the voltage from a battery across a series connection consisting of a known and unknown resistance. The D'Arsonval movement determines the value of the unknown resistance by measuring the fraction of the battery voltage dropped across the known resistance. Since the battery is subject to being worn out, its output voltage must be periodically checked to ensure that it is still capable of performing properly. (The ohmmeter will be discussed in more detail in Chapter 5.)

When used as a dc voltmeter, the VOM has a sensitivity of 20,000 Ω/V and full-scale voltage ranges of 2.5 to 1000 V. As an ac voltmeter, the VOM has a sensitivity of 5000 Ω/V and full-scale ranges of 2.5 to 1000 V. As a dc ammeter, the VOM is capable of measuring currents from about 1 mA to 1 A. Finally, resistance values between 0.1 Ω and 10 MΩ can be measured with the ohmmeter portion of the circuit. Typical accuracies of VOM's are 1–2 percent of full scale, depending on the scale being used.

THERMOCOUPLE ac METERS

The thermocouple meter measures ac and dc quantities by connecting the output signal of a thermocouple to a D'Arsonval movement (Fig. 4-23). The major advantage of thermocouple meters over all other ac meters is that frequencies of up to 50 MHz can be measured with up to 1 percent accuracy. The current in the circuit being measured flows through a heating element and causes the temperature of the heating element to increase. This element in turn heats a thermocouple junction and causes a voltage to

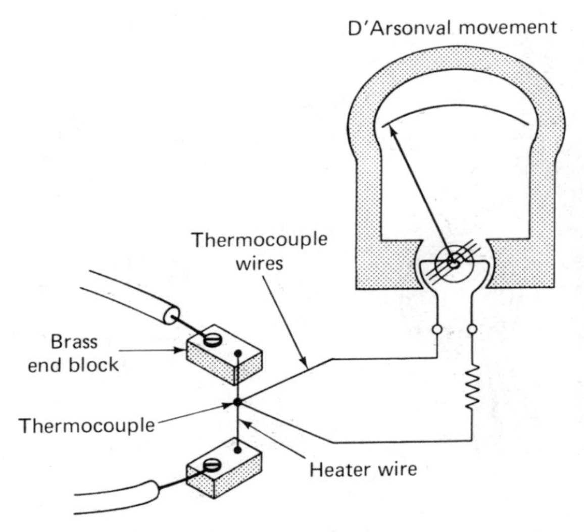

Figure 4-23. Thermocouple ac meter

arise across the junction. The voltage produces a current in the thermocouple wire and this current activates the D'Arsonval movement. Since the heating effect is proportional to i^2R, the resulting current which activates the D'Arsonval movement gives a true rms reading. (i is the current flowing in the heating element and R is the element's resistance.) Current in the 0.5 A to 20 A range and voltages up to 500 V can be measured with this instrument. However, the disadvantages of higher cost and sensitivity to burnout under overloads have limited the widespread use of thermocouple meters. Further discussion on the principles of thermocouples can be found in Chapter 15.

ac CLAMP-ON METERS

The *clamp-on ac meter* is an instrument which is used for measuring ac currents and voltages in a wire without having to break the circuit being measured. The meter makes use of the transformer principle to detect current. (See Chapter 7 for a description of transformer operation.) That is, the clamp-on device of the meter serves as the core of a transformer. The current-carrying wire is the primary winding of the transformer, and a secondary winding is in the meter. The alternating current in the primary is coupled to the secondary winding by the core, and, after being rectified, the current is sensed by a D'Arsonval movement.

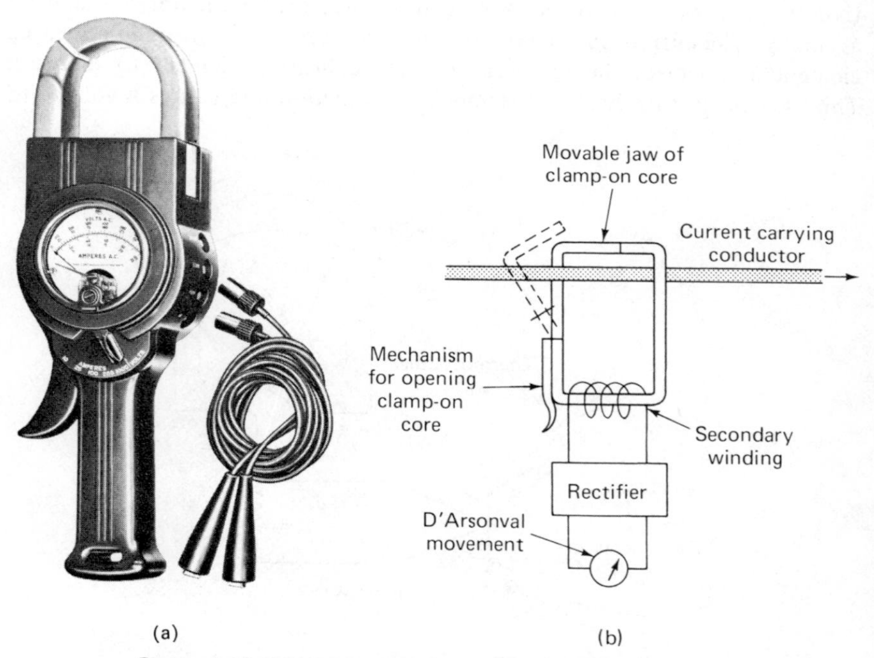

Figure 4-24. Clamp-on ac meter (Courtesy of Weston Instruments, Inc.)

Although the clamp-on meter is very convenient for making rapid ac current and voltage measurements, it is limited to measuring relatively high current levels. The smallest full-scale range on such clamp-on meters is 6 A. However the use of special adaptor cords as shown in Fig. 4-24(a) increases the sensitivity of the meter so that the full-scale sensitivity is increased to 300 mA (0.3 A).

Features of Meter Construction

BEARINGS

To ensure that a meter can respond to the forces which arise from the quantity it is measuring, any friction that would oppose the rotation of their moving member should be kept as small as possible. The moving member of the meter is usually mounted on a shaft that rotates, and such friction would arise at the points where the shaft is supported. To keep the friction to a minimum and still keep the shaft properly centered, jewel bearings are used (Fig. 4-25).

The jewels (usually synthetic sapphire) are used because the steel-shaft and jewel combination provides very low friction. The point of the shaft is usually so small that the pressure exerted on the jewel is very high (in terms of pounds per square inch—30,000 lb/in^2). Such high pressure makes the bearing very susceptible to damage from shocks that cause this load to increase (e.g., roughly laying the meter on the table). Hence great care in handling must be used to keep from cracking or breaking either the pivot or jewel. Some ruggedized instruments have their bearings backed by a spring.

In Fig. 4-25, we note that laying the instrument in such a position as to cause the shaft to be horizontal moves the pivot off-center. This can introduce a small shift in the pointer position during readings. Therefore, for best accuracy, it is advisable to keep the meter in a position in which the shaft is vertical. If the meter is placed so that the pointer moves in a horizontal plane, the shaft will be in a vertical position.

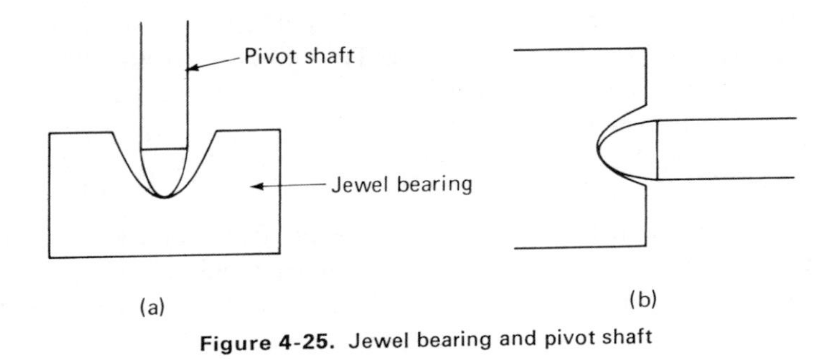

Figure 4-25. Jewel bearing and pivot shaft

TAUT BAND SUPPORT

Another method of supporting a meter movement besides the shaft, jewel bearings, and spring arrangement is the *taut band support*. Here the movement is suspended by two thin metal ribbons or bands (Fig. 4-26). The bands, rather than the springs, provide the electrical connection and restoring torque. An advantage is obtained because there is no friction between moving parts. This method produces highly repeatable measurements. The taut band support is replacing the jewel and pivot bearing in most uses.

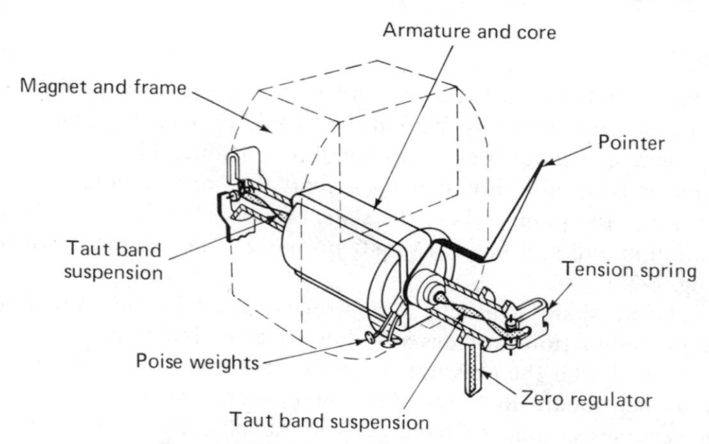

Figure 4-26. Taut band meter movement

DAMPING

If the only friction that acts on a pointer is the bearing friction, the pointer will swing widely before coming to rest (much like a compass needle). This would slow down and add inaccuracies to meter readings. To ensure that the pointer of the meter comes quickly to rest upon deflection, a damping mechanism to stop oscillations is used.

Two common methods of damping are employed. One involves mounting a large but light flap on the end of the pointer, using the air to damp the motion. The other is to construct the frame on which the meter coil is wound of a conducting material. A counterforce is induced in this frame by the motion of the coil in the magnetic field. This slows down the motion near the end of a swing since the counterforce always acts in a direction opposite to the direction of the coil's motion.

SHIELDING

To keep external fields (primarily magnetic) from interfering with the magnetic field of a meter's movements, the meter needs to be shielded. Soft iron or other shielding material is built into the case of the meter, and this should prevent stray external fields from causing errors and interfering with the meter's operation.

How to Use Basic Meters

1. *Ammeters* are always connected in series with the branch whose current is being measured—never in parallel. The ammeter can be destroyed if connected in parallel by mistake. Its low resistance may allow enough current to flow in the meter to burn it out. *Voltmeters* are always connected in parallel with (across) the portion of the circuit whose voltage is being measured.
2. Always make sure that the pointer is set at zero before a meter is connected. If the pointer is not set at zero, adjust the zero-setting screw on the scale face.
3. Do not handle meters roughly. The meter shaft and bearings are easily damaged by sharp knocks or jarring.
4. To protect the meter movements of multirange meters, start all readings of unknown quantities by setting the meters on their highest scale. Take as the final reading the deflection which is nearest to full-scale. This final reading will be the most accurate value.
5. Set portable meters on their backs. This will keep them from being knocked over. It will also keep their readings more accurate by keeping the shaft vertical (previously discussed in the section entitled *Features of Meter Construction*).
6. Correct the readings for any loading effect that is caused by the meter's presence in the circuit (see sections on dc voltmeters and ammeters for a discussion of loading).
7. To give upscale readings, dc meters must be connected so that the meter terminals are connected to points in the test circuit with matching polarities. Reversed polarity connections can lead to movement damage arising from the banging of the pointer against the reverse stop.
8. Ac meters—iron-vane, electrodynamometer, and electrostatic ac meters can be connected independent of polarity considerations.
9. Keep meters (especially iron-vane and dynamometer meters) away from conductors carrying large currents. The magnetic fields associated with the currents can interfere with the meter movement's magnetic fields and introduce errors.
10. VOM—(a) When not in use, always have the function selector switch set to a high dc-volts scale. This will prevent draining of the battery if an accidental shorting of leads occurs. It also protects the rectifier circuit against accidental connection to a dc source. (b) Check the battery to ensure that it is operating above its minimum allowed voltage. (c) Use each meter function as you would use such a meter alone.

11. Meters should be calibrated once a year or according to the manufacturer's specifications. Apply a calibration sticker to the meter stating the date calibration was last performed.

Meter Errors

1. Scale error—Inaccurate markings of the scale during calibration or manufacture. Equally probable along the entire scale.
2. Zero error—Failure to adjust to zero setting before making measurements.
3. Parallax error—Caused by not having line of sight exactly perpendicular to measuring scale. Somewhat eliminated by mirror under scale.
4. Friction error—If the bearing is damaged or worn, its friction may prevent the needle from making a true reading. Can be somewhat eliminated by *gently* tapping on meter while making a measurement.
5. Temperature effects on magnets, springs, and internal resistances. These errors are proportional to the percent of deflection.
6. Error due coil-shaft misalignment on bearing—Reduced by keeping the shaft vertical.
7. Bent pointer or pointer rubbing against scale.
8. Poor accuracy—If a meter is said to be accurate to within some percentage, this usually refers to its full-scale reading. For readings taken at less than full scale, the *actual* percentage error may be much larger.

EXAMPLE 4-7

If a meter is accurate to 2 percent of its 100-mA scale, any reading will only be guaranteed to within ± 2 mA. Thus, if this scale is used to measure 20 mA, the reading may be in error by ± 2 mA or 10 percent of its true value. To minimize this error: (a) always use the reading closest to full-scale for the final reading and (2) make sure the meter is properly calibrated (see Appendix for meter calibration section).

9. Loading-effect error due to introducing a nonideal instrument into a circuit. The disturbance of the circuit by the instrument can be calculated and compensated for in the reading if a more sensitive meter is not available.
10. Specific errors associated with a particular meter's operating principles and design. Examples of these are hysteresis effects in moving-iron-vane movements or waveform errors in nontrue-rms-reading ac instruments. The extent of these errors is determined from a knowledge of the meter and its operation.

Table 4-2 Characteristics of Meters

Type	Frequency Ranges	Voltage Ranges	Input Impedance	Current Ranges	Accuracy
D'Arsonval	Dc only	10 mV–10 kV	1,000 Ω/V to 20,000 Ω/V	1 μA–1000 A	0.1% to 2%
Iron-Vane	25–125 Hz	1–750 V	≈ 100 Ω/V	10 mA–50 A	0.5% to 2%
Electrodynamometer	0–200 Hz	1–300 V	10–30 Ω/V	1–50 A	0.1% to 1%
Electrostatic	0–500 kHz	100 V–100 kV	Virtually Infinite	—	0.5% to 3%
Rectifier (utilizes a D'Arsonval movement)	0–20,000 Hz	2.5–1500 V	20,000 Ω/V (dc) 5,000 Ω/V (ac)	1 mA–1.5 A	1% to 3%
Thermocouple	0–50 MHz	1–500 V	100–500 Ω/V	0.5 A–50 A	1% to 3%
Analog Electronic (Amplifier and D'Arsonval movement)	0–10 MHz (0–500 MHz with special probe)	10^{-8} V–10 kV	10 MΩ or higher	10^{-13} A–50 A	1% to 5%
Digital Electronic	0–1 MHz	1 μV–10 kV	10 MΩ or higher	10^{-13} A–1 A	0.005–0.1% (dc V) 0.1%–2% (ac V and current)

Problems

1. Explain, by describing the effects on the meters and on the circuits being measured, why a voltmeter should never be connected in series, or an ammeter in parallel with the circuit being measured.
2. If a D'Arsonval movement is rated at 1 mA, 60 Ω, what is the current sensitivity of the movement?
3. What functions do the torque springs perform in the D'Arsonval movement?
4. A D'Arsonval movement has a coil form with 50 turns and with dimensions of $l = 2$ cm and $r = 1$ cm. The permanent magnet of the movement supplies a uniform flux density of 0.4 Wb/m^2. If the full-scale current of the movement is 1 mA, find the torque (in lbs) exerted on the coil.
5. If a 10 mA movement has an internal resistance of 4 Ω, calculate the shunt resistance required to increase its range to:
   a) 50 mA
   b) 250 mA
6. Design a multirange ammeter with ranges of 0–10 mA, 0–100 mA, and 0–1500 mA, using a 1 mA movement whose internal resistance is 60 Ω (i.e. show a circuit diagram of the meter you have designed).
7. Why is a "make-before-break" switch used in the design of multirange ammeters?
8. If a 1 mA, 100 Ω movement is inserted into a circuit branch whose resistance is 500 Ω, find the error in the indicated current values due to the insertion of the meter. If the movement is converted into an ammeter whose full-scale range is 10 mA, find the resulting error.
9. Design a voltmeter which uses a 50 mA, 2,000 Ω movement and has ranges of 0–5, 0–50, 0–500 V. What is the sensitivity of this voltmeter? Use a circuit diagram to show the instrument you have designed.
10. If a voltmeter has the same movement as the meter used in Prob. 6, what is the input resistance of the instrument on the following scales: a) 5 V, b) 500 V?
11. A typical vacuum-tube voltmeter has an input resistance of 11 MΩ. If a D'Arsonval movement were used to build a voltmeter, what sensitivity would the movement have to have if the voltmeter were to have the same input resistance as the VTVM when the voltmeter was used on its 0–5 V scale?
12. If a 0–50 V voltmeter has an accuracy of 1% full-scale, between what limits may the actual voltage lie when the meter indicates 12 V? What is the range of error at this reading?
13. Two resistors with values of 12,000 Ω and 6,000 Ω are connected in series across a 50 V voltage source. A voltmeter with a 1,000 Ω/V rating is used to measure the voltage across the 12,000 Ω resistor. Calculate the error in the measured voltage value due to the loading effect of the voltmeter on the circuit.

14. What will be the action of a D'Arsonval movement with a zero-center scale, if it is used to measure quantities in the following ac circuits:
    a) Circuit being driven by a 60 Hz sinusoidal voltage source.
    b) With a slowly varying (less than 5 Hz), but non-periodic source.
15. If a dynamometer ammeter responds to i^2 and a rectifier ammeter to I_{av}, explain how both types of ammeters can be used to indicate the rms values of currents.
16. Predict the readings of a) a dynamometer ammeter and b) a rectifier ammeter if the waveform shown in Fig. P4-1a is applied to the meter.
17. Repeat Prob. 12 for the waveform shown in Fig. P4-1b.

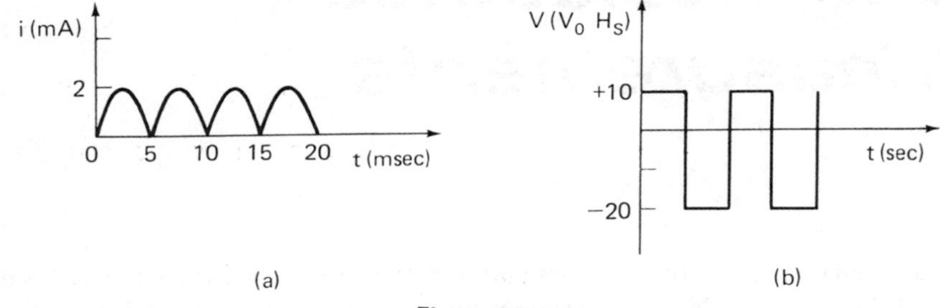

Figure P4-1

18. If a thermocouple ammeter reads 5 A at full-scale, what will be the current passing through the meter when the meter deflection is one-half of full-scale?
19. What is the difference between a taut-band movement and a conventional meter movement?

References

1. RIDER J. F., and PRENSKY, S. D., *How to Use Meters*, 2nd ed. New York. John F. Rider Publisher, Inc., 1960.
2. *General Electric. Manual of Electric Instruments*, 1958.
3. STOUT, M. B., *Basic Electrical Measurements*, 2nd ed. Englewood Cliffs, N.J., Prentice-Hall, Inc., 1960.
4. COOPER, W. D., *Electronic Instrumentation and Measurement Techniques*, Chaps. 4 and 5. Englewood Cliffs, N.J., Prentice-Hall, Inc., 1970.
5. KINNARD, I. F., *Applied Electrical Measurements*, Chap. 4. New York, John Wiley & Sons, Inc., 1956.

5

Resistors and Resistance Measurements

This chapter describes resistors and how their resistance value is measured. First the theory underlying resistance and the part resistors play in circuits is introduced. Next, various common resistor types and structures are examined, along with the limitations on their use. Finally, a discussion and comparison of the techniques and instruments used to measure resistance values are presented.

Resistance and Resistors

Resistance, roughly speaking, describes the tendency of a material to impede the flow of electric charges through it. The unit of measurement of resistance, R, is the ohm (Ω). To understand the physical basis of resistance, we must review the process of charge conduction in a solid.

From Chapter 1, we remember that a current flows in a conductor when a voltage is applied across it. This voltage has associated with it an electric field which accelerates the mobile charge carriers of the conductor (we will restrict this discussion to electrons as the charge carriers). Each individual electron is accelerated by the field until it collides with the atoms of the conductor material. The collisions cause the electrons to change the direction of their motion and transfer some of their energy to the material atoms. Subsequently the electrons are again accelerated until the next collision. This collision-acceleration-collision process continues as long as the voltage is applied. The net effect of the collisions is to deliver energy to the material atoms (lattice). This limits the average forward velocity of the electrons and, consequently, the current flowing in the conductor. The energy transfer process gives rise to the effect of resistance, i.e., by being forced to surrender a part of their energy to the lattice, the progress of the electrons through the material is hindered. In absorbing the energy, the lattice heats up, and the temperature of the material rises.

The material property that describes how much a material impedes current flow is called its *resistivity*, ρ. The resistivities of various materials are listed in Table 5-1. In a good conductor (e.g., copper and silver), the *resistivity*

Table 5-1

Material	Resistivity ρ (ohm-meters)
Silver	1.47×10^{-8}
Copper	1.72×10^{-8}
Aluminum	2.63×10^{-8}
Gold	4.7×10^{-8}
Iron	10.0×10^{-8}
Constantan	49.0×10^{-8}
Nichrome	100.0×10^{-8}
Glass	10^{10}–10^{14}
Wood	10^{8}–10^{11}
Sulfur	10^{15}
Hard rubber	10^{13}–10^{16}

is very low, and the electrons move smoothly without much energy loss. Poor conductors (called insulators) such as wood and glass have a high resistivity, and almost no current flows even under a high voltage.

The resistance, R, of a piece of matter is found from its resistivity and its geometrical dimensions. For example, the resistance of a bar of material with uniform cross-sectional area (and at a constant temperature) is given by

$$R = \rho \frac{l}{A} \tag{5-1}$$

where: R = resistance in ohms,
l = length in meters,
A = cross-sectional area in square meters, and
ρ = resistivity in ohm-meters.

This relation says that a long, thin bar or wire has a greater resistance than a short, thick one of the same material.

EXAMPLE 5-1

A *bus bar* (a large-current-carrying bar) is made of copper; it is 2 meters in length and 1 cm $\times$ $\frac{1}{2}$ cm in cross section. What is the resistance of the bar?

Solution: Using Eq. (5-1), we can find the resistance $R = \rho(l/A)$. From Table (5-1) we see that copper has a resistivity of 1.72×10^{-8} Ω-m. Then

$$R = (1.72 \times 10^{-8}) \times \frac{(2.0)}{(0.01 \times 0.005)} = 6.8 \times 10^{-4}\ \Omega$$

EXAMPLE 5-2

If this same bar is stretched out to form a wire which has the same volume of copper, but which is now 1 mm² in cross section, what will be its resistance?

Solution: Since the volume of copper must be the same as in Example 5-1, this smaller wire is calculated to have a length of 100 m. Then using Eq. (5-1), we find

$$R = \rho\frac{l}{A} = (1.72 \times 10^{-8})\frac{(100)}{(1.0 \times 10^{-6})}$$
$$= 1.72\ \Omega$$

Most wire leads or power lines that transmit electric current are made of materials with very low resistivity; hence, their overall R is usually very low.

If a circuit or device requires the effect that a specific amount of resistance produces (such as limiting the current flowing through it or dissipating energy), an element which increases the overall resistance of the circuit is used. Such an element is called a *resistor*. Table 5-2 lists some uses of resistors in circuit applications.

Table 5-2 Common Uses of Resistors in Circuits

1. Elements which limit the current that can flow in a circuit branch. In such applications they can act as a protector of other elements in the branch, e.g., semiconductor devices or sensitive meter movements.
2. Elements which divide applied voltages so that only the desired voltage appears across a given section of a circuit.
3. Elements which are loads that make use of applied electric power, e.g., electric heating elements and incandescent lamps.
4. Elements which serve as low-resistance paths (as in shunts).
5. Elements which damp (reduce) unwanted oscillations by dissipating the energy of the oscillations.

Resistors are made of materials which conduct electricity but possess a large resistance compared to the resistance of the wires and the contacts. Putting a resistor into a circuit is like putting a toll booth on a busy highway. The resistor limits the current (traffic flow). It also extracts energy flowing in the circuit branch as the toll booths extract money from the drivers.

The instantaneous voltage across a resistor is directly proportional to the current flowing through it. The equation which describes this relation was discovered by George Ohm in his work with dc circuits in 1836. It is given by

$$v = Ri \qquad (5\text{-}2)$$

and is known as *Ohm's law.*

EXAMPLE 5-3

What is the current through a 20-Ω resistor that has a 5-V drop across it?

Solution: $I = \frac{V}{R} = \frac{5}{20} = 0.25 \text{ A}$

EXAMPLE 5-4

What is the potential drop across a 500-Ω resistor with a 1-mA current through it?

Solution: $V = IR = (0.001)\cdot(500) = 0.5 \text{ V}$

If we wish to express how well an element conducts rather than impedes electricity, Ohm's law can instead be written in the form

$$i = Gv \qquad (5\text{-}3)$$

where $G = 1/R$ is called *conductance* and its units are mhos (℧). To say that a circuit element has a low conductance implies that it conducts electricity poorly and has a high resistance. For example, a conductance of 10^{-6} mho (very low conductance) is equivalent to a resistance of 1 megohm.

Ohm's law is a linear relation, which implies that the *resistance* of a resistor does not depend upon the current flowing through it. However, in reality, Ohm's law is only an approximation of the behavior of a resistor under an applied voltage (usually it is an extremely good approximation). Deviations from it arise because the mechanism of current flow is also a function of such variables as temperature, stress on the material, and applied frequency, as well as the magnitude of the instantaneous voltage. How these factors cause a resistor's value to change from the value predicted by Ohm's law is considered in the following sections.

LEAD AND WIRE RESISTANCE

In most electrical measurement and design procedure, the resistance of the wires connecting the various elements of a circuit is neglected compared to the resistance of the circuit elements. This is usually a safe assumption to make and often causes no serious errors, especially if low-resistance elements are not used in the circuit. However, to get an idea of the actual resistance values which are exhibited by various commonly used wire sizes, part of a *wire table* containing this information is presented (Table 5-3).

Wire is manufactured in the U.S. in standard sizes based on a convention called a "wire table." The American Wire Gauge (AWG) sizes in the table correspond to definite cross-sectional areas expressed in circular mils (defined below). A sample of these sizes along with their areas, resistance (in ohms per 1000 feet), and maximum allowable current for standard insulation are given for copper wire (most common wire in use) in Table 5-3.

Table 5-3 American wire gauge (AWG) sizes of copper wire.

Application	AWG #	Area (Circular* Mils)	Ω/1000 ft	Max. Allowable Current (Amperes)
	0000	211,600	0.049	360
Power distribution	00	133,080	0.078	265
	1	83,694	0.124	195
	4	41,470	0.240	125
House main power carriers	6	26,250	0.395	95
	8	16,509	0.620	65
Lighting, outlets, general home use	12	6529	1.588	20
	14	4106.8	2.52	
Television, radio	20	1021.5	10.1	
	22	642.4	16.1	
Telephone instruments	28	159.8	64.9	
	35	31.5	329.0	
	40	9.9	1049.0	

*1 circular mil = 1 CM = $\frac{\pi}{4,000,000} = 7.85 \times 10^{-7}$ in.^{2}
= (diameter of wire in mils)^2 = d^2

From Table 5-3, we see that using household appliance wire (AWG #14) of up to 10 feet will not add any more than 0.025 Ω to the resistance of a circuit. However, we do note that there is a safe maximum current which can be carried by a given wire size. When a large current is expected, an adequately sized wire should be chosen. This must be done to prevent deterioration of the wire insulation by the heat generated from the current passing through the conductor. A larger wire has a smaller resistance and therefore will not generate as much heat for a given current value as a smaller wire.

EXAMPLE 5-5

Find the resistance of a copper wire 3000 ft long with a diameter of 0.1284 in.

Solution: 1 mil is equal to 0.001 inches. A *circular mil* is defined as the area of a circle which is 1 mil in diameter. Thus the diameter of the wire converted to mils is

$$0.1284 \text{ in} = 128.4 \text{ mils}$$

$$\text{Area circular mils} = (\text{diameter in mils})^2$$

So the area in circular mils (cmil) is

$$\text{Area (circular mils)} = A_{cm} = (128.4)^2$$
$$A_{cmil} = 16{,}509 \text{ circular mils}$$

This corresponds to an AWG #8 wire. The resistance of 3000 ft. of this wire is

$$R = (3000\text{ ft}) \times \frac{(0.6283\ \Omega)}{(1000\text{ ft})} = (3 \times 10^3) \times (6.283 \times 10^{-4})$$
$$R = 1.8849\ \Omega$$

CONTACT RESISTANCE

Contact resistance is the resistance arising at the point of connection of wires and circuit element leads. In connecting a circuit together, contact resistance is usually neglected. This can safely be done in most cases because such resistance is usually much smaller than the resistance of the circuit elements. However, when using low-resistance elements such as shunts, or when using a one-ohm resistor to measure current indirectly (as in certain oscilloscope measurements), contact resistance may introduce errors.

For tightly clamped and clean contacts, the resistance is only about 0.001 Ω, and for a well-soldered joint it can be even smaller. The use of either of these two methods for making contacts will usually eliminate the possibility of larger contact resistances (Fig. 5-1).

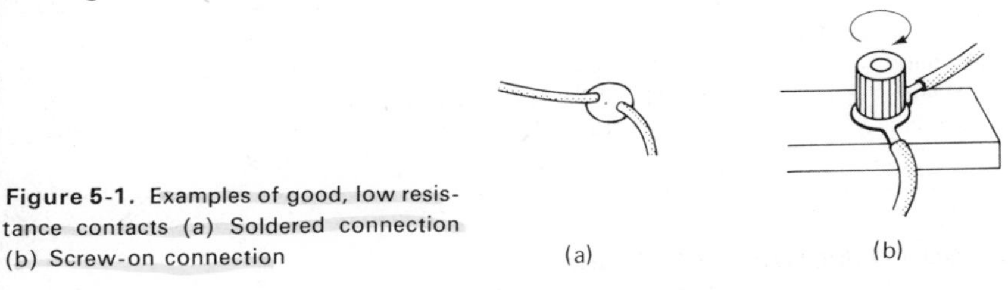

Figure 5-1. Examples of good, low resistance contacts (a) Soldered connection (b) Screw-on connection

Use of other, flimsier contacts such as alligator clips (especially with corroded or dirty surfaces) can introduce contact resistances of 0.1 Ω or more (Fig. 5-2). In certain applications, such extra resistances might cause appreciable errors. It is therefore good practice to use soldered or clean and

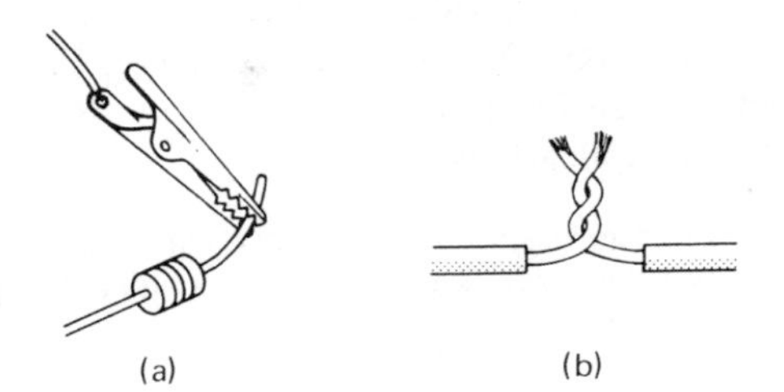

Figure 5-2. Possible sources of contact resistance > 0.1 Ω (a) Alligator clip (b) Loosely joined wires

tightly clamped connections, especially when assembling circuits involving low resistances. Sometimes old switches can also add unexpected resistances to circuits due to wear and corrosion at their contact points. Good switches should have less than 0.01 Ω resistance.

Resistor Types

Resistors are used for many purposes such as electric heaters, telephone equipment, electric and electronic circuit elements, and current-limiting devices. As such, their resistance values and tolerances vary widely. Resistors of 0.1 Ω to many megohms are manufactured. Acceptable tolerances may range from ±20 percent (resistors serving as heating elements) to ±0.001 percent (precision resistors in sensitive measuring instruments). Since no single resistor material or type can be made to encompass all the required ranges and tolerances, many different designs must be employed. The most common of these are discussed in this section. Table 5-4 summarizes the properties of the most common commercially available resistors.

Table 5-4 Characteristics of Various Types of Resistors

Type	Available Range	Tolerance	TC	Max. Power (watts)
Carbon composition	1 Ω to 22 MΩ	5 to 20%	0.1%/°C	2 W
Wirewound	1 Ω to 100 kΩ	0.0005% up	0.0005%/°C	200 W
Metal film	0.1 Ω to 10^{10} Ω	0.005% up	0.0001%/°C	1 W
Carbon film	10 Ω to 100 MΩ	0.5% up	−0.015 to 0.05%/°C	2 W

CARBON COMPOSITION RESISTORS

The carbon composition resistor is the most common resistor used in discrete electric and electronic circuits. Commercially available carbon composition resistors have resistance values which range from 1 Ω to 22 MΩ. They have the advantages of being cheap and reliable, and their stability is high during their lifetimes. However, their tolerances of 5 to 20 percent compare unfavorably with most other resistor types, and their temperature coefficients[1] (TC) are relatively large. The moderate power rating of composition resistors (maximum 2 watts) also prevents them from being used for some applications.

[1]The temperature coefficient (TC) of a resistor is the percentage change in the value of resistance per degree celsius. This quantity is discussed in more detail later in the chapter.

The common form of the composition resistor consists of graphite or other type of carbon embedded in a filler material. Graphite is a moderately good conductor, and by varying the graphite-filler mix, a large range of resistance values can be obtained (the less graphite, the higher the resistance). The body of the resistor has a cylindrical shape, usually brown in hue and with colored bands which code its resistance value. Figure 5-3 shows the construction of a typical carbon composition resistor. The color coding is given in a later section.

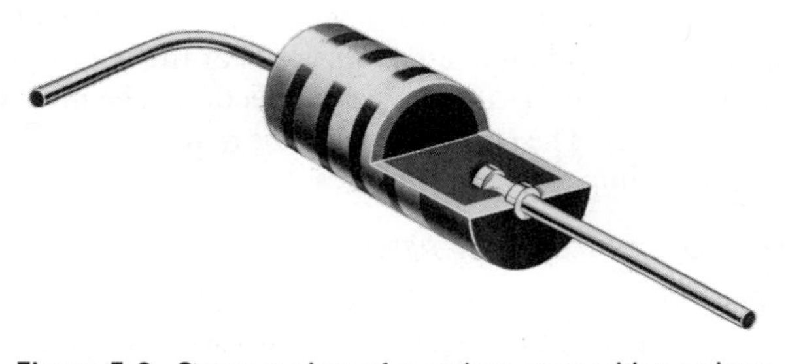

Figure 5-3. Cutaway view of a carbon composition resistor (Courtesy of Allen-Bradley Co.)

WIREWOUND RESISTORS

Wirewound resistors are what their name implies, a length of wire wound about an insulating cylindrical core (Fig. 5-4). Usually wires of materials such as constantan (60 percent copper, 40 percent nickel) and manganin which have high resistivities and low temperature coefficients are used. Because the wire's length, resistivity and size can be carefully controlled, such resistors can be made to be much more accurate than carbon composition resistors. Typical tolerances range from 0.01 percent up to 1.0 percent. The range of resistance values of wirewounds varies from approximately 1 Ω to 1 MΩ. They can also be made for high-power applications of 5- to 200-watt dissipation ratings (tolerance 5 to 10 percent). General purpose wirewound

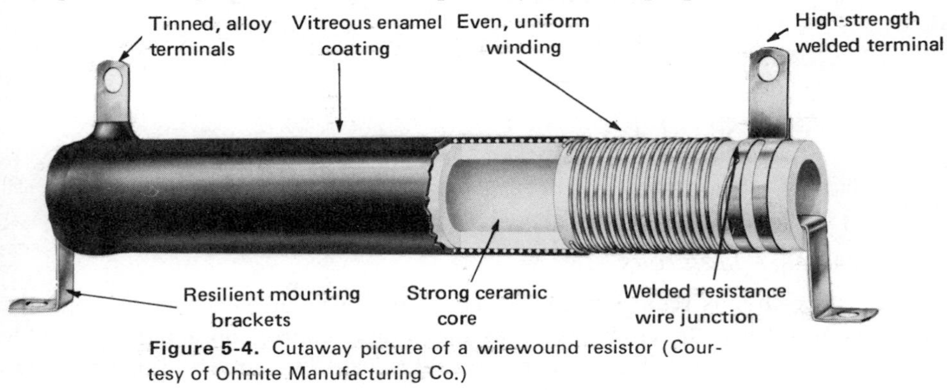

Figure 5-4. Cutaway picture of a wirewound resistor (Courtesy of Ohmite Manufacturing Co.)

resistors are used when carbon composition resistors cannot meet either the tolerance, reliability, or power specifications of a particular application.

If extremely high accuracy is desired (as when making precision measurements), wirewound resistors with special alloys are used to provide long-term stability and small TC. Tolerances as low as ± 0.0005 percent can be achieved.

The main disadvantage of wirewound resistors is the inductance that arises because of their wound, coil-like structures. At high frequencies this often makes the ordinary wirewound resistor unsuitable. To overcome the problem, a bifilar (dual thread) winding is used. One half of the wire is wound in one direction and the other half in the opposite direction. The inductances of the halves cancel each other. The resulting resistor is noninductive but much more expensive than an ordinary wirewound resistor.

METAL FILM AND CARBON FILM RESISTORS

Very thin films of metal can be deposited on insulating materials to provide very high resistance paths. Such metal film resistors can range in value up to 10,000 **MΩ** (10^{10} **Ω**) and are much smaller in size than wirewound resistors. The problems of inductance and wire size, which limit the values of wirewound resistors, are thus overcome. The advantages of high accuracy and low TC associated with using metal as a resistor material are retained. These attributes, coupled with very low noise, make them best suited for use in low-level amplifiers and computers.

The resistor structure consists of two electrodes mounted on an insulating base material with the thin metal film connecting the electrodes. The body is of cylindrical shape (Fig. 5-5).

Carbon film resistors use a deposited layer of carbon rather than metal. This gives lower tolerances and smaller values of resistance than obtained with metal films. However, the carbon film possesses a mildly negative TC which is useful in certain electronic circuits.

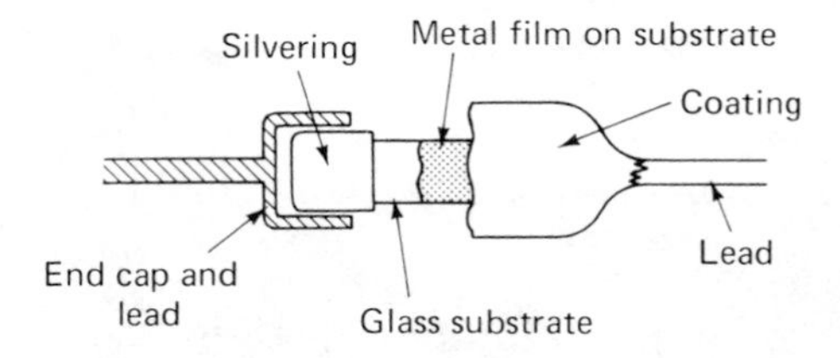

Figure 5-5. Construction of thin film resistor

VARIABLE RESISTORS

For circuits requiring a resistance that can be adjusted while it remains connected in the circuit (such as the volume control on a radio), variable resistors are used. They usually have three leads, two fixed and one movable. If contacts are made to only two leads of the resistor (stationary lead and moving lead), the variable resistor is being employed as a *rheostat* (Fig. 5-6(b)). Rheostats are usually used to limit current flowing in circuit branches. If all three contacts are used in a circuit, it is termed a *potentiometer* or "pot." Pots are often used as voltage dividers to control or vary voltage across a circuit branch (Fig. 5-6(a)).

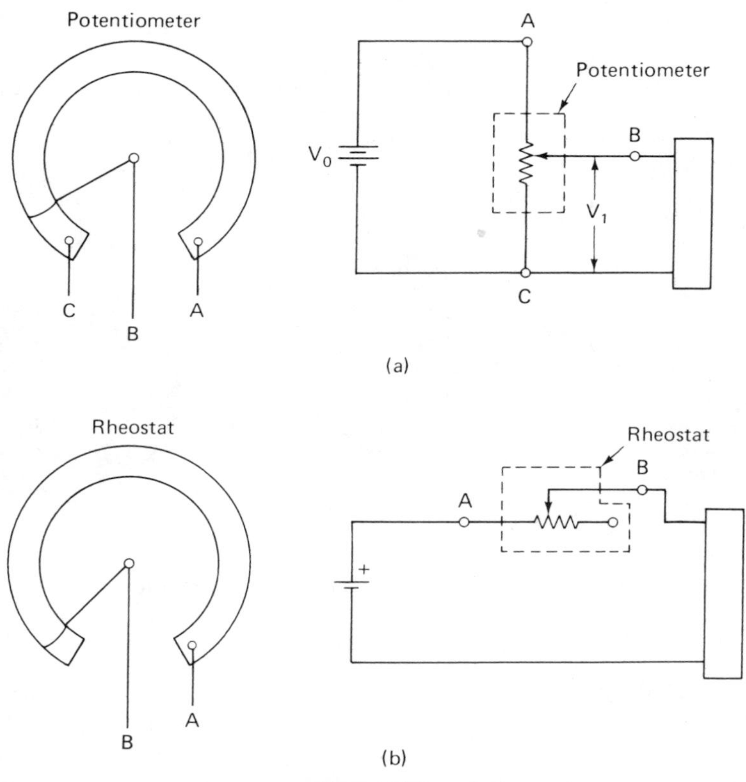

Figure 5-6

A large laboratory potentiometer (often called a slide-wire resistor) is shown in Fig. 5-7. Between the inputs at either end, the wire wound around the insulating cylinder provides a fixed resistance. The sliding contact at the top (sometimes controlled by a screw drive and handwheel) allows a variable resistance to be obtained from the element. Two fuses are used to protect the slide-wire pot from inadvertent overload. However, the accuracy of the slide-wire pot is limited by the wire spacing and the fact that the slide can only make contact with the wire along one line of the wirewound cylinder surface.

For general circuit use, variable resistors are usually made from fixed resistors and a contact capable of rotating on a shaft (Fig. 5-8). The contact is connected to the resistor body in between the fixed terminals. When the

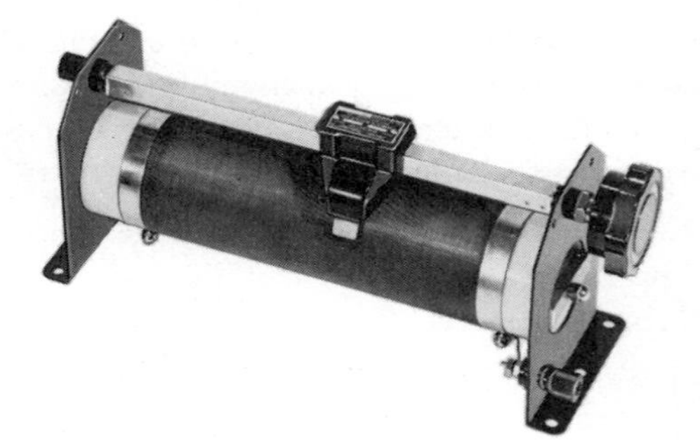

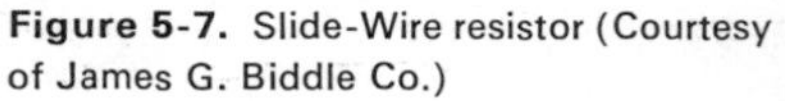

Figure 5-7. Slide-Wire resistor (Courtesy of James G. Biddle Co.)

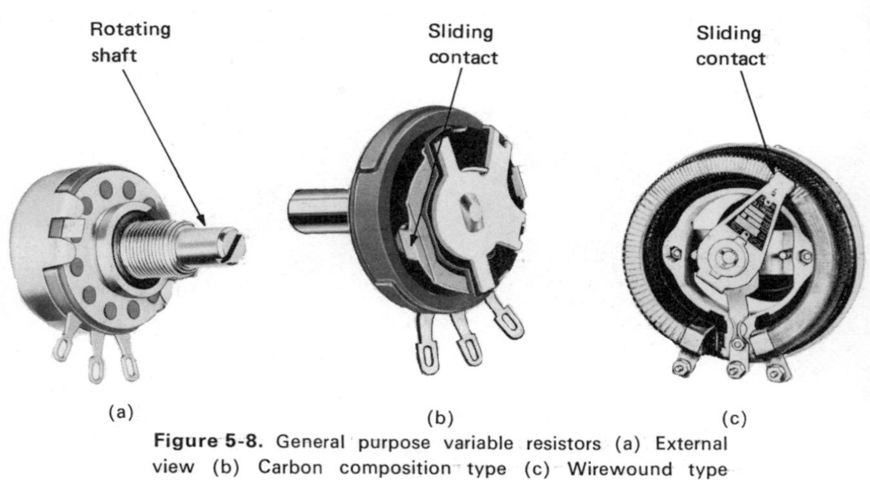

Figure 5-8. General purpose variable resistors (a) External view (b) Carbon composition type (c) Wirewound type (Courtesy of Ohmite Manufacturing Co.)

shaft is rotated, the third terminal is moved along the resistor; hence, the resistance between it and either of the other terminals varies. Usually, a shaft rotation of 270 degrees moves the sliding contact along the full length of the resistance element. The variation of the resistance with shaft rotation is known as the *taper*.

The body of the general purpose variable resistor can be of carbon composition or wirewound type. Ranges of 100 Ω to 1 MΩ for the carbon type and 5 Ω to 50 kΩ for the wirewound type are available. The overall resistance value and power rating are usually stamped on the unit body.

Precision potentiometers have the feature of being accurate in their variation, if not in their overall resistance value. In other words, the linearity of the taper is usually ± 0.05 percent to ± 0.5 percent, while the overall tolerance is usually ± 5 percent. To increase the resolution of a precision potentiometer, it requires between three to ten turns to move the sliding contact along the entire length of the resistance element.

DECADE RESISTORS

Another type of variable resistor found in the laboratory is the *decade resistor*. It gets its name from the fact that each of its switches is connected to a group of resistors whose values differ from the resistor connected to the adjacent switch by a factor of ten (a decade). Each switch can be set to ten positions (marked 0 to 9) and each position will connect a series set of resistors (Fig. 5-9). The switches are all interconnected so that we can obtain a

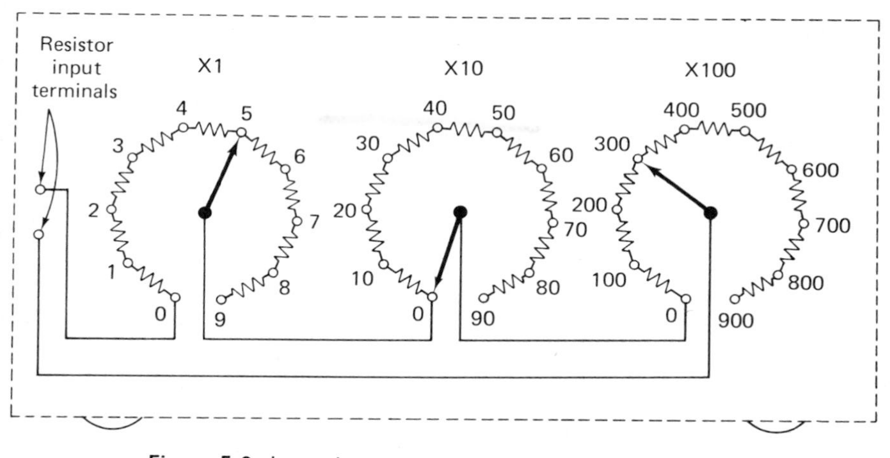

Figure 5-9. Internal connections of a 1-999Ω decade resistor. The switches are set to a position which will yield a resistance of 305Ω.

desired resistance value simply by dialing the switches to the corresponding positions. (See the sample setting of the decade resistor shown in Fig. 5-9). The resistors used in a decade resistor are usually bifilar wirewound resistors. The bifilar winding makes them accurate for frequencies up to 500 kHz. Typical accuracies of each switch position are ± 0.05 percent. Figure 5-10 is a photograph of a typical decade resistor.

Figure 5-10. Decade resistor (Courtesy of Leeds and Northrup Co.)

THE VOLTAGE DIVIDER CIRCUIT

One very simple circuit which is seen over and over again in instrument and measurement applications is the *voltage divider*. It is a circuit which allows a variety of voltages to be obtained from one voltage source.

If several resistors are placed in series and a single voltage is applied across the entire connection, the voltage drop across each of the resistors will be some fraction of the total voltage. If, for example, there are only two resistors in the connection, the total voltage drop will be divided by the two of them in proportion to the ratio of their resistance values, $(V_1/V_2) = (R_1/R_2)$ and $V_1 + V_2 = V_0$. Thus, in a voltage divider circuit:

$$V_2 = V_0 \frac{R_2}{R_1 + R_2} \tag{5-4}$$

In Fig. 5-11(b), a current of

$$I = \frac{V}{R_1 + R_2} = \frac{30}{(5 + 10)} = 2 \text{ A}$$

will flow in the circuit. The voltage across the 10-Ω resistor will be

$$V_1 = IR = 2 \times 10 = 20 \text{ V}$$

and the voltage across the 5-Ω resistor will be

$$V_2 = 2 \times 5 = 10 \text{ V}$$

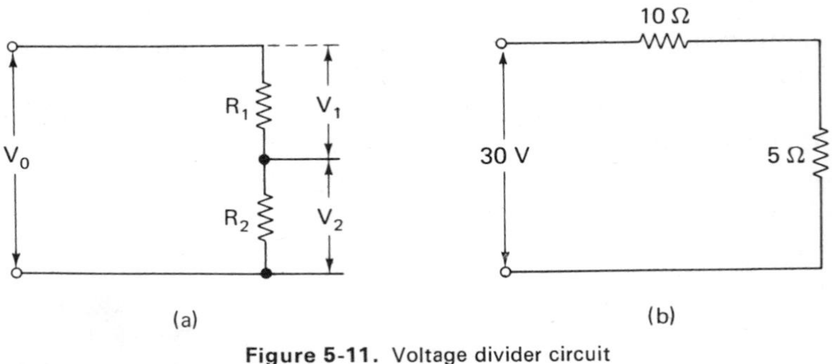

Figure 5-11. Voltage divider circuit

Thus the 30 V of the source will be divided by the two resistors.

If there is an application that requires a 10-V level, and a 30-V source is available, such a voltage divider circuit can be used to provide this 10-V output.

When the voltage requirement is variable, the ratios of the voltage divider circuit can be changed by using a potentiometer as a part of the divider circuit.

Then any fraction of the source voltage can be obtained by changing the resistance of the potentiometer. This is the principle used in the volume control of radios.

Color Coding of Resistors

Most larger resistors have their resistance values and tolerances stamped on their bodies. However, carbon composition resistors and some wire-wounds are too small to use this method of identification. To be able to visually identify the resistance and tolerance of carbon resistors without actually having to measure them, a color code is utilized. Three or four color bands are painted on the resistor body to specify the resistance and tolerance. Figure 5-12 shows a carbon resistor and the formula used for computing the resistance and the tolerance from the colors of the bands.

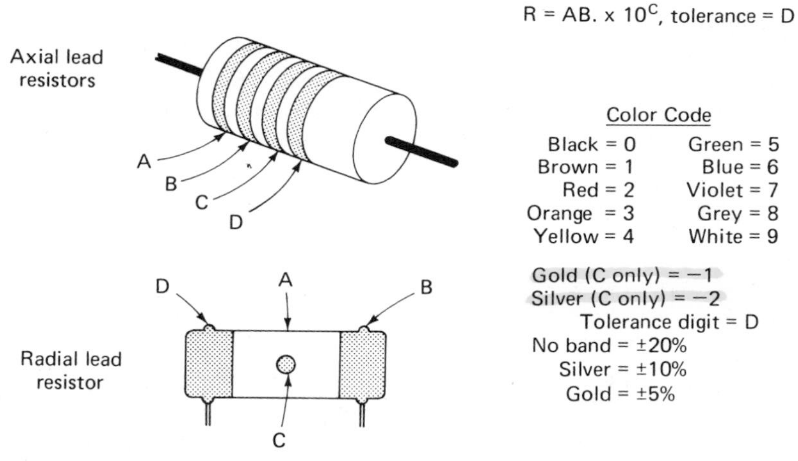

Figure 5-12. Carbon composition resistor color code

EXAMPLE 5-6

Given a resistor with bands:

$$A = \text{blue} \qquad C = \text{orange}$$
$$B = \text{gray} \qquad D = \text{silver}$$

Find the resistance value and tolerance.

Solution:

$$R = 68 \times 10^3, \quad \pm 10\%$$
$$= 68{,}000\ \Omega, \quad \pm 6800\ \Omega$$

EXAMPLE 5-7

Given a resistor with bands:

$$A = \text{yellow} \qquad C = \text{gold}$$
$$B = \text{violet}$$

Find the resistance value and tolerance.

Solution:

$$R = 47 \times 10^{-1}, \quad \pm 20\%$$
$$= 4.7\,\Omega, \quad \pm 0.94\,\Omega$$

The resistance value as indicated by the color bands of carbon composition resistors is called the *nominal* value of resistance. Carbon resistors are manufactured only in a specific set of nominal values. These values are determined according to a formula which states that each nominal value is approximately $(1 + 2N)$ times the value of the preceding nominal value (where N is tolerance of the resistor). By using this formula, the resistance of every resistor manufactured is within the tolerance range of each nominal value. Figure 5-13 shows the nominal values for the 5%, 10% and 20% tolerance resistors.

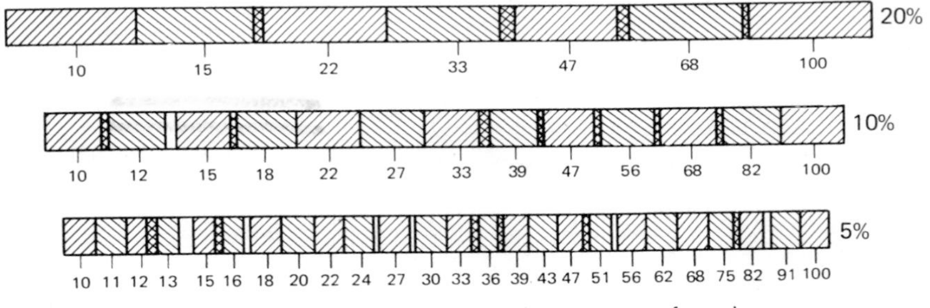

Figure 5-13. Nominal values and tolerance ranges for resistors (logarithmic scale)

Environmental Effects on Resistors

POWER RATING

The power rating of a resistor is the maximum power a resistor can handle before burning out. If this power is exceed, the resulting current flowing in the resistor will transfer (by the process of collisions with the atomic lattice) too much energy to the resistor material. The energy takes the form of heat, and eventually the resistor is destroyed or damaged by some excessive thermal process. Normally, a composition resistor warps and opens (or even explodes), and a wirewound resistor may melt and become an open circuit.

Because severe damage or destruction can result from an overload, one should visually inspect resistors before use. If the resistor body appears warped or discolored, this is an indication that the resistor may have undergone a violation of its maximum power rating in earlier use. Before being used again, such a resistor should be measured to ensure that it still satisfies the coded resistance value.

The maximum power that a carbon composition resistor can handle depends on its size. They are manufactured in $\frac{1}{8}$, $\frac{1}{4}$, $\frac{1}{2}$, 1, and 2 watt ratings. Figure 5-14 shows the actual sizes of commercially available carbon composition resistors and their corresponding power ratings.

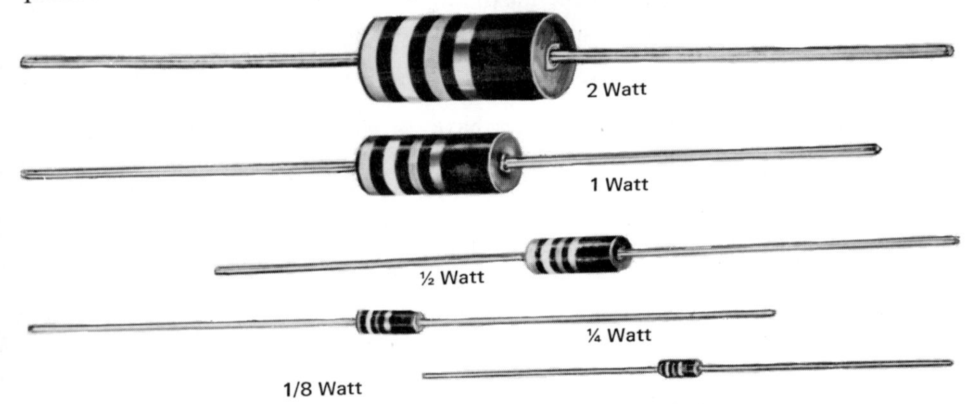

Figure 5-14. (Courtesy of Ohmite Manufacturing Co.)

For applications that require resistors to dissipate more than 2 watts, wirewounds are usually chosen. The power ratings of wirewounds are printed directly on the resistors themselves.

To determine the power rating of a resistor for use in a circuit, one should estimate the maximum voltage or current that will pass through the element and compute the power from

$$P = \frac{V^2}{R} = I^2 R \tag{5-5}$$

For safe, conservative design, assume that the above calculated value might possibly be exceeded by a factor of 4, and then choose your resistor accordingly.

EXAMPLE 5-8

It is estimated that 50 mA of current will pass through the loop of a circuit which has a 100-Ω resistor. The power dissipated by this resistor will be

$$P = I^2 R = (0.05)^2(100) = 0.25 \text{ W}$$

If one wanted to ensure against unexpected overloads, a 1-watt resistor would be a prudent choice.

For very large-valued resistors, the power dissipated by them may be small, even if they are subjected to high voltages. Nevertheless, the high voltage may still produce permanent damage to the resistor. The damage occurs due to the high-energy electrons (excited to this energy by the high voltage) colliding with and distorting the lattice structure of the resistor material. Thus resistors are given a maximum voltage rating which also must not be exceeded.

The ambient temperature of the resistor's environment may also play a part in the calculation of the maximum power that can be dissipated by it. This consideration is discussed in the following section.

TEMPERATURE VARIATION OF RESISTANCE

Temperature variations always have an effect on resistance values. The temperature change can be caused by the external environment or by self-heating effects within the resistor itself. We have seen how the application of too much current or voltage can destroy a resistor. However, even before a resistor burns out, temperature changes cause nondestructive, reversible changes in the resistance value. The variation of resistance with temperature is described by its temperature coefficient (TC). This parameter tells how much of a percentage change in the 25°C resistance value occurs for each degree centigrade of temperature change.

EXAMPLE 5-9

A resistor at 25°C has a resistance of 100 Ω. If the resistance has a temperature coefficient of +0.1 percent per °C, then at 70°C the resistance will be

$$R = 100 + [(+0.001)\cdot(100)\cdot(70 - 25)] = 100 + 4.5 = 104.5\ \Omega$$

In this example, the TC has a plus sign. A positive TC implies that raising the temperature increases the resistance. If the TC had had a negative sign, this would have indicated that the resistance would have decreased as the temperature was raised. Note that most materials possess positive TCs, although some important exceptions possess negative TCs over certain temperature ranges.

If the resistor is placed in a hot ambient environment, it will obviously be able to dissipate less heat for a given current than if it were in cooler surroundings. For example, a given current may burn out a resistor which is operating in an oven at 100°C, while an equal current might not damage the same resistor operating at room temperature (25°C). Thus the resistor must have its maximum power rating decreased, or *derated*, as its ambient temperature is increased. Figure 5-15(a) is a typical derating curve for a carbon composition resistor.

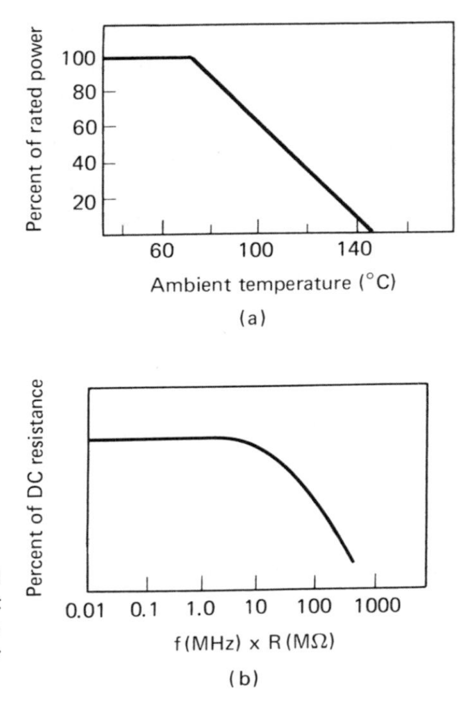

Figure 5-15. Derating curve for a typical carbon composition resistor (a) Ambient temperature (°C) (b) Variation of the resistance of carbon composition resistor vs (frequency) × (resistance)

EXAMPLE 5-10

If a carbon composition resistor is rated at $\frac{1}{2}$ watt, what will be its rating if it is to be used in a 100°C environment?

Solution: From Fig. 5-15, we see that at 100°C a carbon composition resistor will have a rated power that is 63 percent of its room temperature power rating. Thus its maximum power rating is derated to $P = 0.5 \times 0.62 = 0.31$ W.

Finally, it should be noted that, because a resistance changes value with temperature and since a current flow causes a temperature rise, it might be necessary to measure the resistance value of a resistor under actual operating conditions.

HUMIDITY VARIATION OF RESISTORS

High-value resistors (above 100 **MΩ**), wirewound resistors, and composition resistors have their resistance values affected by humidity. The high-value resistor can suffer from increased leakage current due to high humidity conditions and human handling. In other words, lower resistance paths can be established which will cause some of the current to bypass the high resistance path. Thus the overall high-resistance value of the element will change, introducing an error. To minimize the effect of handling, high resistances are usually cleaned in a degreasing solvent before use.

Wirewound resistors can be affected by humidity; however, this change is usually less than 1 percent. In high humidity, the insulation of the wires changes its tension and hence the pressure transmitted to the wire. This cause the change in its resistance value. The effect was first noticed in climates where a large difference in humidity existed between summer and winter. In those standard precision wirewound resistors where the humidity effect is objectionable, the resistors are mounted in sealed containers, thereby eliminating the variation.

In carbon composition resistors, resistance increases with humidity by several percent. To prevent this effect, the resistors must also be sealed.

FREQUENCY EFFECTS

Reactive effects cause changes in a resistor's behavior at high frequencies. Because of stray capacitance and inductance, the apparent resistance deviates from Ohm's law and the dc resistance. Figure 5-15(b) shows the change from dc resistance with the frequency of typical carbon composition resistors.

Measurement of Resistance

VOLTMETER-AMMETER METHOD

The voltmeter-ammeter method is a technique for measuring resistance when one has only voltmeters and ammeters at hand and when an accuracy of 1 or 2 percent is satisfactory. A current is passed through a resistor, and this current is measured by an ammeter. Meanwhile, the voltage across the resistor is monitored by a voltmeter. The resistance of the unknown element is calculated from the ratio of the voltage and the current values read from the meters. The accuracy of the measurement depends on the accuracy of the meters used.

There are two possible ways to connect the meters for this measurement (Fig. 5-16). Assume that we use the connection shown in Fig. 5-16(a) and that the resistance of the voltmeter is very high compared to R_X. Then the voltmeter will draw only a very small current compared to R_X, and we may neglect its loading effect. (See the section on dc voltmeters.) Therefore, this connection is best for low-valued resistance measurements.

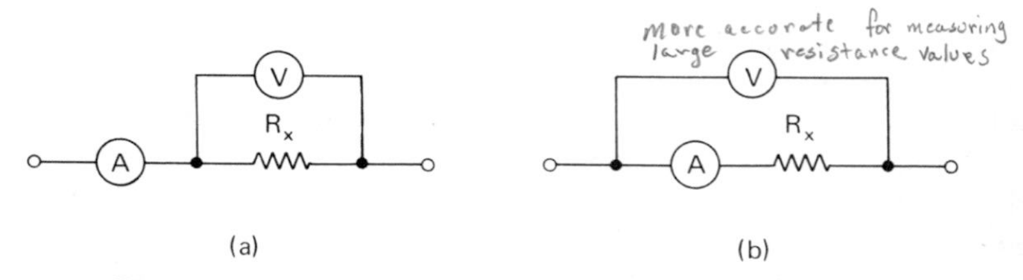

Figure 5-16. Meter connections for measuring resistance with the voltmeter-ammeter method.

On the other hand, consider connection (b) of the same figure. If the value of the ammeter resistance is much smaller than the resistance value to be measured, it hardly disturbs the original current flowing in the resistance. Therefore, connection (b) of Fig. 5-16 is more accurate for measuring large resistance values.

OHMMETERS

The ohmmeter is a simple instrument which applies the fixed voltage of a battery across two resistors in series. One is a resistor of known value and the other is the one being measured. The voltage across the known resistor is measured by a dc voltmeter. The measured voltage causes an indication on the dc voltmeter which is calibrated to display the unknown resistance value directly.

Ohmmeters are useful for making quick measurements of resistance values under many common conditions and ranges. They are used particularly often in servicing communication equipment. Resistance values that can be measured with the ohmmeter vary from milliohms to 50 megohms. However, there are some limitations on their use. Because their best accuracy is about ± 2 percent, they are generally not suitable for highly accurate measurements. Also, certain special precautions must be followed in using ohmmeters to measure circuits with high inductance or capacitance. Finally, because they contain batteries, they should be used only on passive circuits or on circuits which will not be damaged by them. Circuits with active sources connected could contribute currents that would change the voltage/current ratio and might injure the D'Arsonval movement of the ohmmeter. Circuits which contain sensitive devices (such as some semiconductors and fuses) might be burned out by the passage of even the small amount of current put out by the ohmmeter battery.

The circuit of the simple ohmmeter is given in Fig. 5-17. When the ohmmeter probes are connected across the unknown resistance, R_X, the voltage across R_1 will be given by

$$V_1 = V_0 \frac{R_1}{(R_1 + R_X)} \qquad (5\text{-}6)$$

Looking at Fig. 5-17, we note that the resistance (R_v) of the voltmeter portion of the ohmmeter is adjustable. This means that when R_X is zero (probes connected by short circuit), R_v can be adjusted to allow the voltmeter to display a full-scale deflection. A full-scale deflection when $R_X = 0$ indicates that the ohmmeter is calibrated. Since R_v is adjustable, we can continue to keep the ohmmeter calibrated even as aging effects cause variations in the ohmmeter's battery voltage output. Once the ohmmeter is calibrated, a subsequent connection across the unknown resistor gives a reading which is

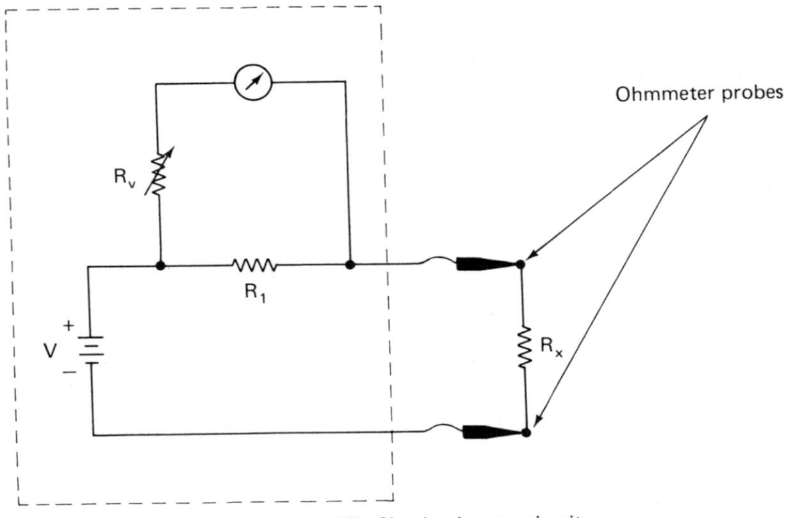

Figure 5-17. Simple ohmeter circuit

less than full-scale and corresponds to the value given by Eq. (5-6). If the scale is properly marked, the meter indicates a reading directly in ohms (Fig. 5-18). Note that, unlike the ammeter or voltmeter, the ohmmeter reads zero at a full-scale deflection and infinity at no deflection. This occurs because, when the probes are not connected, there is no deflection, and the meter is sensing an infinite resistance to current flow.

An ohmmeter's range can be changed by varying its meter sensitivity. This is done by means of a switch that can connect various resistors of different values to replace resistor R_1. Since a half-scale deflection gives the most accurate reading with an ohmmeter, various scales (from the highest on down) should be tried until such a half-scale deflection is approximately reached.

Ohms (R x 100)

30 20 7.5 10 8 6 4 2 0 40 60 100 80 120 1000 500 1500

4 6 2 8 0 10

mA

(Open = ∞) (Short)

Figure 5-18. Nonlinear ohms scale used on ohmmeters

USING THE OHMMETER

1. If the resistance to be measured is known approximately, switch the ohmmeter to the scale that will indicate this value most accurately. If the resistance is not known at all, set the ohmmeter to the highest resistance scale.
2. Before connecting the probes of the ohmmeter across the unknown resistor, touch them together to close the battery circuit. The adjustment knob should be turned until the resulting needle position indicates an exact full-scale deflection. (This corresponds to a zero-ohms reading.)
3. Disconnect the probes from each other and put them across the resistance to be measured (first making sure all electric power to the resistor being measured is turned off). Switch the scale settings until approximately a half-scale deflection is achieved, (making sure to adjust the zero of each scale). This scale will yield the most accurate result. The resistance can be read directly from the needle deflection on the proper scale.
4. Shut the ohmmeter off to keep the battery from draining.

RESISTANCE BRIDGES

A *bridge* is the name used to denote a special class of measuring circuits. They are most often used for making measurements of resistance, capacitance, and inductance. Bridges are used for resistance measurements when a very accurate determination of a particular resistance value is required. The most well known and widely used resistance bridge is the *Wheatstone bridge.* It is used for measuring resistance values above one ohm. Another type, the *Kelvin bridge*, is used for measuring resistances below one ohm. Our discussion will be concerned primarily with Wheastone-bridge measurements. Most commercial Wheatstone bridges are accurate to approximately 0.1 percent. Thus the values of resistance obtained from the bridge are far more accurate than the values obtained from the ohmmeter or the voltmeter-ammeter method.

The circuit of the Wheatstone bridge is shown in Fig. 5-19, where R_X is the resistance to be measured. The bridge works on the principle that no current will flow through the very sensitive D'Arsonval galvanometer connecting

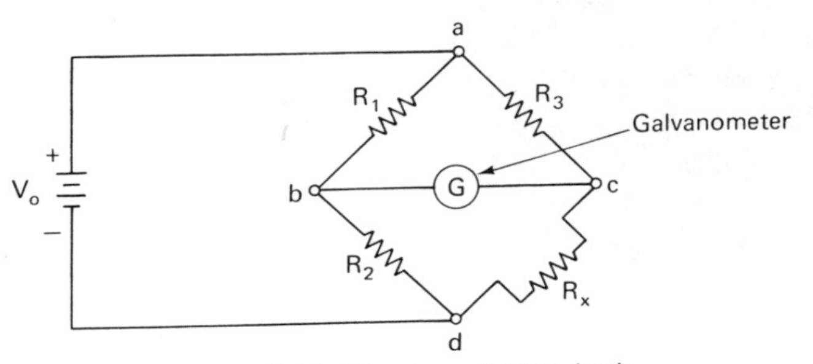

Figure 5-19. Wheatstone bridge circuit

points b and c of the bridge circuit if there is no potential difference between them. When no current flows, the bridge is said to be *balanced.* The balanced condition is achieved if the voltage V_0 is divided in path *abd* by resistors R_1 and R_2 in the same ratio as in path *acd* by resistors R_3 and R_X. Then points b and c will be at the same potential. Thus the conduction of no current flow through the galvanometer implies

$$\frac{R_X}{R_3} = \frac{R_2}{R_1} \tag{5-7}$$

Now if R_X is unknown and R_1, R_2, and R_3 are known, we can find R_X from:

$$R_X = R_3\left(\frac{R_2}{R_1}\right) \tag{5-8}$$

In practical bridges, the ratio of R_2 to R_1 is controlled by a switch that changes this ratio by decades (i.e. factors of ten). Thus the ratio R_2/R_1 can be set to 10^{-3}, 10^{-2}, 10^{-1}, 1, 10, 10^2, and 10^3. R_3 is a continuously adjustable variable resistor. When a null is achieved, the resistance can be read directly off the dials because these dial settings correspond to the variables of Eq. (5-8). Figure 5-20 is a photograph of a Wheatstone bridge.

Since the resistance value of a resistor is known to vary with frequency, resistors used in high-frequency applications should be measured at the frequency of use. When such measurements are performed, an ac source rather than a battery is used. A number of detectors, including the oscilloscope or even earphones, are available to determine a null or balanced condition.

Figure 5-20. Wheatstone bridge (Courtesy of Beckman Instruments, Inc., Cedar Grove Operations)

HOW TO USE THE WHEATSTONE BRIDGE

1. Connect the unknown resistor, R_x, to the terminals of the bridge with good, tight contacts. This will minimize contact resistance.
2. Set the scale of the galvanometer to the least sensitive setting. (A variable shunt resistor is connected across the galvanometer to allow variation of its sensitivity.) This will prevent damage to the D'Arsonval movement if the bridge is severely unbalanced.
3. Adjust the variable resistor dials until a null is reached. (Zero deflection of the galvanometer needle.)
4. Move to a more sensitive scale setting, and null again.
5. Continue until the most sensitive scale setting is reached.
6. Calculate the resistance from

$$R_X = R_3\left(\frac{R_2}{R_1}\right)$$

or read from the dial settings.

EXAMPLE 5-11

A resistor is measured using a Wheatstone bridge and a null is reached for $R_2 = 100\ \Omega$, $R_1 = 1000\ \Omega$, and $R_3 = 120.3\ \Omega$. The fixed resistors are known to be accurate within $\pm 0.02\%$ and the variable resistor within $\pm 0.04\%$. What is the unknown resistance?

Solution:

$$R_X = R_3 \times \frac{R_2}{R_1} = (120.3) \times (1 \pm 0.0004) \times \frac{(100) \times (1 \pm 0.0002)}{(1000) \times (1 \pm 0.0002)}$$

In the worst case, the errors of R_3 and R_2 will be in the plus direction, and the errors of R_1 will be in the minus direction. Then the errors will add to give

$$R_X = (120.3) \times (1 + 0.0004) \times \frac{(0.1) \times (1 + 0.0002)}{(1 - 0.0002)}$$

$$= (12.03) \times (1 \pm 0.0008)$$

$$= 12.03 \pm 0.01\ \Omega$$

ERRORS OF THE BRIDGE

The possible errors that arise from using the bridge include the following:

1. Discrepancies between the true and stated resistance values in the three known branches of the bridge circuit. This error can be estimated from the resistor tolerances.
2. Changes in the known resistance values due to self-heating effects.
3. Thermal voltages in the bridge or galvanometer circuits caused by different materials in contact and at slightly different temperatures.

4. Balance-point error caused by lack of galvanometer sensitivity.
5. Lead and contact resistances introduced when making low-resistance measurements.

Finally, care should be exercised in keeping the voltage across the arms low enough so that permanent damage due to self-heating is not caused in the precision resistors of the bridge.

COMMERCIAL RESISTANCE BRIDGES

Several resistance bridges are available commercially. An example of one is shown in Fig. 5-21. These instruments are basically similar to the Wheatstone bridge, but some also have the capability of measuring capacitance and inductance values. To measure resistance, the dc source and dc detector are used. Several scales are provided for a wide range. A summary of their capabilities is given as part of Table 5-5.

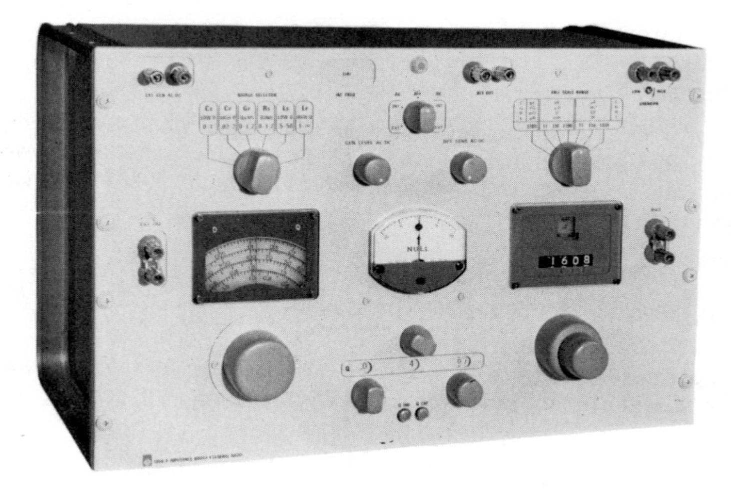

Figure 5-21. Commercial resistance bridge (Courtesy of General Radio Corp.)

SUBSTITUTION METHOD

The technique for making an extremely accurate determination of an unknown resistance involves comparing its resistance to the resistance of a high-precision resistor. The unknown resistor is first used to balance a resistance bridge. The unknown is then replaced by an *adjustable* precision resistor. The precision resistor is varied until the null is found again. The value of the unknown equals the value of the precision resistor at the null point. By this method, the errors of the bridge are bypassed, and the error depends on the accuracy of the adjustable resistor only.

Table 5-5 Table of Resistance Measuring Techniques

Method	Approximate Resistance Ranges	Advantages and Uses	Limitations
Voltmeter-ammeter	1 Ω to 1 MΩ	Easy measurement if voltmeter and ammeter are only meters available.	Limited to about 1% accuracy.
Ohmmeter	0.1 Ω to 5 × 10^7 Ω	1. Rapid measurement. 2. Small instrument. 3. Often part of a VOM or VTVM.	Limited to about 2% accuracy. Cannot be used in all circuits.
Wheatstone bridge	1 Ω to 10 MΩ	1. Accurate to 0.01%. 2. Common laboratory instrument.	More expensive and not as small or capable of making rapid measurements as the ohmmeter.
Commercial bridges	HP 4260-A: 10 mΩ to 10 MΩ GR 1650-B: 1 mΩ to 1.1 MΩ	Portable, rugged, and accurate. (HP 4260-A and GR 1650-B are accurate to 1%).	Same limitations as Wheatstone-bridge methods
Substitution method	1 Ω to 10 MΩ	Most accurate method available for the measurement of resistance (up to 0.0001%). Basically a method which employs the Wheatstone bridge.	Requires special resistance standards. This level of accuracy is usually not required.
Megohmmeter	1 kΩ to 1 TΩ	Most suitable for very high resistance measurements.	Not suitable for measurements below 1 kΩ.
Kelvin bridge[1]	0.1 μΩ to 100 Ω	Accurate to 2–5%. Used for wire and switch measurements.	Only for very low resistance values.
Milliohmmeter[1]	0.1 μΩ to 1 kΩ	Accurate to 2–5% wire and switch measurements.	Specialized instrument.
Digital ohmmeter[2]	0.01 Ω to 10 MΩ	1. Rapid measurement. 2. Digital readout (0.1 to 1% accuracy).	More expensive than simple ohmmeter.

[1] Not discussed in this text.

[2] See Chap. 12 (Section on Digital Multimeters) for a more detailed discussion of Digital Ohmmeters.

MEGOHMMETER

Measuring extremely large resistance values (such as leakage resistance of capacitors, insulation resistance of cables, and resistance of vacuum tubes) presents a special problem. High voltages must be produced to allow even a small current flow to be detected. In addition, a voltmeter with a very high internal resistance must be used to prevent error. The megohmmeter is a device which fulfills these requirements and can measure resistance values up to 10^{12} ohms. Its operation is similar to that of a simple ohmmeter, except that it has a high voltage supply and a more sensitive current detector.

Problems

1. How is the resistance of a wire changed by
   a) doubling its length?
   b) tripling its cross-sectional area?
   c) replacing copper with aluminum?
2. Calculate the resistance of a copper wire whose diameter is 1 mm and whose length is 3 m. Repeat the calculation for a nichrome wire of the same dimensions.
3. An aluminum wire whose length is 40 m has a resistance of 2.5 Ω. Find its diameter in inches.
4. Convert the following quantities from inches to circular mils. Approximately what gauge numbers would wires with these diameters have?
   a) 0.036 in
   b) 0.18 in
   c) 0.025 in
5. What is the area, in circular mils, of a conductor whose cross-sectional dimensions are 0.1 in by 0.2 in?
6. If a 120 V source is connected across an 80 Ω resistor, how much current will flow in the resistor?
7. A resistor whose value is 0.35 MΩ has a current of 0.5 mA flowing through it. What is the voltage across the resistor?
8. If the armature of an automobile starting motor has a resistance of 0.033 Ω, and the automobile uses a 12 V battery, find the current drawn by the starting motor when it is connected to the battery. What is the conductance of the starting motor?
9. What is the unit used to describe the temperature coefficient of a resistor? Explain the meaning of the unit.
10. If a resistor wound of nichrome wire has a resistance of 60 Ω at 50°C, find the resistance of the resistor at 70°C. (The temperature coefficient of nichrome is +0.0004.)
11. Give the color code which would be used to identify the following composition resistors.

a) 6800 Ω, ±20% c) 3.9 Ω, ±10%
b) 510 Ω, ±5% d) 100 Ω, ±20%

12. A 110 W, 120 V tungsten incandescent lamp has a resistance of 10.5 Ω when no current flows through it. What is the resistance of the lamp when it is connected to a 120 V source?

13. Two resistors with unmarked values are received. However, it is known that one is a 250 Ω, 5 W resistor and the other is a 50 Ω, 25 W resistor. How can the 50 Ω, 25 W resistor be identified by inspection?

14. If a 5000 Ω resistor carries a current of 25 mA, how much power does it dissipate? What should be its power rating for safe design?

15. Design a voltage divider whose possible voltage outputs are 1.0 V, 2.5 V, 5.0 V, and 15 V. Assume that a 15 V battery is the source of voltage and that no current is drawn from the output terminals.

16. Explain the difference between a *rheostat* and a *potentiometer*.

17. In reference to the voltmeter-ammeter method for measuring resistance, explain why the connection of Fig. 5-16a is more accurate for the measurement of low-valued resistors than the connection of Fig. 5-16b. (Hint: consider the effects of the internal resistances of both meters on the measurement.)

18. List some of the factors that limit the accuracy of the Wheatstone bridge.

19. Find the value of the unknown resistance in the balanced Wheatstone bridge shown in Fig. P5-1.

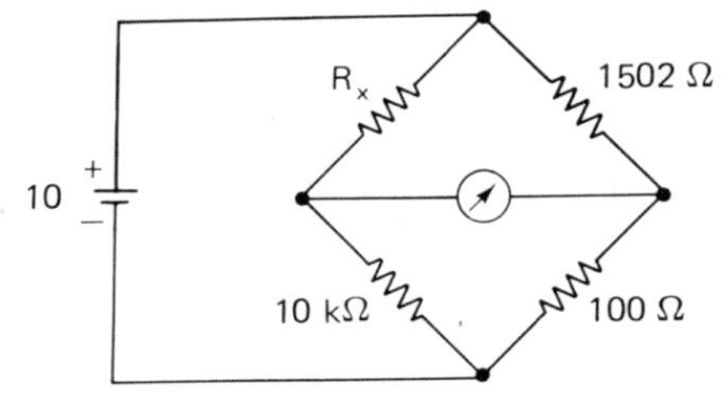

Figure P5-1

20. The Wheatstone bridge of Fig. P5-2 has $R_1 = 250\ \Omega$, $R_2 = 750\ \Omega$, and R_3 is a decade box with steps from 100 Ω to 0.1 Ω. R_1 and R_2 are known to within ±0.02% and the resistors of the decade box are known to within ±0.05%. If R_3 is set to 153.7 Ω when a balance is found, determine a) R_X and b) the percentage error which exists in the calculated value of R_X.

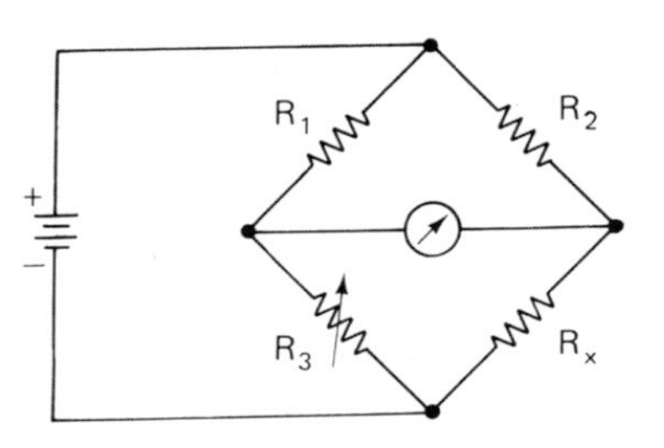

Figure P5-2

References

1. DeFrance, J. J., *Electrical Fundamentals*, 3rd ed., Chaps. 2, 3, 6, and 17. Englewood Cliffs, N. J., Prentice-Hall, Inc., 1969.

2. Stout, M. B., *Basic Electrical Measurements*, 2nd ed., Chaps. 4 and 5. Englewood Cliffs, N. J., Prentice-Hall, Inc., 1960.

3. Kinnard, I. F., *Applied Electrical Measurements*, Chap. 5. New York, John Wiley & Sons, Inc., 1956.

6

Capacitors and Capacitance Measurements

In contrast to the resistor whose ideal static and dynamic characteristics are identical, the electrical characteristics of the capacitor and inductor differ under static and dynamic operating conditions. That is, capacitors and inductors will exhibit different behavior if subjected to dc (static) rather than ac (dynamic) currents and voltages. The objective of this chapter is to examine capacitors and how their static and dynamic characteristics are used in various circuit applications. We will see how different dielectrics and structural configurations are utilized to construct capacitors in a wide range of values and tolerances. Finally, some methods used to measure capacitance will be discussed. The next chapter will be devoted to inductors.

Capacitance and Capacitors

Material bodies which possess opposite electric charges will be attracted to one another by a force whose strength is found from Coulomb's law. To help represent this force, an electric field and a voltage between these bodies can be calculated. It has been observed that, for each particular configuration of two charged bodies in which the shape of the bodies and their separation remains fixed, the ratio of the charge to the voltage existing between them is a constant. This observation is expressed mathematically as

$$\frac{q}{v} = C \tag{6-1}$$

The constant, C, is known as the *capacitance* of the particular geometrical configuration. To put it another way, *capacitance* refers to the amount of charge the configuration can store for each volt of potential difference that exists between the two bodies (Fig. 6-1).

If a circuit element is built so that it deliberately possesses a particular capacitance value, this element is called a *capacitor*. The unit of capacitance is the farad (F), and it is expressed as

$$1 \text{ farad} = \frac{1 \text{ coulomb of charge stored}}{1 \text{ volt}} \tag{6-2}$$

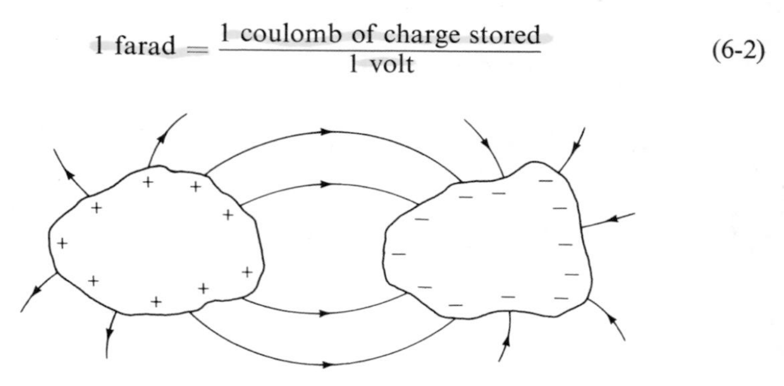

Figure 6-1. Two bodies which are separated by a fixed distance (and are not connected by any conducting path) will store a constant amount of charge for each volt of potential difference between them.

One coulomb is a very large amount of charge, and the quantity of charge stored for each volt in most real capacitors is much smaller than a coulomb. This makes the farad too unwieldy to describe the capacitance of actual capacitors. As a result, it is more common to see the capacitance of particular configurations and capacitors expressed in picofarads (1 pF $= 10^{-12}$ F) or microfarads (1 μF $= 10^{-6}$ F). For example, the large capacitors used in power supply filters have capacitance values of 10 to 1000 μF. The small-valued capacitors used in radio communication instruments have capacitance values of 25–500 pF.

The circuit symbol used for the capacitor is either ⊣(or ⊣⊢ or ⊣̸((the last represents variable capacitors). The special configuration of two closely spaced, parallel metal plates is used to construct almost all circuit elements that are used as capacitors. Such capacitors are called *parallel-plate capacitors*, and an example of their form is shown in Fig. 6-2. The capacitance value of parallel-plate structures is found from the expression

Capacitance of a parallel plate capacitor

$$C = \frac{K\epsilon_o A}{d} \tag{6-3}$$

where K is the relative dielectric constant, ϵ_o is the permittivity of free space (and is constant of value $\epsilon_o = 8.85 \times 10^{-12}$ farads per meter), A is the area of the plates (in square meters), and d is the distance between the plates

$\varepsilon_o = 8.85 \times 10^{-12}$ farads/meter

(in meters). From Eq. (6-3) we can see that, to increase the value of the capacitance of a parallel-plate structure, we can either increase its plate area or dielectric constant value, or decrease the distance between the plates.

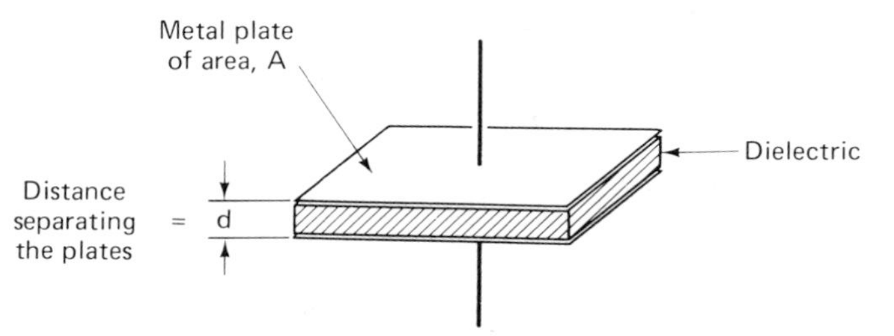

Figure 6-2. Parallel plate capacitor

EXAMPLE 6-1

We are given a parallel-plate capacitor which has plates of 10 cm on a side and a separation of 2 mm between the plates. The dielectric between the plates is air. For this structure, find (a) the capacitance value, (b) the electric field strength between the plates, and (c) the charge on each plate for a 200-V potential difference.

Solution: (a) To find the capacitance, we convert all the dimensions of length to meters and use Eq. (6-3). K for air is approximately 1.0. Thus

$$C = \frac{K\epsilon_o A}{d} = \frac{(1.0) \times (8.5 \times 10^{-12}) \times (0.1)^2}{2 \times 10^{-3}}$$

$$= 42.5 \times 10^{-12} \text{ F} = 42.5 \text{ pF}$$

(b) The electric field, E, is found from

$$E = \frac{V}{d} = \frac{200}{2 \times 10^{-3}} = 10^5 \text{ V/m}$$

(c) The charge per plate is found from Eq. (6-1):

$$q = Cv = (42.5 \times 10^{-12}) \times (200) = 8.5 \times 10^{-9} \text{ coulombs}$$

Capacitance may also exist between conductors of other shapes and non-parallel separations (e.g., Fig. 6-1). When irregularly shaped configurations give rise to capacitance, more elaborate computational techniques than Eq. (6-3) must be used to calculate their resulting capacitance values. Such calculations will not be covered by this text.

DIELECTRICS

A dielectric is an insulating material placed between the plates of a capacitor to increase the capacitance value. Many different capacitance values can be obtained from two parallel plates of the same size and separation through the use of various dielectrics. Figure 6-3 explains the physics of this capacitance increase.

The relative dielectric constant, K, introduced in Eq. (6-3), is the parameter which indicates how much a particular dielectric inserted between a capacitor's plates can increase the capacitance relative to vacuum. Table 6-1 gives the dielectric constants of various materials used in capacitors.

Table 6-1

Dielectric	$K = \epsilon/\epsilon_o$ (Average Value)	Dielectric Strength (V/cm)
Vacuum	1.0	
Air	1.0006	3×10^4
Polystyrene	2.5	6×10^5
Paper, paraffin	4.0	5×10^5
Mica	5.0	2×10^6
Oxide films	5–25	
Ceramic (low-loss)	8.0	8×10^4
Ceramic (high-K)	100–1000	
Water	80	

EXAMPLE 6-2

For the capacitor of Example 6-1, find the capacitance value if the following dielectrics are used: (a) paper and (b) high-K ceramic (with $K = 100$).

Solution: (a) Paper, $K = 4.0$, $C = 4 \times 42.5$ pF $= 170$ pF.
(b) Ceramic (high-K), $K = 100$, $C = 4250$ pF $= 0.0042$ μF.

The dielectric constant, however, is not the only parameter that must be considered when choosing a dielectric for a particular capacitance application. The choice of dielectric also depends upon such factors as the dielectric loss, the leakage resistance, and the dielectric strength. Often a dielectric may be able to provide a capacitor with the proper capacitance value, but will fail to satisfy the other specifications. All of these additional parameters are measures of how much an actual capacitor will depart from the behavior of an ideal element. They are discussed in more detail in the section, *Capacitor Circuit Models and Losses.*

ENERGY STORED IN A CAPACITOR

When a voltage source is applied across the plates of an uncharged capacitor, electrons flow away from one plate and accumulate on the other. The total charge, q, eventually deposited on the plates is found from

$$q = Cv \tag{6-4}$$

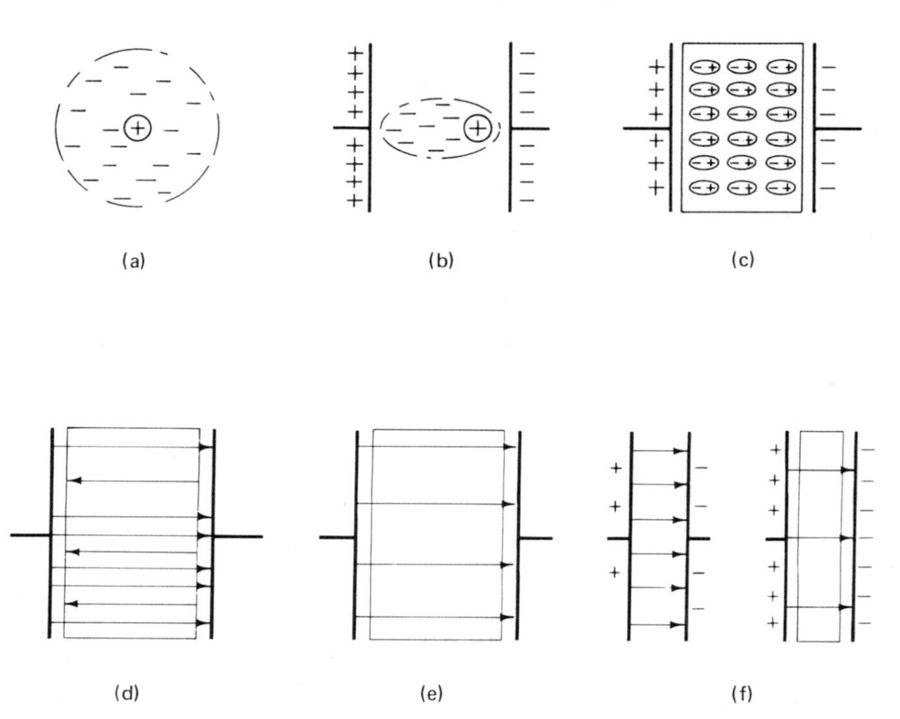

Figure 6-3. The effect of dielectrics inserted between the plates of capacitors
(a) An atom has a positively charged nucleus which is surrounded by a "cloud" of negative charge made up of electrons. If the atom is put in an electric field (such as that which exists between two charged metal plates), the electrons of the atom will be attracted to the positive plate and the nucleus toward the negative plate.
(b) If the attraction is not strong enough to tear any electrons away from the parent atom, it can still cause the oppositely charged parts to stretch slightly away from each other. An electric field will now exist between the slightly separated positive and negative parts of the atom.
(c) In a dielectric material which is inserted between two such plates, this effect occurs in all its atoms. Since the dielectrics are insulators, the electrons of the atoms remain attached to their respective atomic systems. However, the positive and negative components of each atom are pulled slightly apart.
(d) We see that this results in a slight shift in the position of the positive charges toward one plate of the capacitor and the negative charges toward the other. Between the separated charges an electric field exists. Note, however, that its direction is opposite to the original field.
(e) Since the two fields act in opposite directions, the effect is to cancel part of the strength of the original field (which existed due to the charges on the capacitor plates). Hence the material is termed a "dielectric" (which means opposing (di), electric field (electric). A smaller net electric field thus exists between the plates.
(f) Since the strength of the electric field between two plates is $E = V/d$, a smaller voltage will exist for an equal charge on the plates separated by the dielectric. Capacitance is defined as $C = q/v$. Thus if a dielectric is placed between two charged plates, this configuration will have a higher capacitance than two equally charged plates with no dielectric between them.

where C is the capacitance of the capacitor and v is the voltage of the source. If the source is removed and the capacitor terminals are kept unconnected, the charge is trapped on the surface of the capacitor plates (Fig. 6-4).

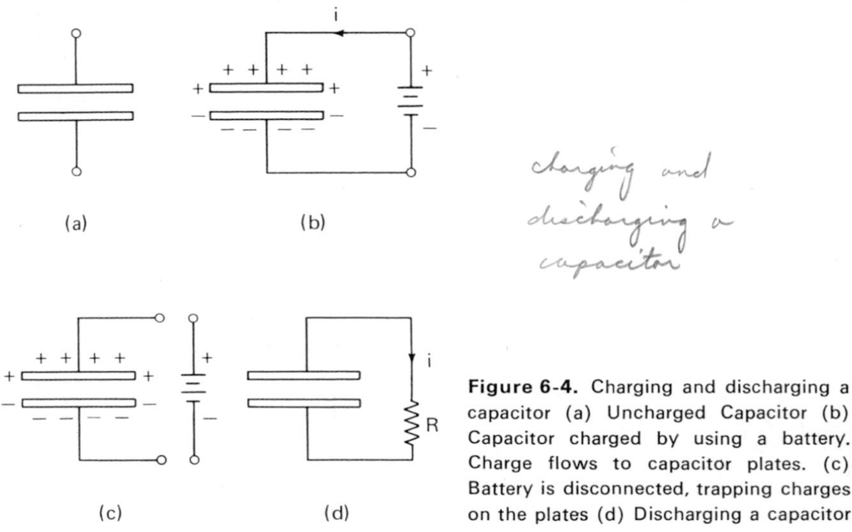

Figure 6-4. Charging and discharging a capacitor (a) Uncharged Capacitor (b) Capacitor charged by using a battery. Charge flows to capacitor plates. (c) Battery is disconnected, trapping charges on the plates (d) Discharging a capacitor

The transfer of charge from one capacitor plate to the other during charging requires an expenditure of energy because once some negative charges accumulate on the negative plate, they act to repel additional charges trying to flow there. Thus, work must be done against this repulsive force. The energy required to do the work is stored as potential energy by the capacitor and can be recovered by allowing the excess negative charges to flow from the negatively charged plate back to the positively charged plate. If a conducting path is introduced between the charged capacitor plates, the charges will flow until no more potential difference exists between the plates. The energy existing in a charged capacitor is also said to be *stored* in the electric field between the plates. The mathematical expression of this stored energy is given by

$$w = \tfrac{1}{2}Cv^2 \tag{6-5}$$

where w is the energy in joules and v is the voltage between the capacitor plates.

EXAMPLE 6-3

A 10-μF capacitor is charged 500 V. What is the energy stored by the capacitor?

Solution: $w = \tfrac{1}{2}(10 \times 10^{-6})(500)^2 = \tfrac{1}{2}(2.5)$ joules $= 1.25$ joules

CAPACITOR SAFETY

As indicated in the previous section, a charged capacitor stores energy. If the capacitor has a large capacitance value and is charged to a high voltage, the quantity of energy stored can become quite large. During discharge, the energy is released by the current flowing in the connection between the plates. If this discharge accidentally takes place through a human conducting path, the resulting electric shock can be painfully nasty or sometimes even fatal. Since a charged capacitor looks no different from an uncharged one, a charged capacitor represents a disguised safety hazard. This means if a capacitor is ever charged during use, it must be discharged before being handled or put back on the shelf.

The discharge of the capacitor should always be made through an appropriate resistor. By simply short-circuiting its leads we can easily damage a sound capacitor. Therefore, such methods as connecting the capacitor terminals of a charged capacitor together with a screwdriver blade are not acceptable ways of discharging capacitors.

STRAY CAPACITANCE

As noted earlier, capacitance can and does exist between conductors which are at different potentials, regardless of their shape. Various configurations of circuit elements and leads which are at different potentials often exhibit such capacitance. Usually this capacitance effect is unplanned and unwanted because it appears as an extra capacitance element in a circuit or system. For this reason, it is usually referred to as *stray capacitance*. Sometimes stray capacitance effects are small and can be neglected; at other times the effects may be relatively large and can cause significant changes in a circuit's behavior. For example, at high frequencies the stray capacitance can shunt large amounts of signal energy which should actually be transferred to other points in the circuit.[1] When stray capacitance is significant, it must either be reduced or its magnitude must be included into the analysis of the circuit or measurement-system design. Examples of situations that can give rise to stray capacitance are shown in Fig. 6-5.

Figure 6-5(a) shows two current-carrying wires which are also at different potentials. A capacitive effect is established by the difference of potential between them. Figure 6-5(b) shows how a capacitive effect occurs between the turns of a coil in an inductor. Because of the proximity and voltage drop between neighboring turns (a small but measurable drop), a stray capacitance results. A detailed discussion of the effects of stray capacitance on measurements and how their values can be estimated was undertaken in Chapter 3.

[1] In the next section we will see why a capacitor appears as a low-impedance path at high frequencies.

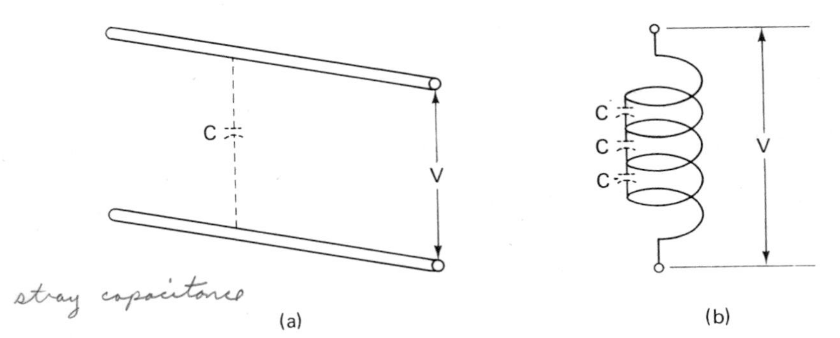

Figure 6-5. Stray capacitance effects (a) Stray capacitance between two current-carrying conductors at different potentials (b) Stray capacitance arising between the neighboring turns of a coil.

CAPACITORS IN ELECTRIC CIRCUITS

The most important circuit properties of capacitors involve their charging characteristics and their impedance to ac currents. Both of these characteristics involve the dynamic aspects of a capacitor's behavior.

Consider first the capacitor's response to charging and discharging. When a voltage is applied across a capacitor, current flows to its plates. However, a voltage cannot exist across the capacitor itself until some of the charge has been deposited on the plates (according to $v = q/C$). Since it takes a finite time for the charge to flow onto the plates, the full voltage impressed on a capacitor will not appear across it instantaneously. This effect means that a capacitor is an element which resists abrupt or instantaneous changes in its voltage. The voltage across it must change smoothly with time (although this smooth change can be made to take place rather rapidly). However, there is no effect which prevents abrupt variations in the charge flow from occurring instantaneously. This means that the amount of charge flowing into or out of a capacitor can change instantaneously.

If an ac voltage is applied across the terminals of a capacitor, an ac current will flow in the branch in which the capacitor is connected. This can be seen if we take the time derivative of Eq. (6-4):

$$Cv = q$$

or

$$C\frac{dv}{dt} = \frac{dq}{dt} = i \tag{6-6}$$

Although no electrons actually move through the capacitor itself, they surge back and forth between the plates and move back and forth in the circuit connected to the capacitor. However, the other elements within the circuit react in the same way as if the ac current was actually passing *through* the capacitor.

As expressed by Eq. (6-6), the magnitude of current (as given by its rms value) flowing in a circuit branch containing a capacitor will increase as the frequency of the applied voltage increases.[2] This current increases even if the magnitude of the applied voltage (also expressed as an rms value) remains constant. Since the ratio of V/I decreases as ω increases, the impedance of the capacitor must be dropping with increasing ω.[3] Mathematically, the relation which expresses how the impedance of the capacitor depends on the applied frequency and the value of capacitance is

$$X_C = \frac{1}{\omega C} \tag{6-7}$$

X_C is called the *capacitive reactance* of a capacitor, and its units are ohms (just like the units of resistance). Note that Eq. (6-7) states that $X_C = \infty$ for dc voltages. This result is in accord with the principle that ideal capacitors appear as open circuits to dc quantities.

EXAMPLE 6-4

Find the capacitive reactance, X_C, of a 50-pF capacitor at 60 Hz and 100,000 Hz. Then find X_C of a 50-μF capacitor at these same frequencies.

Solution: (a) For the 50-pF capacitor:

At 60 Hz: $$X_C = \frac{1}{2\pi(60)(5 \times 10^{-11})} = \frac{10^8}{2} = 5 \times 10^7\ \Omega$$

At 100 kHz: $$X_C = \frac{1}{2\pi(10^5)(5 \times 10^{-11})} = \frac{10^6}{31.4} = 3 \times 10^4\ \Omega$$

(b) For the 50-μF capacitor:

At 60 Hz: $$X_C = \frac{1}{(2\pi)(60)(5 \times 10^{-5})} = \frac{10^2}{2.0} = 50\ \Omega$$

At 100 kHz: $$X_C = \frac{1}{(2\pi)\,(10^5)(5 \times 10^{-5})} = 0.03\ \Omega$$

We see that, as the frequency of the signal applied to a capacitor becomes very high, the impedance of most capacitors (and even small stray-capacitive effects) can become quite small. This means that ac currents at high frequencies will begin utilizing these low-impedance paths. If such low-impedance

[2] For a voltage $v = A \sin \omega t$, $dv/dt = (A\omega) \cos \omega t$. Thus as the frequency, ω, increases, so does the amplitude of dv/dt.

[3] Remember that impedance (in ohms) is defined as $Z = V/I$. When the voltage and the current are sinusoidal ac signals, V and I of this equation represent their rms values.

paths exist in parallel with impedances that remain large at high frequencies, the ac current will bypass the high-impedance paths and flow in the capacitor path instead. This *shunt effect* can be a useful characteristic in some applications and a limitation in others.

Now that we have become acquainted with the important principles of capacitor operation, we can examine Table 6-2 and see how capacitors are used in various electrical applications. The applications clearly show that the dynamic circuit properties of capacitors are utilized as much as, if not more than, their static properties.

Table 6-2 Examples of Capacitor Use in Circuit Applications

1. *Dc-blocking elements.* When a capacitor is placed in series with a circuit branch, the dc components of the current in the branch are prevented from flowing. However, ac quantities are not completely blocked. Blocking capacitors are used in amplifier, rectifier, and oscillator circuits.
2. *Bypass element for ac quantities.* When a capacitor is placed in parallel with a large-valued resistor, the capacitor can form a low-impedance path for ac quantities. Dc is still forced to use the high-resistance path because the capacitor remains a virtually open circuit for dc. Capacitors are used in this way in amplifier circuits.
3. *Energy storage elements.* Used in circuits that provide energy to electronic photoflash units, electron accelerators, and laser flash lamps. Energy can be stored slowly during the charging of the capacitor, and it can be released quickly by rapid discharging.
4. Use of the capacitor's transient charging and discharging characteristics in pulse generating, timing, and analog computer circuits.
5. Used in power-supply filters to reduce fluctuations of output waveforms.
6. Used in oscillators as part of the oscillation-producing circuits.
7. *Transducer elements.* The change of some physical variable can be used to change the capacitance value of a capacitor structure .The variation is thereby converted to an electrical variation.
8. Increase the efficiency of power transmission systems by increasing their power factor.*

*For further information on how capacitors can be used to change the efficiency of power transmission systems (and for a definition of the term *power factor*), see Chapter 10.

Capacitor Circuit Models and Losses

An ideal capacitor element stores but does not dissipate energy. It is a lossless element. However, a real capacitor always has some losses connected with its operation. Let us examine the structure of real capacitors to discover the sources of the major loss mechanisms.

If the dielectric separating the capacitor plates were a perfect insulator and if the leads and plates were made of perfectly conducting materials, there would be no energy dissipated by the capacitor during charging and discharging. However, since real dielectrics are not perfect insulators, they do cause some energy loss when a capacitor is operating in a circuit. This *dielectric loss* depends on how imperfect the dielectric is and on the frequency of the applied voltage.

In the special case of an applied dc voltage, a small current flows through the capacitor because of the few free charge carriers that exist in the dielectric. (The more the dielectric resembles a perfect insulator, the fewer are the free charge carriers.) Such a current is called *leakage current.* Polystyrene and mylar dielectrics possess the lowest leakage currents. Electrolytic capacitors have some of the highest leakage currents. Humidity and defects in the capacitor's encapsulation or packaging sometimes lead to additional leakage currents.

The other major losses of a capacitor involve so-called *resistance losses* or *plate losses.* These are due to the resistance of the material making up the plates and leads of the capacitor. At high frequencies the capacitor is partially charged and discharged at a high rate. Each time current flows into or out of a capacitor, it must flow through these conductors and lose some energy. In addition, at high frequencies the resistance of conductors can be much higher than their dc resistance value. Therefore, at such high frequencies the resistance loss effect can become quite significant.

These plate losses, together with the dielectric losses, show up as heat generated during a capacitor's operation. They must therefore be held to some reasonable level in order to avoid damaging the element by excessive heating.

The overall losses of an actual capacitor can be taken into account when creating an equivalent-circuit model of a capacitor for use in circuit analysis. One commonly used model is a resistor connected in parallel with an ideal capacitor. It is referred to as the *parallel model* (Fig. 6-6(a)). The leakage current of a capacitor can be thought of as flowing through the resistor of this model. The lower the leakage current that exists for a given voltage, the larger is the *leakage resistance*, R_p. (A high leakage resistance is considered to be 100 MΩ or more. A low leakage resistance would be 1 MΩ or less.)

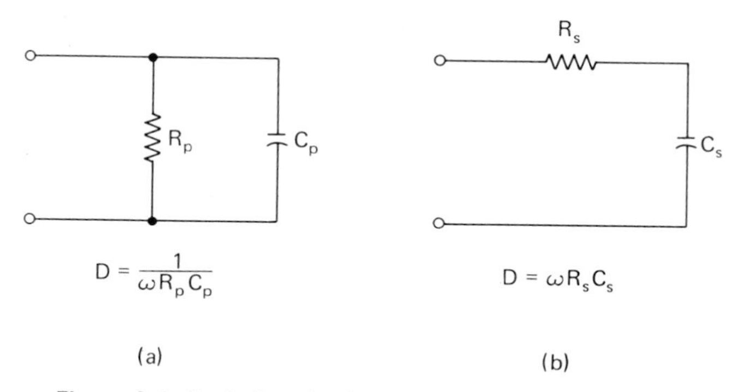

Figure 6-6. Equivalent circuit models of capacitors (a) Parallel (b) Series

Another model which is also used to represent capacitors is the *series model;* it is shown in Fig. 6-6(b). It is not as easy to associate leakage current with this model, but the model is more useful than the parallel model for analyzing certain circuits. It is mentioned here because it is used in some of the circuits of the capacitance-measuring instruments which will be discussed in a later section.

Both the parallel and the series equivalent-circuit models of a capacitor are frequency dependent. This means that the value of the elements used in them is liable to vary with the applied frequency. However, most capacitors have a range of frequencies over which C and R remain approximately constant (see Table 6-3). As long as the capacitor is used in this range, the simple models given in Fig. 6-6 can be used with confidence. (The values of C and R are determined by measurement.)

The dissipation or loss factor, D, is also frequency dependent. This factor is defined as the ratio of the conductance to the capacitive reactance of a capacitor. For the parallel model, D is found from

$$D = \frac{1}{\omega C_p R_p}\bigg|_{\omega} \tag{6-8}$$

Here ω is the frequency of the applied voltage, and C_p and R_p are the values of the elements of the parallel model measured at the applied frequency. For the series model, D is found from

$$D = \omega C_s R_s |_{\omega} \tag{6-9}$$

where C_s and R_s are the values of the capacitance and resistance elements of this model measured at the applied frequency.

In an ideal capacitor, D would be zero, with R_p being infinite or R_s zero. In an actual capacitor, the larger the value of D, the larger the overall loss. Values of D vary from about 0.1 in electrolytic capacitors to about 10^{-4} in polystyrene types.

In addition to the losses of a capacitor, there are other parameters which represent a departure in electrical behavior from that of an ideal capacitor structure. The most important of these is the *dielectric breakdown.* When the voltage across a dielectric exceeds a certain value, the bonds restraining the bound electrons of the material atoms are finally torn asunder. This results in a high current flowing through the capacitor, and it is called *dielectric breakdown.* The value of the electric field at which the breakdown takes place is called the *dielectric strength* of the material. Table 6-1 also gives the dielectric strength (in V/cm) of the materials listed in the table.

The maximum voltage a given capacitor can withstand is the product of its dielectric strength and the thickness of its dielectric layer. This is called the *breakdown voltage* of the capacitor.

EXAMPLE 6-5

What is the maximum voltage that can be applied across a 0.1-μF capacitor which has a plate area of 0.2 m^2? The dielectric is mica.

Solution: The thickness of the dielectric of this capacitor is first found from Eq. (6-3).

$$C = \frac{K\epsilon_o A}{d} = \frac{(5.0)(8.85 \times 10^{-12})(0.2)}{d}$$

or

$$d = \frac{(5.0)(8.85 \times 10^{-12})(0.2)}{(1 \times 10^{-7})}$$

$$d = 8.85 \times 10^{-5} = 0.088 \text{ mm}$$

The dielectric strength of mica, as given by Table 6-1, is 2×10^6 V/cm. Then the maximum voltage that can be applied to this capacitor is

$$V_{max} = 1.7 \times 10^4 \text{ V}$$

Capacitor Types

The dielectric which separates the plates of a capacitor primarily determines the capacitance value, the leakage current, and the breakdown voltage of the capacitor. Consequently, capacitors are generally classified according to the materials that are used as their dielectrics.

The objective in developing a capacitor design is to have the largest capacitance value in the smallest possible volume. In addition, the capacitance should not change with time, voltage, or mechanical stress, and it should have a minimum of losses.

For a given plate area and dielectric type, capacitance can be increased only by making the dielectric as thin as possible. Unfortunately, making the dielectric thinner reduces the maximum voltage that can be applied across the capacitor. If this maximum voltage is exceeded, breakdown and rupture of the dielectric occurs. The same effect also causes higher leakage in larger capacitance units. Thus an essential compromise must be made between high capacitance and the ability to withstand high voltages.

The metal-dielectric-metal sandwich that makes up a fixed capacitor can be rolled into a tube, folded, or otherwise reduced in overall size and sealed in an environmentally protected package. The resulting capacitors can there-

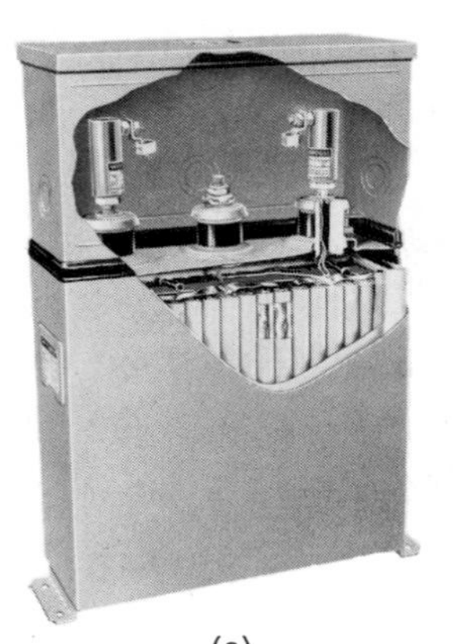

(a)

(b)

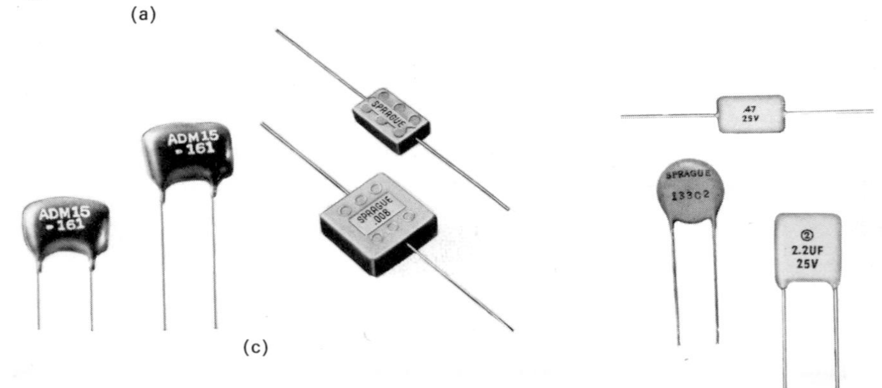

(c)

(d)

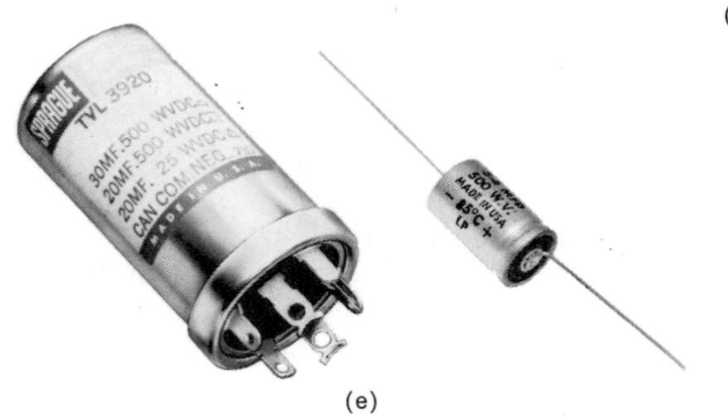

(e)

Figure 6-7. Common types of capacitors (a) Oil-filled paper capacitor (Courtesy of Westinghouse Electric Corp.) (b) Cutaway view of plastic film capacitor (Courtesy of Sprague Electric Co.) (c) Mica units (molded and dipped types) (Courtesy of Sprague Electric Co. and Aerovox) (d) Ceramic capacitors (Courtesy of Sprague Electric Co.) (e) Electrolytic capacitors (Courtesy of Cornell-Dubiller Corp. and Sprague Electric Co.)

fore vary widely in shape and size. Figure 6-7 shows some various fixed-capacitor packages.

MICA CAPACITORS

Mica is a transparent, high-dielectric-strength mineral that is easily separated into uniform sheets as thin as 0.0001 inch. It has a high breakdown voltage and is almost totally chemically inert. These characteristics produce capacitors which are very stable under temperature and electric stress and with the passage of time.

Mica capacitors are built in round, rectangular, or irregular shapes. They are constructed by sandwiching layers of metal foil and mica, as shown in Fig. 6-8(a). Sometimes silver is deposited on the mica in lieu of metal foil. The resulting stack of metal and mica sheets is firmly clamped and encapsulated in a plastic package.

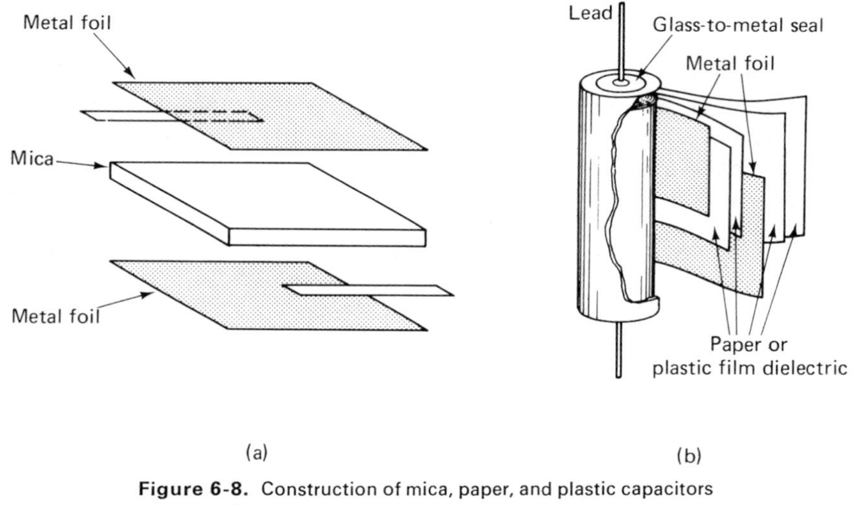

Figure 6-8. Construction of mica, paper, and plastic capacitors
(a) Mica Capacitors
(b) Paper and plastic capacitors

Mica capacitors possess very small leakage current and dissipation factors. Available capacitance ranges are 1 pF to 0.1 μF, with tolerances of ± 1 to ± 20 percent. The capacitance is limited to this relatively small upper value because mica is not flexible enough to be rolled into tubes. As a result, the size of the mica capacitor structures cannot be markedly reduced.

Mica capacitors are employed as precision capacitors because of their small tolerances and high stability under temperature. They are also used in such high-frequency applications as oscillator tuning and filter construction, where small capacitance values and low dissipation factors are desirable.

Finally, high-voltage applications are often best served by mica capacitors.

Mica capacitors have no polarity preference and their capacitance value and other characteristics are often indicated by a color-code scheme printed on their package.

CERAMIC CAPACITORS

There are two different types of ceramic capacitors being built: the low-loss, low-dielectric-constant type, and the high-dielectric-constant type. The low-loss types have a very high leakage resistance (1000 MΩ) and can be used in high-frequency applications almost as well as mica capacitors.

The high-dielectric-constant types provide a large capacitance value in a small volume. However, their value of capacitance can change strongly with variations of temperature, dc voltage, and frequency. This is due to the fact that the dielectric constants of high-K capacitors are highly dependent on these variables. Thus, this type of capacitor is only suitable if an exact capacitance is not required (such as in circuit coupling or bypass applications). Capacitance values of the high-K types range from 100 pF to 0.1 μF. A typical tolerance range is $+100$ percent to -20 percent of its stated value.

The disc type of construction used to build ceramic capacitors is shown in Fig. 6-9. A ceramic disc or plate is coated with metal on both faces. Leads are attached to the metal and the resultant capacitor is packaged in a coating of plastic or ceramic to protect it from moisture and other environmental conditions. The capacitance value is printed directly on the body, or a color code is used. Ceramic capacitors possess no required voltage polarity.

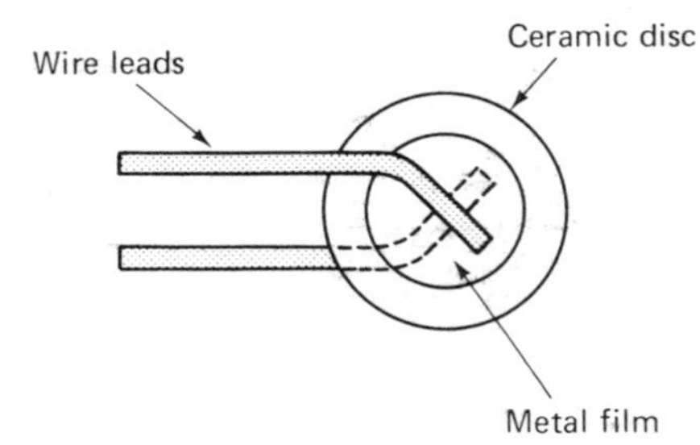

Figure 6-9. Ceramic disc capacitor construction

PAPER CAPACITORS

Paper capacitors are the most widely used type of capacitors. Their popularity is due to their low cost and the fact that they can be built in a broad range of capacitance values (500 pF to 50 μF). Furthermore, they can be designed to withstand very high voltages. However, the leakage currents of paper capacitors are high and their tolerances are relatively poor (± 10 to 20 percent). These limitations restrict their use in some applications. If size permits, the capacitance value and voltage are usually printed on the capacitor body. For small units, a color code is used. When the color code is not used, a band (usually black) is often printed on the tube nearest the lead that is connected to the outer metal sheet. This lead should always be connected to the circuit lead of lower potential.

Many paper capacitors are of a cylindrical shape because they are made by rolling a sandwich of metal and impregnated paper sheets into a tube. Axial leads are attached to each metal sheet, and the tube is encapsulated in waxed paper or plastic (Fig. 6-8(b)).

Various substances such as oil, wax, or plastic are used to soak the paper. If paper deposited with thin metal films is used rather than separate metal sheets, the volume per unit of capacitance can be reduced by 50 percent and the leakage current reduced by 90 percent. Unfortunately, this creates a resulting structure which is more prone to rupture by high-voltage transients.

Special oil-filled paper capacitors are made which have high capacitance values and high breakdown voltages. These types are usually mounted in metal cases and have ceramic insulators surrounding the input leads. Such high-voltage paper capacitors are used primarily in certain power-supply and transmitter circuits. Figure 6-7(a) shows a photograph of a typical oil-filled paper capacitor.

PLASTIC-FILM CAPACITORS

Plastic-film capacitors are constructed in basically the same way as paper capacitors, except that a thin sheet of plastic (such as Mylar, Teflon, or polyethylene) is used as the dielectric. This dielectric improves the properties of the capacitor by minimizing leakage currents, even at temperatures of up to 150–200°C. Their other characteristics are similar to those of paper units. However, the cost is higher for plastic units, so they are not usually used except when a paper capacitor cannot meet the design specifications. Commercial plastic-film capacitors are manufactured in ranges between 500 pF and 10 μF.

ELECTROLYTIC CAPACITORS

Electrolytic capacitors are usually made of aluminum or tantalum.[4] The basic structure of the aluminum electrolytic capacitor consists of two aluminum foils, one of which is coated by an extremely thin oxide (Fig. 6-10). The oxide is grown on the metal by a process of applying a voltage to the capacitor; the process is called *forming*. The thickness of the oxide depends on the forming voltage. Between the foils is an electrolytic solution soaked into paper. This electrolyte is a conductor and serves as an extension of the nonoxidized metal foil. Since it is a fluid, the electrolyte can butt up directly

[4] Aluminum and tantalum are used for the plates of electrolytic capacitors because they form oxides with very high dielectric strengths.

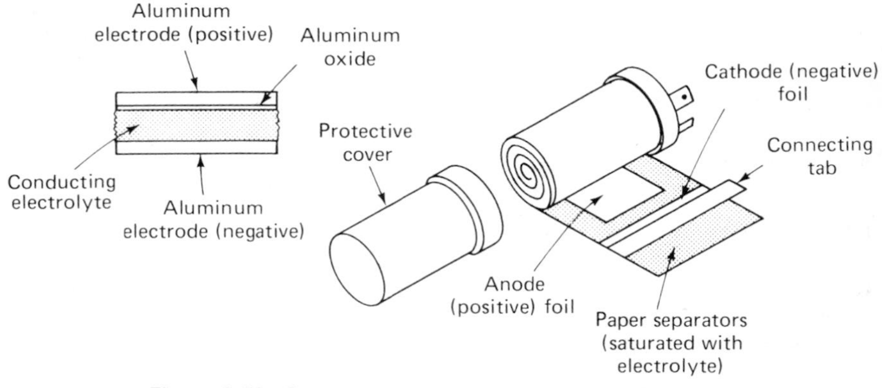

Figure 6-10. Construction of aluminum electrolytic capacitors

against the oxide dielectric. The two oppositely charged plates are then effectively separated by only an extremely thin oxide film which possesses an extremely high dielectric constant.

Once the oxide is formed, the foils are rolled into a tube, and the piece of foil without the oxide is connected to the capacitor's exterior package. This lead serves as the negative connection to the capacitor. The other lead is marked by a "+" on the capacitor body and *must* be connected to the positive terminal of the circuit in which it is used.

It needs to be strongly emphasized that the electrolytic capacitor should only be connected in a circuit with the proper polarities. If one connects the positive lead of the capacitor to the negative lead of a circuit, chemical action by the electrolyte will rupture the oxide dielectric and *destroy* the capacitor. (With reversed polarity, the oxide no longer acts like an insulator. As a result, a substantial leakage current can flow and disintegrate the oxide.[5]) In addition, as for other capacitors, the rated voltage must not be exceeded. For the largest capacitance values, the maximum voltage will be small because the oxide layer is so thin.

Electrolytic capacitors have the largest capacitance values per volume of element of any capacitor type. But they also possess very large leakage current values. These properties limit their use to special applications. For example, in transistor circuits, large capacitances in a small volume are desirable, but leakage currents or exact capacitance values are not necessarily critical. Thus electrolytic capacitors are suitable for some of these circuits. Electrolytic capacitors are available in values that range from 1 to 500,000 μF. However, their corresponding leakage resistances are only about 1 MΩ.

[5]If an oxide is grown on both metal foils of an electrolytic capacitor, the problem of proper polarity connections does not exist. However, the capacitance-to-volume ratio of the element is also reduced by half. These types of electrolytic capacitors are called *nonpolar electrolytic capacitors*, and they are not commonly used.

VARIABLE CAPACITORS

As with resistors, it is often necessary to be able to vary the value of a capacitor while it remains connected in a circuit. For example, it may be desired to tune the circuit of a radio receiver or an oscillator. *Variable capacitors* are available to fulfill such application requirements.

The *air-variable* capacitor is one such common variable type. It is constructed by mounting a set of metal plates (usually aluminum) on a shaft, and meshing them with a comparably shaped set of fixed metal plates (see Fig. 6-11). As the shaft is rotated, either more or less area (depending on the direction of rotation) between the adjacent and oppositely charged plates is created. This variation of the area changes the capacitance. (The larger the interleaved area, the greater the capacitance.) By designing the shapes of the plates appropriately, various capacitance-versus-shaft-position curves can be achieved. For example, a linear or a square-root variation of capacitance can be obtained. Because the dielectric is air, the separation between the plates must be kept fairly large to ensure that they do not touch and discharge. (If dust or conducting debris falls in between capacitor plates, it can cause arcing and changes in capacitance values. Thus air-variable capacitors must be kept clean.) This limits capacitance values of air-variable capacitors to about 500 pF. Because air capacitors have such low leakage, they are used to build precision adjustable capacitors which act as standards for measuring small values of capacitance.

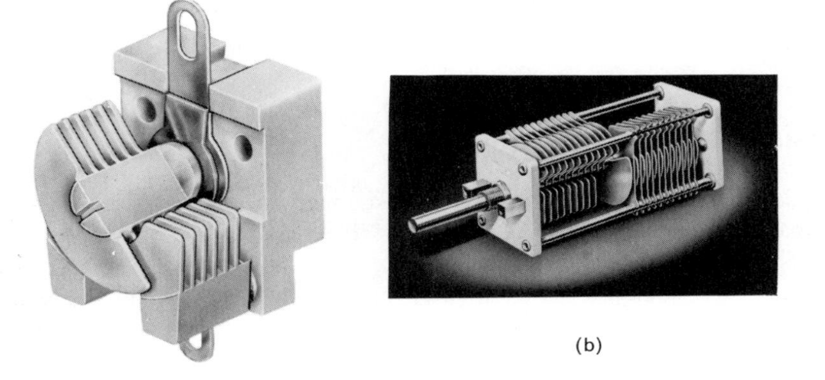

(a) (b)

Figure 6-11. Air variable capacitors (Courtesy of E. F. Johnson Co. and James Millen Mfg. Co., Inc.)

The trimmer capacitor is also a variable capacitor, but it is used primarily in circuits that need only one-time or infrequent tuning adjustments (as in the setting of the frequency range of a tuned amplifier). The trimmer capacitor is usually a mica capacitor that has a screw which clamps the metal-mica sheets. When the screw is tightened, the separation between the plates (and thus the capacitance) is adjusted. The overall range of trimmer capacitors is about 15–500 pF. Each individual unit has a small variable range, i.e., between 5–40 pF or 20–100 pF.

Table 6-3* Various Capacitor Types and Their Characteristics

Dielectric	Available Capacitance Values	Tolerances (%)	Leakage Resistance (MΩ)	Maximum Voltage Ranges	Useful Frequency Ranges (Hz)
Mica (silvered)	1 pF–0.1 μF	±1 to ±20	1000	500–75 kV	10^3–10^{10}
Ceramic (low-loss)	1 pF–0.001 μF	±5 to ±20	1000	6000 V	10^3–10^{10}
Ceramic (high-*K*)	100 pF–0.1 μF	+100 to −20	30–100	100 V or smaller	10^3–10^8
Paper (oil-soaked)	1000 pF–50 μF	±10 to ±20	100	100 V to 100 kV	100–10^6
Polystyrene	500 pF–10 μF	±0.5	10,000	1000 V or smaller	0–10^{10}
Mylar	5000 pF–10 μF	±20	10,000	100 V to 600 V	100–10^6
Electrolytic	1 μF–0.5 F	+100 to −20	1	500 V or smaller	10–10^4
Air-variable	10 pF (unmeshed) to 500 pF (meshed)	±0.1		500 V	

*Adapted from B. D. Wedlock and J. K. Roberge, *Electronic Components and Measurements* (Englewood Cliffs, N.J.: Prentice-Hall, Inc., 1969), p. 96.

Color Coding of Capacitors

Capacitance values at one time were stamped directly on all the bodies of capacitors. However, the popularity of the color-coding scheme used on resistors led to the development of color-coding systems for capacitors. Today color coding is used on many capacitors which have small packages. It is seen most commonly on tubular paper, mica, and ceramic units. Figure 6-12 gives the color codes used for each of these various capacitor types.

TEMPERATURE EFFECTS ON CAPACITORS

Capacitors are somewhat sensitive to changes in their ambient temperature. As a general rule, their capacitance values increase with increasing temperatures. However, some types of ceramic and polystyrene capacitors are specifically designed to decrease in capacitance value as their temperature increases.

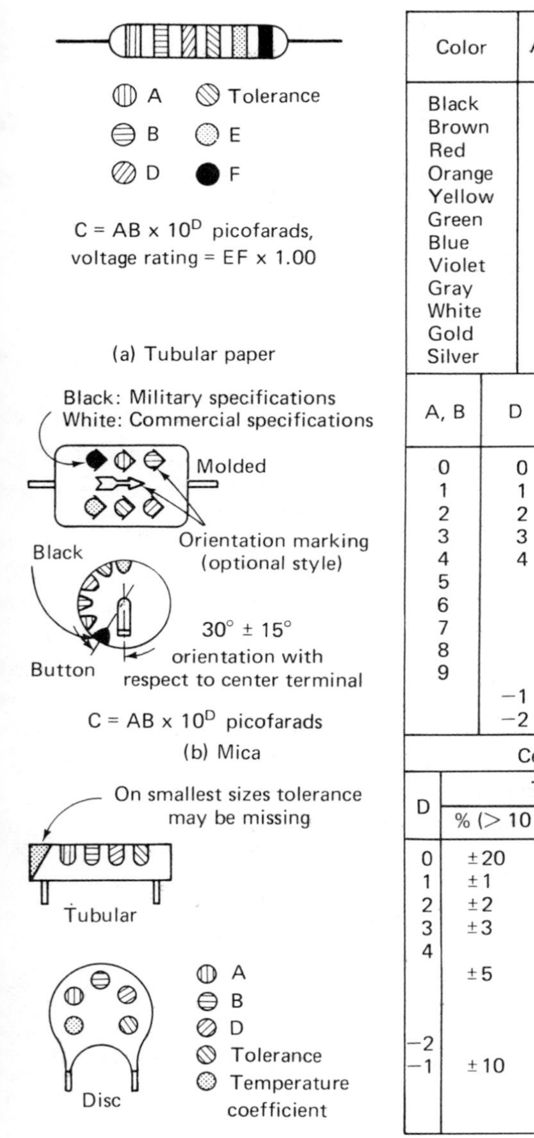

$C = AB \times 10^D$ picofarads

(c) Ceramic

Color	A, B, D	Tubular: Tolerance	Tubular: E, F
Black	0	20	0
Brown	1		1
Red	2		2
Orange	3	30	3
Yellow	4	40	4
Green	5	5	5
Blue	6		6
Violet	7		7
Gray	8		8
White	9	10	9
Gold			
Silver			

A, B	D	Mica: Tolerance %	Mica: Temp. coeff. ppm/°C
0	0	±20 M	
1	1	±1 F	±500
2	2	±2 G	±200
3	3		±100
4	4		−20 to +100
5		±5 J	0 to +70
6			
7			
8			
9			
	−1	±0.5 E	
	−2	±10 K	

Ceramic (A, B same as Mica)

D	Tolerance: % ($>$ 10 pF)		Tolerance: pF ($<$ 10 pF)		Temp. coeff. ppm/°C
0	±20	M	±2.0	G	0
1	±1	F	±0.1	B	− 33
2	±2	G			− 75
3	±3	H			−150
4					−220
	±5	J	±0.5	D	−330
					−470
					−750
−2			±0.25	C	−1500 to +150
−1	±10	K	±1.0	F	−750 to +100

Figure 6-12. Capacitor color codes (a) Tubular paper (b) Mica (c) Ceramic (Adapted from B. D. Wedlock and J. K. Roberge, *Electronic Components and Measurement*, Englewood Cliffs, N.J.: Prentice-Hall, Inc., 1969, p. 15.)

The percentage of change in capacitance per degree centigrade temperature change is called the *temperature coefficient* (TC) of a capacitor. The reference value is the value of the capacitance measured at 25°C. Thus a 50-μF capacitor with a +0.01 percent TC will increase its capacitance value 0.005 μF for each degree of temperature rise above 25°C.

In addition to changing the value of capacitance, increasing temperature lowers the rated breakdown voltage of capacitors. Conversely, decreasing the external temperature makes capacitors more susceptible to internal losses and potentially destructive heating effects.

Capacitance Measurement

Although capacitance can be measured by instruments such as the electrometer and by indirect methods such as measuring *RC*-circuit time constants, these methods are both hampered by lack of accuracy. Consequently, most capacitance measurements are made using bridge circuits. Capacitance measurements made with bridges yield very accurate results.

The bridge circuit was introduced in Chapter 5 in the form of the Wheatstone bridge for measuring resistance. Recall that a condition of balance in the bridge circuit established a null reading in the bridge detector. At balance, the value of the unknown resistance could be computed from a knowledge of the other resistance values in the circuit.

The methods for measuring capacitance and inductance by the use of bridges are also based on the principle of establishing a null condition in a bridge circuit. The unknown value is calculated from the other elements of the circuit at balance. In this section we will examine those bridge circuits which are suitable for determining capacitance.

BRIDGE CIRCUITS FOR MEASURING CAPACITANCE VALUES

The condition for balance of the Wheatstone bridge examined in Chapter 5 was found to be

$$R_X R_1 = R_2 R_3$$

where R_X was the unknown value of resistance and R_1, R_2, and R_3 were the known values. If the resistances of the Wheatstone bridge are replaced by impedances of both a resistive and reactive nature (Fig. 6-13), and if an ac voltage is applied between points A and B of the circuit, the balance equation is, in general:

$$Z_X Z_1 = Z_2 Z_3 \tag{6-10}$$

Since any impedance, Z, can be expressed as a complex number

$$Z = R + jX \tag{6-11}$$

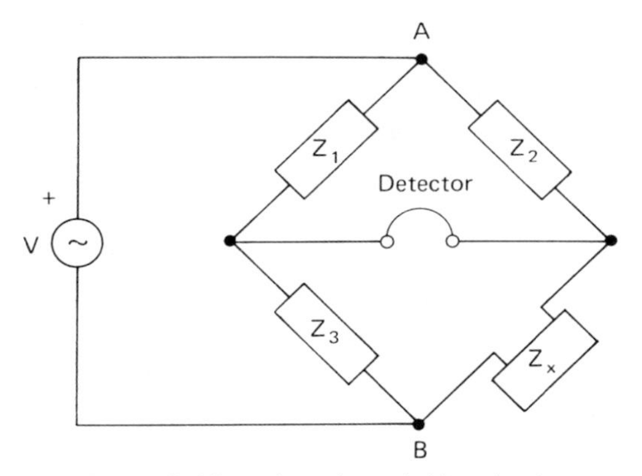

Figure 6-13. ac impedance bridge circuit

then deriving an unknown Z_X from the condition of balance is not quite as easy as finding R_X in a purely resistive bridge. In fact, for a null condition to be achieved, it is necessary to specify two matching conditions—one for the resistive part of Z_X and one for the reactive part

$$Z_X = R_X + jX_X = \frac{Z_2 Z_3}{Z_1} = \mathrm{Re}\,\frac{(Z_2 Z_3)}{Z_1} + \mathrm{Im}\,\frac{(Z_2 Z_3)}{Z_1} \tag{6-12}$$

Two types of capacitance bridge circuits are commonly employed for measuring the capacitance values and the dissipation factors of a capacitor. If the dissipation factor of a capacitor is small ($0.001 < D < 0.1$), the *series-capacitance comparison bridge* is used (Fig. 6-14(a)). If D is larger than this ($0.05 < D < 50$), the *parallel-capacitance comparison bridge* is used (Fig. 6-14(b)).

For the *series-capacitance comparison bridge* the impedances of Eq. (6-10) are $Z_1 = R_1$, $Z_2 = R_2$, $Z_3 = R_3 - j(1/\omega C_3)$ and $Z_s = R_s - j(1/\omega C_s)$. Substituting these impedances into Eq. (6-10) and separating the imaginary and real values, we find that when a null is established, R_s and C_s are calculated from the following relations:

$$R_s = \frac{R_2 R_3}{R_1} \quad \text{and} \quad C_s = C_3 \frac{R_1}{R_2} \tag{6-13}$$

C_s is the value of the capacitance we want to determine.

Note that if R_2 and C_3 are chosen as the fixed quantities, and R_3 and R_1, as the variable elements, we can achieve a null condition by varying R_3 and R_1 and then calculating R_s and C_s. In this circuit the only variable elements are resistors, and no reactive elements have to be adjusted to achieve a condition of balance. This aids the user of the bridge in approaching the null value rapidly. In addition to C, the dissipation factor (D) is also indicated by the

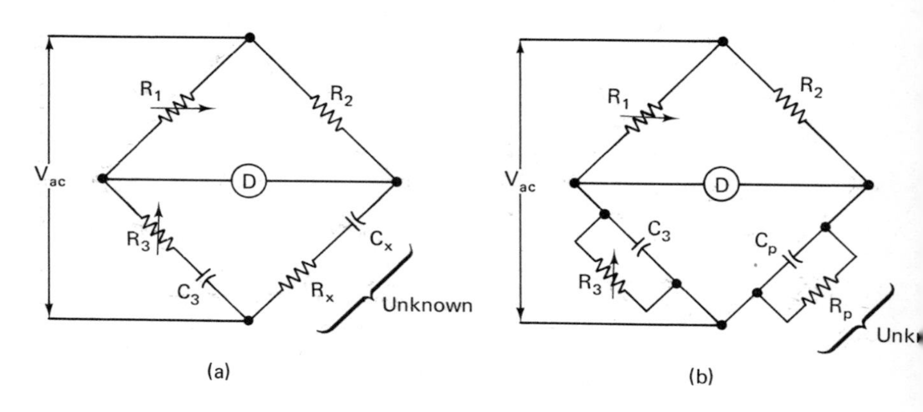

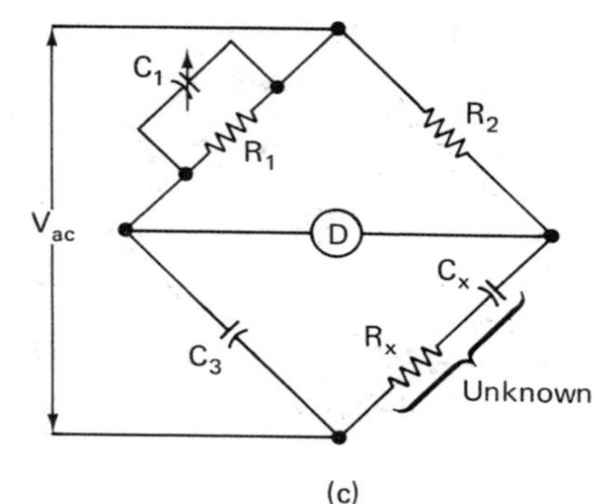

Figure 6-14. Bridge circuits for measuring capacitance (a) Series comparison bridge (b) Parallel comparison bridge (c) (c) Schering bridge

bridge. Usually a dial on the meter is calibrated so that the value of D is automatically calculated and directly indicated.

The condition of balance of the parallel-capacitance comparison bridge is

$$R_p = \frac{R_2 R_3}{R_1} \quad \text{and} \quad C_p = \frac{C_3 R_1}{R_2} \tag{6-14}$$

In this bridge R_2 and C_3 are fixed and R_3 and R_1 are variable. Both C_p and D are read directly from the bridge settings at the condition of balance.

For measuring capacitors in circuits where the phase angle is very nearly 90°, the Schering bridge offers more accurate readings than either of the capacitance comparison circuits. This bridge is shown in Fig. 6-14(c).[6]

[6]For more details on the Schering bridge, see M. B. Stout, *Basic Electrical Measurements*, 2nd ed. (Englewood Cliffs, N.J.: Prentice-Hall, Inc., 1960), pp. 268–269.

COMMERCIAL CAPACITANCE BRIDGES

There are two general forms in which capacitance bridges are manufactured commercially. The first type is an instrument which is capable of measuring only capacitance. It usually contains two capacitance comparison bridges—a series-comparison type and a parallel-capacitance type. Most such bridges contain an internal ac source at a single fixed frequency. A few others provide multiple internal frequencies. However, almost all contain provisions for connecting additional external ac sources so that the instrument can be used at other frequencies as well. Typical capacitance values which can be measured using capacitance bridges range from 1 pF to 1000 μF to accuracies of 1 percent. Some extremely accurate bridges can measure capacitance values up to ± 0.1 percent in accuracy. Examples of this type of bridge are shown in Fig. 6-15.

The other type of bridge that is sold is called the universal bridge and is capable of measuring R and L as well as C. It usually has five or six built-in bridge circuits with appropriate switching to connect them. For measuring capacitance, two of these bridges are the series-comparison and parallel-comparison bridges. They also usually contain a fixed internal ac source as

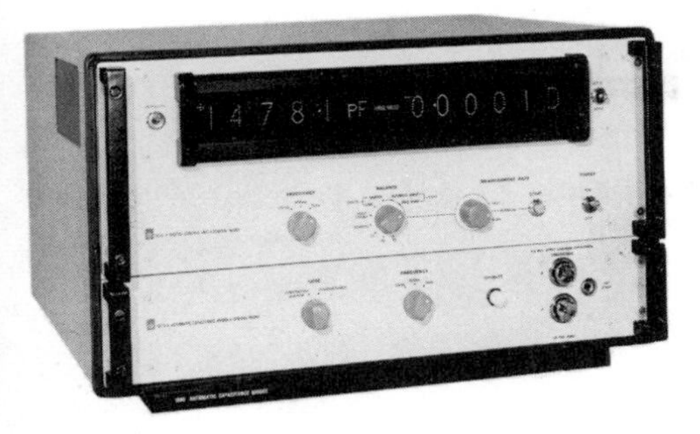

(a)

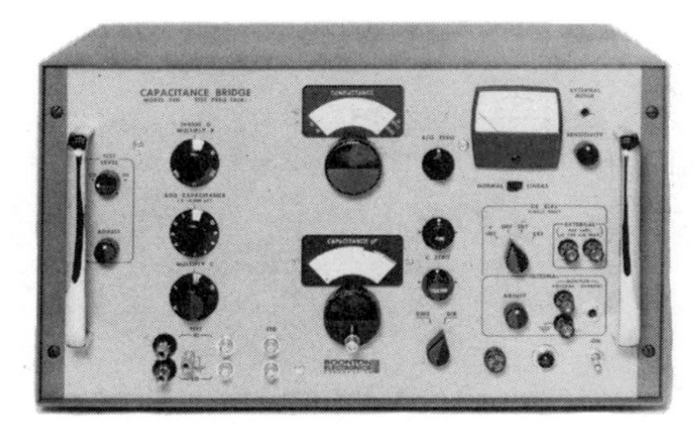

(b)

Figure 6-15. Commercial capacitance bridges (a) GR-1680 (b) Boonton 75D (Courtesy of General Radio Co. and Boonton Electronics, Inc.)

well as having the option of connecting additional external ac signals. The operation and overall specifications of the universal bridge will be examined more thoroughly in Chapter 7 in the discussion on inductance measurements. For measuring capacitance, however, their ranges are about the same as for the capacitance bridge alone.

CAPACITANCE MEASUREMENT USING AN AC VOLTMETER

Although capacitance measurements made with a capacitance bridge are quite accurate, this instrument is not always available when an unknown capacitor needs to be measured. Therefore, a method which uses a high impedance ac voltmeter (e.g., with an input impedance of 10–11 MΩ) to determine the unknown capacitance is presented here. This type of measurement is limited to measuring capacitors with values of 0.001 μF or more, and it is good to only about 10 percent accuracy (because of uncertainties in the applied frequency and voltage and because of meter inaccuracies).

The unknown capacitor is connected in series with a resistor and the combination is put across the 115-V, 60-Hz power line (Fig. 6-16). Then the voltage across each element is measured separately. We first find I (an rms value) from

$$I = \frac{V_R}{R} \tag{6-15}$$

where R is the resistance of the resistor and V_R is the rms voltage measured across the resistor. Then we find C from

$$V_C = IX_C = \frac{I}{\omega C} = \frac{I}{2\pi f C}$$

or

$$C = \frac{I}{2\pi f V_C} \tag{6-16}$$

where V_C is the rms voltage measured across the capacitor.

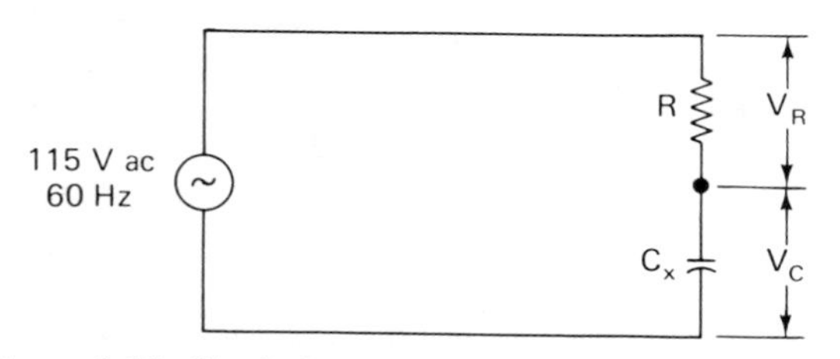

Figure 6-16. Circuit for measuring capacitance with an ac voltmeter

EXAMPLE 6-6

We connect a 4000-Ω resistor in series with an unknown capacitor across the 115-V, 60-Hz power-line voltage. An ac voltmeter measures V_R to be 80 V and V_C to be 100 V. What is the value of C?

Solution: First find I from Eq. (6-15):

$$I = \frac{80\text{ V}}{4000\ \Omega} = 0.02\text{ A} = 20\text{ mA}$$

Then use Eq. (6-16) to find C:

$$C = \frac{0.02}{(6.28) \times (60) \times 100} = \frac{0.02}{(37{,}700)} = 0.7\ \mu\text{F}$$

Problems

1. Find the capacitance of a capacitor structure if a charge of 120 μC is deposited on the plates of the structure when 15 V are applied across it.
2. How much charge is deposited on the plates of a 500 pF capacitor if 300 V are applied across it?
3. A parallel plate capacitor has a capacitance of 10 μF. What will be the effect on the capacitance of this component if
   a) the area of the plates is doubled?
   b) the spacing between the plates is increased from 1 mm to 3 mm?
   c) if the oil-soaked paper dielectric is replaced by a mica dielectric?
4. If the capacitance of an air capacitor is 600 pF and an unknown dielectric is placed between the plates, its capacitance increases to 3000 pF. Identify the unknown material used as the dielectric.
5. If capacitors are connected together in parallel with one another, the capacitance of the entire connection is given by the expression:

   $$C_t = C_1 + C_2 + C_3 + \cdots + C_n$$

   If the capacitors are connected in series, the capacitance is given by the expression:

   $$\frac{1}{C_t} = \frac{1}{C_1} + \frac{1}{C_2} + \cdots + \frac{1}{C_n}$$

   For capacitors whose individual values are 6 μF, 10 μF, and 4 μF, find the capacitance value of the three capacitors connected a) in parallel, b) in series.
6. Find the total capacitance value of the connection shown in Fig. P6-1.

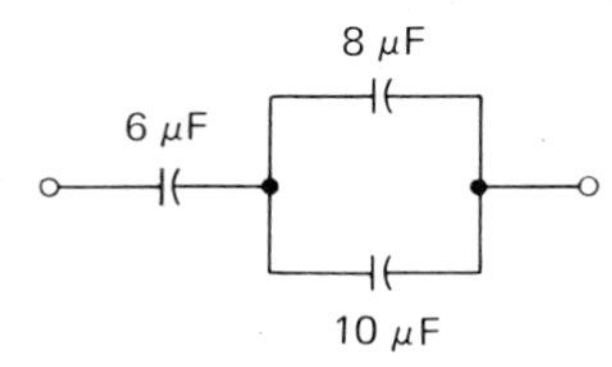

Figure P6-1

7. What capacitance value must be added in series to a capacitor whose value is 900 pF in order to reduce the total capacitance to 300 pF?

8. Describe how the shaft rotation of a variable capacitor causes the capacitance to change.

9. Shielded cable, which is often used to carry low level signals, has a specific capacitance value per unit of length. Describe how the capacitance effect within the cable arises.

10. If each of the following events in a circuit branch containing a capacitor occurs, what effect does it have on the value of X_c of the branch?
    a) Capacitance value is increased by a factor of 3.
    b) Radian frequency is doubled.
    c) Capacitance is reduced to one-tenth of its original value.

11. What is the rms value of the current that flows in the circuit shown in Fig. P6-2?

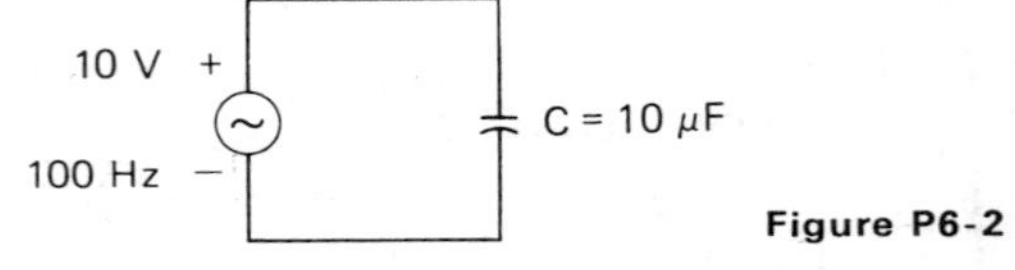

Figure P6-2

12. At what frequency will a 100 pF capacitor appear to have a reactance of 10,000 Ω?

13. Explain why it appears that current "flows through" a capacitor in ac circuits.

14. Select the type (or types) of capacitors which would most likely best satisfy the following design requirements:
    a) large value of capacitance in a very small volume,
    b) low cost, 0.1 μF capacitor,
    c) capacitor to be used as a standard capacitor for comparison measurements,
    d) capacitor which can withstand high voltages.

References

1. MULLIN, W. F., *ABC's of Capacitors*. Indianapolis, Ind., Howard W. Sams & Co., Inc., 1967.

2. BOYLESTAD, R., *Introductory Circuit Analysis*, Chap. 8. Columbus, Ohio, Charles E. Merrill Books, Inc., 1968.

3. STOUT, M. B., *Basic Electrical Measurements*, 2nd ed., Chaps. 9, 10, 12, 13, and 14. Englewood Cliffs, N.J., Prentice-Hall, Inc., 1960.

4. COOPER, W. D., *Electronic Instrumentation and Measurement Techniques*, Chap. 8. Englewood Cliffs, N.J., Prentice-Hall, Inc., 1970.

7

Inductors and Transformers

In Chapters 5 and 6 we examined two of the three passive elements of electric circuits—the resistor and the capacitor. We saw that resistors are circuit elements that dissipate electrical energy and maintain the same ideal circuit properties under both ac and dc operating conditions. Capacitors were seen to be circuit elements that store energy in their electric fields and have drastically different properties when interacting with ac instead of dc quantities. The third of the three ideal passive elements is the *inductor*. Inductors are elements that store energy in magnetic fields and also possess different circuit properties when excited by ac rather than dc quantities. In fact, the quantity of inductance relies on a condition of changing current as a part of its definition.

In this chapter, we will examine the structure, circuit properties, and methods of measuring real inductors. In addition, since inductors rely on magnetic effects for their operation (and on magnetic materials to enhance their electrical characteristics), a brief discussion concerning magnetic fields and properties of magnetic materials will also be included. The operation of transformers and relays will also be presented since they too are devices whose functioning relies on magnetic effects.

Inductance and Inductors

If charge flows in a conductor, there will be a magnetic field associated with this current in the space surrounding the conductor. When the charge flow ceases, the magnetic field decreases to zero. These phenomena demonstrate that the source of magnetic fields is a flow of charge, or current.[1]

When current flows in a wire conductor, a magnetic field with a configuration like that shown in Fig. 7-1(a) surrounds the wire. If this wire is bent into a loop, the magnetic field of the loop has a form like that shown in Fig. 7-1(b). Finally, if the wire is wound into the shape of a *coil* (Fig. 7-1(c)), the magnetic fields of the individual loops combine to form a composite magnetic field. (A coil is also called a *solenoid*, from the Greek word—tubelike.) The field of the coil has the same pattern as the field of a permanent bar magnet (with the field at one end corresponding to a north pole and the field at the other end, a south pole). The lines which represent the magnetic fields in these illustrations are called *flux lines*.

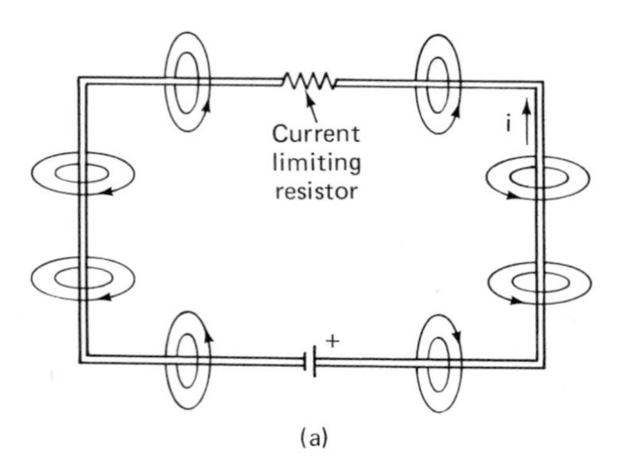

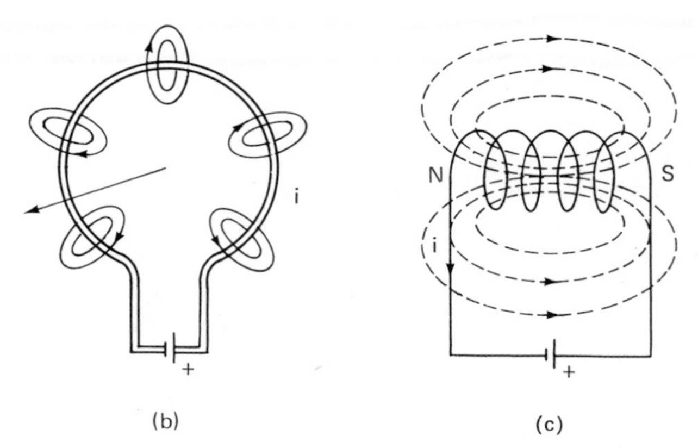

Figure 7-1. (a) Magnetic field around a current-carrying wire (b) Magnetic field of a circular loop of wire (c) Magnetic field of a coil which is carrying current

[1] In permanent magnets, the "organized" motion of electrons about their atoms constitutes this source current.

The magnetic fields surrounding current-carrying conductors have energy stored in them. The amount of energy depends upon the magnitude of the current which is the source of the field. This energy is "deposited" into the magnetic field when the current that causes the field is increasing. This same energy is returned to the charges when the current in the conductor decreases. (A part of the energy may also be radiated away in the form of electromagnetic waves.)

Another phenomenon associated with energy storage by a magnetic field is the fact that a voltage is induced in a conductor if the conductor is being *cut* by a magnetic field. (The word *cut* means that the flux lines of a magnetic field move through the region of space occupied by the conductor. In other words, flux lines cut a conductor if the conductor and the magnetic field are in *relative motion*. Either the conductor can be moving in a stationary field, or the field can be moving, or the field can be time-varying in the space around a fixed conductor.) When a current is initiated in a coil, the magnetic field created by the current cuts the coil itself and induces a voltage which appears across the ends. (As the magnetic field of each loop of the coil increases, it spreads out into the surrounding space. Thus, in the process of spreading, its flux lines cut the other turns of the coil.) This fact indicates that a coil becomes a *source of voltage* during the time in which a current or magnetic field within it is changing. (Once the current reaches a steady value, there is no more voltage across the coil.) Mathematically, this effect of induced voltage in a coil is expressed as

$$v = N\frac{d\phi}{dt} \tag{7-1}$$

where ϕ is the *magnetic flux* of the field (in webers) and N is the number of turns in the coil. Equation (7-1) is known as *Faraday's law*, and it forms the basis of the operation of motors and generators as well as inductors.

Since the magnetic flux, ϕ, in the coil is proportional to the current which causes it, we can write

$$\phi = Ki \tag{7-2}$$

where K is some constant of proportionality. Then we can rewrite Eq. (7-1) as

$$v = NK\frac{di}{dt} = L\frac{di}{dt} \tag{7-3}$$

where $L = NK$.

The quantity represented by L is called the *self-inductance* (or just the *inductance*) of the coil. When an element is deliberately designed to possess a certain value of inductance in a circuit, the element is known as an *inductor*. Most inductors are built in the shape of coils because the maximum inductance per volume of element is developed by the coil configuration.

The unit of inductance is the *henry* (H). One henry is defined as the inductance necessary to induce a voltage of one volt in an element when the current in it is changing at a rate of one ampere per second. Inductors used in electronic applications have values ranging from a high of about 30 H (i.e., the "heavyweight" or large-valued filter chokes[2]) to a low value of approximately 50 μH (i.e., the small "peaking" coils used in television circuitry). The circuit symbols of the inductor are shown in Fig. 7-2.

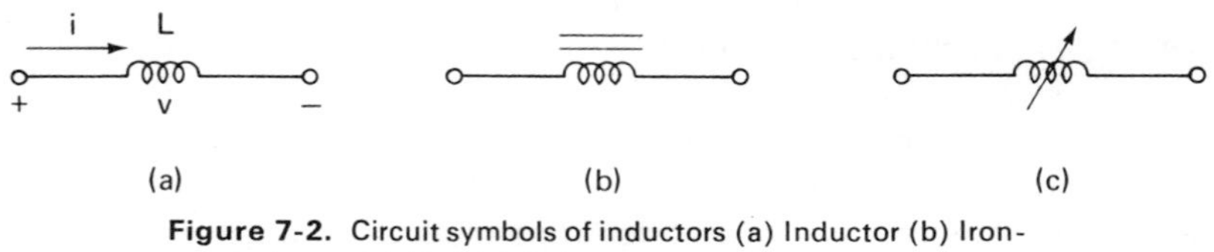

Figure 7-2. Circuit symbols of inductors (a) Inductor (b) Iron-core inductor (c) Variable inductor

INDUCTORS IN ELECTRIC CIRCUITS

When a dc current is being passed through an ideal inductor, the inductor presents no impedance to the flow of the current. (An ideal inductor is defined as being made of an ideal conductor and therefore contains no resistance.) Thus, there will be no voltage drop across an ideal inductor when a dc current is flowing in it. However, if an ac (or changing) current is applied to an inductor, a voltage is induced in the element. The rms value of this voltage increases as the frequency of the applied curent increases. If the applied ac current, i, is sinusoidal, these phenomena have the following quantitative implications: First, the expression for i can be written as

$$i = I_o \sin \omega t \tag{7-4}$$

Then,

$$\frac{di}{dt} = \omega I_o \cos \omega t \tag{7-5}$$

Since the rms value of $I_o \cos \omega t$ equals the rms value of $I_o \sin \omega t$, the rms value of di/dt can be written

$$\left|\frac{di}{dt}\right|_{\text{rms}} = \omega I \tag{7-6}$$

where I is the rms value of $i = I_o \sin \omega t$.

[2]*Choke* is a term used to describe inductors which are placed in a circuit to present a high impedance to frequencies above a specified frequency range. At the same time, the choke should not appreciably affect the flow of dc current. The way that inductors "choke off" such high frequencies is elaborated upon in a later section of the chapter.

Now Eq. (7-3) can be written in terms of the rms values of v and i as

$$L = \frac{V}{\omega I} \tag{7-7}$$

Since the ratio of V/I yields the impedance of an element, we can therefore write

$$X_L = \omega L = \frac{V}{I} \tag{7-8}$$

The quantity X_L is known as *inductive reactance*. The units of inductive reactance are ohms (Ω). We see from Eq. (7-8) that when $\omega = 0$, $X_L = 0$. This conclusion is in accord with the principle that an ideal inductor should offer no impedance to dc currents. We also see that X_L increases as ω increases. Thus inductors act as elements of higher impedance to high-frequency signals and as elements with lower impedance to low-frequency signals.

EXAMPLE 7-1

Find the inductive reactance of a 1.0-H and a 50-μH inductor at 60 Hz, 100 KHz, and 10 MHz, respectively.

Solution: (a) The inductive reactance of a coil is found from Eq. (7-8). For the 1-H coil:

At 60 Hz, $X_L = \omega L = 2\pi f L$
$= (6.28) \times (60) \times (1)$
$= (6.28) \times (60) = 377\ \Omega$

At 100 kHz, $X_L = 6.28 \times 10^5\ \Omega$

At 10 MHz, $X_L = 6.28 \times 10^7\ \Omega$

(b) For the 50-μH coil:

At 60 Hz, $X_L = (6.28) \times (60) \times (5 \times 10^{-5}) = 1.8 \times 10^{-2}\ \Omega$

At 100 kHz, $X_L = (6.28) \times (10^5) \times (5 \times 10^{-5}) = 31.4\ \Omega$

At 10 MHz, $X_L = (6.28 \times 10^7) \times (5 \times 10^{-5}) = 3.14 \times 10^3\ \Omega$

Table 7-1 lists some uses of inductors in electric circuits.

ENERGY STORED IN AN INDUCTOR

When the current flowing in an inductor reaches a steady value, its magnetic field also stabilizes and stores a steady value of energy. The energy stored by the magnetic field of an inductor is computed from the expression,

$$w = \frac{1}{2} L i^2 \tag{7-9}$$

where i is the value of the current flowing in the inductor.

This expression indicates that the energy stored by an inductor increases as the square of the magnitude of the current. As in ideal capacitors, the stored energy is returned to the circuit if its source current is halted.

Table 7-1 Uses of Inductors in Circuit Applications

Inductors are used in electric circuits as:

1. Elements which block high-frequency signals but allow low-frequency signals to pass without significant attenuation. Such *choke coils* can be placed in series with the inputs to 60-Hz power supplies. Any high-frequency (hf) signals are thereby prevented from entering the power-supply circuit.
2. Elements used in combination with capacitors to allow the passage of selected frequencies only (e.g., frequency-selective filters).
3. Elements used to produce high values of voltage in such applications as automobile spark plugs.
4. Elements which store energy in magnetic fields as part of oscillator circuits.
5. Elements which act as transducers to indicate such quantities as position and velocity. (See Chapter 15 for further details on position-sensing transducers.)
6. Devices which produce magnetic fields for deflecting the electrom beam of television-type cathode ray tubes. (Oscilloscope CRTs use an electrostatic deflection system instead.)

Properties of Magnetic Materials

Inductors and transformers are essentially coils of wire which utilize self-inductance and mutual inductance effects to function as electric circuit elements. To increase the self-inductance of a given inductor structure, the coil can be wound around a piece of magnetic material called a *core*. (In a *magnetic material*, a stronger magnetic effect is created by a current than would be created in a vacuum by the same current.) Likewise, the coils of a transformer are wound on a core of magnetic material to increase the effect of mutual inductance. In this way, magnetic materials play an analogous role in inductors and transformers that dielectric materials play in capacitors.

However, unlike dielectric materials (which are relatively loss-free), magnetic materials also cause other effects when they are used in the construction of inductors or transformers. Magnetic materials are more complex and are subject to significant loss effects and nonlinear behavior.

To be able to explain how a magnetic material changes the properties of inductors and transformers, we will present a brief discussion on the properties of magnetic fields and materials. We will start by introducing the units used to describe magnetic effects.

There are two sets of quantities used to describe magnetic fields—the magnetic flux density (B) and magnetic intensity (H).[3] The strength of the magnetic field is described by B, and its units are webers per square meter (Wb/m^2). The tendency of moving charges to create a magnetic effect is described by H, and its units are amperes/meter (A/m). In a vacuum H and B are related by the expression

$$B = \mu_o H \tag{7-10}$$

The constant μ_o in Eq. (7-10) is known as the *permeability of a vacuum* and its value is $\mu_o = 4\pi \times 10^{-7}$ henries/meter (or Wb/A-m).

In a vacuum, the linear relation of Eq. (7-10) is always valid because μ_o is constant. However, for magnetic materials, the permeability μ is different from μ_o, furthermore, it varies nonlinearly with H. In fact, it is usually necessary to use a graphical relation rather than an equation to demonstrate the relation between B and H in magnetic materials. Such graphs are called *magnetization* or *B-H* curves. Figure 7-3 shows how B and H are related in a typical magnetic material.

The graph in Fig. 7-3 shows that as we increase H (and hence i), the magnetic field strength B increases approximately linearly until some specified

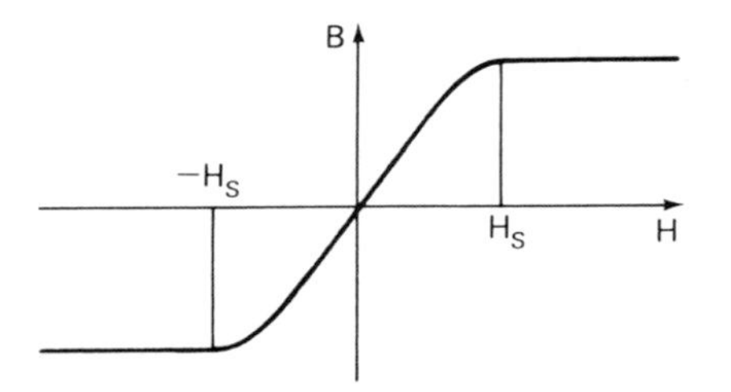

Figure 7-3. Magnetization curve of a typical magnetic material

[3]It may be questioned at this point why two quantities which describe magnetic effects must be used. The reason is that each one describes a different aspect of magnetic action. The magnetic intensity, H, defines the magnetic effect generated by moving electric charges regardless of the surroundings in which they move. That is, a current of a specific magnitude will always cause a condition which is described by H. H is always proportional to i.

The magnetic flux density, B, describes the strength of a magnetic effect as caused by i as well as that caused by the material in which the magnetic intensity exists. Thus the value of B depends not only on the magnitude of i, but also on the material in which the field is created.

For example, a current whose magnitude is I_1 will cause a magnetic intensity H_1 and a magnetic flux density B_1 to exist in nearby nonmagnetic materials. However, if the material near the current-carrying conductor is a magnetic material, the resulting strength of B will be much greater than B_1, although H_1 will have the same value as in the nonmagnetic material.

magnitude of H is reached (i.e., $|H_s|$). Between the values of $-H_s$ and $+H_s$, the relation

$$B = \mu H \tag{7-11}$$

is approximately valid. Therefore this operating region is called the *linear region* of the magnetic material. μ in Eq. (7-11) is then the permeability of the material for values of H that are smaller than $|H_s|$. In this linear region, the relative permeability $\mu_r = \mu/\mu_o$ of common inductor core materials are given in Table 7-2.

Table 7-2

Material	Relative Permeability of Inductor Core Materials (μ_r)	Saturation, B_M, (Wb/m^2)
Silicon steel	10^3–10^4	1.6
Nickel-iron alloy	10^4–10^5	1.5
Powdered nickel-iron alloy	10–100	0.6
Ferrites	10^3	0.3

Since the permeability of magnetic materials is much higher than the permeability of air, the flux density (B) in a coil whose core is made of magnetic material will also be much greater. Now if the *magnetic flux* (ϕ) is uniform over some area of space, A, then ϕ and B are related by the expression

$$B\,(Wb/m^2) = \frac{\phi\ (Wb)}{A\ (m^2)} \tag{7-12}$$

Thus the ratio of the flux established in the ferromagnetic core to the flux established in the air core of the same size, is also given by μ_r.

Since a higher flux, μ, will exist in the iron-core inductor, such an inductor will have a greater inductance value than that of a comparable air-core type.

When the magnitudes of H become greater than H_s, the magnetic material described in Fig. 7-3 starts to *saturate*. This means that B no longer increases in magnitude as rapidly as in the linear region.[4] Therefore, as i increases beyond the linear region, H will still increase, and so will B. However, the rate of increase of B eventually approaches the rate at which it would increase in a vacuum.

[4]In magnetizing a ferromagnetic material, a reorientation of small permanent-magnet-like regions within the material takes place. These regions are characteristic of magnetic materials and are called *domains*. At first only a small current is needed to begin reorienting these domains. As they become aligned, their magnetic effect adds to the magnetic effect caused by the current i. However, once most of the domains have become properly oriented by the current, an additional current cannot cause further reorientation. Thus the material is no longer able to contribute to the strengthening of the field. This state occurs when the value of H_s is exceeded and is the reason the name "saturation" is used for the effect.

In many electronic applications, saturation leads to undesirable circuit effects because of the nonlinearity which occurs in the relation between *B* and *H*. Therefore, various measures designed to prevent saturation are used. One common method of counteracting core saturation is to leave a small air gap in the magnetic core. With an air gap in the core, the resulting flux (for a given current) will be much lower in the core. This reduces the likelihood of *B* reaching a saturation value.

HYSTERESIS AND EDDY CURRENTS IN MAGNETIC MATERIALS

It was noted in the previous section that coils are usually wound around cores of magnetic materials to increase their value of inductance. In this section we will examine the other effects a core has on inductors.

When a current is first initiated in a coil with magnetic core, *B* and *H* start out as being zero. As *i* is increased, *B* and *H* will rise in value according to a curve like that shown in Fig. 7-3. If, after increasing *H* from zero to some value H_1, we reduce *H* to zero (by reducing the current flowing to zero), there will still be some magnetic flux remaining in the material. This remaining magnetic effect (*remanence*) occurs because some of the tiny domains in the magnetic material remain oriented in such a way that they continue to cause a weak magnetic flux to exist. The fact that some magnetic flux remains when *H* is returned to zero can be expressed by saying that *B lags behind H.* As a consequence, this lagging effect is called *hysteresis* (from the Greek word *hysterein*—to lag).

Now let us examine what will occur if the curve of *B* vs *H* is traversed in the following way: First, *H* is started from zero and is increased to a point of saturation. Then *H* is decreased until a point of negative saturation is reached. Finally, *H* is increased again until the core reaches positive saturation.

We see that a path as shown in Fig. 7-4(a) is traced as a consequence of this repetitive increasing and decreasing of *H*. The traversed path is called a *hysteresis loop.* The area enclosed by the loop represents dissipated energy because there is an inelastic action involved when the magnetic domains are reoriented in direction by the magnetic effect of the current. The dissipated energy manifests itself as heating of the core. As a result, if an ac current is applied to an inductor with a magnetic core, the number of times the ac current is reversed represents the number of times the hysteresis loop is traversed. Since the same energy is dissipated during each cycle, a high-frequency signal can be subjected to severe energy loss from such hysteresis effects. In fact, hysteresis losses often limit many core materials from being used in high-frequency applications. Since the area of the loop is directly proportional to the magnitude of the energy loss, a smaller loss per cycle will exist if the amplitude of *i* (which causes *H*) is kept small. Figure 7-4(b) shows how this effect occurs.

Another loss effect which becomes significant in the cores of inductors is

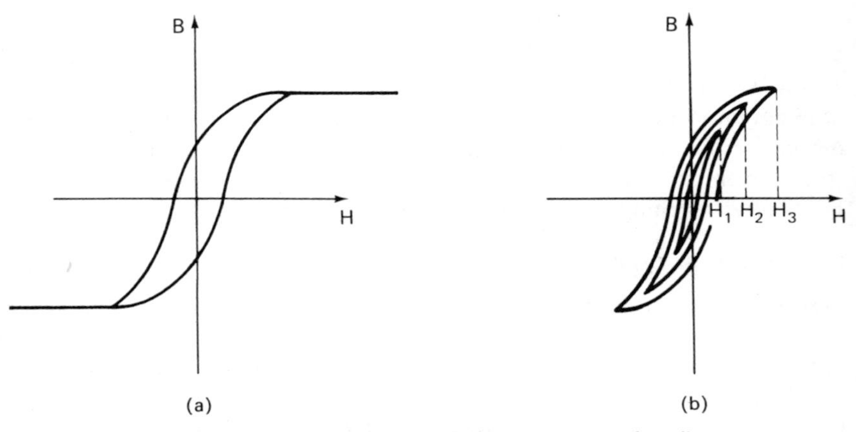

Figure 7-4. (a) Hysteresis loop (b) Hysteresis loops of smaller areas result as H_{max} of each loop is made smaller

that due to induced *eddy currents.* Since the core material of an inductor is usually an electric conductor as well as a magnetic conductor, the time-varying current of the surrounding coil induces electric currents in the core (as well as a magnetic field). Because such currents swirl around throughout the entire core structure, they are called *eddy currents.* (Eddy currents in rivers are currents that separate themselves from the main current and tend to swirl around outcroppings of land or rocks.) This current flow in the core is subject to a resistance heating loss according to

$$P_{\text{eddy current loss}} \, P_{ECL} = I_{\text{EC}}^2 R \tag{7-13}$$

where R is the resistance of the path in the core and I_{EC} is the rms value of the eddy currents. In addition to being an energy loss as far as the signal is concerned, this effect also heats up the core.

To minimize eddy current losses, inductor cores are usually constructed of very thin metal sheets (i.e., ≈ 0.02 inch thick) that are insulated from one another by a thin coat of varnish. These thin sheets are called *laminations* and they restrict the flow of eddy currents to each sheet alone. Since each lamination is much thinner than the entire core, the resistance that appears to any flowing eddy currents is much greater. (In addition the laminations are oriented so that currents are at a minimum in the plane of the lamination.) The overall eddy current flow is thus drastically reduced and P_{ECL} is decreased.

MODELS FOR LOSSES IN ACTUAL INDUCTORS

The eddy current and hysteresis losses present in iron-core inductors are called *core losses.* There is also a loss which arises from the resistance of the wires with which the coil is wound. These losses are called *copper losses.*

Copper loss is always present but can be minimized by using as large a wire as space and cost limitations will allow. As noted earlier, the hysteresis loss can be minimized by operating the core at low levels of magnetic intensity (low H), while eddy current losses are reduced by the use of laminations, powdered iron, or ferrites[5] as the core.

Nevertheless, inductor losses cannot be completely eliminated. This means that the value of the losses must be estimated or measured in order to be able to predict how they will cause the operation of a real inductor to change from its ideal behavior. Various models which take the losses into account can be used to represent an inductor for such analytical purposes. One such model is shown in Fig. 7-5.

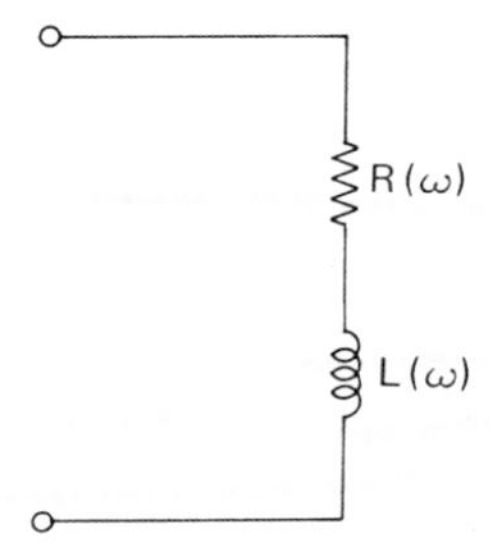

Figure 7-5. Circuit model of an inductor

In this model all the losses are lumped together as one resistor whose resistance value will depend on the frequency at which the inductor is operated. In general, the impedance of an inductor, Z, to any sinusoidal signal of frequency, ω, which is calculated from this model can be written as the sum of the resistive and inductive effects:

$$Z(\omega) = R(\omega) + jX_L(\omega) \tag{7-14}$$

where

$$X_L(\omega) = \omega L(\omega)$$

However, the resistance of an inductor is rarely specified in a manufacturer's data. Instead, a factor called the *quality factor*, Q, is used. It tells what the ratio of the inductive reactance to the resistance of an inductor will be at a specific frequency and is expressed as

$$Q = \frac{\omega L(\omega)}{R(\omega)} \tag{7-15}$$

If $R(\omega)$ were equal to zero, we would have an ideal inductor, and Q would be infinite. Thus, the larger the value of Q, the more ideal is the inductor. (The best-made inductors have Q's of about 1000.) However, it is necessary to

[5] *Ferrites* are high-permeability materials made of metallic oxides (such as iron, zinc, manganese, and nickel). Because the ferrites are oxides, they have very high resistivities.

specify the frequency and the amplitude of a test signal as well as the Q of an inductor in order to make meaningful comparisons between various elements. When measurements of inductance at specific frequencies are made, a measurement of Q is also desirable. Although common inductance-measuring instruments also allow Q to be measured, Q-meters are usually used when it is necessary to make very accurate measurements of Q.

Inductor Structures

Inductors are constructed by winding wire in various coil configurations. This restricts the magnetic field to the physical space around the inductor and creates the largest inductance effect per volume of element. (For the closely wound toroidal coil, the magnetic field is almost wholly confined to the space enclosed by the winding.)

The major factors which determine the magnitude of the inductance of a coil are:

1. The number of coil turns,
2. The type and shape of the core material, and to a lesser extent,
3. The diameter and spacing of the turns.

The coils are usually wound around cores of ferromagnetic material because this makes magnetic flux density within the wound coil area vastly greater than if the core is air. (We saw earlier how this effect was a consequence of the atomic structure of ferromagnetic materials.) The larger flux density allows an increase in the inductance of the structure. But this type of core also makes the inductor subject to eddy current and hysteresis losses.

For inductors shaped like those shown in Fig. 7-6, an approximate value of inductance can be calculated from Eq. (7-16) (as long as the current is not so large that the linear region of B vs H curve is exceeded).

$$L = \frac{\mu_r N^2 A}{l} \tag{7-16}$$

In Eq. (7-16), L is the inductance in henries, μ_r is the relative permeability of the core, N is the number of turns,[6] A is the area of one turn, and l is the length of the coil.

EXAMPLE 7-2

Given an inductor such as the one shown in Fig. 7-6(b), with $N = 100$ turns, $l = 6$ cm, and $r = 0.5$ cm. For cores of (a) air and (b) iron, find the inductance of each coil. (Choose $\mu_{r_{\text{iron}}} = 1000$.)

[6]The term N^2 is used to calculate L in most practical inductor structures because of the configuration of the coil winding. The coil is wound so that the magnetic flux of each loop is allowed to cut or intersect each of the other loops. In this manner, the changing current in each turn will cause an inductive effect in all of the other turns.

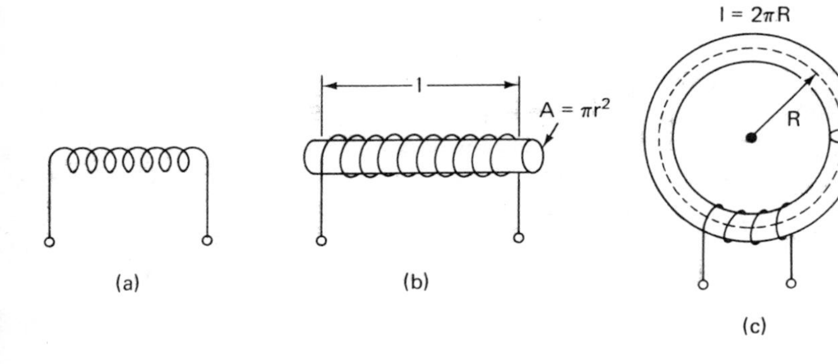

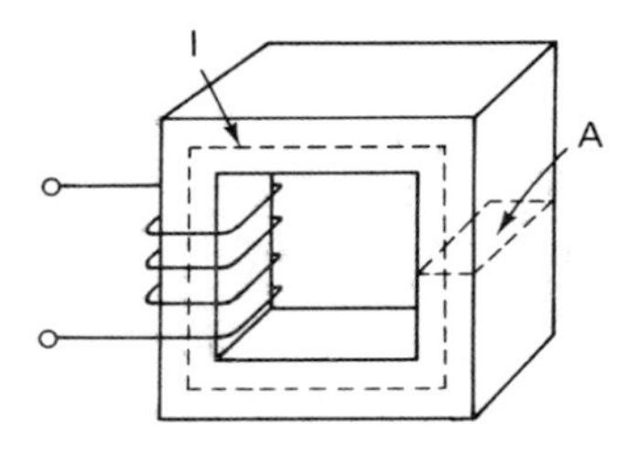

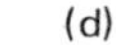
(d)

Figure 7-6. Various inductor configurations

Solution: (a) $L = \dfrac{\mu_o N^2 A}{l}, \qquad l = 6 \text{ cm} = 0.06 \text{ m}$

$$A = \pi r^2 = \pi(0.005 \text{ m})^2$$

$$= \pi(2.5 \times 10^{-5}) \cong 8 \times 10^{-5} \text{ m}^2$$

$$\mu_o = 4\pi \times 10^{-7} = 12.6 \times 10^{-7} \text{ H/m}$$

Then

$$L = \frac{(12.6 \times 10^{-7}) \times (10^4) \times (8 \times 10^{-5})}{(6 \times 10^{-2})}$$

$$= 1.6 \times 10^{-5} \text{ H}$$

$$L_{\text{air}} = 16 \ \mu\text{H}$$

(b) $L_{\text{iron}} = \mu_r L_{\text{air}} = 1000 \times 16 \ \mu\text{H} = 16 \text{ mH}$

LOW-FREQUENCY INDUCTORS

For the low-frequency inductors which are generally used in filters and audio-frequency chokes (and which therefore carry frequencies from 20–20,000 Hz), a large value of X_L is usually required. Since $X_L = \omega L$, at low frequencies a large inductive reactance requires an element which has a large inductance value. Inductors of 5 henries or more are common for these applications.

In low-frequency inductors, N must be large and the core material must have a high permeability. Hence, laminated iron or silicon-steel (to reduce core losses) is usually chosen for the core. At very low frequencies (between 25 and 400 Hz), the eddy currents generated are not very great, and hysteresis losses are also small. Therefore, the low-cost silicon-steel cores with laminations can be used with good results. Figure 7-7 shows a typical af inductor.

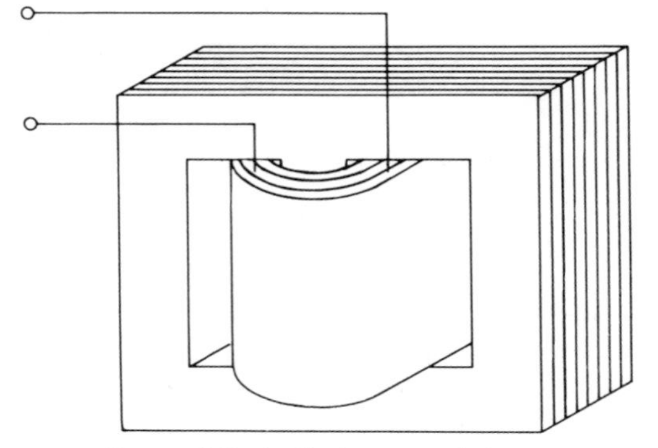

Figure 7-7. Audio-frequency inductor

HIGH-FREQUENCY INDUCTORS

For high-frequency applications (i.e., 500 kHz to 100 MHz), the inductance values required of inductors are usually much smaller than the inductance values of af inductors ($\approx 10^{-6}$–10^{-3} H) because a larger reactance effect results from the high-frequency current in the inductor. As a consequence, a much smaller number of turns is required for such inductors. However, in these elements, the high-frequency effects make core losses and stray capacitance much more critical than in low-frequency inductors. The laminated core is no longer satisfactory because even the laminations do not prevent large eddy current flows at these frequencies. Hence different core materials must be used. Air cores, powdered-iron *slugs*, and ferrite cores are the most common choices. The powdered-iron and ferrite cores can be made in any shape desired (Fig. 7-8).

Powdered-iron cores are made of fine iron granules separated from each other but bound together by insulating material. The current paths in each of these granules are so tiny that eddy currents can be virtually curtailed. Thus powdered-iron cores can be used at very high frequencies (several hundred MHz). Cores from this material are also very stable under temperature variations and are therefore used to construct highly accurate inductors. Their biggest drawback is that their relative permeabilities are quite low ($\mu_r \approx 10$–100).

Ferrites are high-permeability ceramic materials made of metallic oxides which have extremely high resistivities and hence small eddy current losses.

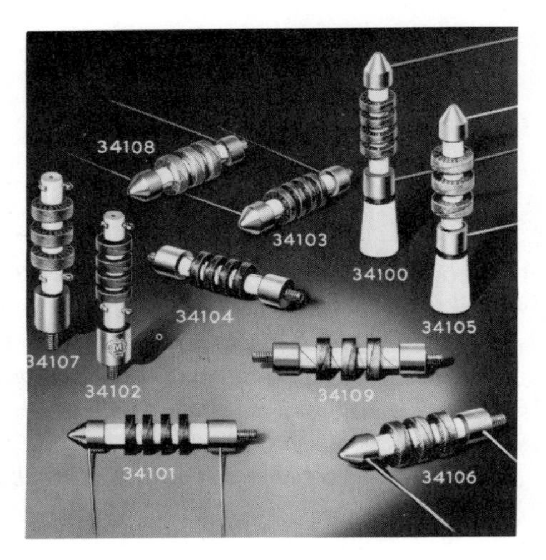

Figure 7-8. Radio-frequency inductors (Courtesy of James Millen Mfg. Co., Inc.)

Ferrites are also used in constructing accurate inductors for frequencies that range from 1 kHz to 10 MHz (Fig. 7-9). They are not used in low-frequency coils because their permeabilities are not as large as those of iron or silicon steel.

Figure 7-9. Ferrite core shapes (Courtesy of Ferroxcube Corp.)

One unique shape of core used in rf inductors is the *pot core*. The *pot-core* inductor has its core both inside and outside the coil. This ensures that the coil is completely surrounded by the core material and provides a highly efficient path for the magnetic flux. The core is constructed by shaping ferrite or powdered iron into a cup. By winding the coil around a rod of ferrite and then slipping the cups over it, we have the form as shown in Fig. 7-10. The pot core also provides an excellent shield against noise.

When a current flows in any real inductor, a small but measurable voltage drop exists between neighboring turns of its coil. Thus a capacitive effect also exists between them. To limit the effects of this unwanted capacitance (which can become significant at high frequencies), several steps can be taken.

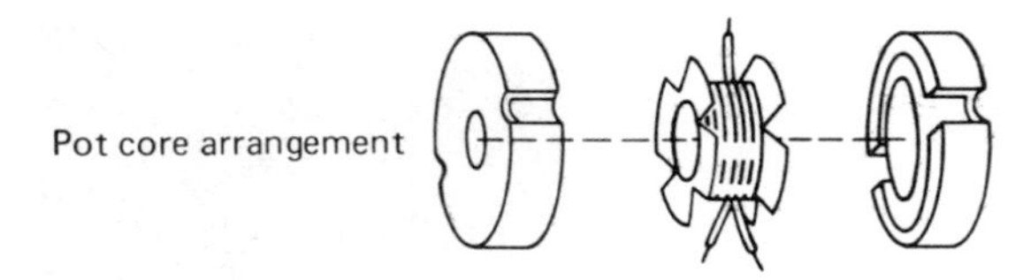

Figure 7-10. Pot-core arrangement

If the inductance value required by an application is small, the coil can be wound in a single layer. This type of winding has the least capacitance but also produces small values of inductance.

Another method used to reduce capacitive effects is to wind the coil in sections called *pies* along one core (Fig. 7-8). In pie-wound inductors a zigzag winding ensures that adjacent layers are not parallel and the capacitance of each wound element is added in series (thus reducing the overall capacitance of the element).

TOROIDAL INDUCTORS

The *toroid* or donut-shaped inductor is a highly efficient inductor because of its shape. Virtually all of the flux of a toroid coil is enclosed by the winding, and hence a large value of inductance per volume of space is achieved. In addition, toroidal inductors are relatively immune to stray magnetic fields. The disadvantage of toroidal inductors is that they are more expensive to wind because of their circular form.

Toroidal inductors are manufactured in various ways. One way is to roll a long narrow strip into a ring-shaped core (Fig. 7-11). This produces a core that is, in effect, laminated. Another manufacturing method is to stack flat circular "washers" to a desired thickness to provide the core. If environmental protection of the core is desired, the toroid core may be coated with silicone rubber and the entire inductor encapsulated in epoxy or molded Bakelite.

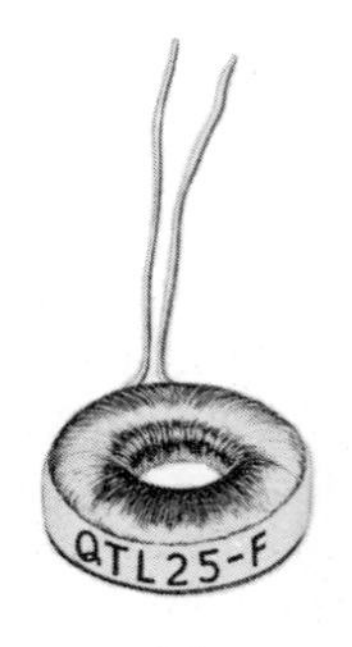

(a) (b)

Figure 7-11. Toroidal inductors (a) Tape-wound toroid (b) 25 mH Toroidal inductor (Courtesy of Microtran Co., Inc.)

VARIABLE INDUCTORS

Some applications call for variable rather than fixed inductors. Tuning circuits, phase shifting, and switching of bands in amplifiers sometimes require a variable inductance. Such inductors can be made in different ways. Figure 7-12 shows how inductance is varied in several commercial elements. The inductor shown in Fig. 7-12(a) can be varied by switching from one tap on the coil to another. In Fig. 7-12(b) a movable core is employed. As more of the core is inserted into the coil, the inductance increases. By appropriately varying the spacing of the coil windings, we can obtain a relatively linear variation of inductance with core insertion.

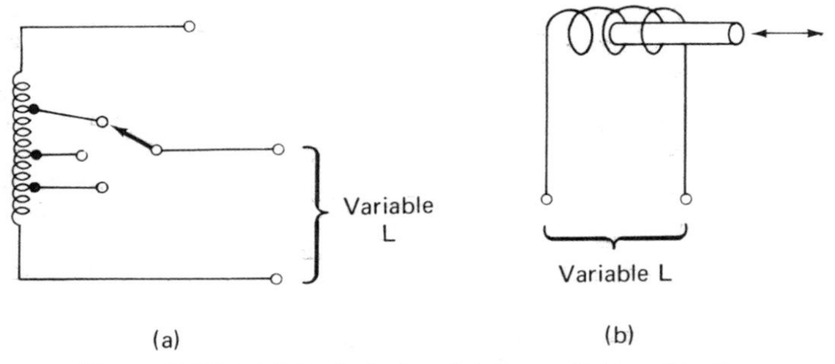

Figure 7-12. Variable inductors (a) Tap switching (b) Movable-core type

Measurement of Inductance

In this section we will discuss two methods for measuring inductance. The first method utilizes ac bridge circuits (which are somewhat similar to those used in measuring capacitance). This method yields very accurate values of the inductance under test. Special inductance-bridge instruments are available to perform this type of measurement. The second method utilizes an ac voltmeter and the voltage from the power line. The results of this method are much less accurate than those obtained from bridge measurements, but the method is useful for determining a value of inductance when an inductance bridge is not available.

INDUCTANCE BRIDGES

There are two types of bridge circuits most commonly used to determine inductance. The first, the *Maxwell bridge* is best suited for measuring inductors which have a low Q (i.e., $1 < Q < 10$). The second is the *Hay bridge*, and it determines L most accurately when the Q of an inductor is high (i.e., $10 < Q < 1000$).

The Maxwell bridge is shown in Fig. 7-13(a). It measures unknown induc-

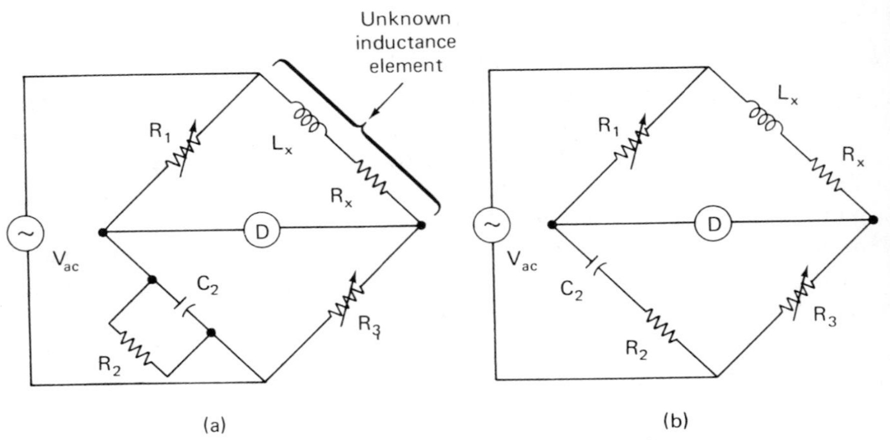

Figure 7-13. (a) Maxwell bridge (b) Hay bridge

tance by comparison to a standard capacitance. The use of a capacitance as the standard element is advantageous because a capacitor is a compact element and is easy to shield.

A condition of balance in a Maxwell Bridge exists when

$$L_X = R_1 R_3 C_2 \tag{7-17a}$$

and

$$R_X = \frac{R_1 R_3}{R_2} \tag{7-17b}$$

In these equations, L_X is the value of the unknown inductance and R_X is the corresponding resistance value of the element. We see from these equations that by choosing C_2 and R_3 as constants, we only have to vary R_1 and R_2 until a null is achieved. However, since R_1 appears in both Eq. (7-17a) and Eq. (7-17b), finding the two conditions of balance requires several adjustments. The common procedure for establishing a balance is to first determine L_X. Then the balance of R_X is sought. In finding the condition of balance for R_X, we invariably disturb the condition of balance for L_X. We must then return to find L_X again. After making several adjustments, we finally reach a balance for both conditions simultaneously.

The Hay bridge shown in Fig. 7-13(b) is best for measuring the inductance of high-Q inductors. It also uses a standard capacitor as a comparison element for determining the unknown L. However, the capacitor is in a series with a resistor in one arm of the bridge, rather than in parallel. The equations of balance for the Hay bridge are

$$L_X = \frac{R_2 R_3 C_1}{1 + \omega^2 C_1 R_1^2} \tag{7-18a}$$

and

$$R_X = \frac{\omega^2 C_1{}^2 R_1 R_2 R_3}{1 + \omega^2 C_1{}^2 R_1{}^2} \tag{7-18b}$$

These equations appear more complex than the balance conditions for the other bridges considered. Also, the equations for L_X and R_X appear to be dependent on ω. However, for the cases when $Q > 10$, the term with ω becomes less than $\frac{1}{100}$ and hence can be neglected. In such cases, the equation for L_X becomes

$$L_X = R_2 R_3 C_1 \tag{7-19}$$

For this reason, the Hay bridge is not as accurate if the value of Q of the inductor being measured is less than 10.

COMMERCIAL INDUCTANCE BRIDGES

Commercial inductance bridges are available which measure L with values from nanohenries up to 1100 H. The basic accuracies of these instruments range from 0.1 percent to 1 percent. The Maxwell and Hay bridges are usually used as the bridge circuits which determine L. Figure 7-14 shows a typical inductance bridge.

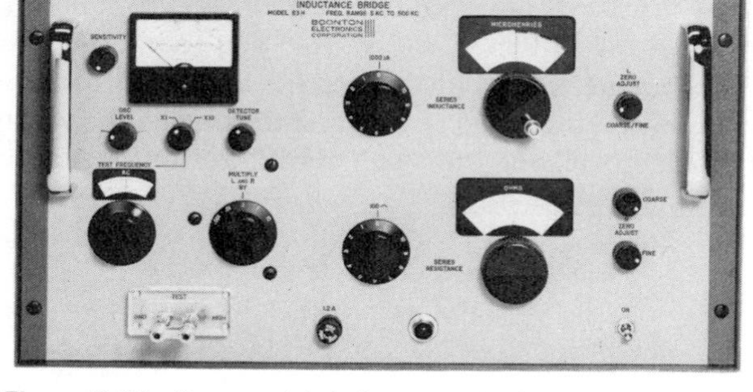

Figure 7-14. Commercial inductance bridge (Courtesy of Boonton Electronic Corp.)

UNIVERSAL IMPEDANCE BRIDGES

The *universal impedance bridge* is an instrument designed to be able to measure R, L, and C over a wide range of values. To be able to carry out these functions, the instrument has five or six built-in bridge circuits. These five bridge circuits are the Wheatstone bridge, the series and parallel capacitance bridges, the Maxwell bridge, and the Hay bridge. The specifications of three commercially available universal impedance bridges are given in Table 7-3. Figure 7-15 shows a photograph of two of the models listed in the table.

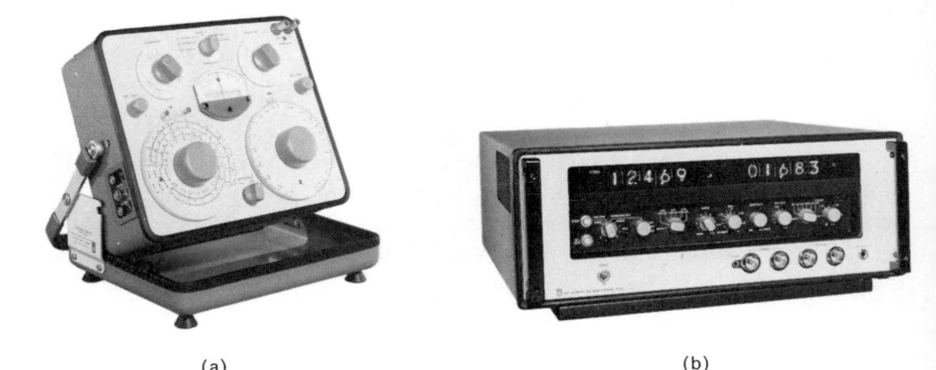

(a) (b)

Figure 7-15. Universal impedance bridges (a) GR-1650B (b) Automatic *RLC* bridge GR-1883 (Courtesy of General Radio Co.)

MEASUREMENT OF INDUCTANCE WITH AN AC VOLTMETER

A quick measurement of inductance can be made by using an ac voltmeter and the 115-V, ac voltage from the power line. The method is not nearly as accurate as measuring inductance with a bridge, but it does produce values that are adequate for many applications. Its chief advantage is that the measurement can be made with instruments that are commonly available in the electronics laboratory.

The method consists of connecting the unknown inductance in series with a variable resistor R (as shown in Fig. 7-16). The total ac voltage is applied across the series connection, and the voltage across both elements is equalized. The equalization is carried out by first measuring the voltage across the inductor and then connecting the meter across R_1. The resistance is adjusted

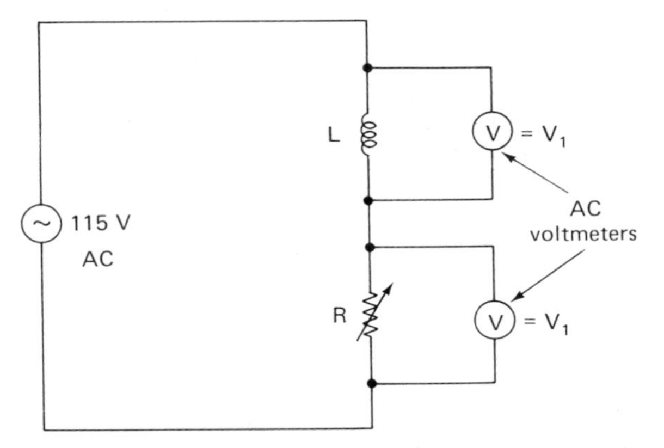

Figure 7-16. Measurement of an unknown inductance with ac voltmeters

until the voltage across it is equal to the voltage measured across the inductor.

Table 7-3 Summary of the Specifications of Three Typical Universal Impedance Bridges

	R	C	L	Percent Accuracy	D	Q	Int. Source Frequency
GR 1650-B	1 mΩ– 1.1 MΩ	1 pF– 1100 μF	1 μH– 1100 H	1%	0.001–50	0.02–1000	DC, 1 kHz
John Fluke 710B	10 mΩ– 12 MΩ	1 pF– 1200 μF	1 μH– 1200 H	1%	0–1.05	0–1000	DC, 1 kHz
Hewlett-Packard 4260-A	10 mΩ– 10 MΩ	1 pF– 1000 μF	1 μH– 1000 H	1%	0.001–50	0.002–1000	DC, 1 kHz

When both voltages are equal, the impedances of the inductor and the resistor are also equal. Therefore we can equate them and use the resulting relation to calculate the value of the inductance. The relation for finding L is

$$L = \frac{\sqrt{R^2 - r^2}}{2\pi f} \tag{7-20}$$

where R is the value of the adjustable resistor at balance, r is the measured value of the dc resistance of the coil, and f is 60 Hz (or any other conveniently available frequency).

EXAMPLE 7-3

An unknown inductor is connected in series with an adjustable resistor as shown in Fig. 7-16. The dc resistance of the inductor had previously been measured to be 100 Ω. When the series connection is hooked up across the 60-Hz power line, a condition of equal voltage across both elements exists when R is set to 3200 Ω. Find the value of the inductance.

Solution: Using Eq. (7-20) we find that

$$L = \frac{\sqrt{R^2 - r^2}}{2\pi f} = \frac{(3200^2 - 100^2)^{1/2}}{6.28 \times 60} = \frac{3150}{377} = 8.4 \text{ H}$$

Q MEASUREMENTS

The quality factor, Q, of a coil was defined as the ratio of the reactance of the coil to its resistance, or

$$Q = \frac{\omega L}{R} \tag{7-15}$$

where ω is the test frequency ($\omega = 2\pi f$) and R is the effective series resistance of the coil. The effective resistance often differs markedly from the dc resistance because of such ac-associated resistance effects as eddy currents and the skin effect. Therefore, R varies in a highly complex manner with frequency. For this reason Q is rarely calculated by determining R and L. Instead, R is determined indirectly from a measurement of Q.

One way in which Q can be measured is by using the commercial inductance bridge. But since the circuits of inductance bridges are rarely capable of producing an accurate measurement when Q is high, special meters designed to yield accurate values are built for measuring Q.

The circuit of the common Q-meter is shown in Fig. 7-17. A variable-frequency voltage source (with a very low impedance) is used to drive a series connection of a capacitor and the inductor under test. Since the resistance of the inductor is in series with its inductance, this connection makes a series

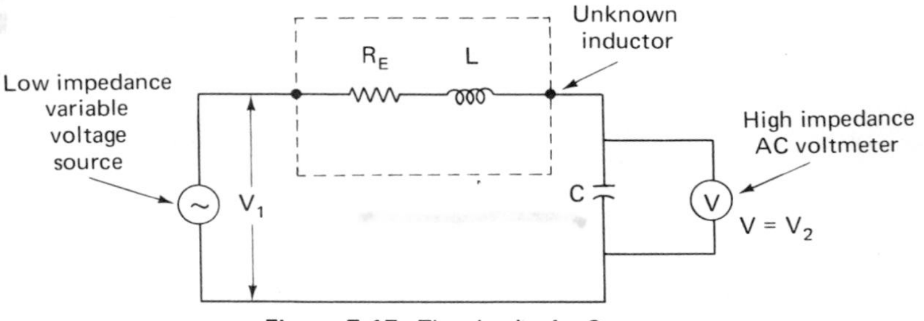

Figure 7-17. The circuit of a Q-meter

RLC network. By varying the frequency of the voltage source, we can find the resonant frequency, ω_0, of the circuit. At resonance the capacitive reactance and the inductive reactance of the circuit cancel, leaving only the effective resistance, R_E, of the inductor to limit the current. Thus, at resonance the current in the entire circuit (rms value) will be found from

$$|I| = \frac{|V_1|}{R_E} \tag{7-21}$$

Since the current must have the same magnitude in the entire circuit, the current in the capacitor, I_C, will also be equal to

$$|I| = |I_C| = \omega_o C |V_2| \tag{7-22}$$

Therefore, in connection with the rapid increase of I at resonance, a high-impedance voltmeter placed across the capacitor will detect a sharp increase in $|V_2|$ at resonance. We also know that at the resonant frequency, ω_0

$$\omega_o^2 = \frac{1}{LC} \tag{7-23}$$

$\omega_0 \rightarrow$ resonance

Therefore, by combining Eqs. (7-21), (7-22), and (7-23), we see that

$$\left|\frac{V_2}{V_1}\right| = \frac{\omega_o L}{R_E} = Q \tag{7-24}$$

In commercially available Q-meters, a continuously variable oscillator with a very low output impedance acts as the voltage source. The meter also contains a high-impedance VTVM and a variable capacitor. The scale of the VTVM is calibrated to indicate Q directly. A separate scale is often provided for low Q (i.e., between 0–10). The frequency range of the oscillators is typically 50 kHz to 50 MHz, with special oscillators available for lower and higher frequencies. The accuracy of such Q-meters varies from about 1 to 5 percent, depending on the magnitude of Q and the resonant frequency. A more detailed account of the operation and use of Q-meters can be found in such texts as Cooper[7] or in the operating manuals accompanying each instrument.

Transformers

Transformers are devices which are designed to transfer electric energy from one circuit to another. They achieve this transfer by utilizing a magnetic field which intersects both circuits. In addition to performing such energy transfers, transformers are also capable of delivering a different value of ac current or voltage at their output terminals than the value applied to their input terminals.

The ability of transformers to step up or step down ac voltages and currents is used in electric power distribution systems and in ac metering applications. This property is also utilized in electronic circuits for providing high voltage levels or for matching impedances. The capability of transferring energy by using magnetic fields allows transformers to be used to create isolation connections. (Isolation connections are connections in which no physical joining between the circuits is necessary.) To understand how a transformer performs all these functions, we must examine the principles of its operation in more detail.

The transformer operates by using the electrical phenomenon of *mutual inductance*. Mutual inductance is the effect that occurs when the magnetic field of one element also influences other elements in its vicinity. The result of such magnetic coupling is that currents and voltages are induced in the nearby elements. Although mutual inductance may be an undesirable effect in some cases, the operation of a transformer depends upon using this effect to its fullest extent.

The transformer consists of two coils (called the *primary* and the *second-*

[7]William D. Cooper, *Electronic Instruments and Measurement Techniques* (Englewood Cliffs, N.J.: Prentice-Hall, Inc., 1970), pp. 292–302.

ary) wound around a common core of magnetic material (Fig. 7-18). If a current flows in the primary winding, it sets up a magnetic field which is largely restricted to the magnetic core around which the primary is wound. If another winding (called the *secondary*) is also wound on the same core, the magnetic field will also link the secondary winding. If the current in the primary is steady (dc), it will not affect the secondary coil because the magnetic field will also be constant. In particular, no current will flow in the secondary coil.[8]

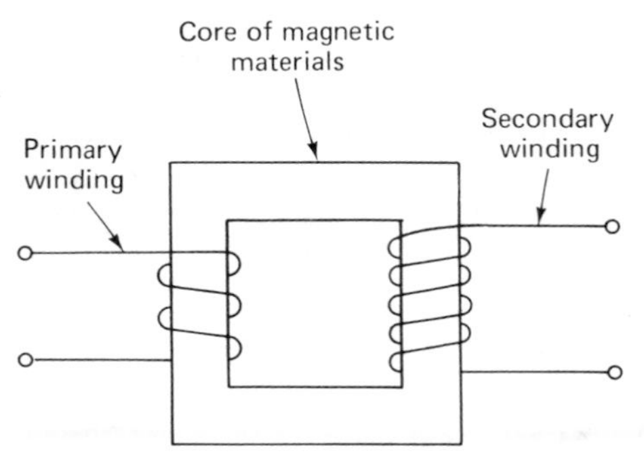

Figure 7-18. Diagram of a transformer

If the current in the primary is a changing (ac) rather than a steady current, the magnetic field in the core will also change. Since a changing magnetic field appears to a conductor as a moving magnetic field, the free charges in the conductor of the secondary coil experience a force. Since they are mobile, these free charges will move under the influence of the force, and a current will flow. In this manner, a changing current in the primary will cause a current to flow in the secondary of a transformer.

The current flow induced in the secondary also has a voltage associated with it. Faraday's law says that the magnitude of the voltage induced by the changing of magnetic flux in a coil of N turns is given by

$$v = NK\frac{d\phi}{dt} = M\frac{di}{dt} \tag{7-25}$$

where M is called the mutual inductance that exists between the coils.

In the ideal transformer, all the magnetic flux created by the primary coil also links the secondary. Then the voltage at the terminals of the secondary is dependent on the rate of change of current in the primary:

$$V_2 = M\frac{di_1}{dt} \tag{7-26}$$

[8]There must be either a motion of charges in a magnetic field or a moving or time-varying magnetic field in order for the charges to experience a magnetic force.

The ratio of the number of turns on the primary to the number on the secondary is an important quantity and is written as:

$$\text{Turns ratio} = \frac{N_p}{N_s} \tag{7-27}$$

In this equation, N_p is the number of turns on the primary and N_s is the number of turns on the secondary. It is the *turns ratio* that determines how much a transformer steps up or steps down a voltage (see Fig. 7-19).

In fact, the ratio of the voltage across the primary (V_p) to the voltage across the secondary (V_s) is equal to the turns ratio

$$\frac{V_p}{V_s} = \frac{N_p}{N_s} \tag{7-28}$$

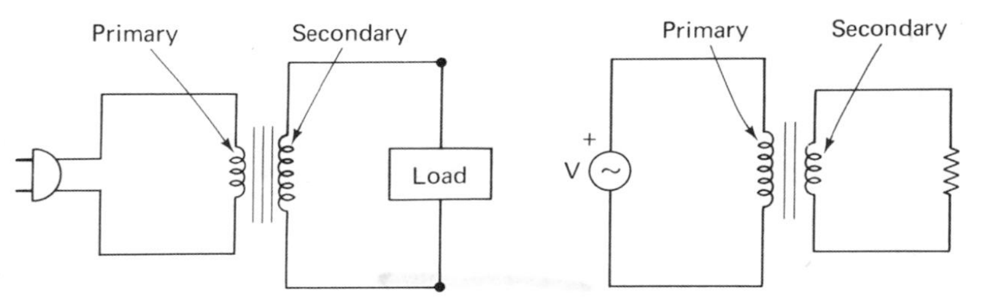

Figure 7-19. (a) Voltage step-up transformer $N_p < N_s$ (b) Voltage step-down transformer $N_p > N_s$

EXAMPLE 7-4

We are given a transformer with a 1 : 10 step-up ratio. If this transformer is connected to the power line, what will be its output voltage?

Solution: Using Eq. (7-28) and the fact that the power line voltage is about 115 V, we have

$$\frac{V_p}{V_s} = \frac{N_p}{N_s}$$

$$V_s = V_p \times \frac{N_s}{N_p} = 115 \times \frac{10}{1}$$

$$V_s = 1150 \text{ V}$$

In practice, the value of the output voltage of the transformer examined in Example 7-4 will be somewhat smaller than the calculated result. There are always some resistance and core losses that exist and reduce the ideal output from the secondary coil. This loss can be anticipated by the transformer designer and he can add a few turns to the secondary to ensure that the input-to-output voltage ratio will be as specified.

Although the transformer can increase or decrease voltage, it does this by correspondingly decreasing or increasing the current. This means that a transformer which has a step-up voltage ratio also has a step-down current ratio. For example, a transformer that steps up voltage by a factor of five steps down the current in the secondary to a value of one-fifth the current in the primary. Mathematically, this is expressed by

$$\frac{I_p}{I_s} = \frac{N_s}{N_p} \tag{7-29}$$

where I_p and I_s are the currents in the primary and secondary coils.

In addition to stepping up and stepping down voltages and currents, a transformer can make the impedance of a load which is connected to its secondary appear to have a different value. If such an impedance is measured from the primary winding, it will appear changed by an amount that depends on the construction of the transformer. This allows the transformer to be used as an impedance-matching device. For example, a load resistance connected across the secondary winding of a transformer can appear to have a different resistance value if it is looked at through the coils of the primary. Quantitatively, the load resistance R_L appears to be equal to R_{eq}, as calculated by

$$R_{\text{eq}} = \left(\frac{N_p^2}{N_s^2}\right) R_L \tag{7-30}$$

EXAMPLE 7-5

What does the impedance of a 10-Ω load connected across the secondary appear to be if we are looking at it from the primary winding of an 8: 1 step-down transformer?

Solution: From Eq. (7-30):

$$R_p = \left(\frac{N_p}{N_s}\right)^2 R_s$$

$$= (64)(10)$$

$$R_p = R_{eq} = 640\ \Omega$$

As pointed out in Chapter 3, a typical application of impedance matching occurs when we desire to match the impedance of the output of an amplifier to that of a loudspeaker. The transformer makes the low impedance of the loudspeaker appear to be much higher, so that it matches the higher output impedance of the amplifier. Therefore, a maximum amount of power is transferred to the speaker.

In addition to matching impedances, the transformer can also connect two parts of a circuit without an *electrical* connection. Since only *magnetic* effects actually link the two parts, they remain *isolated electrically* from one another. This is a very useful property in such applications as restricting a high dc voltage level to one part of a system. When an isolation transformer is used, the ac component of a signal can be coupled between two parts of a circuit, while the dc level is kept from being transferred. (Remember that dc quantities are not passed by a transformer.)

NAMEPLATE DATA AND DOT CONVENTION

The symbols used for transformers are shown in Fig. 7-20. The polarities of the coils of the transformer are indicated by the dot convention. They show that the dot-marked terminals have the same voltage polarity at every instant of an ac cycle and are used in helping to connect a transformer properly to a circuit. (i.e., when the current flows into the dotted side of the primary, the dotted side of the secondary will have a positive voltage relative to the undotted end.)

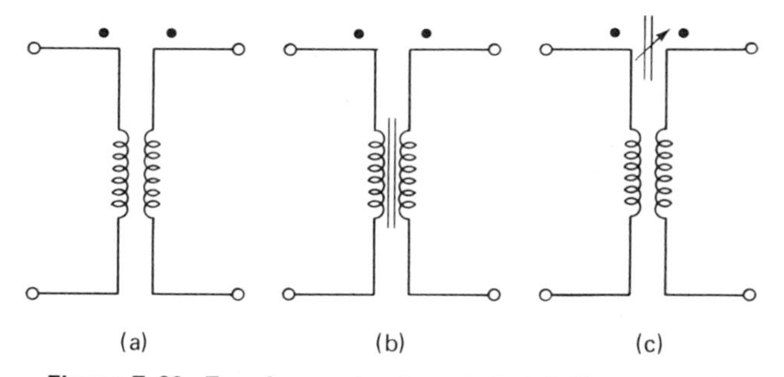

Figure 7-20. Transformer circuit symbols (a) Air-core transformer (b) Iron-core transformer (c) Variable-core transformer

On most transformers there is a *nameplate* which specifies their normal operating conditions. A typical nameplate contains data of the following sort:

Transformer: 5 kVA, 4400/220 V, 60 Hz

The information on this nameplate indicates that 4400 V and 220 V are the maximum voltages that can be applied to each of the windings. That is, if the 4400-V winding is used as the primary, a maximum of 4400 V can be applied to the transformer. Likewise, when the 220-V winding is used as the primary, a maximum of 220 V can be applied to it. Furthermore, this information indicates that the turns ratio is 20: 1. The 60-Hz value refers to the fact that the transformer is operating at the knee of its magnetization curve when 4400 V or 220 V are applied to the respective windings. If the frequency of the current applied to the primary is decreased far below

60 Hz, the flux needed to maintain the primary voltage, V_p, would have to rise. Since the transformer is already operating on the knee of its *B-H* curve, the current in the primary would have to increase greatly to maintain the value of this flux. Such a high current could burn out the coil wire insulation. Thus, the applied frequency should not be allowed to drop far below the frequency stated in the nameplate.

The kVA rating specifies how much output can be continuously maintained without excessively heating the transformer.[9]

LOSSES IN TRANSFORMERS

Although there are losses associated with the operation of actual transformers, they are usually not severe if the unit is well-designed. In fact, the ideal transformer model described in this chapter closely approximates the operation of actual, commercially available transformers. However, for completeness, we will list some of the mechanisms that prevent actual transformers from acting exactly like ideal devices. These major loss effects are the following:

1. Loss due to the resistance of the wires that make up the primary and the secondary windings.
2. Leakage—Not all the magnetic flux created by the primary winding links the secondary winding. (In a well-designed transformer, leakage remains quite small.)
3. Core storage—The material of the core stores magnetic energy (which is then not transferred from one circuit to the other).
4. Core losses—Losses due to hysteresis and eddy currents arising in the transformer core.
5. Core nonlinearity—Since the core material of a transformer saturates at some maximum flux density (B_M), this places a limit on the maximum voltage that can be induced in any given winding.

Types of Transformers

Transformers come in many shapes and sizes, depending on the specific application in which they are to be used. However, most transformers share the property of being wound on a ferromagnetic core and possessing one primary and one or more secondary windings. If there is a large transformer ratio, the high-current winding is usually wound of heavy-gauge wire to reduce resistance losses. The low-current winding is then wound of fine wire. For purposes of shielding, most transformers are enclosed in a suitable

[9]The reason why the kVA rather than the power rating (kW) is used is explained in the section on ac power in Chapter 10.

ferromagnetic or copper housing (depending on the frequency the transformer is designed to handle).

Most common transformers are referred to by either the frequency range they can handle or by the type of circuit application for which they are best suited. As a result, the categories of transformers most likely encountered in measurement work include:

1. Audio-frequency (af) transformers.
2. Radio-frequency (rf) transformers.
3. Power transformers.
4. Autotransformers.
5. Instrument transformers.

AUDIO-FREQUENCY TRANSFORMERS

Audio-frequency (af) transformers are used to couple signals whose frequencies fall in the 20–20,000-Hz range. Af transformers are used in microphones, loudspeakers, and audio amplifiers. Since the frequencies of af transformers are relatively low, losses due to eddy currents and hysteresis are not very high. Therefore silicon-steel laminations are suitable for use in their cores. Af transformers are usually relatively large units and Fig. 7-21 shows a common style of their construction.

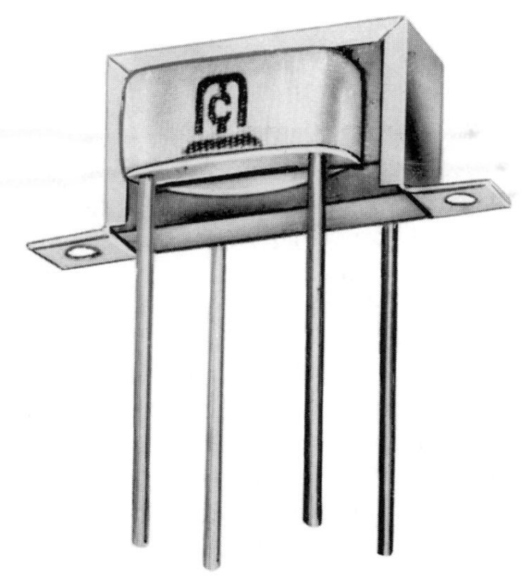

Figure 7-21. Open frame audio-frequency transformer (Courtesy of Microtran Co., Inc.)

RADIO-FREQUENCY TRANSFORMERS

When transformers are to be used to couple high-frequency signals (i.e., rf frequencies), the losses due to eddy currents would become too high if laminated cores were used. Thus, as in high-frequency inductors, powdered-iron or ferrite cores are employed. In some rf transformers, the shape of the core can be made in a threaded form (like a screw). The core can then be turned into or out of the transformer for tuning applications.

For very high frequencies (greater than 50 MHz), the losses in any type of magnetic core may become so great as to render the core more of a hindrance than an aid. In such cases air cores are often used. (Even though they are called air cores, the windings may really be wound around ceramic material.) Rf transformers are usually shielded to reduce undesirable coupling and feedback effects. Figure 7-22 shows some typical rf transformers with and without their shields.

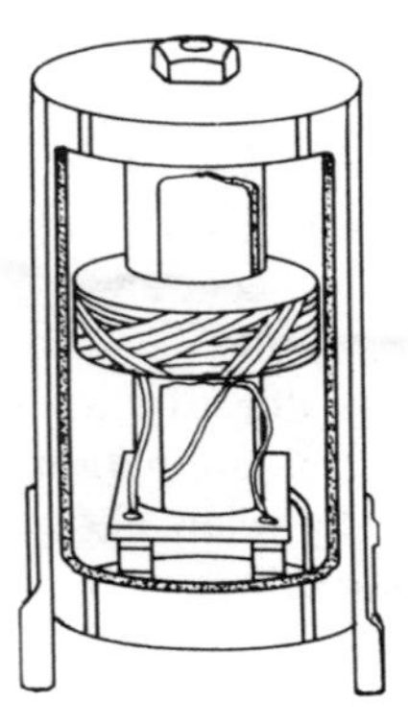

Figure 7-22. Radio-frequency transformer

POWER TRANSFORMERS

Power transformers are used to step up and supply high voltages to the various parts of communication equipment or measuring instruments drawing 60-Hz power. Usually a single power transformer can feed several different elements that require such voltages. Thus power transformers can have one primary and several secondary windings.

Figure 7-23(b) shows a schematic of such a power transformer connected on its primary side to the 115-V, 60-Hz power line. There are several secondaries—two for 350 V, one for 10 V, and one for 60 V. The wire in these transformers is enamel- or plastic-insulated to prevent conduction or arcing between coil turns (particularly in the high-voltage secondary windings). Power transformers are also shielded to prevent the magnetic field which they generate from producing 60-Hz hum signals in nearby elements. Sometimes an electrostatic shield is also employed between the primary and secondary coils to prevent rf and noise voltages from entering through the power lines.

ISOLATION TRANSFORMERS

Isolation transformers are used to couple parts of circuits without electrical connections. The turns ratio for an isolation transformer is usually 1 : 1,

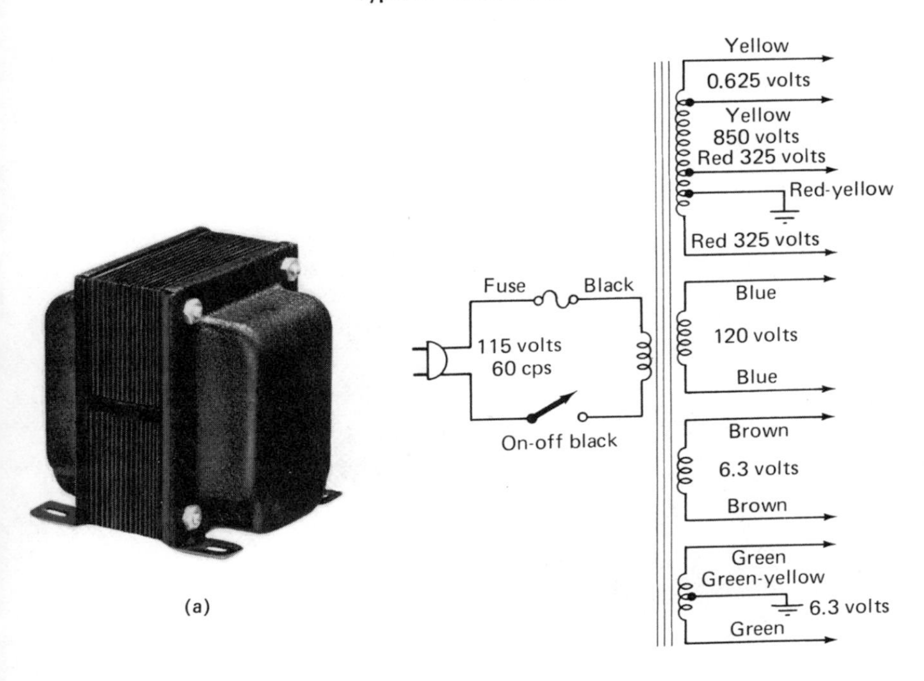

Figure 7-23. Power transformer used in an oscilloscope (a) Power transformer photograph (b) Schematic (Courtesy of Microtran Co., Inc.)

and it is rated in *watts* to indicate the magnitude of the load that can be connected safely to its secondary. (Isolation transformers contain no common lead as do autotransformers, which are described in the following section.)

AUTOTRANSFORMERS

Autotransformers are an exception to the rule that no electrical connection exists between the primary and secondary windings of a transformer. In the autotransformer, the same coil serves as both the primary and the secondary winding. The single winding has a *tap* which can be connected anywhere along the length of the winding. If the transformer is to be used to step down voltages, the entire length of the coil is used as the primary (Fig. 7-24(b)). The part between the tap and the bottom end acts as the secondary. If step-up action is required, the entire coil is used as the secondary (Fig. 7-24(a)). Special variable autotransformers known as *variacs* or *powerstats* are also available for use where manual regulation may be required. Figure 7-25 is a cutaway photograph of a variac.

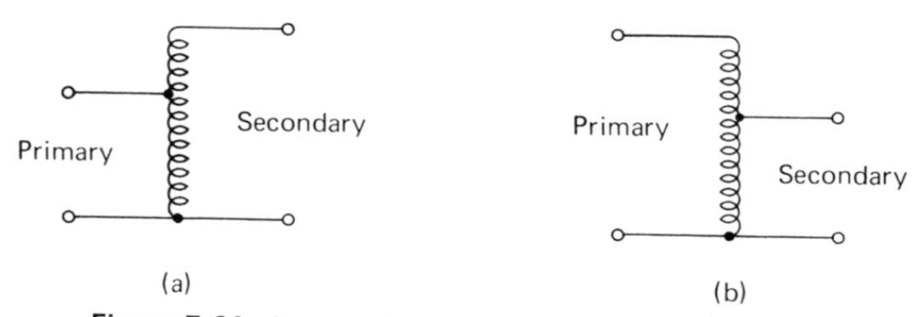

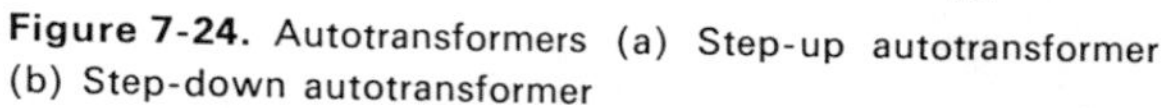

Figure 7-24. Autotransformers (a) Step-up autotransformer (b) Step-down autotransformer

Figure 7-25. Cutaway of a Variac (Courtesy of General Radio Co.)

INSTRUMENT TRANSFORMERS

Transformers that are utilized in connection with instruments are referred to as *instrument transformers.* Current transformers are the most common type of instrument transformer, although voltage transformers are also occasionally seen. Instrument transformers perform two principal tasks. They (a) extend the range of an instrument and (b) isolate an instrument from high-voltage lines.

Ac ammeters use current transformers rather than shunts to extend their ranges because shunts are less accurate in dividing ac currents. The reactance associated with a meter and a shunt causes the ac currents to be divided differently at different frequencies. When a current transformer is used, this difficulty does not occur.

For high-voltage ac measurements, it is better to use an instrument transformer to step down the voltage rather than a high-valued series resistor in a voltmeter. To see why this is so, consider the problem of trying to monitor high-voltage transmission lines (which may be carrying 300,000 V or more).

It may be necessary to meter these high voltages, as well as the somewhat lower voltages of generators, on switchboard meters. To prevent these high voltages from constituting a safety hazard to the switchboard watchers, the voltages must be stepped down to a much lower value. Instrument transformers are used to accomplish such reductions.

A typical instrument transformer is shown in Fig. 7-26. By using its various binding-post terminals, we can obtain four different ratios of input to output currents. In addition, the primary wire can be placed through the hole in the case to obtain additional current step-down ratios. For example, in the Weston model shown in Fig. 7-26, one turn looped through the opening gives a current ratio of 800/5 between the primary and secondary windings. Two loops through the hole makes the ratio 400/5, and so on.

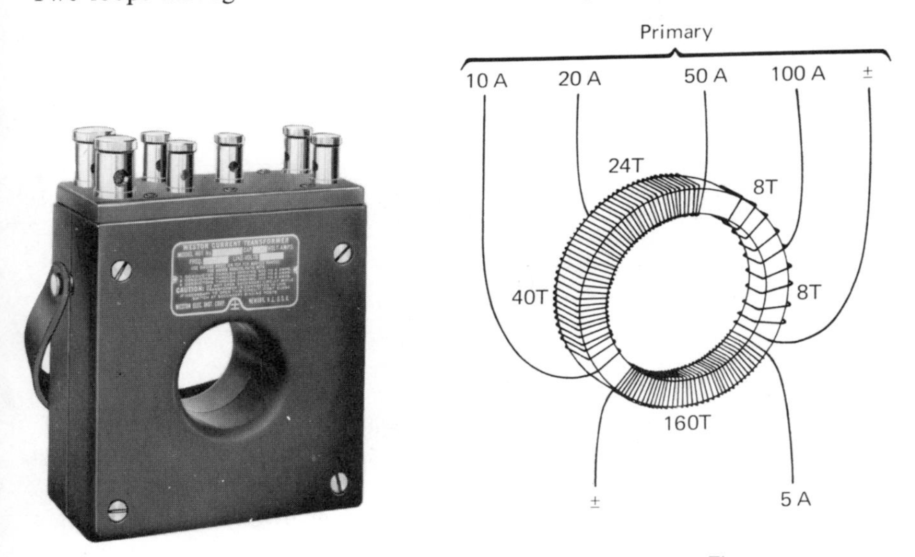

Figure 7-26. Current transformer to extend ac ranges. The secondary winding for the basic instrument range (5 A) is shown at the bottom winding and the four extended current ranges are obtained from taps on the primary winding, shown at the top. (Courtesy of Weston Instruments, Inc.)

PRECAUTIONS TO BE FOLLOWED WHEN USING INSTRUMENT TRANSFORMERS

Never connect the primary of current transformer unless the secondary is already connected to some circuit (e.g., ammeter, wattmeter, etc.). Not only does an open secondary become a highly dangerous piece of equipment, but the transformer itself may be ruined if it is left open-circuited. If the secondary is open, a condition of very high voltage exists across it, even though a current transformer is considered to be a low-voltage device. This can easily lead to an electrical accident.

Electromagnets and Relays

As mentioned earlier, when a coil is wound around a straight core of magnetic material and a steady current is applied to the coil, the magnetic field of the entire device has the same pattern as the field of a permanent magnet (Fig. 7-27). Since the magnetic field of this device only arises when an electric current is passed through the coil, the device is called an *electromagnet*.

The strength of the field of an electromagnet can be varied by changing either the current or the number of turns of the coil. In any case, a relatively weak current in the winding can cause a strong magnetic field if the wire is wound around a core of good magnetic material. (The discussion on magnetic materials earlier in the chapter dealt with the relation between the magnetic field strength, B, and the current, i.)

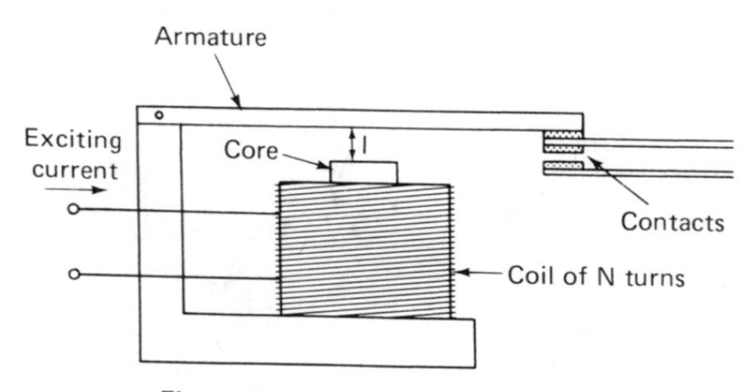

Figure 7-27. Electromagnetic relay

Since a bar magnet exerts a force on nearby ferromagnetic materials, such an attractive force will also be created by an electromagnet when a current is passed through its windings. If this force is used to attract and move a pivoted piece of metal called an *armature* and if the motion of this armature is used to open and close electric contacts, the assembly is called an *electromagnetic relay*. (In most relays, when the current is stopped, the spring action of the armature causes the contacts to return to their original positions.)

The nomenclature of relays follows that of the switch nomenclature described in Chapter 3. That is, if the relay has one armature, it is known as a single-pole relay. If relays have two, three, or more armatures, they are called *double-pole*, *triple-pole*, *or multipole relays*, respectively. If each armature merely opens or closes one circuit, the relay is called a *single-throw* type. On the other hand, when the armature is double-acting (i.e., when it breaks one circuit while simultaneously making another), it is called a *double-throw relay*. Like switches, relays are thus classified as single-pole, single-throw; double-pole, double-throw; etc. The contacts of the relay that are open when no current is passing through the relay are called *normally open* (NO) contacts. Those that are closed when no current is passing are called *normally closed* (NC) contacts. Figure 7-28 shows two common relay models.

(a) (b)

Figure 7-28. Electromagnetic relays (Courtesy of Potter and Brumfield)

Relays are used in a tremendous number of industrial and electronic applications. They play a key role in the operation of automated machinery. Some common consumer applications of relays include the voltage regulators in automobiles and the release mechanism which opens and closes doors activated by door buzzers. Relays also are responsible for the clicking noises heard when dishwashers, air conditioners, and washing machines are being shifted from one cycle to another during their operation. Figure 7-29 shows another application of how an "electric eye" is used with a relay to open and close a door.

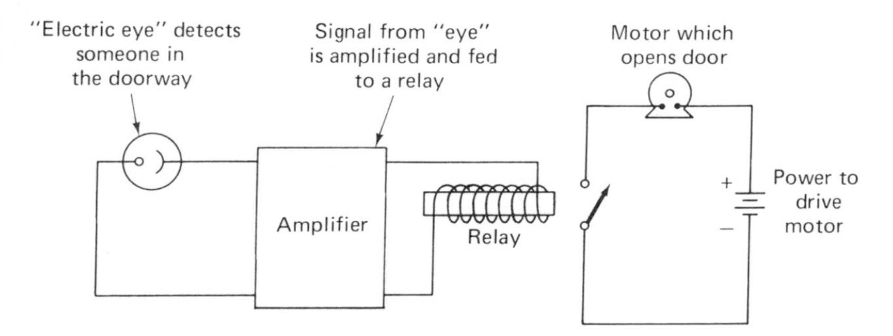

Figure 7-29. Circuit for "electric-eye" controller of doorway opening

The force that is developed by the electromagnet on the armature located at a distance l from one end of the magnet is given by

$$f = \frac{B^2 A}{2\mu_o} = \mu_o \frac{N^2 i^2 A}{2l^2} \tag{7-31}$$

where: A = the area of the gap (in meters squared),
N = number of turns in the coil,
μ_o = permeability of free space = $4\pi \times 10^{-7}$ H/m,
l = length of the gap (in meters), and
i = current in amperes.

EXAMPLE 7-6

A relay consists of an electromagnet of 4000 turns, wound on a core whose cross section is 1 cm². The gap when the contacts are open is 3 mm, and it is 1 mm when they are closed. A current of 50 mA flowing in the coil is necessary to activate the relay and close the contacts. How much force is exerted on the armature when the relay is activated?

Solution: Using Eq. (7-31) we can calculate the force acting on the armature (in newtons):

$$f = \frac{\mu_o N^2 i^2 A}{2l^2}$$

$$= \frac{4\pi \times 10^{-7}(4 \times 10^3)^2(5 \times 10^{-2})^2(0.01)^2}{2 \times 9 \times 10^{-6}\ \text{m}}$$

$$f = 0.27\ \text{N}$$

Since 1 newton = 0.224 lb, then

$$f = 0.27 \times 0.22 = 0.06\ \text{lb} \approx 1\ \text{oz}$$

Another commonly used type of relay is the *reed relay* shown in Fig. 7-30. In the single-pole, single-throw type shown in this figure, two ferromagnetic

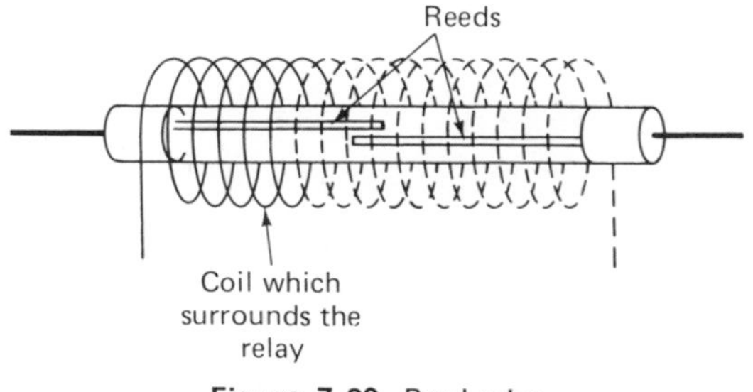

Figure 7-30. Reed relay

reeds are separated by a small distance. These reeds are mounted in a hermetically sealed glass tube, and a coil surrounds the glass tube. When a current is passed through the coil, the magnetic field of the coil magnetizes the two reeds. One reed is thereby made a north pole, and the other becomes a south pole. When enough attractive force is developed, the reeds are drawn together and a connection is established. When the current is halted, spring action due to the stiffness of the reeds forces them apart again. The contacts of the reeds are gold-plated to provide low-resistance contacts.

Problems

1. Explain why the current flowing in an inductor does not rise immediately to its maximum value when a voltage is applied across the inductor.
2. If the flux linking a coil which has 100 turns changes at a rate of 0.170 Wb/sec, what is the induced voltage across the coil?
3. What is meant by "cutting lines of force"? Describe three ways in which a conductor and a magnetic field are in relative motion.
4. a) An induced voltage of 3 V is developed in a coil when the current in it is changing at a rate of 0.5 A/sec. What is the inductance of the coil?
   b) What would be the inductance of the above coil if the induced voltage were:
   1) 10 V
   2) 0.03 V
   3) 0.0005 V
5. If a coil with 100 turns has an inductance of 250 μH, how many turns of the coil must be removed for it to have an inductance of 200 μH?
6. For the electromagnet shown in Fig. P7-1, find the flux density in the core.

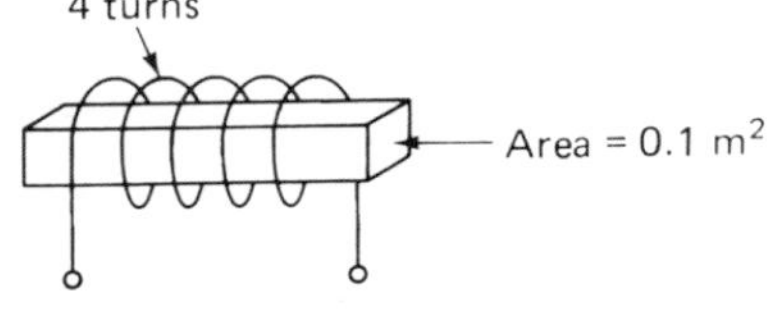

Figure P7-1 $\phi = 4 \times 10^{-4}$ Wb/m^2

7. Find the inductance of the inductor shown in Fig. P7-2.

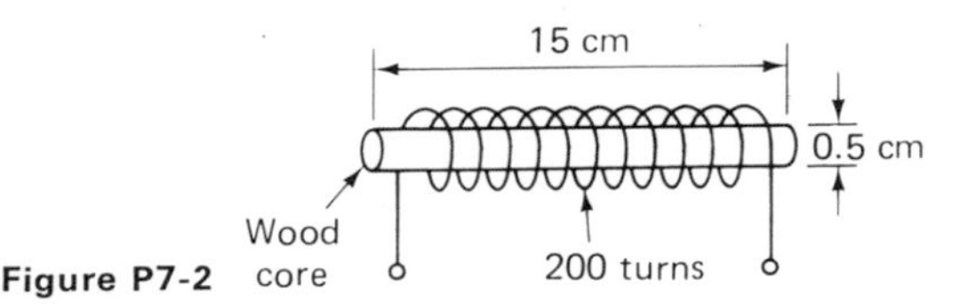

Figure P7-2

8. Find the inductance of the inductor shown in Fig. P7-3.

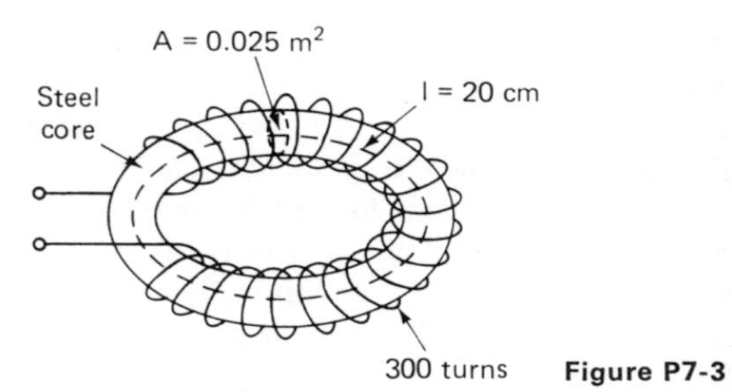

Figure P7-3

9. What is the effect of the following actions on the inductance of a coil?
   a) Reducing the number of turns by one half.
   b) Using an iron core rather than an air core.
10. Explain how an inductor stores energy.
11. What is "hysteresis"? Explain when hysteresis losses can become serious in inductors and transformers.
12. List and explain the operation of three types of variable inductors.
13. What are "eddy currents" and how are they produced? How does a laminated core in a transformer reduce eddy currents?
14. Explain why eddy current losses can be high at radio frequencies. What types of cores minimize these losses?
15. What is the reactance of a 50 mH coil at a) 2500 Hz, b) 2500 kHz?
16. At what frequency will a 500 μH coil have a reactance of 1500 Ω?
17. What happens to the value of inductive reactance if
    a) the inductance is tripled?
    b) the frequency is doubled?
    c) the frequency is reduced to one-fifth?
18. Why are inductors and transformers often enclosed in metal containers? What types of metals are usually used for such containers?
19. What is the significance of the Q of an inductor?
20. If a 350 μH inductor has a resistance of 75 Ω at 5 MHz, what is the Q of the coil?
21. For the transformer shown in Fig. P7-4, find the magnitude V_2 of the induced voltage.

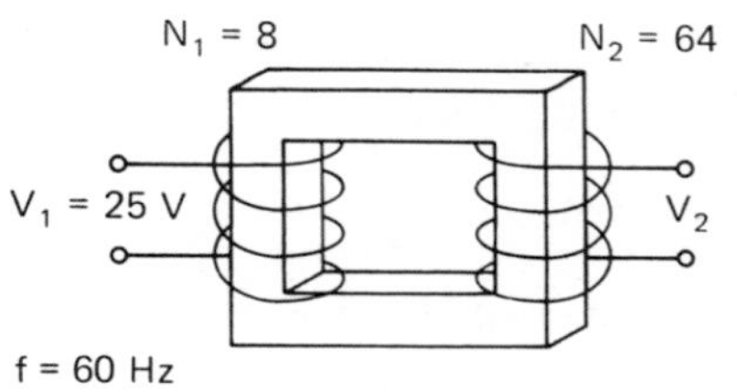

Figure P7-4

22. Repeat Problem 21 for $N_1 = 240$ and $N_2 = 30$.
23. Give a short explanation of the effect of mutual inductance.
24. An ideal transformer is rated at 10 kVA, 2400/120 V, 60 Hz.
    a) Find the transformation ratio if 120 V is the secondary voltage.
    b) Find the current rating of the primary if 120 V is the secondary voltage.
    c) Repeat parts a) and b) if 2400 V is the secondary voltage.
25. Describe the operation of a *relay*.

References

1. BUKSTEIN, E., *ABC's of Coils and Transformers*. Indianapolis, Ind., Howard W. Sams & Co., Inc., 1968.
2. KINNARD, I. F., *Applied Electrical Measurements*, Chap. 8. New York, John Wiley & Sons, Inc., 1956.
3. DEFRANCE, J. J., *Electrical Fundamentals*, 3rd ed., Chaps. 18 and 19. Englewood Cliffs, N.J., Prentice-Hall, Inc., 1969.
4. COOPER, W. D., *Electronic Instrumentation and Measurement Techniques*, Chap. 8. Englewood Cliffs, N.J., Prentice-Hall, Inc., 1970.
5. BOYLESTAD, R., *Introductory Circuit Analysis*, Chaps. 10 and 22. Columbus, Ohio, Charles E. Merrill Books, Inc., 1968.
6. STOUT, M. B., *Basic Electrical Measurements*, 2nd ed., Chaps. 9, 10, 12, 13, 14, and 15. Englewood Cliffs, N.J., Prentice-Hall, Inc., 1960.

8

Batteries, dc Power Supplies, and Standard Cells

All circuits and electrical equipment require sources of electric power. Some require ac power, others dc power. Electronic amplifiers that use vacuum tubes or transistors all require *dc power* to allow them to amplify electric signals (whether these signals are in dc or ac form). Since amplifiers are utilized in oscilloscopes, electronic voltmeters, radio transmitters and receivers, and a host of other electronic equipment, dc power sources are required for operating such instruments. However, the electric power commonly delivered from generating stations to the home and laboratory is 60-Hz ac power (see Chapter 10 for more details on how electric power is distributed). In order to get dc power for the electronic devices described, another source must be used or the available ac power must be converted to dc form. The *battery* is the most common alternative source of dc power. It utilizes energy from electrochemical reactions to supply this power. On the other hand, the device used to convert the 60-Hz line power to dc power at the desired current and voltage levels is called a *dc power supply*. The power supply can be constructed as a separate piece of laboratory equipment. It may also be designed to be an integral component of equipment that requires dc power but is operated from the ac power line.

Solar cells are also sources of dc power. They convert solar energy to dc power directly. The conversion process makes use of certain semiconductor properties which allow the dc power to be generated.

The *standard cell* is a special type of battery which is used as a voltage standard or reference in making accurate voltage measurements. It is designed to be used only for such reference applications and is *never* to be used as a power source. The precise voltage output of the standard cell would be upset if even a small amount of current were passed through it. Nevertheless, if used correctly, standard cells provide extremely stable and long-lasting sources of constant voltage. *Zener diodes* can also be used as voltage standards. In some applications requiring voltage references, they are replacing standard cells. However, Zener diodes are not actual energy sources. They require an external energy source to provide their voltage output.

Batteries

Batteries were the first available sources of dc power and were therefore used almost exclusively to power early electronic circuits (amplifiers, radios, etc.). As other dc sources were developed, batteries were replaced because of their expense, limited life, and large size. For example, early radio amplifiers needed 90 volts or more to power their vacuum tubes. Such high voltages required the use of many cells in series—an expensive, heavy arrangement. Consequently vacuum-tube power supplies became almost universally used to power electronic equipment.

Recently, the development of the transistor and improved battery technology have begun to make batteries popular again. Far less voltage is required to drive circuits which use transistors instead of vacuum tubes (i.e., 2–10 V vs 90 V). In addition, new advances in battery technology have led to longer battery life and more power per volume of battery. These factors now make batteries attractive for powering portable instruments and for use in equipment located where a central line voltage is not available. In some applications, they are also employed as reserve power sources in the event of accidental power-line failures.

The battery functions by releasing electrochemical energy and converting it to dc electrical energy. The basic principle of battery operation relies on the relationship between its two different metal *electrodes* and the *electrolyte* in which they are immersed. Figure 8-1 gives a summary of the process involved in the functioning of a battery.

The potential difference due to the charge difference that exists between two unconnected electrodes of a battery varies according to the materials used for the electrodes and the electrolyte. For example, in the zinc-carbon cell, the potential difference developed is about 1.5 V, while in the lead storage battery, a 2.2-V potential difference exists.

The reactions taking place at an electrode sometimes yield a gas as their product. If this gas forms bubbles that cling to the electrode surface, they can have a deleterious effect on the battery operation. Gas is a poor conductor, and if it surrounds the electrode, it will prevent further ions from wandering near to it. The chemical reactions and hence the current flow in the battery will slow down. Such gas formation is called *polarization* and is counteracted by substances called *depolarizers*. Depolarizers react with the gas to form a liquid, keeping the electrolyte in its well-conducting state.

Each configuration of two electrodes and electrolyte is called a *cell*.

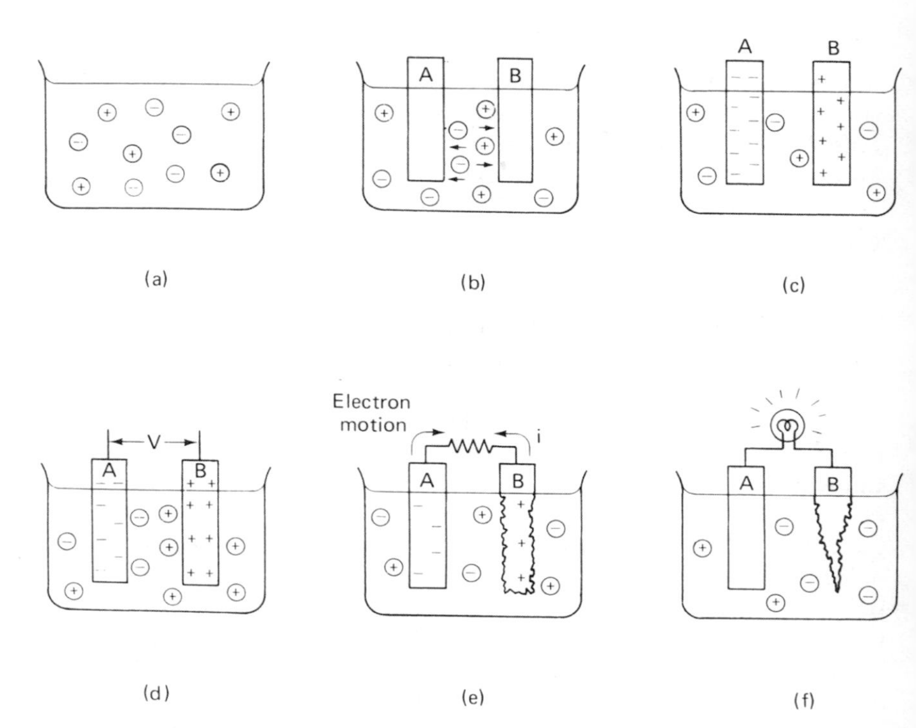

Figure 8-1. Principles of battery operation (a) An electrolyte is a liquid which breaks up some of its molecules into positive and negative ions. The positive ions are short of electrons; the negative ions possess extra electrons. The entire liquid remains electrically neutral as the positive and negative ions cancel each other's effects. (b) To construct a battery cell, we insert two rods of special materials into the electrolyte. These rods are called the *electrodes* of the battery. Their chemical properties are such that one of them (B) reacts with the negative ions of the electrolyte and the other (A) reacts with the positive ions. (c) The reactions of the negative ions with electrode A result in electrons being deposited into the electrode, giving it a negative charge. The reactions at electrode B remove electrons and make it positively charged. The ions come into contact with the electrodes through the random motion of the liquid molecules. (d) If no external electrical connection is made to the two electrodes, the growing charge difference between the electrodes leads to a potential difference (voltage) and an electric field between them. The electric field is directed so that the ions needed to sustain the reactions are repelled by their respective electrodes. This eventually makes the chemical reactions stop. (e) If the electrodes are externally connected, electrons will flow from the negative electrode to the positive one. This will reduce the field between the electrodes and ions can again come into contact with their desired electrodes. The current-feeding reactions can continue. (f) The energy released by the chemical reactions is transferred to the electrons. This energy is used to power the load connected to the battery (i.e., a lightbulb of a flashlight). The reactions continue as long as material in the electrodes and electrolyte is available.

Batteries are made of one or more cells. (The term battery is derived from "battery of cells.") The circuit symbol for a single-cell battery is $\overset{+}{\dashv}\vdash$. Batteries consisting of two or more cells in series are shown as $\overset{+}{\dashv}||\vdash$. When two or more cells are connected in series, the voltage across the entire connection is equal to the sum of the voltages of each of the individual cells. When cells of equal voltage are connected in parallel, the voltage output of the assembly will still be the same as the voltage of one cell. However, the current output of the assembly will be equal to the sum of the individual current outputs.

Cells are categorized in different groupings. One distinction is made between *wet* cells and *dry* cells. The difference between the two types is that the electrolyte in wet cells is completely liquid, whereas the electrolyte in dry cells is mixed with other substances to form a semiliquid or moist paste. The common zinc-carbon flashlight battery is a dry cell which has an electrolyte of ammonium chloride and zinc chloride. The lead storage (automobile) battery is a wet-cell battery which uses sulfuric acid (H_2SO_4) diluted in water as its electrolyte.

A further distinction is made between *primary* and *secondary* cells. In a *primary cell*, a decomposed electrode cannot be restored to use again by charging. The chemical reactions that led to its decomposition are irreversible. A *secondary cell* possesses the property that once the battery is discharged and its electrodes are partially decomposed into other compounds, it can be restored to its initial chemical state by means of another external energy source (i.e., a power-supply battery charger). The external source recharges the battery by passing an electric current between the terminals of the battery in a direction opposite to the battery's normal current-flow direction.

The size of the battery determines the total amount of energy it can deliver (the more electrolyte and electrode material, the longer the energy-generating reactions can be sustained). The quantity termed the *capacity* of a battery indicates the number of *ampere-hours* (A-hr) a battery can deliver before its terminal voltage drops below some designated level. A battery with a 100-A-hr rating will theoretically deliver a steady current of 100 A for one hour, 10 A for ten hours, etc.

The actual output of a battery is also affected by the temperature and rate of its discharge. A very rapid discharge, a drop in temperature, or long storage before use will lead to a decrease in the actual A-hr rating. For example, an automobile battery rated at 50 A-hr at 2A should deliver 2A for 25 hours. If the discharge rate is increased to 25 A, the A-hr rating may fall by as much as 25 percent to 38 A-hr. A drop in temperature from 80°F to 30°F will also drop the A-hr rating of such a battery by 20 percent.

Another way to measure a battery's capacity is by specifying the total amount of energy (in joules) it can provide before its voltage drops below a specified value. Using this measure, a D-type (flashlight) zinc-carbon cell can deliver 2×10^4 joules under conditions of moderate discharge (i.e., no more than about 100 mA is drawn from the battery at any time). An automobile battery can provide about one hundred times as much energy as a zinc-carbon flashlight battery.

Most batteries have a limited *shelf life*. This quantity is usually defined as the period of time required to reduce battery voltage to a specified percentage (usually 90 percent) of its original voltage value if the battery is not used.

A graph showing the discharge characteristics of various D-size batteries versus time is shown in Fig. 8-2.

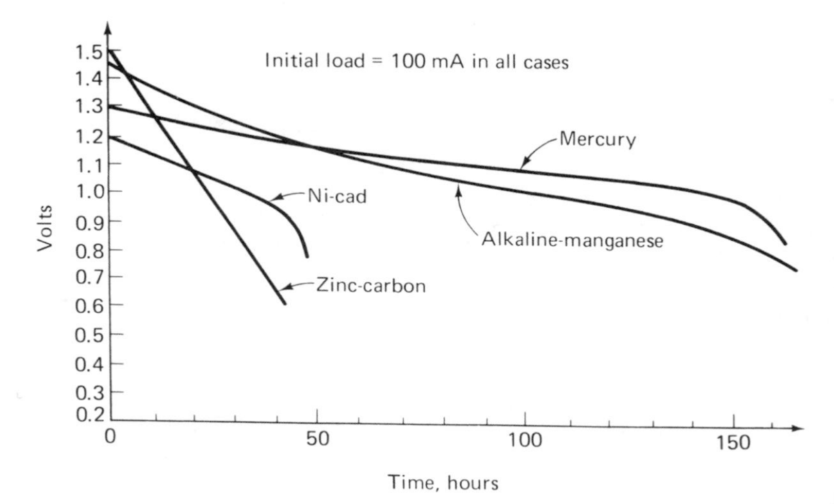

Figure 8-2. Discharge characteristics of D-size cells (Courtesy of B. D. Wedlock and J. K. Roberge, *Electronic Components and Measurements*, Englewood Cliffs, N.J.: Prentice-Hall, Inc., 1969, p. 136)

BATTERY INTERNAL RESISTANCE

The battery is used as a voltage source and is treated as such in analyzing a circuit. However, the electrical characteristics of actual batteries do not exactly equal those of an ideal voltage source. The main difference between them is that an ideal voltage source is defined as having zero resistance, whereas an actual battery always exhibits some resistance. This resistance exists because the electric current generated in a battery must first flow in the electrolyte and electrodes before reaching the external circuit. Since these component parts of a battery are not perfect conductors, the current is presented with an impedance before it gets to the external circuit. This impedance is called the *internal resistance* of the battery. In the circuit model used to represent actual batteries, the internal resistance is shown as a resistor in series with the ideal voltage source (Fig. 8-3(a)).

The value of the internal resistance varies with the condition and age of the battery. For example, the internal resistance of a freshly prepared zinc-carbon dry cell is about 0.05 Ω, while the internal resistance of the same cell that has aged a year without use may be 100 Ω or more. When the internal resistance of a battery becomes too large, the voltage drop across this resistance becomes so great as to render the battery useless as a voltage source.

To measure the internal resistance of a battery experimentally, we can use a circuit such as the one shown in Fig. 8-3(b). The resistor R should have a resistance of about 10 Ω and a power rating of $\frac{1}{2}$ W (or more). For accurate results, a digital voltmeter should be used to perform the measurement (although any other voltmeter can also be used).

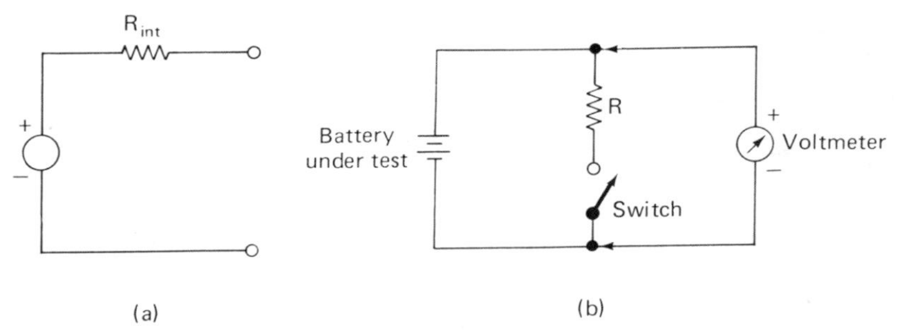

Figure 8-3. (a) Equivalent-circuit model of a battery (b) Circuit used for measuring the internal resistance of a battery

The voltage across the battery terminals with the switch open is measured first. This voltage is called the *open-circuit voltage* of the battery (V_{oc}). Then the switch is momentarily closed and the voltage is measured again. The second voltage value is referred to as the *loaded voltage*, V_L, (or terminal voltage). The internal resistance of the battery R_{int} is then found from the expression,

$$R_{int} = R\left[\frac{V_{oc}}{V_L} - 1\right] \tag{8-1}$$

EXAMPLE 8-1

If a battery connected as shown in Fig. 8-3(b), is found to have a V_{oc} of 1.50 V and a V_L of 1.41 V, what is the internal resistance of the battery (assuming $R = 10\ \Omega$)?

Solution: Using Eq. (8-1), we obtain

$$\begin{aligned} R_{int} &= R\left[\frac{V_{oc}}{V_L} - 1\right] = 10\left[\frac{1.50}{1.41} - 1\right] \\ &= 10[1.06 - 1] \\ &= 0.6\ \Omega \end{aligned}$$

EXAMPLE 8-2

If the same battery has aged and now has an internal resistance of 100 Ω, what will be the value of V_L if we use the measuring circuit of Fig. 8-3 (with $R = 10\ \Omega$)?

Solution: The open-circuit voltage of the battery V_{oc} remains at 1.50 V. Then using Eq. (8-1) and solving for V_L, we obtain

$$R_{int} = R\left(\frac{V_{oc}}{V_L} - 1\right)$$

$$V_L = \frac{V_{oc}}{(1 + R_{int}/R)}$$

or

$$V_L = \frac{1.5}{(1 + 100/10)} = \frac{1.5}{11}$$

$$V_L = 0.14\text{ V}$$

Common Battery Types

ZINC-CARBON CELL

The zinc-carbon cell is a primary dry cell developed by Georges Le Clanche in 1868. When freshly prepared, it has a terminal voltage of about 1.55 V. Dry-cell batteries of the zinc-carbon type are manufactured with voltage values of 1.5, 3, 6, 7.5, 22.5, 45, 67, and 90 V. Such batteries are very commonly used because of their low cost and general applicability. However, they do suffer from the following disadvantages. First, their capacity is very dependent on the rate of discharge. Second, they are hampered by temperature limitations (at temperatures below freezing they are quite ineffective).

The electrodes of this battery (Fig. 8-4) are made of carbon and zinc.

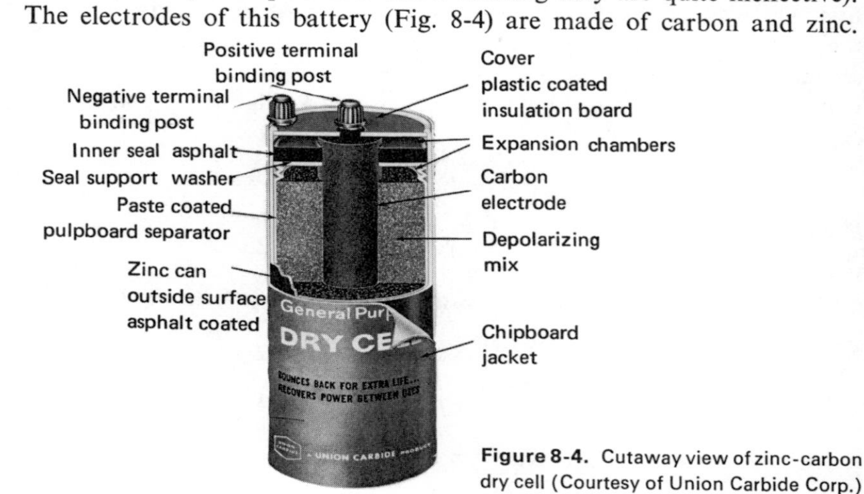

Figure 8-4. Cutaway view of zinc-carbon dry cell (Courtesy of Union Carbide Corp.)

The electrolyte is a paste made of ammonium chloride, NH_4Cl, and zinc chloride, $ZnCl_2$. Hydrogen gas forms around the carbon electrode during discharge, and it is the job of the depolarizer (manganese dioxide, MnO_2) surrounding the carbon electrode to remove it. The MnO_2 absorbs the H_2^+ and converts it to water, thus preventing the H_2^+ from seriously reducing the battery life. The zinc electrode meanwhile decomposes and forms $ZnCl_2$. When the zinc is all consumed, the battery ceases to supply energy.

Unfortunately, the zinc continues to slowly decompose even if the terminals are unconnected. This action causes eventual deterioration of the battery output, and after about one or two years of storage, the battery becomes useless. Furthermore, since the zinc electrode forms part of the outer wall of the cell, its dissolution weakens the cell structure. As the evolved hydrogen-gas pressure builds up, it can rupture a weak cell and spill its corrosive contents. Therefore equipment should never be stored with such dry cells installed.

A zinc-carbon cell operates best under intermittent use. During "off" periods the depolarizer is able to reduce the hydrogen gas that forms around the carbon electrode. Thus after some "rest" the cell again presents a lower internal resistance to current flow.

LEAD STORAGE BATTERY

The lead storage battery, which is commonly used in automobiles, is made of secondary wet cells. An example of such a battery is shown in Fig. 8-5. Each lead cell has a terminal voltage of about 2.2 V when fully charged. The 6- and 12-volt requirements of automobile electrical systems are met by placing three or six cells in series. Because the lead battery is made of secondary cells, it can be recharged to its initial state hundreds of times after partial discharge. However, lead batteries are quite bulky and require considerable maintenance in order to give proper service.

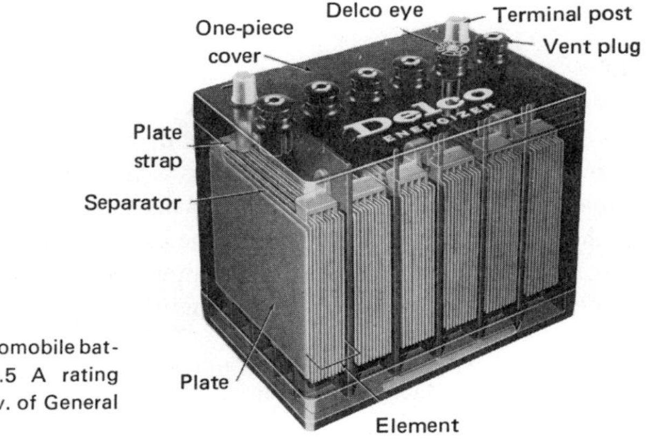

Figure 8-5. Lead storage automobile battery; 12 volt, 70A-hr at 3.5 A rating (Courtesy of Delco-Remy, Div. of General Motors)

One electrode of their cells is lead, Pb (in a porous, sponge-like form), and the other is lead peroxide, PbO_2 (Fig. 8-6). The electrolyte is sulfuric acid (H_2SO_4) diluted in distilled water. The electrolyte decomposes into positive ions of hydrogen H^+ and negative ions of sulfate $SO_4^=$. The Pb electrode combines with the sulfate ions to form lead sulfate $PbSO_4$. As a result of this reaction, electrons are deposited in the Pb electrode. If there is an external connection made between the Pb and PbO_2 electrodes, these electrons flow through the connection to the PbO_2 electrode. The PbO_2 uses the electrons to react with the sulfate ions of the electrolyte. This reaction also forms lead sulfate ($PbSO_4$) as well as oxygen ions ($O_2^=$). The oxygen ions combine with the hydrogen ions of the electrolyte to form additional water. Because the electrolyte loses sulfate ions and gains water, the concentration of the sulfuric acid decreases. Thus a nearly discharged battery will have a very diluted electrolyte. The *specific gravity*[1] of the electrolyte is a measure of its sulfuric-acid-to-water ratio. By determining the specific gravity of its electrolyte, we can tell how much a lead battery is discharged. A fully charged lead battery has a specific gravity of 1.30, while a discharged battery has a specific gravity of less than 1.05. (Pure water has a specific gravity of 1.00.)

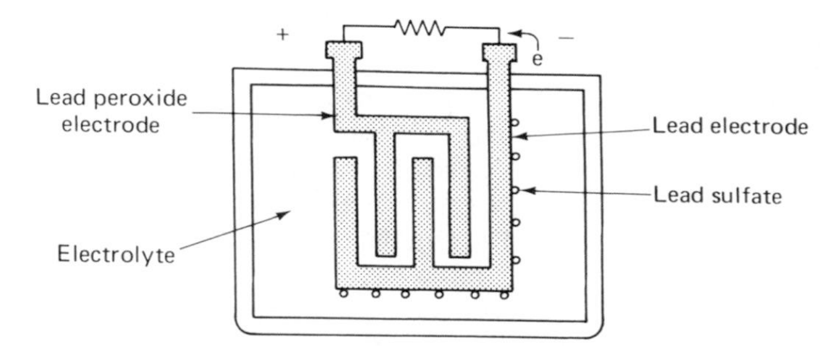

Figure 8-6. Schematic of lead storage cell

When the lead battery is discharged, it can be recharged again. The action of recharging decomposes the lead sulfate on the electrodes and restores them again to lead and lead peroxide. Meanwhile the electrolyte also regains its full strength. However, if the lead sulfate is allowed to remain on the battery plates for a long time, it tends to harden and will no longer be convertible to lead and lead oxide. Thus a lead battery in storage should be kept either fully charged (no sulfate on the plates) or empty of electrolyte. Because an unused battery will discharge internally within a few months, a stored battery containing electrolyte must be periodically recharged.

[1] Specific gravity is defined as the ratio of the weight of a given volume of a substance to an equal volume of pure water.

LEAD BATTERY CHARGING AND CARE

In order to recharge a secondary cell, current must be passed between the terminals of the cell in a direction opposite to the direction of current flow that delivers power to the load. This fact points out that an energy source must be able to supply a unidirectional current in order to charge a battery. Thus, an ac generator would not be suitable as a charger if used alone. (The output of an ac generator is sinusoidal and therefore changes direction sixty times per second.) However, if a rectifier is used with the ac generator output, the ac sinusoid can be converted to a unidirectional current. In this way the rectifier allows the ac generator (and thereby the ac power delivered to consumers) to serve as an energy source of a battery charger.

During the charging of a battery, the current level must not be so excessive as to evolve gas and deteriorate the electrodes. However, when a battery is completely discharged, it can stand a higher charging rate than when it is almost fully charged. If one wants to charge a battery quickly, a high charging rate can be applied until the battery is 85 percent charged. Then a reduced finishing rate can be used to complete the charging process. Most battery chargers are regulated so that they apply current in approximately this manner.

Since hydrogen and oxygen are evolved during charging, the battery should be well ventilated. The two gases together make an explosive mixture. Furthermore, only distilled water should be used to replace battery water. Iron and chlorine found in common tap water will decrease the performance and life of the battery.

MERCURY BATTERY

The mercury battery shown in Fig. 8-7(a) is a primary cell which has the advantage of keeping a relatively constant terminal voltage throughout its useful life. This voltage stability makes the mercury battery especially attractive for use in those devices which require a specific voltage to operate properly. Occasionally, mercury batteries are also used as voltage references in measurement circuits. When a battery is exhausted, its internal resistance rises drastically (within two to three hours of the end of its 150-hour life). The capacity in ampere-hours exceeds that of a comparably sized zinc-carbon cell by about a factor of 3. The chief disadvantages of mercury batteries are their greatly decreased capacity at temperatures below 40°F and their high price (about five times the cost of an alkaline battery of equal size).

One electrode of the mercury battery is made of mercuric oxide and carbon, and the other of zinc amalgam. The electrolyte is sodium or potassium hydroxide. The resulting chemical reactions produce a battery voltage of either 1.35 V or 1.40 V.

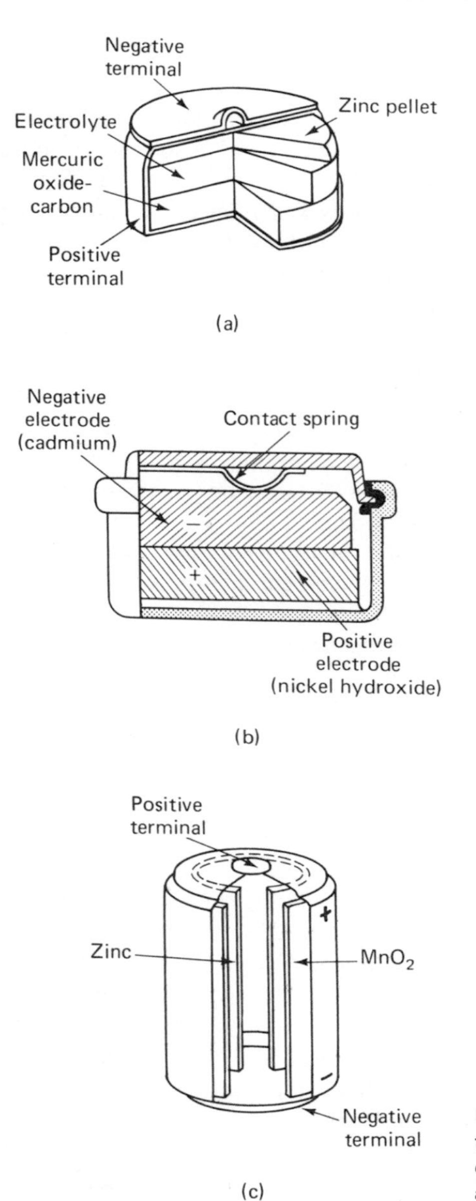

Figure 8-7. Construction of various battery types (a) Mercury cell (b) Nickel-cadmium button-type cell (c) Manganese-alkaline cell

NICKEL-CADMIUM BATTERY

Nickel-cadmium batteries (Fig. 8-7(b)) are secondary cells that are easily recharged with a simple battery charger. They are very well-suited for use as power sources in battery-operated appliances. Under normal use, these batteries can be recharged several hundred times, and they are also quite unharmed by large loads, overcharging, being left discharged, and low temperatures. Their terminal voltage is about 1.25 V, and they can be made in many different sizes and capacities. Although the initial cost of the nickel-cadmium battery is quite high (about 25 times the price of zinc-carbon batteries of comparable size), their cost per use drops to a very low sum as they are recharged many times. For example, D-size Ni-Cad cells and a charger can provide a flashlight with indefinite battery life. Their electrodes are nickel hydroxide and cadmium, and potassium hydroxide is used as the electrolyte.

MANGANESE-ALKALINE BATTERY

Like the nickel-cadmium battery, the manganese-alkaline battery (Fig. 8-7(c)) represents one of the recent developments in battery technology. It can be made as either a primary or secondary cell, but the primary type is more common. This battery is very similar in construction and terminal voltage (1.5 V) to the carbon-zinc cell, but it uses an electrolyte (potassium nitrate) that is strongly basic rather than acidic.

The electrodes of the manganese-alkaline battery are powdered zinc and manganese dioxide. Together, these components give the alkaline cell a lower internal resistance, longer shelf-life and larger energy capacity than that of the carbon-zinc cell. In addition, alkaline cells can operate at temperatures as low as −40°F. Although their initial cost is greater, the larger capacities of alkaline cells make their cost per ampere-hour lower than that of carbon-zinc cells.

Solar Cell

Solar batteries convert light energy directly to electric energy in the form of dc power. They are called solar cells because their primary source of energy is usually meant to be solar radiation. Solar cells are used as power sources largely in space electronics. However, several other applications such as powering rural telephone systems and transistor radios have also been developed. Solar cells are also used as light sensors in some photographic light meters. In this application their output current is used to deflect a D'Arsonval galvanometer.

Solar cells convert light energy to electric energy through a special semiconductor diode device. This diode is designed to generate current and voltage when illuminated with light. Because of the optical properties of the materials used to construct the diodes, they are only moderately efficient in converting solar energy to an electrical form (10 to 15 percent efficiencies are typical). That is, the ratio of the power output of a solar cell to the power of the light incident upon it is relatively small. Because the current and voltage output of each individual cell is also quite low, solar cells are usually connected together in series or parallel combinations. In this way, some satellite batteries have been able to generate power outputs of 10–15 watts per pound of battery.

Research on improving the operating characteristics and applications of solar cells is continuing. The relative abundance of their chief ingredient (silicon), their ease of manufacture, and the vast amount of energy available from solar radiation make them important candidates for expanded use in future energy-conversion techniques.

dc Power Supplies

In order to make the 115-V, 60-Hz electric power suitable for driving dc electrical equipment, a *dc power supply* is used. The name may be misleading because the power supply does not actually generate power; it only converts ac power to an approximate dc voltage or current. Most dc power supplies are made to provide constant voltages rather than currents. This is because most applications require sources that have electrical characteristics which resemble those of voltage sources rather than current sources. Once dc voltage power supplies are set, they keep a constant voltage output even as the current requirement changes. Such constant-voltage supplies are commercially available with ranges from 0 to 4 volts up to 0 to 3000 volts. To see how such power supplies operate, let us examine Fig. 8-8.

From this figure we see that the incoming 115-V ac voltage is first stepped up or down by a transformer (see Chapter 7 for a description of transformer operation). The initial voltage change enables the power supply to deliver

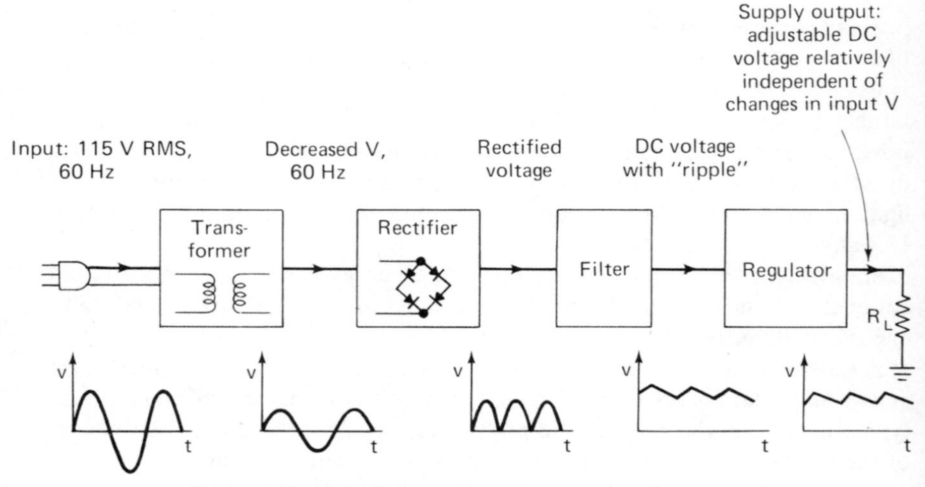

Figure 8-8. Block diagram of dc-voltage power-supply operation

voltages which may be greatly different from the 115-V power-line voltage. Next, the transformed voltage (still at 60 Hz) is fed to a rectifier (discussed in Chapter 4 on Rectifier ac Meters). The rectifier converts the ac voltage to a rectified dc voltage. (The output of the rectifier shown in Fig. 8-8 is a full-rectified waveform, although half-wave and peak rectifiers are also available). The rectified waveform is then fed to a filter which smooths out the variation or *ripple* in the rectified waveform. Filters can eliminate most, but not all, of the ripple in a signal. Therefore the output of a filter is a dc quantity with a small residual ac component superimposed on it (Fig. 8-9). However, if sufficient filtering has been used, the magnitude of the remaining ac component is small enough that it will not impair the operation of the device which is drawing power from the supply. The magnitude of the remaining ripple voltage in the output of a power supply is expressed in terms of its rms value. Typical commercial supplies have ripple voltages in their outputs which range from less than 0.1 mV to about 10 mV.

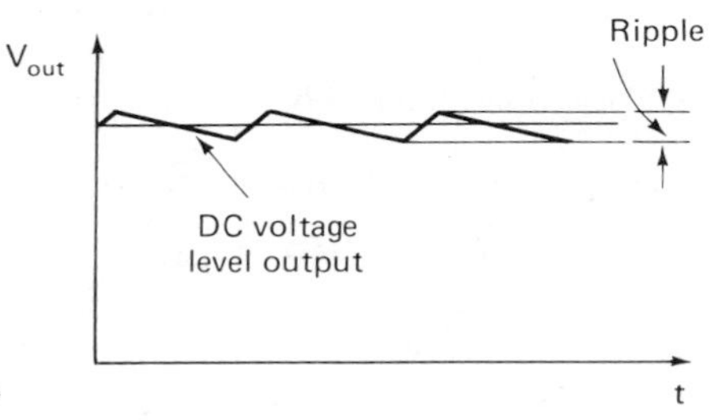

Figure 8-9. dc waveform with ripple

The filtered waveform could now be employed as the dc output of the power supply, but it becomes more useful with the addition of one more modification. As is, the value of the voltage output is fixed by the input voltage level and the transformer construction. There is no way to adjust the magnitude of the output level. Furthermore, any variations in the power line voltage or load current could vary the dc output level. For instruments which depend upon a constant voltage level, this variation (possibly greater than the variation due to the remaining ripple) might be too large to allow proper functioning. For such reasons, a regulation device must be installed to allow adjustments of the output level and help keep the output value constant once the level is chosen. With the help of the *regulator*, the supply output is a dc voltage that is adjustable over the designed range.

The ability of a power supply to keep a voltage constant once an output level is chosen is referred to as its *regulation* ability. This quantity is expressed by the percentage of change the output will undergo for each one percent of change of the input voltage. Another way of expressing regulation is to specify the maximum percentage change in output voltage for a particular change in input voltage (usually from 105 V to 125 V). Typical values of regulation ability in various power supplies allow about 0.05 percent change in the output for a variation of 20 volts (105 V to 125 V) at the input.

Note that if the power supply is to be used as a voltage source, it should have an output impedance (V_{out}/I_{out}) as close to zero as possible. Conversely, those power supplies that are designed to be constant-current sources should have an almost infinite output impedance. Typical commercial power supplies acting as voltage sources have output impedances of less than 0.01 Ω at 60 Hz.

SAFE OPERATION AND CURRENT LIMITING OF POWER SUPPLIES

When using a constant-voltage power supply, the current requirement of the load can vary. If the required current becomes too high, the power flowing in the regulator of the supply ($V_{out} \times I_{load}$) may reach a value so large that the supply regulator could be burned out. To guard against such an occurrence, *electronic current limiters* are included in the regulator circuits. They limit the maximum current flowing in the output, regardless of the output voltage required of the supply. By doing this, they ensure that the power supply is being operated safely (in the sense that it is kept from burning out).

ADDITIONAL CONTROL FEATURES OF POWER SUPPLIES

There are some applications that require a power supply to provide power to a point that is at a fairly distant location from the supply itself. Thus, wires must carry the power from the supply to that point. When high currents are required by such a distant load, the voltage at the load may differ from the voltage at the output terminals of the power supply. This change occurs because the wires connecting the supply output to the load may drop a measurable percentage of the supply voltage before it gets to the load. Therefore, the load voltage will be dependent on the connections to the load as well as the output of the supply. The regulation at the load may not be as good as at the power supply output. To overcome this problem, a voltage reading can be taken at the load and this value can be fed back to the supply by an additional pair of wires. Similar to the way that the voltage output of the supply is regulated internally, this voltage monitor at the load can control the output of the supply. It will maintain the desired voltage at the load regardless of the drop that occurs in the connections. This feature is available in some supplies and is called *remote sensing*.

DC POWER SUPPLY SPECIFICATIONS

dc Output—Describes the range of dc voltages or currents available from a particular power supply.

ac Input—Describes the characteristics of the ac voltage required to drive the power supply. Usually the required ac input is 115 V ac ± 10 percent,

50–63 Hz. However, some supplies are built to also operate over other voltages and frequencies.

Load Regulation—The change in the dc output voltage resulting from a change in the load resistance from zero (short circuit) to the value that results in the supply's maximum rated output voltage.

Line Regulation—The change in the dc output voltage of the supply resulting from a change in the input line voltage from its lowest to its highest values (usually the maximum change allowed is from 105 V to 125 V).

Ripple and Noise—Describes the rms value of the ac component which remains unfiltered and superimposed on the dc output.

Output Impedance—For a supply designed as a constant-voltage source, the output impedance should be very small ($\rightarrow 0$). For a constant-current supply, this output impedance should be very large ($\rightarrow \infty$).

Temperature Rating—The temperature range over which the supply can operate and remain within its capabilities.

HOW TO USE A POWER SUPPLY

Figure 8-10 shows a typical power supply and the functions of the controls on its case. Let us examine this figure to see how a power supply is used. The supply should first be plugged into a three-terminal power outlet. The *Power* switch turns on the supply. The supply output voltage is obtained by

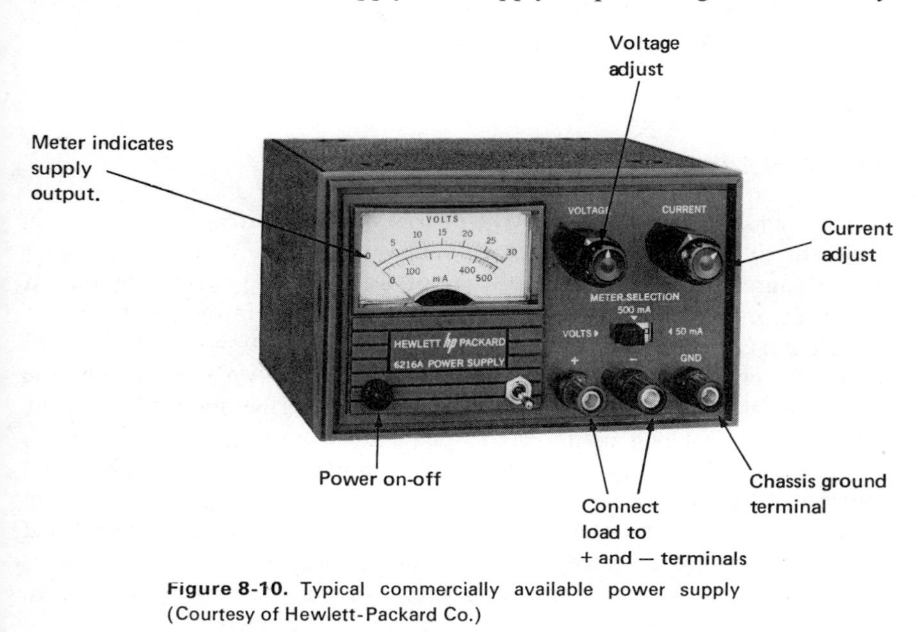

Figure 8-10. Typical commercially available power supply (Courtesy of Hewlett-Packard Co.)

connections to the "+" and "−" terminals of the supply. The equivalent circuit of the supply as seen from these terminals is shown in Fig. 8-11(a).

The equivalent circuit shows that the ground terminal is connected to the third-wire of the power-line cord through the chassis, but it is not connected internally to either the + or − terminal. If a load is connected between the + and ⏚ terminals alone, the output voltage of this connection will be zero. To get a voltage output relative to ground, the ground terminal must be separately connected to either the + or − terminal (depending on the polarity desired). See Fig. 8-11(b) and (c). With a proper connection made to the supply terminals, the output voltage is adjusted by means of the *Voltage Adjust* control. This output is indicated on the meter scale. The range of the meter scale is controlled by the *Meter Range* switch. The *Short Circuit Current* control indicates the maximum output current of the supply.

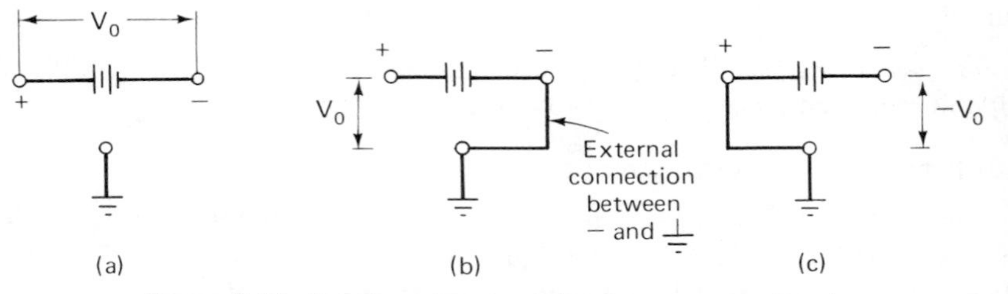

Figure 8-11. How to make connections to power supply output (a) Ungrounded voltage V_o between + and − terminals (b) With extra ground wire connected as shown, a positive voltage V_o relative to ground exists between + and − terminals (c) With extra ground wire connected as shown, a negative voltage V_o relative to ground exists between + and − terminals

Standard Cells and Zener Diodes

The standard cell is a special type of battery which provides an extremely accurate voltage reference for measurements, but is *never* to be used as a power source. It can maintain a fixed voltage over long periods of time if cared for properly. The standard cell is primarily used in the potentiometer circuit as a calibration source and its role in this application will be discussed in a later chapter dealing with potentiometers. Here we will examine the principles of its operation and precautions that must be observed when the standard cell is being used.

The standard cell was developed by Weston and adopted by the National Bureau of Standards as the standard for measuring voltages. It is built in two versions, the saturated (or normal) Weston cell and the unsaturated standard cell. The saturated cell is the more accurate of the two and yields a voltage which is within 1 μV of its stated value at 20°C. The unsaturated cell is more portable and better suited as a working reference.

The voltage of the saturated Weston cell at 20°C is 1.01830 V. This voltage is so reproducible that the legal definition of a volt in this country is 1/1.01830 of the potential of this cell at 20°C. A picture of the standard cell is shown in Fig. 8-12. The positive electrode of the cell is mercury and the negative one is a cadmium-mercury amalgam. The electrolyte is a solution of cadmium sulfate. These materials are put into an H-shaped glass tube. In the saturated cell, the electrolyte contains a surplus of cadmium sulfate and therefore crystals of cadmium sulfate cover the electrodes. The saturated cell is subject to voltage changes of $-40\ \mu$V per 1°C temperature change. Hence it is important to maintain it at 20°C to get an accurate measurement. However, it is very stable with time.

The unsaturated cell is similar to the saturated one, except that the electrolyte is unsaturated and contains no cadmium sulfate crystals at room temperatures. In addition, it has a retainer over the electrodes so that it can be made portable. (The normal cell is hard to transport.) It exhibits virtually no voltage variation with small changes in temperature near 20°C, but it does change its voltage about $-3\ \mu$V per month. The voltage output of each

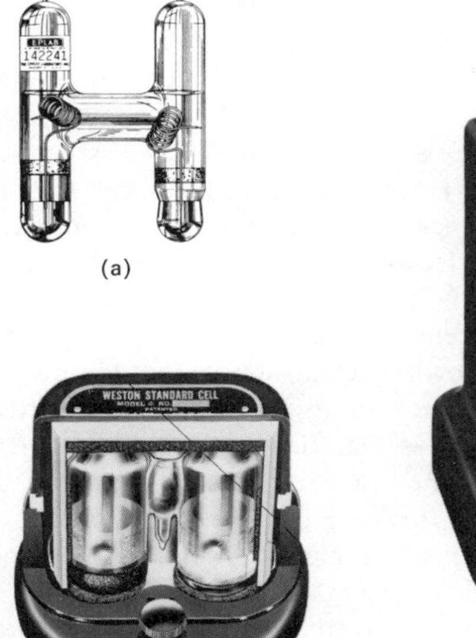

(a)

(b)

(c)

Figure 8-12. The standard cell (a) Standard-cell construction (Courtesy of The Eppley Laboratory, Inc.) (b) Cutaway view of Weston standard cell (Courtesy of Daystrom, Inc., Weston Instrument Division) (c) Standard-cell mounting (Courtesy of The Eppley Laboratory, Inc.)

particular unsaturated cell lies at some specific value between 1.0180 V and 1.0200 V.

PRECAUTIONS IN THE USE OF STANDARD CELLS

The most important rule which must be obeyed when using a standard cell is to avoid drawing current from it during measurements. In a potentiometer, it is used to detect a null, and no current needs to flow. If more than 100 μA is *ever* drawn from the cell, even momentarily, the exact voltage that it maintains may be disturbed. Since it is used as a standard against which other voltages are checked, any suspicion that its voltage has been changed renders it useless. The following guidelines indicate other precautions which must be observed to obtain proper voltage readings from the cell over a long lifetime.

1. Do not expose the cell to temperatures of less than 4°C or greater than 40°C.
2. Keep all parts of the cell at the same temperature.
3. *Never* allow currents of more than 0.0001 A (100 μA) to pass through the cell.
4. Recheck the voltage of the cell against another cell every one or two years.

ZENER DIODES

Zener diodes are diodes which are designed to break down[2] when subjected to a specific reverse-bias voltage. If less reverse voltage than that required for breakdown is applied, the Zener diode behaves like an ordinary diode and conducts virtually no current. However, after breakdown has started, the voltage across the diode remains constant regardless of how much current is passing through it (until a current level is reached that will burn up the diode). The physics of the breakdown mechanism involves the properties of semiconducting materials and will not be discussed here. Nevertheless, the important point for our discussion is that the Zener diode in breakdown maintains a constant voltage as the current passing through it changes. This characteristic can be used to provide regulation for a power supply or to act as an accurate voltage reference. The circuit symbol for the Zener diode is given as —⊳ƶ— .

Zener diodes are built to break down at a specific voltage. Depending upon design, this breakdown voltage may range from 2 to several hundred volts. Therefore Zeners can be fabricated to suit almost all supply output voltages.

When used as a voltage reference, the Zener diode can be made as accurate as a secondary standard cell, but not quite as accurate as a primary cell.

[2]When a diode begins to conduct a large current under reverse bias, it is said to have broken down.

However, their construction makes them much more rugged and far less vulnerable to current overload damage than either of the standard-cell configurations. Thus they are suitable for use as standards in applications whose accuracy requirements are not as high as those of the normal standard cell.

The voltage-reference device is constructed by placing a Zener diode in a temperature-controlled oven (Fig. 8-13). A precision voltage divider is used to obtain an accurate voltage output from the breakdown voltage, V_o. The temperature of the oven can be controlled to within $\pm 0.03°C$ if the exter-

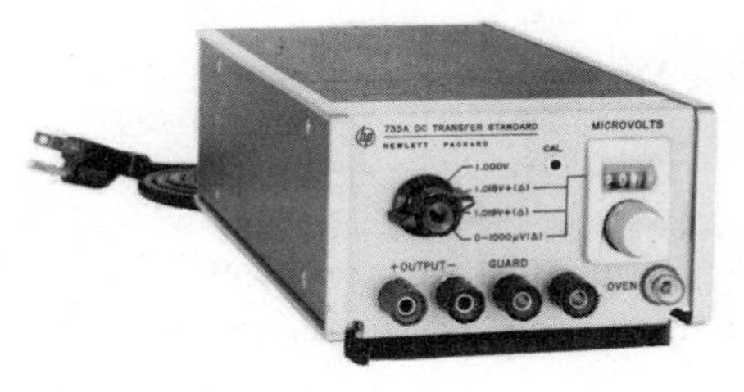

Figure 8-13. A dc transfer standard that can be used as a 1.000-V reference source, a standard-cell comparison instrument, and a 0-1,000 μV dc source

nal temperature is between 0° and 50°C. (It is necessary to provide this temperature control because the breakdown voltage will vary with temperature.)

In addition to providing an accurate voltage reference, such instruments can usually also act as precision variable-voltage sources. In these cases they are provided with a variable voltage divider. An application for this latter capability is to act as an adjustable voltage source that can vary the level of the power-supply output.

Problems

1. Give a definition of the following terms: a) ion, b) electrolyte, c) electrode.
2. What is the difference between
   a) a primary and a secondary cell,
   b) a wet and a dry cell,
   c) a cell and a battery.
3. What is the A-hr rating of a battery that can provide 700 mA for 72 hr?
4. What current will a battery with an A-hr rating of 100 theoretically provide for 20 hr?
5. What is meant by polarization of a cell? How is polarization reduced in the zinc-carbon cell?
6. Three cells whose open-circuit voltages are each 1.30 V are connected in series. What is the open-circuit voltage of the series connection? If the same three cells were connected in parallel, what would be the open-circuit voltage of the parallel connection?

7. What are two functions of the zinc can in the zinc-carbon dry cell shown in Fig. 8-4? What is the disadvantage of this type of construction?
8. What type of cells would be best suited for the following applications:
   a) constant voltage, light loads,
   b) low temperatures (below 0°F),
   c) portable appliances,
   d) long life, moderate initial cost,
   e) high-current values?
9. A cell has an open-circuit voltage of 1.50 V and a terminal voltage of 1.46 V at a load of 200 mA. What is the internal resistance of the cell?
10. An automobile battery has an open-circuit voltage 12.60 V and a total internal resistance of 0.012 Ω. What is the terminal voltage of the cell when it is delivering 200 A to the starter motor?
11. What is the purpose of a power supply, and why must ac power be rectified for use in electronic circuits?
12. What is the *ripple factor*? How is it determined?
13. What is meant by the regulation of a power supply?
14. If a power supply has a no-load voltage output of 100 V and a 98 V output under full load, what is the percentage regulation of the supply?
15. List the major precautions which must be observed when using a standard cell.
16. A standard cell is found to have an open-circuit voltage of 1.01892 V. When a 1 MΩ resistor is connected across the cell terminals the terminal voltage of the cell drops to 1.01874 V. Find the internal resistance of the cell.
17. How much does the open-circuit voltage of the normal standard cell change from its value at 20°C if it is operated at 0°C?

References

1. DeFrance, J. J., *Electrical Fundamentals*, 3rd ed., Chaps. 11 and 12. Englewood Cliffs, N.J., Prentice-Hall, Inc., 1969.
2. Malmstadt, H. V., Enke, C. G., and Toren, E. C., *Electronics for Scientists*, Chap. 2. New York, Benjamin, 1962.
3. *Power Supply Handbook*. Berkeley Heights, N.J., Hewlett-Packard Company, New Jersey Division, 1970.
4. Stout, M. B., *Basic Electrical Measurements*, 2nd ed., Chap. 6. Englewood Cliffs, N.J., Prentice-Hall, Inc., 1960.
5. *Zener Diode Handbook*. International Rectifier Corp., El Segundo, Calif.

9

The Oscilloscope

When measuring an electrical quantity or a quantity which is converted to an electrical form, the measuring instrument must somehow display the measured result. Two of the more common mechanisms used by instruments to provide the display are found in analog[1] meters and oscillographs. Analog meters (both electromechanical and electronic) use a pointer which is moved along a scale to indicate the value of the measured quantity. The oscillograph and *x-y* recorder employ a moving pen assembly which is deflected along one axis while either the paper or the pen assembly is moved along the other. However, both of these display methods are limited by mechanical inertia when measuring high-frequency signals. The mass of the components which make up the display devices prevents them from rapidly changing direction in response to changes in the applied signal. As a result, meter movements can only follow instantaneous variations up to a few cycles per second (Hz). Likewise, high-speed oscillographs are limited to displaying signals below 500 Hz. However, very many signals of interest in electrical applications have frequencies which are far higher than these limits. If the waveforms of such higher-frequency signals are to be measured or examined, other display mechanisms not hindered by inertia must be found.

The display mechanism available in the cathode ray oscilloscope (CRO) is such a device. Because of this unique display mechanism, CRO instruments are being made which can follow signals with frequencies up to 500,000,000 Hz (500 MHz). In fact, even higher frequencies are being displayed by using

[1]An instrument which uses a meter movement to display the quantity being measured along a continuous scale is one type of analog instrument.

the sampling oscilloscope (which is a variation of the basic oscilloscope instrument).

The display device which allows such high-speed variations to be observed is the *cathode ray tube* (a close relative of the television cathode ray tube). The tube generates a thin beam of electrons (the cathode ray) within itself. This beam is directed so that it strikes a fluorescent screen which covers one end of the tube. Wherever the beam strikes the screen, a spot of visible light is emitted. As the beam is moved across the screen, it "paints" a trace of its path. Since the beam is made of electrons which are electrically charged particles, it can be quickly and accurately deflected by appropriate electric or magnetic fields placed in its path. In addition, because electrons are very light, the beam is hardly hindered by inertia. It can respond almost instantaneously to the rapid variations of high-frequency signals. This capability also allows the CRT to display virtually any type of waveshape on the oscilloscope screen. The fields which cause the electron beam deflections are created along the path of the beam by *deflection plates*. The strengths of the fields are determined by the voltages applied to the plates, thereby making the amount of deflection directly proportional to the applied voltage. This indicates that the scope[2] display depends upon the voltages impressed upon the CRT plates. It also follows from this conclusion that the oscilloscope is really a voltmeter—a voltmeter with a super-high-speed display mechanism.

Additional components of the oscilloscope extend the capabilities of its CRT so that the image on its screen is not merely a voltage display. Depending upon the mode of operation being used, the pattern displayed on the scope screen is also a *graph* of the voltage variation with time or the graph of the voltage variation of one signal versus another. This may be a useful point to remember if confusing patterns are seen on a scope screen. Furthermore, voltage is not the only quantity that can be measured. By properly interpreting the characteristics of the display, we can use the oscilloscope to indicate current, time, frequency, and phase difference. Finally, by using the oscilloscope to monitor the output of various transducers, we can also measure a large variety of nonelectrical quantities.

Even from this brief introduction, it is easy to see the wide range of measurement applications made possible by an oscilloscope. In fact, the oscilloscope is probably the most versatile and useful single instrument invented for electrical measurement work.

In this chapter we will examine the operation of basic oscilloscopes in some detail. The discussion will be aimed at giving the reader a sense of familiarity with the major scope subsystems, as well as the ability to use a scope to measure the more common electrical quantities just described. In line with this objective, some practical suggestions concerning scope operations and

[2]The word *scope* is commonly used as an abbreviation for oscilloscope.

limitations will also be given. To tie some of the general discussions of oscilloscope use to actual instruments, the detailed operation of two commonly used laboratory oscilloscopes will also be described. Higher frequency and special purpose scopes will be mentioned only briefly.

Oscilloscope Subsystems

Our discussion will begin with an examination of the major subsystems which make up a basic oscilloscope. By learning how each of these subsystems function separately, one can understand how they operate when connected together to make an oscilloscope. The subsystems found in a basic oscilloscope are the following:

1. Cathode ray tube.
2. Vertical and horizontal amplifiers.
3. Time-base circuitry.
4. Power supplies.

CATHODE RAY TUBE

The heart of the oscilloscope, the cathode ray tube (CRT) is shown in Fig. 9-1. The tube itself is a sealed glass vessel with an electron gun and deflection system mounted inside one end and a fluorescent screen on the other. The air is removed from the tube, leaving a high vacuum. This high vacuum is required because the fine electron beam produced within the tube would be

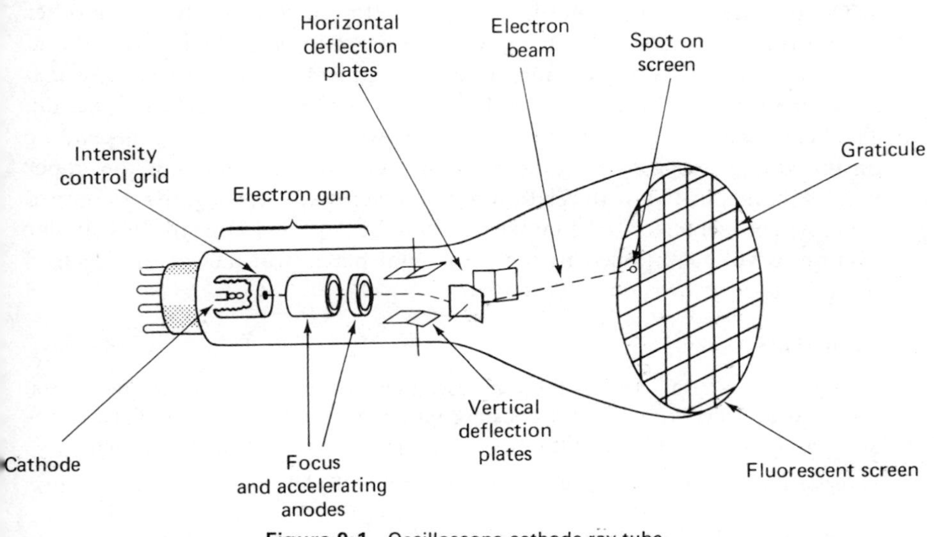

Figure 9-1. Oscilloscope cathode ray tube

scattered by collision with any gas molecules in its path. Many such collisions would destroy the pencil-thin character of the beam.

It is the function of the *electron gun* to produce the electron beam. The gun consists of a thermionic cathode (a cathode made of a material which emits electrons upon being heated), various accelerating electrodes, and controls for focus and intensity. When the cathode is heated to a high temperature, it begins to emit electrons. Some of these electrons pass through a small hole in the *intensity control grid* that surrounds the cathode. If a negative voltage is applied to this grid, only a limited number of the emitted electrons can pass through the hole. This number can be controlled by varying the magnitude of the voltage. The intensity of the spot of light where the beam strikes the fluorescent screen depends on the number of electrons in the beam.

The emerging electrons are compressed by the focus and accelerating anodes of the gun to form a tight beam. An electrostatic focusing scheme is used in oscilloscope CRTs to achieve this compression. The same electrostatic fields also direct the electrons along the axis of the beam and accelerate them forward toward the fluorescent screen.

After leaving the electron gun, the focused and accelerated beam passes between two sets of *deflection plates*. If there is no voltage difference between the plates, the beam continues straight through and strikes the fluorescent screen at its center. If there is a voltage difference between either or both sets of these plates, the beam will be deflected from its straight path. The amount of deflection is determined by the magnitudes of the voltage differences. In typical oscilloscopes, between twenty and fifty volts rms must be applied to a set of plates to deflect the spot one centimeter.

The two sets of deflection plates are placed perpendicular to one another so that they can independently control the beam in both the horizontal and vertical directions. For example, a voltage applied to the *vertical deflection plates* changes the direction of the electron beam only in the vertical direction. The beam can be deflected upward or downward by these plates, depending on the voltage polarity existing between them. If the voltage of the upper plate is made positive with relation to the lower plate, the negatively charged beam will be attracted to the upper plate and be deflected upward. In a similar fashion, a voltage applied to the horizontal plates will deflect the electron beam to the left or right.

EXAMPLE 9-1

Figure 9-2 is a front view of the oscilloscope screen which shows the horizontal and vertical deflection plates. If constant voltages are applied to the plates in the manner shown, in which quadrant will the electron beam strike the screen?

Solution: Second quadrant, because the upper and the left-hand plates are positive.

Example 9-1 shows how the position of the electron beam on the screen can be changed when no external signal is applied to the scope. By adjusting

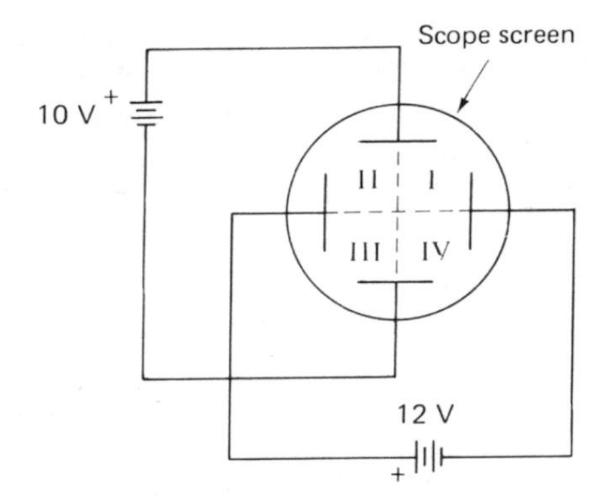

Figure 9-2

the dc voltages applied to the plates, we can shift the center of the image displayed on the screen to any point. These voltages originate in the internal power supplies of the scope and are adjusted by the *Position* controls on the scope face.

The *fluorescent screen* of the CRT is coated with a phosphor. At the point where the electron beam strikes the screen, the phosphor emits a spot of visible light. Most phosphors continue to emit light for a short time after the beam has stopped striking it. Thus, if the electron beam is repeatedly moved across the screen along the same path, and this retracing is performed rapidly enough, the image "painted" on the scope screen will appear to be a solid line.

The length of time it takes the intensity of the spot to decrease to 10 percent of its original brightness is called the *persistence* of the phosphor. The value of the persistence varies according to the phosphor type. For laboratory scopes, a green phosphor with medium persistance provides a steady image of a repeated trace.

Heat as well as light is generated when an electron beam strikes the screen. In fact, 90 percent of the electron beam energy is converted to heat and only 10 percent to visible light. Therefore, care must be taken to prevent the beam from burning holes in the phosphor coating. This is done by keeping the beam *intensity* set to a small value, especially when the spot is stationary. (However, in any case, do not allow the spot to remain stationary for a long time.)

The *graticule* is the set of horizontal and vertical lines permanently scribed on the screen of the CRT. The graticule acts as a scale for measuring the quantities displayed by the CRT. Most oscilloscope graticule lines are spaced one centimeter apart.

THE OSCILLOSCOPE AMPLIFIERS

The voltages applied to the inputs of an oscilloscope are often too small to cause any noticeable deflection of the electron beam if applied directly to the deflection plates of the CRT. Therefore, before being applied to the plates, the input voltages are fed to amplifiers which increase their magnitudes.[3] For example, if the signal voltage to the vertical input of the scope is V_v and the amplifier changes this value by a factor of K, the voltage applied to the deflection plates is KV_v. The connections to the deflection plate inputs through these amplifiers is shown in Fig. 9-3.

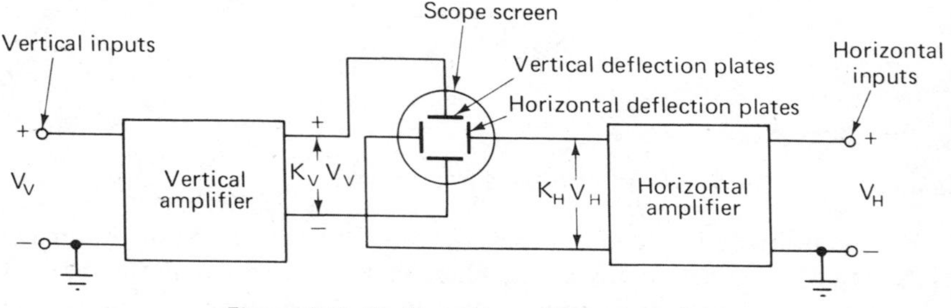

Figure 9-3. Oscilloscope amplifier connections

The amount of amplification which each amplifier provides is selected by the sensitivity (*V/Div.*) controls of the scope. To make the scope versatile, the sensitivity control is designed to be set to a wide range of discrete amplification levels. These levels are calibrated to produce the level of amplification specified by the control setting (to within 1–3 percent, depending on the scope). However, the *Variable V/Div.* knob on such scopes must be set to the *calibrate* position in order for the calibration to be accurate.

Most laboratory oscilloscope amplifiers are direct-coupled amplifiers. The term *direct-coupled* refers to the fact that an amplifier is capable of amplifying dc as well as ac signals. However, there is still a maximum frequency limit beyond which direct-coupled amplifiers become ineffective. The bandwidth of each oscilloscope specifies this maximum frequency.[4] Direct-coupled amplifiers and their associated terminology will be discussed in more detail in Chapter 16. However, the various types of *amplifier inputs* and *signal couplings* available in connection with oscilloscope amplifiers need to be introduced here.

There are two types of inputs through which a signal can be connected to an oscilloscope. They are the *single-ended input* and the *differential input*. Their names are derived from the type of amplifier for which they form an input. For our purposes, it will only be necessary to describe the capabilities

[3] Actually, as part of the amplifying process, all input voltages are reduced to approximately the same voltage level before being fed into the amplifier. Calibrated attenuators associated with the amplifier circuitry provide this attenuation.

[4] The maximum frequency an oscilloscope can handle effectively is defined as the frequency at which the amplitude of the voltage indicated by the scope drops to 0.707 (−3 dB) of the value of the true voltage applied to its input. The bandwidth is the frequency interval from dc to this maximum frequency value.

of oscilloscopes equipped with each type of input; we can ignore the details of their associated amplifiers.

Single-ended inputs have only one input terminal besides the ground terminal at each amplifier channel. Most basic scopes have single-ended inputs as shown in Fig. 9-4(a). Only voltages relative to ground can be measured with a single-ended input (unless the scope is "floated" above earth ground by using a three-to-two-wire adapter. More will be said about this in a later section).

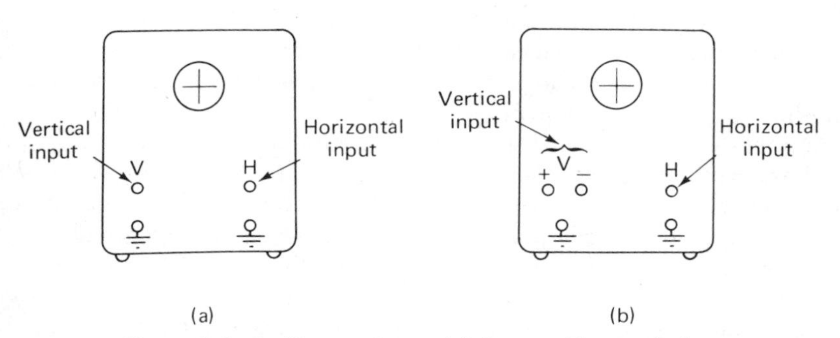

Figure 9-4. Oscilloscope inputs (a) Scope with two single-ended inputs (b) Scope with differential input on vertical amplifier and single-ended input on horizontal amplifier

The *differential input* has two terminals besides the ground terminal at each amplifier channel. An example of this type of input is shown on the vertical amplifier channel of the scope in Fig. 9-4(b). With a differential input, the voltage between two nongrounded points in a circuit can be measured without floating the scope. Each point is merely connected to one of the two differential input terminals. The amplifier electronically subtracts the voltage levels applied at the two terminals and displays the difference on the screen. In addition, unwanted interference and noise in the signal are reduced by the use of the differential amplifier and input. However, a differential amplifier is more complex than a single-ended amplifier, and hence oscilloscopes equipped with differential inputs are more costly.

To measure a voltage relative to ground with a differential input, only one of the input terminals (and the ground terminal) needs to be used. The other terminal is left unconnected and its internal circuitry is grounded. In this type of application, the differential input is being used as a single-ended input.

Once the signal has been connected to an oscilloscope through its inputs, it may be desired to feed only a part of the signal to the CRT plates for display. For instance, the signal may contain both ac and dc components. Sometimes only the ac components are of interest; at other times the entire signal needs to be displayed. The controls of many oscilloscopes allow us to choose which part of the signal components will be coupled to the amplifier circuitry for subsequent display. For example, a typical input to a vertical amplifier is shown in Fig. 9-5.

When the DC position of the *Vertical-Input Coupling Switch* is selected, the input terminal couples (connects) the entire signal to the vertical amplifier (both the ac and dc components). On the other hand, if the AC position is chosen, we see that the input branch to the amplifier now contains a capacitor in series. This capacitor appears as an open circuit to dc components and hence blocks them from entering the amplifier circuitry. Likewise, very-low-frequency ac components are also blocked (below 2–10 Hz, depending upon the scope design). Thus a signal with both ac and dc components applied to an input which is set for ac coupling will be displayed without its dc and very-low-frequency components.

Ac input coupling is used only when a high-frequency ac signal is to be displayed without its dc component. This may be a useful option if the magnitude of the ac component of a signal is much smaller than the dc level. If the signal with both components were displayed, any changes in the larger dc component might drive the entire signal off the screen. By eliminating the dc component, the ac part of such a signal can be more easily observed. The dc input coupling is generally appropriate for displaying most other types of signals.

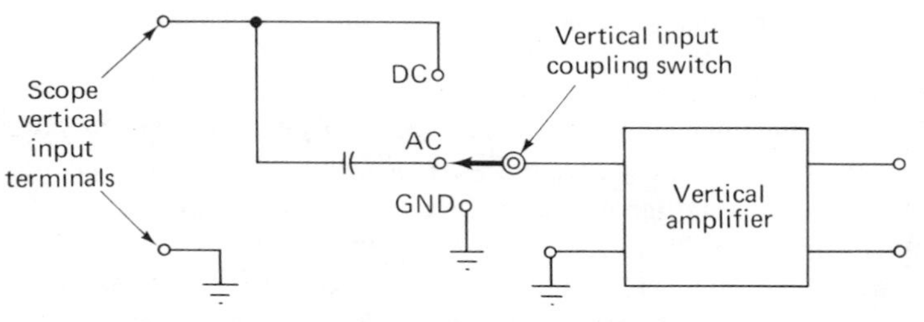

Figure 9-5. Functioning of the input coupling switch

The GND position of the input coupling switch of Fig. 9-5 grounds the amplifier's internal circuitry. It does *not* convert the input terminal to a ground point for an input signal. The purpose of this feature is to remove any voltages applied to the CRT deflection plates and allow the electron beam to be recentered. This option can be useful if one desires to center the beam without removing the input leads to the scope. If the input to a scope is a differential input, both of the nongrounded terminals usually have a separate input coupling switch.

Some very inexpensive scope models have amplifiers that are not direct-coupled. In these scopes, dc signals or dc components cannot be displayed. Only ac coupling of the input signal exists, and no input coupling switch is needed.

TIME-BASE CIRCUITRY

The most common application of an oscilloscope is the display of signal variation versus time. To generate this type of display, a voltage which makes the horizontal position of the beam proportional to time must be applied to the horizontal plates of the scope. In addition, this same voltage must be repetitively applied to the horizontal plates so that the beam can retrace the same path rapidly enough to make the moving spot of light appear to be a solid line. Finally, this voltage must be synchronized with the signal being displayed in such a way that the same path actually *is* retraced and a steady image results on the scope screen.

The *time-base* circuitry of the oscilloscope has the job of producing such a repetitive and sychronized voltage signal. To see how it performs this function, let us examine the principles of its operation with the following questions in mind. First, what kind of signal must the time base produce in order to make the horizontal position of the beam be proportional to time? Second, how is this signal repetitively generated? Finally, how is this signal synchronized with the signal to be displayed on the scope screen?

The signal generated by the time base is called a *sweep waveform*,[5] and the rate at which this signal causes the beam to move horizontally across the screen is calibrated in terms of a specific time for a given distance of spot travel. That is why this subassembly is called the time base.

The shape of one cycle of the sweep waveform is shown in Fig. 9-6, where V_H is the voltage applied to the horizontal plates of the CRT. If the spot of the electron beam is located at the left edge of the screen when $t = 0$, the increasing voltage of the sweep waveform will cause the spot to be pulled horizontally across the screen. At the end of T_1 seconds, the spot will have been moved across the full length of the screen. During the time from T_1 to T_2, V_H will decrease to zero, and the spot will be returned quickly to the left edge of the screen. Since V_H increases linearly with time from $t = 0$ to $t = T_1$, the position of the spot during this time interval will be proportional to the time elapsed from the beginning of the sweep waveform. The *Time/Div.* switch on the scope face determines how much time it takes for the sweep waveform to move the spot across one division of the screen. If no

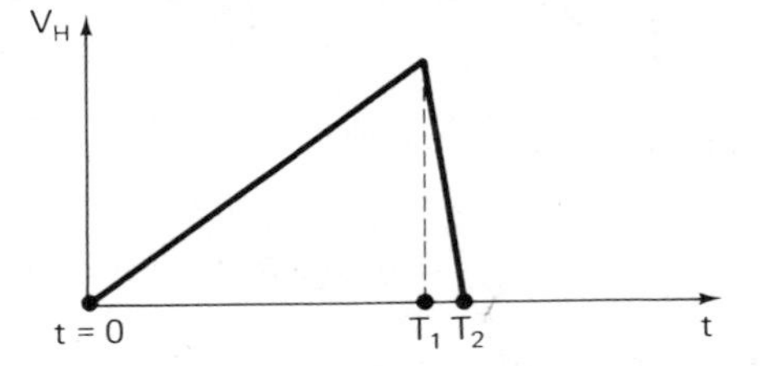

Figure 9-6. One cycle of the sweep waveform

[5] Because of its shape, this waveform is also sometimes called a *sawtooth* or a *ramp* waveform.

external signal is applied to the vertical plates, the sweep waveform will cause the spot to trace a horizontal line on the scope screen. If there is a vertical input signal, the sweep waveform will make the scope display a V-versus-time plot. Figure 9-7 shows how the time variation of a signal is displayed with the help of the sweep waveform signal.

During the short time the spot is being returned from the right edge of the screen to its starting position, additional circuitry within the scope is used to shut off the beam. This action prevents the beam from leaving a trace during the return trip and is known as *retrace blanking*.

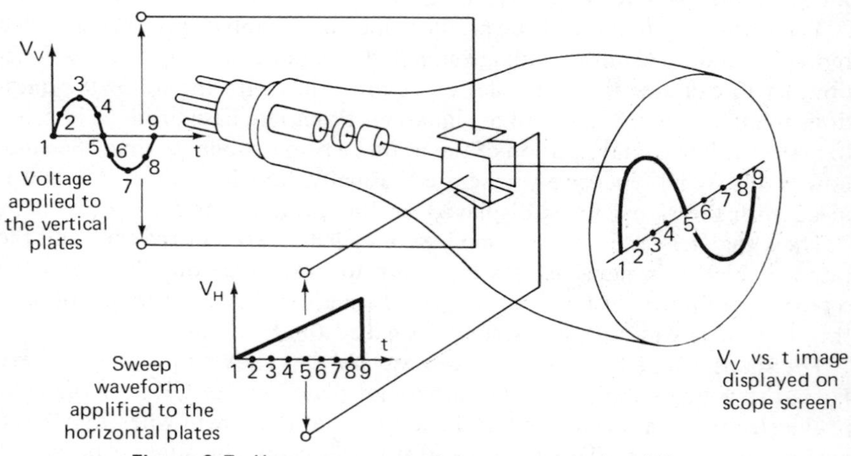

Figure 9-7. How a sweep waveform generates a plot of V_v vs t

Let us now look at a block diagram of the time-base circuitry to understand how the sweep waveform is generated (Fig. 9-8). We see that a signal called the *triggering signal* is first fed to the pulse generator of the time base.

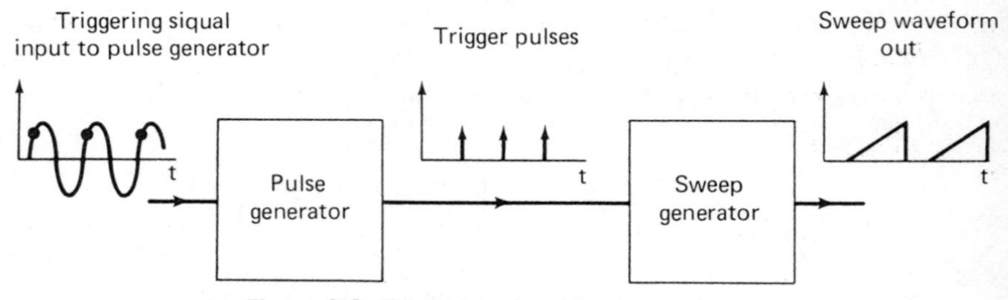

Figure 9-8. Triggering signal input to pulse generator

Every time this triggering signal crosses a preselected slope and voltage level condition, the pulse generator emits a pulse. The emitted pulse triggers the sweep generator to begin producing one cycle of the sweep waveform. Figure 9-9 shows how the triggering signal, emitted pulses, and sweep waveform are

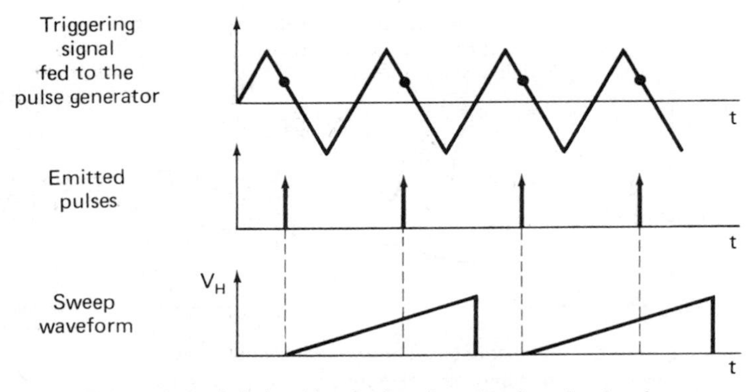

Figure 9-9. Relationship of triggering signal, emitted pulses and sweep waveform in time

related in time. We note that not all pulses from the pulse generator cause the sweep generator to generate a sweep waveform cycle.

If the sweep generator receives a pulse during the middle of one cycle, the pulse is ignored. This allows the scope to display more than one period of the input signal without having itself retrigger a new sweep waveform. At the end of each cycle, the sweep generator stops its output and awaits the arrival of the next pulse before producing a new sweep waveform.

The point on the triggering signal at which the pulse generator emits a pulse is controlled by the *Trigger Slope* and *Trigger Level* switches of the scope. The Trigger Slope switch allows one to choose whether the slope of the triggering signal should be plus or minus when the pulse generator emits a pulse (Fig. 9-10). Likewise, the Trigger Level switch determines the value (sign and magnitude) of the triggering voltage at which a pulse is generated. For example, if the Trigger Level switch is set to 0 and the Trigger Slope set to *plus*, a pulse (and hence a sweep waveform) will be triggered when the triggering signal passes through 0 on a positive-going slope (points 1 and 6 of the curve shown in Fig. 9-10).

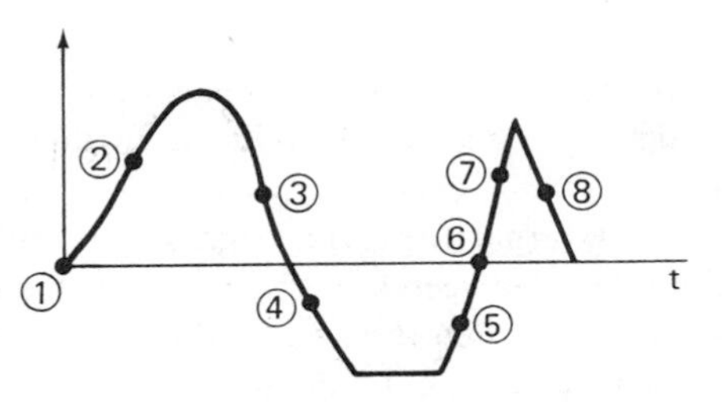

Figure 9-10. Points 1, 2, 5, 6, and 7 are examples of points which have a *positive* slope. Points 3, 4, and 8 are points of *negative* slope

We mentioned earlier that the final condition necessary for a stable display of a time-varying signal is that the sweep waveform must be started at the same point of the input signal waveform as the point at which the previous sweep waveform was started (Fig. 9-11).

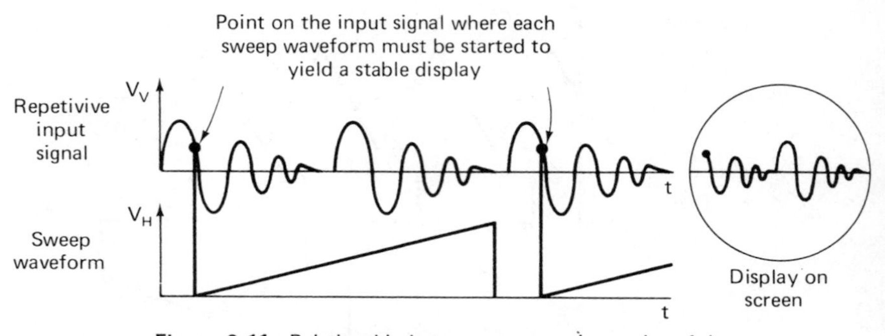

Figure 9-11. Relationship between consecutive cycles of the sweep waveform and vertical input signal when a stable image is displayed

Since the triggering signal is the agent which causes the sweep waveform to be started, the triggering signal and the display on the scope screen must be synchronized to achieve a stable image. This synchronization is easy to achieve if the input signal also acts as its own trigger signal. In this case the input signal and the trigger signal are already synchronized, being one and the same signal. As a result, the sweep waveform is started by the input signal itself, and the first point of the image on the scope screen will be equal to the point on the input signal where its slope and level triggers the sweep waveform. In an actual scope, this type of triggering is called *internal triggering* because a part of the input signal is tapped from the vertical amplifier and used as the trigger signal. The only requirement is that the input signal be large enough to produce a 0.5 Div deflection at the sensitivity level chosen. Internal triggering is the most commonly used type of triggering.

If the trigger signal is externally applied, the frequency of the external signal f_E must be related to the frequency of the input signal being displayed, f_S, by Eq. (9-1) in order for the display to be stable.

$$f_S = Nf_E \tag{9-1}$$

Here N must be either an integer or a fraction such as $\frac{1}{2}$, $\frac{1}{3}$, $\frac{1}{4}$, etc. The external triggering signal is introduced through the input terminals marked *Ext. Trig.*

An external triggering signal is useful in the measurement of the phase difference between two sine waves of the same frequency (discussed in a later section) or when the amplitude of the input signal is too small to trigger the pulse generator. On most scopes another feature allows the 60-Hz line voltage to be the triggering signal. This type of triggering is actually another form of external triggering. Such *line triggering* is useful for observing waveforms which bear a fixed relationship to the line frequency. The choice of the source of the trigger signal is controlled by the *Trigger Source* switch on oscilloscopes.

The Trigger Level switch usually contains two extra positions (AUTO and FREE RUN) which should also be explained at this point. When the Trigger Level switch is in the AUTO position, the pulse generator is set to be triggered every time the triggering signal crosses zero. The triggering signal is also ac-coupled to the pulse generator. In order to be able to trigger the pulse generator, the trigger signal must have enough ac amplitude to be able to cause a deflection of one half a division. However, if no triggering signal is received by the pulse generator (or if the signal is too weak), the AUTO position continues to trigger a sweep waveform automatically (although the sweep is no longer synchronized to the weak signal). This feature allows the user to generate a horizontal trace without feeding in a triggering signal. The AUTO triggering mode is best for most triggered sweep displays. As a result, the Trigger Level switch is usually left set to the AUTO position.

The FREE RUN position disconnects the pulse generator from the sweep generator circuit and the sweep generator begins to produce sweep waveforms, one after another, with no triggering. In other words, the sweep generator does not wait for pulses to produce consecutive cycles. As soon as one cycle is finished, a new one is started. The sweep rate is still controlled by the *Sweep Time/Div.* switch. Thus a horizontal trace can also be produced in the FREE RUN position without using a triggering signal. It is useful for generating a base line when dc levels are being observed, since no triggering pulse is required to generate a sweep.

EXAMPLE 9-2

A sine wave of 400 Hz is to be displayed so that four complete cycles are to appear on a scope screen which has ten horizontal divisions. To what settings should the Trigger Source and Sweep Time/Div. be set to allow this pattern to be displayed?

Solution: Trigger source should be set to INT. The time of one cycle (i.e., period) of the waveform is

$$T = \frac{1}{f} = \frac{1}{400} = 2.5 \text{ ms}$$

The time required to display four periods would be $4T = 10$ ms.

Thus the sweep time required would be 1 ms/Div.

Note that in this example the internal triggering signal causes the pulse generator to emit four pulses during one sweep waveform cycle. The last three of these four are ignored by the sweep generator as it is already in the middle of producing one cycle of the waveform. In addition, if it is acceptable to have the display started as the sine wave crosses the zero level, the Trigger

Level switch can be set to either AUTO or the ZERO position. If it is desired to start the display at some other voltage level along the waveform, the Trigger Level switch needs to be set to this value.

SCOPE POWER SUPPLIES

The oscilloscope has two power supplies which are used to provide dc voltages for its various subassemblies. One is a high-voltage power supply which is capable of delivering up to 10,000 V and is used to drive the cathode ray tube. The other is a low-voltage power supply which provides power for the scope amplifiers and triggering circuitry. The maximum low-voltage supply output is about 350 V.

How an Oscilloscope Displays a Signal

There are two basic ways in which an oscilloscope is used to display electrical signals—the *X-Y* mode and the triggered-sweep mode. The *X-Y mode* displays the graph of the variation of two external signals—one versus the other. The *triggered-sweep mode* displays the variation of only one external signal as it varies with *time*. The triggered-sweep mode is by far the most commonly used of the two modes.

When the scope is being used in the *X-Y* mode, an external signal applied to the horizontal plates controls the deflection of the electron beam to the right and left (*X*-direction). Simultaneously, an external signal applied to the vertical plates moves the beam up and down (*Y*-direction). If both applied voltage signals change slowly with time, the display on the screen will be a slowly moving spot of light. If both signals are periodic, the moving spot will retrace the same path on the screen over and over again. If the frequencies of the periodic signals are high enough, the repeating trace will appear to be a steady pattern painted by solid lines of light on the screen. (Frequencies of 50 Hz or more will cause the trace to appear as a nonflickering solid line due to the persistance of vision and the persistance of the phosphor.) Figure 9-12 shows the circuitry involved when the *X-Y* mode of the scope is being used.

In the triggered-sweep mode, the time base generates a sweep waveform at accurately timed intervals. The sweep waveform (Fig. 9-6) causes the electron beam to be swept horizontally across the screen at a uniform rate. Simultaneously, the vertical input signal deflects the electron beam up and down. Because the horizontal position of the beam is proportional to time, the resulting display is an image of the vertical input signal as it varies with time. In order for this image to be a steady pattern on the screen, the sweep waveform must be started at exactly the same point on the vertical input waveform as the previous sweep waveform was started. In addition, for the pattern to appear to be a solid line, the sweep waveform must be repeated at least 50 times per second.

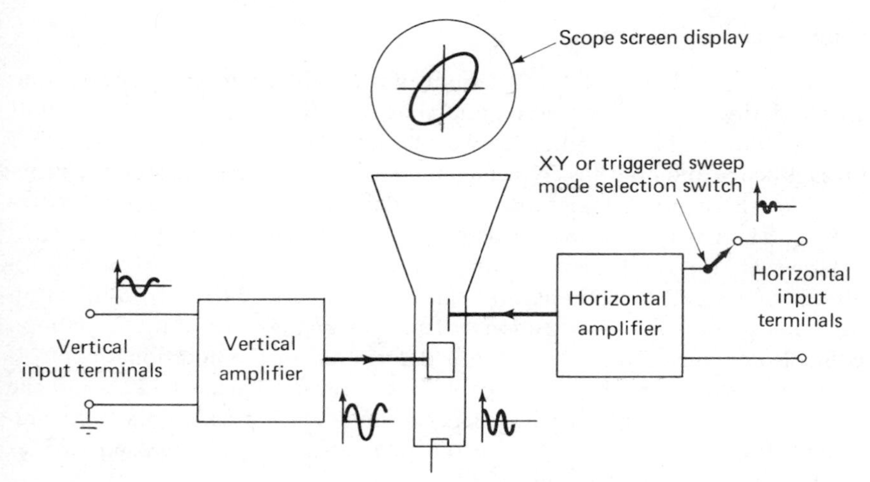

Figure 9-12. $X - Y$-mode operation

The *sweep rate* or the length of time it takes the beam to sweep horizontally across one division of the scope screen is determined by the circuitry of the sweep generator. Figure 9-13 shows the scope circuitry involved during the triggered-sweep mode of operation.

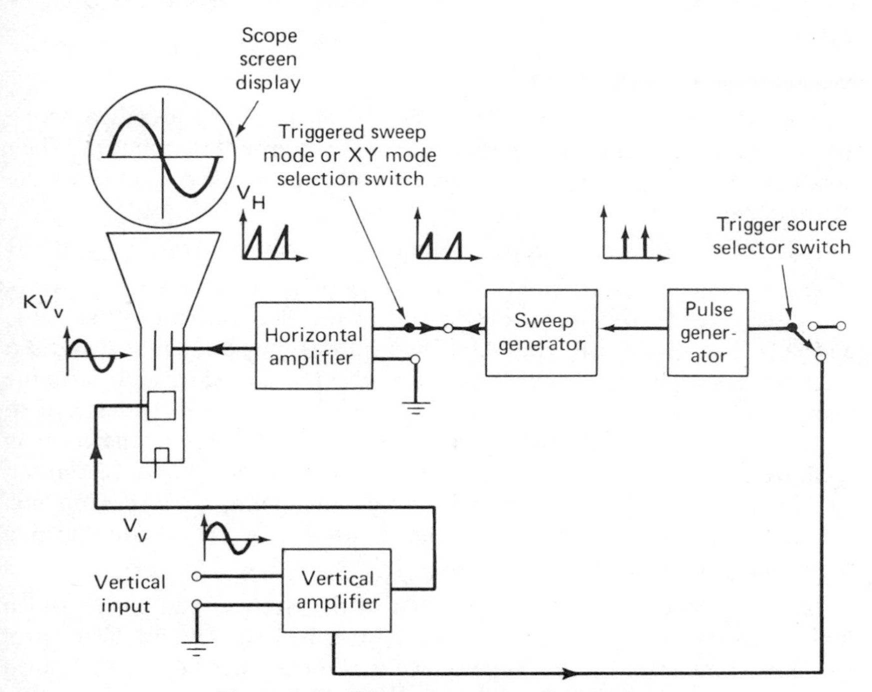

Figure 9-13. Triggered-sweep-mode operation

Oscilloscope Controls

The knobs and switches on the front of an oscilloscope can appear to be a bewildering array to the user unfamiliar with the instrument. This initial confusion might best be relieved by explaining the function of every control on each scope and how it is connected to the scope's inner subsystems. Unfortunately, there are so many different models and manufacturers of oscilloscopes that it is not feasible to describe every control of every model.

However, there are some controls which are used quite universally on almost all scopes, even though they may be found in different positions on different instruments. This section will start by giving a list of these common controls and their functions. Next, a description of the additional controls found on two basic laboratory scopes—the Hewlett-Packard 122A and the Tektronix 561B equipped with the 2A63 and 2B67 plug-ins—will be presented. Finally, a list of miscellaneous controls which are used on various other elementary scopes will be given.

If the scope being used by the reader is not one of the above two models, the best guide to its controls may be the instruction manual accompanying the scope. However, if the manual is not available or if it is not sufficiently clear, the reader can scan the lists of controls presented here and probably find an explanation of most of the control functions of the scope (that is, with the exception of the controls of special purpose or very-high-frequency scopes).

COMMON SCOPE CONTROLS

Although the following list describes the most commonly used scope controls, their names may be slightly different on a given scope model. When more than one name is apt to be used, an attempt is made to also list the alternate names.

Power—Turns the oscilloscope on and off (after it has been plugged in).

Intensity—Controls the brightness of the scope trace. This knob provides a connection to the intensity control grid of the electron gun in the CRT. When the knob is turned clockwise, the grid repelling voltage is decreased and more electrons can emerge from the hole in the cathode grid to form the beam. A larger number of electrons in the beam causes a brighter spot to appear on the screen. *Caution*: Care must be taken to prevent the electron beam from burning spots on the screen. A stationary spot should be kept on very low intensity. If the intensity is kept high, the spot must be kept moving. If a "halo" appears around the spot, the intensity is *too high*! Before turning the scope on, turn down the intensity.

Focus—The focus control is connected to the electron gun anode which compresses the emerging electron beam into a fine point. When this control is adjusted, the trace on the scope screen is made as sharp and well-defined as possible.

Astigmatism—This control works in conjunction with the focus control to improve the focus over the entire CRT. On some scopes an astigmatism knob is provided. On others, this control requires a screwdriver for adjustment or is placed inside the chassis.

Position—The position knobs are used to shift the trace or the center of the displayed image around the screen. The position knobs provide this control by adjusting the dc voltages applied to the deflection plates of the CRT.

Vertical Position—Controls the vertical centering of the trace.

Horizontal Position—Controls the horizontal centering of the trace.

Scale Illumination—Provides illumination for the graticule. The etched lines of the graticule are brightened by light applied from the edge of the screen, thereby producing no glare to interfere with the displayed image.

Vertical Sensitivity or V/Div. or V/cm—Determines the necessary value of voltage that must be applied to the vertical inputs in order to deflect the beam one division (or cm). This control connects a stepped *attenuator* or a *potentiometer* to the scope amplifier and allows the vertical sensitivity to be controlled by discrete steps. The settings of the HP 122A range from 10 mV/cm to 10 V/cm. The Tektronix 561B has scales from 10 mV/Div. to 20 V/Div.

Variable V/Div. (usually a red dial)—Allows a smooth (rather than a stepped) variation of the vertical sensitivity. This knob *must* be set to the calibrated position (usually fully clockwise past the click stop) for the vertical sensitivity of the scope to be equal to the value marked by the Vertical Sensitivity switch. Some inexpensive scopes possess only a variable V/Div. control.

Sweep Time or Time/Div.—Controls the time it takes for the spot to move horizontally across one division of the screen when the triggered-sweep mode is used. A very small Time/Div. setting indicates a very short sweep time. The HP 122A has sweep times from 5 μs/cm. to 200 ms/cm. The Tektronix 561B with a 2B67 plug-in has sweep times from 1 μs/cm to 5 s/cm.

Variable Time/Div. (usually a red dial)—This vernier control allows a continuous but noncalibrated Time/Div. sweep rate to be chosen. Some very inexpensive scopes only have a continuously variable Time/Div. control.

Source or Trigger Source or Sync Selector—Selects the source of the triggering signal. By using this control, one chooses the type of signal being used to synchronize the horizontal sweep waveform with the vertical input signal. The possible selections usually include:

1. *Internal.* The output of the vertical amplifier is used to trigger the sweep. This choice makes the input signal control the triggering. For most applications, this type of triggering is appropriate.
2. *Line.* This position selects the 60-Hz line voltage as the triggering signal. Line triggering is useful when the frequency of the vertical input signal is related to the line frequency.

3. *Ext.* When this position is used, an external signal must be applied to trigger the sweep waveform. This signal must be connected to the *External Trig.* input. The external trigger signal must have a compatible frequency with vertical input signal in order to get a stable display.

Sweep Magnifier—This control allows one to decrease the time per division of a sweep waveform. However, the reduction is accomplished by magnifying a portion of the sweep waveform rather than by changing the time constant of the internal circuits that generate it.

Trigger Slope—This switch determines whether the pulse circuit in the time base will respond to a triggering signal of positive or negative slope. This topic was discussed in more detail in the time-base section (see Fig. 9-10).

HEWLETT-PACKARD 122A

The HP 122A was chosen for description because it is an older unit used in many basic labs. It is also a dual-trace oscilloscope and the terminology of the controls associated with this feature is applicable to many other dual-trace models. The dual-trace capability means that the scope can simultaneously display the traces of two separate input signals by time-sharing one electron gun and deflection system between two vertical input channels. If only one channel of the two channels available is used, the dual-trace scope acts like an ordinary single-trace instrument. The controls of the HP 122A are shown in Fig. 9-14.

The bandwidth of the 122A is 200 kHz, and the choice of vertical sensitivities ranges from 10 mV/cm. to 10 V/cm. Its input can be either ac- or dc-coupled, and a differential input is available (but only when the 10 mV/Div. sensitivity is used). In order to employ either of the vertical input channels in a differential input mode, the grounding strap connecting the lower two input terminals must be disconnected and the vertical sensitivity switch must be set to the 10 mV/Div. position (see Fig. 9-17). The vertical sensitivity steps are calibrated to within ± 3 percent and the time-base accuracy is ± 5 percent. The input impedance of the vertical amplifiers is 1 MΩ shunted by a capacitance of 50 pF. The graticule is designed to be parallax-free.

The control knobs of the HP 122A are either black or red. The red printing on the scope face refers to the red control knobs; the black printing refers to the black control knobs. The controls pertinent to this instrument are the following:

Vertical Presentation—Selects the type of display which the scope will exhibit. (This control is sometimes called *Mode* on other dual-trace scopes.)

1. *Position A*. Output from channel *A* is displayed.
2. *Position B*. Output of channel *B* is displayed.

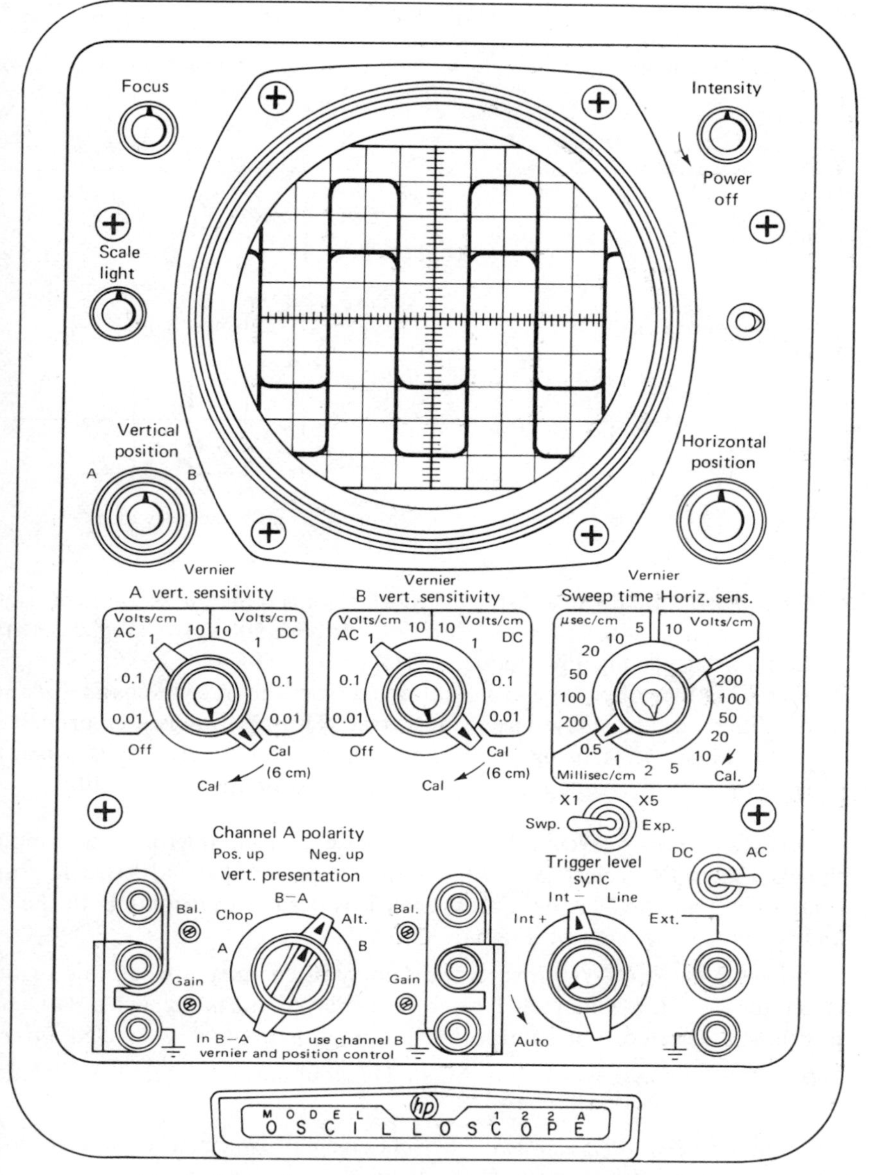

Figure 9-14. Face of Hewlett-Packard 122A oscilloscope (Courtesy of Hewlett-Packard Co.)

3. *Alt*. Signals from both channels are displayed by alternately showing one channel output after the other. The change from one channel to the other is made at the end of each sweep (Fig. 9-15(a)). ALT. is best for displaying higher-frequency signals ($>$ 2 kHz).

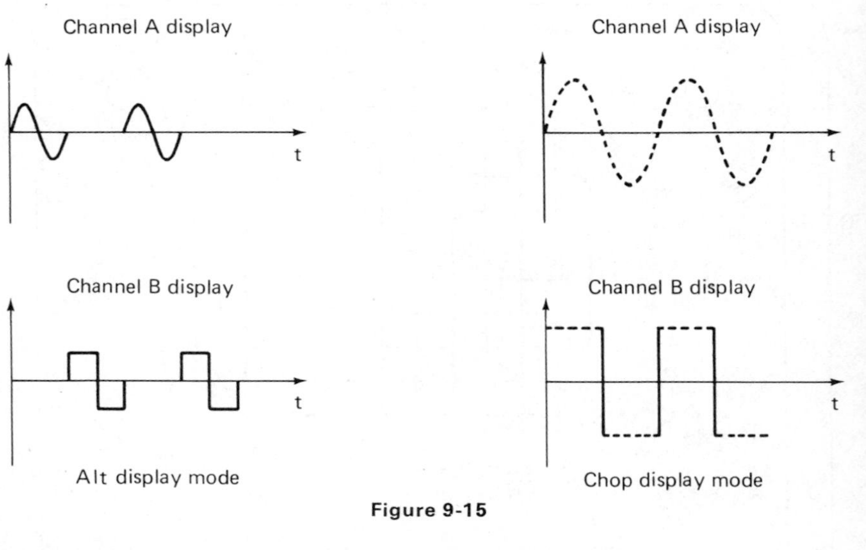

Figure 9-15

4. *Chop*. Also a dual-trace display of both signals. One-ms sections of each signal are displayed at a repetition rate of about 500 kHz/s (Fig. 9-15(b)). Best for displaying low-frequency signals (< 2 kHz).
5. *B-A*. The trace of channel *A* is subtracted from the trace of channel *B* and the difference is displayed as a single trace. This position is used primarily when the scope is being used as a difference voltmeter, that is, when it is comparing the voltages of two ungrounded points in a circuit.

Vertical Sensitivity—(V/cm AC) positions select ac coupling for vertical input signals. (V/cm DC) selects dc coupling for vertical input signals. Use dc positions for signals of 2 Hz or less or when it is desired to display both the dc and the ac components of a signal.

Channel-A Polarity (*red*)—Inverts the display of channel *A* when this switch is set to *Neg. Up*. Channel-*A* trace is displayed normally when *Pos. Up* position is selected. On some other scopes this control is labeled *Invert*.

DC-AC (*toggle switch*)—Selects AC or DC coupling for external horizontal input signals.

×1 Swp–×5 Swp (*toggle switch*)—Selects normal sweep when set on ×1. Sweep is expanded five times when set to ×5.

Trigger Level (*red*)—When this knob is rotated fully counterclockwise, AUTO triggering occurs. This means that the sweep is triggered internally approximately when the vertical input signal crosses its zero level. When triggering at some other level than zero is desired, turn knob clockwise. If no triggering signal is applied, an internal sweep of approximately fifty sweeps per second will continue to be generated if the switch is in the AUTO position.

Sync—Selects the source of the trigger signal as described in *Common Controls* section. However, Int. + triggers the sweep on a positive slope of an internal triggering signal while Int. – triggers the sweep on a negative slope.

Sweep Time and Horizontal Sens—*Sweep Time* positions determine the rate at which the electron beam is swept across the scope screen by the sweep waveform. These positions are used when the scope is being operated in the triggered-sweep mode. *Horizontal Sens* positions choose the V/Div. sensitivity of the horizontal amplifier when the *X-Y* mode of operation is used.

TEKTRONIX 561B

The Tektronix 561B with a 2A63 differential amplifier vertical plug-in unit and a 2B67 time-base horizontal plug-in will be described in this section. (Fig. 9-16) The 561B was chosen because it is a modern laboratory oscilloscope which is made extremely versatile by virtue of its *plug-in* construction. The plug-in feature refers to the fact that other types of vertical amplifiers and time-base units can be installed into the 561B by a simple replacement of interchangeable units (see also Fig. 9-39). We will discuss the 2A63 plug-in because it is an example of a vertical amplifier with a differential input. The 2B67 is chosen because it is a general purpose time-base unit.

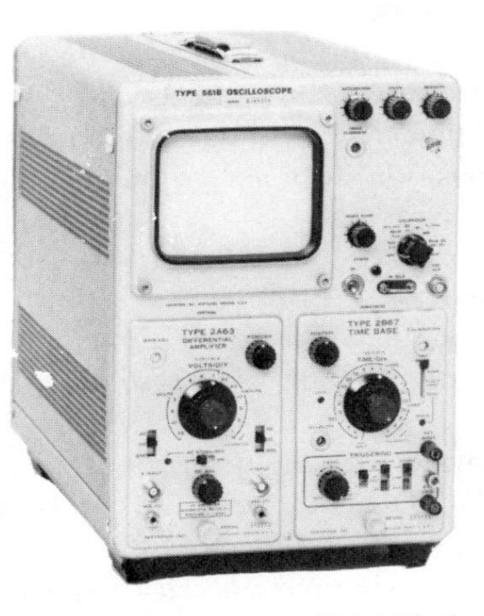

Figure 9-16. Tektronix 561B oscilloscope (with 2A63 differential amplifier and 2B67 time base) (Reproduced with permission of Tektronix, Inc.)

With these plug-ins, the Tektronix 561B has a dc-to 300-kHz bandwidth and vertical voltage sensitivities which range from 1 mV/Div. to 20 V/Div. However, the horizontal sensitivity is not adjustable; it is set to a fixed 1 V/Div. The amplifier gains are calibrated to within ±3 percent. Single-sweep as well as normal repetitive operation is also available.

In this section, the most commonly used controls which appear on the 561B scope face will be listed. However, a few will not be covered. For example, the *AC Stabilized*, *Alignment*, and *Calibration* are special controls which are used when making very accurate readings with the scope. Because of their rather esoteric nature, they need not be discussed in this text. If information on their functions is desired, consult the instruction manual of the appropriate plug-in unit.

TEKTRONIX 561B CONTROLS

Input +, *Input* ——These terminals make up the input points to the vertical amplifier. If the scope is used to measure a signal voltage relative to ground, only the Input + terminal needs to be used. In this case, the Input − terminal coupling switch should be set to the GND position. However, if the voltage between two nongrounded points in a circuit is being measured, the two points are connected to the Input + and Input − terminals, respectively.

AC, *DC*, *GND*—These switches select the type of coupling between the input signal and the inputs of the scope amplifiers (see Fig. 9-5).

1. AC position. Dc and low-frequency components (less than 5 Hz) of the input signal are blocked from entering the vertical amplifier input. Therefore, only the ac components above 5 Hz can pass and be displayed by the scope.
2. DC position. All components of the input signal can pass to the amplifier input and be displayed.
3. GND. The vertical amplifier input circuit is grounded and no voltage appears across the vertical deflection plates. This switch is used to center the scope trace without disconnecting the scope input leads. However, this position does *not* ground the applied signal.

Level—Selects the voltage level of the triggering signal which causes the sweep to trigger. If this control is set to AUTO, the sweep will automatically trigger, even if no triggering signal is present. If a triggering signal is being applied, this position will cause it to trigger the sweep as the signal crosses zero. When set to FREE RUN, the repetitive sweep of the trace is *not* synchronized with any triggering signal that may be applied.

Mode—NORMAL position selects the repetitive sweep operation of the oscilloscope. SINGLE SWEEP selects single-sweep operation. The sweep is ready to be triggered once (as indicated by the Ready Light). After being triggered once, the switch must be set to RESET before it will retrigger. Returning the switch to the NORMAL position will allow repetitive triggering to start again.

Coupling—This control is similar to the vertical input coupling switch, except that it chooses the coupling available for the trigger signal. If one wants the triggering circuit to respond to the entire triggering signal (whether it be external or internal), the DC position should be chosen. If only the time-varying component of the signal is to be applied to the triggering circuit, the AC SLOW position is chosen. When only the high-frequency components of a signal are to be used to trigger the sweep, use AC FAST.

Time/Div.—This control connects the horizontal amplifier to either the output of the sweep generator or an external signal. To allow an external signal to be connected, set this switch to EXT. INPUT.

DC Bal.—This control allows adjustment of the output of the amplifier to zero when the amplifier inputs are grounded. Often the amplifiers may drift so that a dc offset appears even if the amplifiers are grounded. This control corrects such drift.

Cal. Out—This terminal provides a calibrated squarewave waveform whose magnitude is selected by the position of the calibrator switch. The square wave can be used to calibrate the vertical amplifiers and check probe compensation.

MISCELLANEOUS CONTROLS

These controls are sometimes found on other scope models.

Gain Adj. (Gain Adjustment)—Used to adjust the amplifier gains to their proper value if scope plug-in units are interchanged. Often an internally placed adjustment control.

Volt. Reg.—Used to set the voltage output of the oscilloscope's internal power supplies.

Sweep Freq.—Calibrates horizontal sweep oscillator frequencies to match sweep-selector switch positions.

Freq. Comp.—Adjusts amplifier and attenuator elements for wide-band response (both vertical and horizontal circuits).

Hum Balance—Used in some scopes to cancel power-supply hum.

Z-Axis Input—Some scopes have an extra input for allowing modulation of the beam intensity by an input signal.

Beam Finder—When activated, this control will find and center a trace that has drifted off-screen.

How to Operate an Oscilloscope

The previous sections of this chapter introduced the functions of the oscilloscope subsystems and the controls on the scope face. They also discussed the two modes in which an oscilloscope is used to display the quantities being measured. This information can now be used to learn how to operate an oscilloscope. Proper operation includes making correct connections to the instrument, knowing how to turn it on, and how to display an accurate trace of the signal being measured. It also involves knowing how to measure such quantities as voltage, current, time, and frequency.

We will approach the subject by first examining the test leads and grounds of an oscilloscope and seeing how they are used to connect the scope to the circuit under test. Then various techniques used for measuring the quantities listed above will be presented.

OSCILLOSCOPE TEST LEADS AND PROBES

A probe is a device which is used to transfer the signal from a circuit being tested to the measuring instrument (in this case an oscilloscope). An ideal probe should not disturb the circuit under test or the performance of the scope.

The simplest type of probe is a test lead. *Test leads* can be merely convenient lengths of wire which connect the oscilloscope input to the test point. At one end, such leads usually have spade lugs, banana plugs, or other tips which fit the input terminals of the scope. At the other end, there may be alligator clips or other convenient means for connecting the probe to the circuit being measured. An additional refinement to many such simple probes is an electrostatic-shield cable that surrounds the test lead. This shield is meant to be connected to a ground terminal of the scope. The purpose of the shield is to protect the test lead from picking up unwanted hum and noise signals. Other types of probes which perform special functions, such as increasing the input impedance of the scope or measuring current, will be discussed in later sections of this chapter.

SCOPE GROUND TERMINALS

Figure 9-17 shows the input terminals of the HP 122A and Tektronix 561B including the ground terminals.[6] All of the ground terminals of oscilloscopes

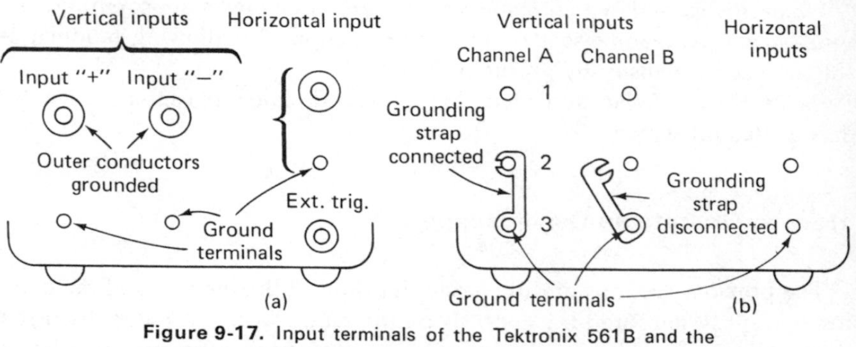

Figure 9-17. Input terminals of the Tektronix 561B and the HP 122A oscilloscopes (a) Tektronix 561B input terminals (b) HP 122A input terminals

[6]Note that on the HP 122A, terminals 2 and 3 on both vertical input channels are grounded to the chassis as long as the grounding strap is connected. The only time the strap is disconnected is when the channel is being used in its *differential input* capability. In that case, terminal 2 is no longer connected to terminal 3 and is therefore not grounded.

are connected together because they are all wired directly to the scope case or chassis. Since the chassis of the scope is connected to the building ground (and hence to earth ground) by the third wire of the line cord, all the ground terminals are also connected to earth ground. On scopes whose input terminals have both an inner and outer conductor, the outer conductors are also connected to the chassis. (This feature facilitates grounding the shielding conductor which surrounds the input cables.) Because there are so many ground terminals on an oscilloscope, care must be taken to prevent shorting out the measured circuit by inadvertently connecting a lead from a hot point of the circuit to a ground terminal.

MAKING CONNECTIONS TO AN OSCILLOSCOPE

A voltage measurement always involves placing the voltmeter across the two points being measured. Therefore, when making a measurement with an oscilloscope, at least two leads must be connected from the circuit under test to the scope inputs. The number of leads and the type of connection depends upon the type of amplifier input and whether the voltage is being measured relative to ground or some other nongrounded level. The following rules should spell out how connections are made in the proper manner.

1. If some point in the circuit being measured is earth grounded, connect the scope ground to this circuit ground with a separate lead.
2. If the voltage of the point in question is being measured relative to ground, then one additional connection to the scope must be made. Depending on whether the scope input is single-ended or differential, this connection is made in the following way:
   (a) Single-ended input—Connect the probe to the point in question of the test circuit and to the nonground input terminal of the oscilloscope.
   (b) Differential input—Connect the probe to the point of interest in the circuit and to the *plus* input terminal of the differential input. Ground the *minus* terminal of the input by setting its input coupling switch to the GROUND position or by ensuring that the grounding strap to that terminal is connected. (The grounding procedure depends upon the design of the particular input involved.)
3. If the voltage being measured by a scope is a voltage between two nongrounded points in a circuit, the method of connection also depends on the type of input available on the scope being used.
   (a) Single-ended inputs—With a single-ended input, the measurement of ungrounded voltages involves a method which is quite hazardous (and may sometimes lead to erroneous results). This method involves the use of the three-to-two-wire adapter on the scope power-line plug (Fig. 9-18). This adapter (also called a *cheater plug*) disconnects the third wire of the three wire cord from the building ground and thereby also disconnects the scope ground terminals and chassis from the building ground. (See

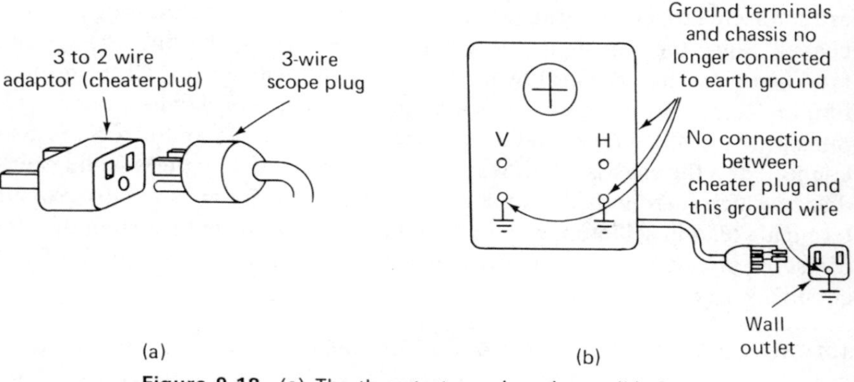

Figure 9-18. (a) The three-to-two-wire adapter (b) Scope with cheater plug attached to its three-wire plug. This disconnects its ground terminals and chassis from earth ground.

Chapter 3 for a review of grounds and grounding.) Since the ground terminals of the scope are no longer connected to earth ground, a nongrounded voltage can be connected between the two vertical input terminals, and the scope will display the voltage difference between them.

The possible hazard that can arise when using this method involves the fact that the chassis of the scope assumes the same voltage level above ground as the voltage connected to any of its ground terminals. If the ground terminal is connected to a point in the circuit which is at 115 V above ground, *the entire chassis of the scope* is like an exposed 115-V power-line wire sitting on the workbench. If a user somehow touches the scope and an earth ground point, a potentially lethal shock can occur. Therefore, the use of cheater plugs is mentioned in this section primarily to warn the user of the danger associated with their use and *not* as a recommendation.

On some dual-trace scopes with single-ended inputs, another method for measuring nongrounded voltages (which is not hazardous) is sometimes possible. The scope must have a feature that allows it to display the difference of the voltage levels fed to its two input channels (*B-A* position on Presentation control of HP 122A). Then the voltage of one nongrounded point in the test circuit can be fed to channel *A* and the other to channel *B*. With the Presentation control set to the subtracting position, the scope will display the voltage difference between the two points without having to resort to a cheater plug.

(b) Differential input—Connect one point being measured to the + input terminal of the differential input and the other to the − input. The scope will electronically subtract the one voltage from the other and display the difference of the voltages between the two points.

TURNING ON THE OSCILLOSCOPE

1. Before plugging in the scope, be sure that the Power switch is off and that the Focus and Intensity controls are set to their lowest positions.
2. Set Vertical and Horizontal Position controls to approximately their midpoint positions.
3. Set the Sweep Time/Div. switch to EXT. INPUT (or on the HP 122A, to a Horiz. Sens. setting of 1 V/cm).
4. Plug in power cord. If the scope chassis is to be earth grounded, use just the three-wire plug. If the scope chassis is to be purposely isolated from ground (i.e., floated), use a three-to-two-wire adapter plug.
5. Turn Power switch to ON.
6. Wait about 2–5 minutes until the scope has warmed up. Then increase the intensity until a spot just appears on the screen. If the spot does not appear, use the Position controls to locate it. *Always* keep the intensity of the spot as low as possible.
7. Use the Focus control to get a sharp fine spot.
8. Center the spot using the Position controls.
9. Set the Trigger Level switch to AUTO.
10. Set the Trigger Source (or Sync) switch to LINE and the Sweep Time/Div. switch to 1 ms/Div. (or less) to generate a solid horizontal trace. Center the trace along the horizontal centerline.
11. The scope is now ready for use.

VOLTAGE MEASUREMENTS

When connecting a scope to a circuit to measure voltage, follow the rules listed in the section on *How to Make Connections to the Scope*. Voltage can be measured from the resulting display as described next.

The oscilloscope is primarily a voltmeter. If it is used in the triggered-sweep mode, it displays the time variation of the voltage being applied to its vertical input. The height of the vertical deflection of the displayed trace combined with the setting of the V/Div. switch yields the *peak-to-peak* voltage of the input signal. For example, the waveform shown on the screen as in Fig. 9-19 has a vertical excursion of four divisions. If the V/Div. switch is set to 0.1 V/Div. (and the Variable (red) V/Div. switch to the *calibrate* position), the peak-to-peak voltage being displayed is 0.4 V.

To get an accurate reading from the display, position the trace so that its bottom edge is aligned with one of the lines of the graticule. Also position one of the peaks near to the vertical centerline.

If the waveform being examined is a *sinusoid*, one can convert the peak-to-peak reading to get an rms reading by using the relation:

$$V_{\text{rms}} = 0.35\ \text{V}_{\text{peak-to-peak}} \tag{9-2}$$

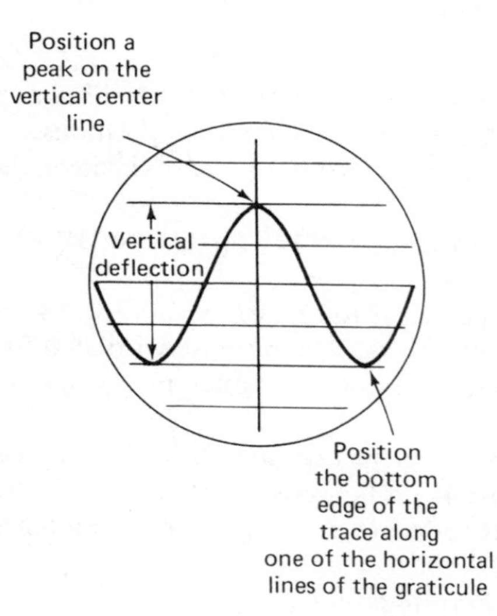

Figure 9-19. Measuring voltage from a scope display

The relations between peak and rms values of some other waveforms are given in Fig. 1-18.

CURRENT MEASUREMENTS USING A TEST RESISTOR

Although the oscilloscope really measures voltage, indirect measurements of current can also be made. One way to do this is to pass the current through a known test resistor and measure the resulting voltage drop. Figure 9-20 shows how such a connection can be made. R is often chosen to be a 1-Ω, noninductive resistor. Since $v = iR$, a 1-Ω resistor eliminates calculations and

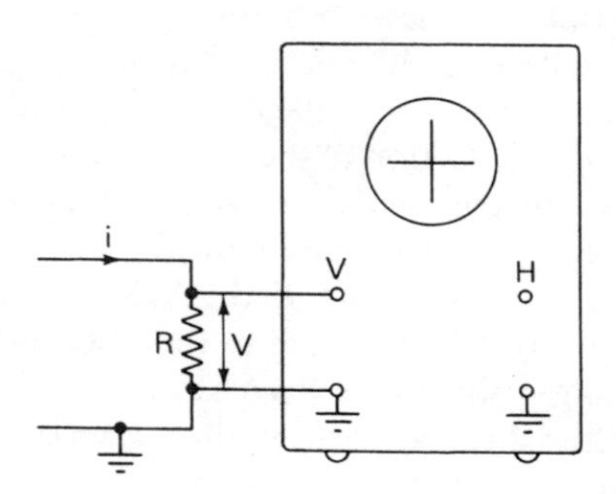

Figure 9-20. Measuring current by using a test resistor

yields the current immediately. When this connection is used, the power rating of the test resistor should be large enough to handle the anticipated power resulting from the current flow. Note that, unless a differential input is available or unless the scope ground is being floated, this method requires that one side of the resistor be grounded. Therefore, current cannot always be measured using this method.

CURRENT PROBES

An additional method used to measure current that does not require the insertion of a test resistor requires the use of a special *current probe* (shown in Fig. 9-21). The probe is clipped onto a current-carrying conductor and the voltage produced in the probe is proportional to current flowing in the conductor. There are two types of probes used to sense current. One type makes use of the *transformer effect* and the other, the *Hall effect*.

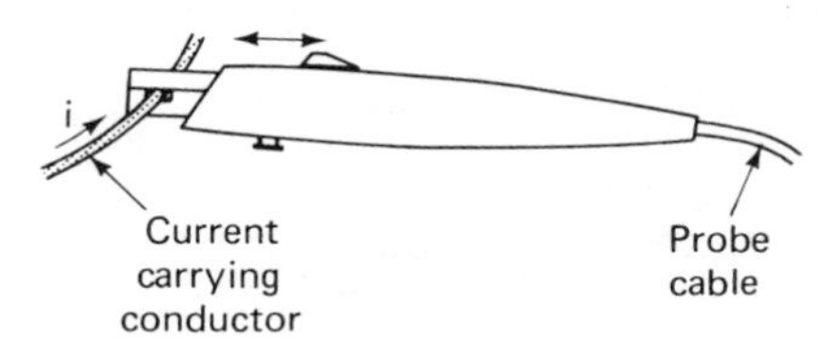

Figure 9-21. Current probe clamped around current-carrying conductor

The principle of transformer operation discussed in Chapter 7 can be directly applied to understanding the operation of the transformer-type current probe. (See Chap. 7, on transformer operation, if necessary.) Because a transformer requires ac excitation, this probe can only be used to measure ac currents. The typical frequency range over which transformer-type current probes are effective extends from about 800 Hz to several MHz.

The Hall effect is discussed in more detail in Chapter 15 on Transducers. The Hall effect does not depend on ac signals and so current probes based on this effect can measure dc as well as ac currents.

MEASUREMENTS OF TIME

When used in the triggered-sweep mode, the time-base circuit of a scope is used to provide sweep waveforms with various values of sweep times (s/Div.). If a signal is displayed when the scope is set to a specific sweep time per division, the number of horizontal divisions between two points along the signal waveform is a measure of the time elapsed. The following relation can be used to calculate the time from such a reading:

$$\text{time} = \begin{array}{l}\text{horizontal distance}\\ \text{between points of}\\ \text{display}\end{array} \times \begin{array}{l}\text{horizontal sweep}\\ \text{setting}\\ \end{array}$$

$$= d \times \text{s/Div.} \tag{9-3}$$

EXAMPLE 9-3

The horizontal distance between points 1 and 2 of the signal waveform shown in Fig. 9-22 is 5 Div. The horizontal sweep is set to 0.5 ms/Div. What is the time duration between points 1 and 2?

Solution: Using Eq. (9-3), we obtain

$$t = 5 \times 0.5 \text{ ms/Div} = 0.0025 \text{ s} = 2.5 \text{ ms}$$

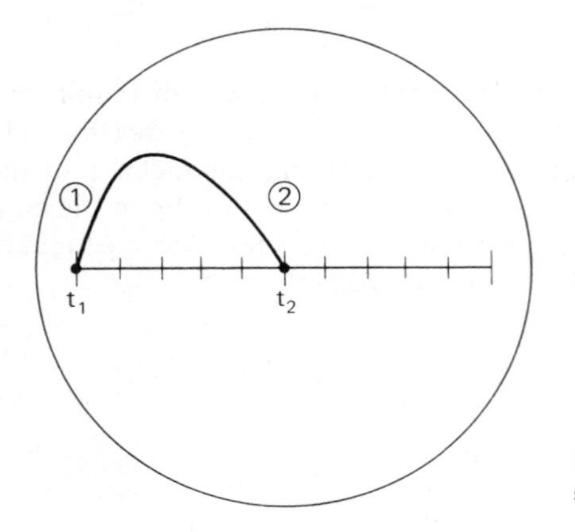

Figure 9-22. Measuring time from the scope display

For accurate measurements, position at least one point of the waveform of interest on the horizontal centerline of the screen.

FREQUENCY MEASUREMENTS (Triggered-Sweep Method)

Measuring the frequency, f, of periodic waveforms using the triggered-sweep mode is essentially the same technique that is used for measuring time. However, one additional calculation must be made for determining f. The frequency of a waveform is the number of cycles per second. Therefore

$$f = \frac{1}{T} \tag{9-4}$$

where T is the time of one cycle, or the period. To find f we measure the time of one period and use Eq. (9-4).

EXAMPLE 9-4

If a periodic function displayed on the scope screen has a distance of 4 cm between the beginning and end of a cycle, and if the Time/Div. control is set to 1 ms/Div., what is the frequency of the waveform?

Solution: First find the time duration of one waveform:

$$\begin{aligned} t &= \text{horizontal distance} \times \text{horizontal sweep setting} \\ &= 4 \text{ Div.} \times 0.001 \text{ s/Div.} \\ &= 0.004 \text{ s} \end{aligned}$$

Since in this case $t = T$, use

$$f = \frac{1}{T} = \frac{1}{0.004} = 250 \text{ Hz}$$

PHASE MEASUREMENTS (Triggered-Sweep Method)

The phase difference between two waveforms of the same frequency can be found by using the triggered-sweep method and the Lissajous figures method. In this section the triggered-sweep method is discussed.

The triggered-sweep method for determining phase difference compares the phases of two signals by using one signal as the reference. The shift in the position of the second signal compared to the reference signal can be used to calculate the phase difference between the signals.

To make the measurement, the phase of one signal is chosen as zero and the scope display is calibrated to indicate this choice. The calibration procedure involves setting the scope to External Trigger, the Level to zero, and the Slope to + so that the sweep triggers when a trigger signal crosses zero with a plus slope. The first signal, A, is connected then to both the vertical inputs and the external trigger terminals. The waveform displayed by the scope is like the waveform shown in Fig. 9-23. Next, the vertical input signal is changed from

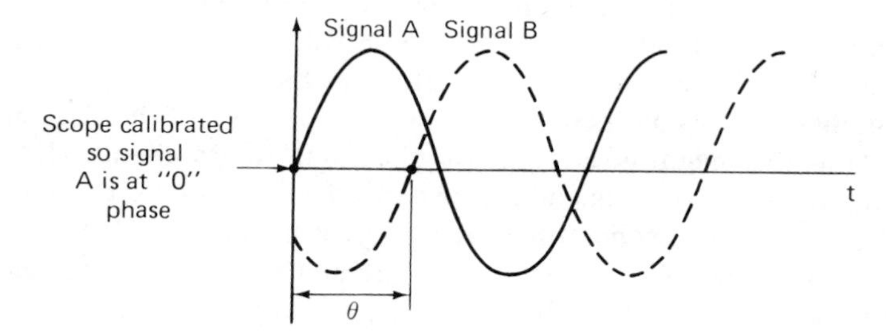

Figure 9-23. Phase-difference measurement using the triggered-sweep method

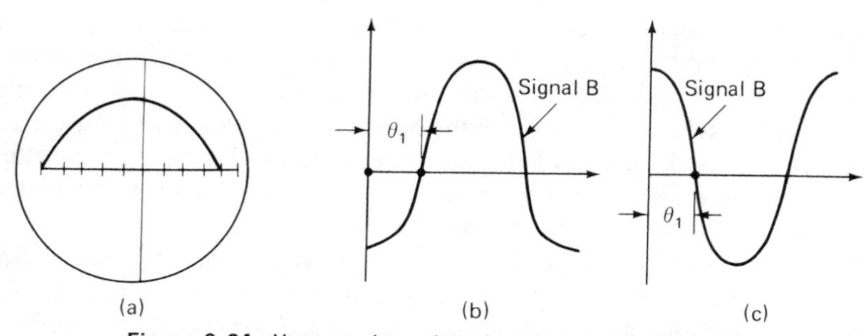

Figure 9-24. How to determine the phase angle from the triggered-sweep display (a) Calibration of horizontal axis so that 180° equals nine divisions (b) Phase of signal B is equal to $-\theta$ for this type of position shift (c) Phase of signal B is $\theta = 180° - \theta$ for this type of position shift

signal A to signal B. Signal A remains connected as the external triggering signal. Thus, if signal A triggers a sweep when signal B is not at the same level and slope, the display of signal B will be shifted in position along the horizontal (time) axis. To calibrate the time axis so that it corresponds to 20°/Div., use the *Variable Sweep Time* control to adjust the display of the waveform so that half a cycle of signal A corresponds to nine divisions (Fig. 9-24(a)). Then the phase shift can be found by measuring the distance to the first zero-scrossing of signal B (Fig. 9-24(b) and (c)).

LISSAJOUS FIGURES

If two sine waves are simultaneously fed to an oscilloscope (one to the vertical inputs and the other to the horizontal inputs) and the scope is set to operate in the X-Y mode, the resulting display on the scope screen is referred to as a *Lissajous pattern*. If the two sine waves are of the same frequency and phase, the Lissajous pattern will be a diagonal line. If the sine waves are of the same frequency but out of phase by 90°, the pattern will be an ellipse (if the amplitudes are also equal, the ellipse will instead be a circle). Figure 9-25 shows how Lissajous patterns result from the input of two sine waves.

The numbered dots on these figures trace the position of the electron beam as time and the magnitudes of the applied sine waves change. If the two signals are not of equal frequencies, the pattern will not be a diagonal, ellipse, or circle, but it will be some other unusual gyrating pattern. Thus, if the frequency of one signal is known, the other can be found by varying the known frequency source until a steady Lissajous pattern is displayed.

FREQUENCY MEASUREMENTS USING THE X-Y MODE

Since the sweep times of the sweep waveform are usually calibrated to within 5 percent of their stated values, frequency measurements using the triggered-sweep method can be in error up to this amount. However, if an accurate, adjustable frequency source is applied to the horizontal input of a scope, an unknown frequency can be determined much more accurately by comparison (Fig. 9-26). This is done by varying the frequency of the accurate frequency source until either a Lissajous pattern of a circle or an ellipse appears on the screen. The appearance of the steady Lissajous pattern indicates that the frequencies of both applied signals are equal.

If it is not possible to adjust the frequency of the source to obtain a circle or ellipse, the known frequency should be adjusted until a stationary Lissajous pattern with a number of loops is reached. The ratio of the number of horizontal to vertical loops of the stationary pattern yields the unknown frequency (Fig. 9-27).

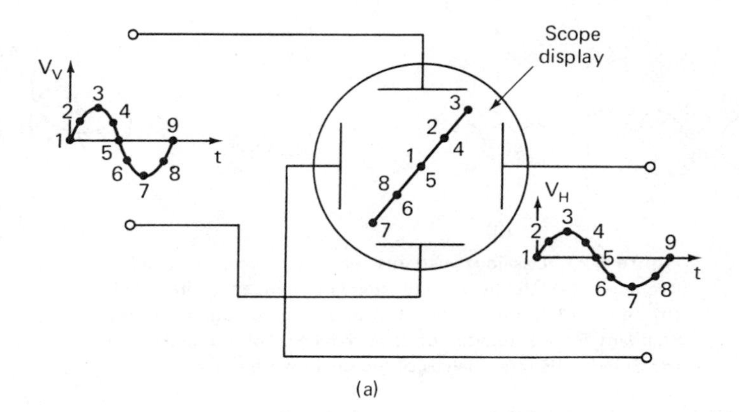

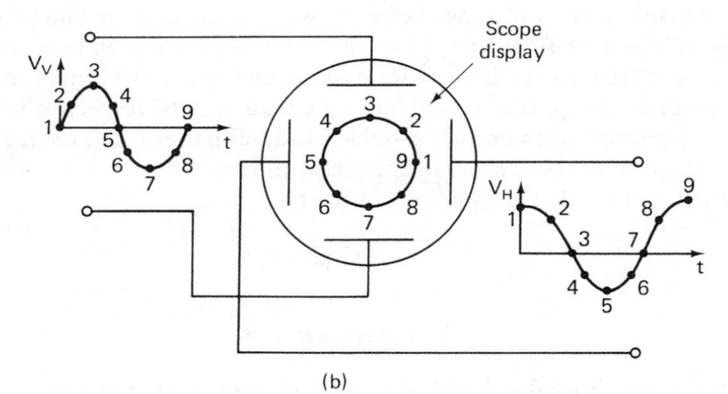

Figure 9-25. How Lissajous patterns are generated (a) Sine waves of equal frequency and phase applied to both horizontal and vertical plates (b) Sine waves of equal frequency and amplitude but with a phase difference of 90° applied to both the horizontal and vertical plates.

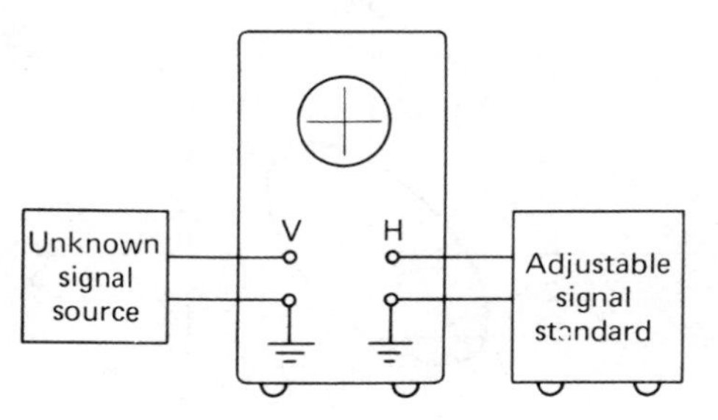

Figure 9-26. Connections for measuring an unknown frequency by comparison to a known frequency

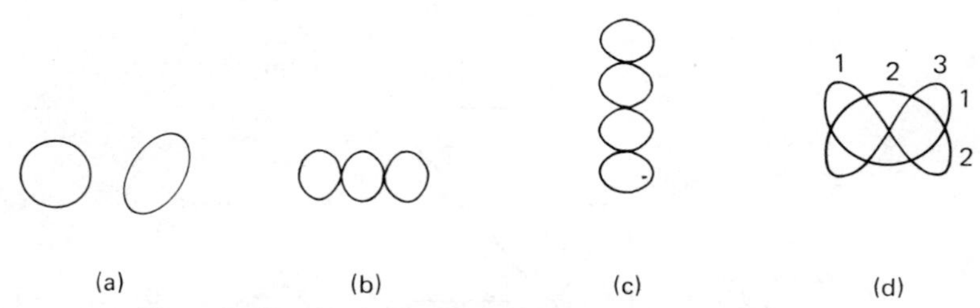

Figure 9-27. Lissajous patterns obtained when measuring frequency (a) Unknown and standard frequency are equal (b) Standard is three times the frequency of unknown (c) Standard is one quarter of the frequency of the unknown (d) Standard is three-halves of the unknown frequency

PHASE MEASUREMENT USING LISSAJOUS PATTERNS

To measure phase difference between two sine waves, they must by definition be of the same frequency. (The phase difference between two sine waves of different frequencies is meaningless). Therefore, if two equal-frequency sine waves are fed to the vertical and horizontal inputs, respectively, the display on the scope screen will be a stable Lissajous pattern. The characteristics of the shape of the pattern allow the phase difference between the two signals to be determined. If the equations of the two waves are

$$X = C \sin \omega t \tag{9-5}$$

and

$$Y = B \sin (\omega t + \theta) \tag{9-6}$$

the phase difference θ is found from the Lissajous pattern by the equation

$$\frac{A}{B} = \sin \theta \tag{9-7}$$

where A is the point where the ellipse crosses the Y axis (Fig. 9-28).

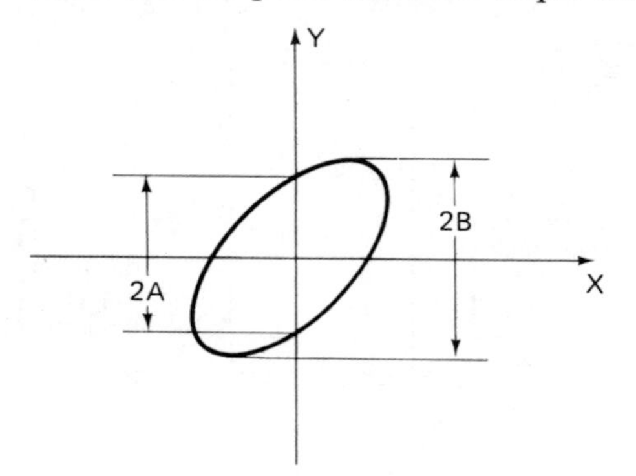

Figure 9-28. Determining phase angle from Lissajous patterns

Note that if the display is a diagonal line, $A = 0$ and $A/B = 0$; therefore, $0 = \sin(0°)$. This means the two sine waves are in phase. If the display is a circle, $A/B = 1$ and $1 = \sin(90°)$. This indicates the two signals are 90° out of phase.

Oscilloscope Limitations

OSCILLOSCOPE LOADING AND PROBE USE

In Chapter 3 we discussed the general concept of input impedance and how a measuring instrument can change the values of the current and voltage in the circuit it is measuring. When a measuring instrument disturbs the circuit under test by drawing current from it, the effect is known as *loading*. The loading effect of a voltmeter manifests itself by causing the voltmeter to indicate a smaller voltage value than the actual voltage being measured. The extent of the deviation from the actual voltage is found from the equation

$$V = V_o \frac{R_M}{R_M + R_T} \tag{9-8}$$

where V is the voltage indicated by the voltmeter, V_o is the actual voltage, R_M is the input impedance of the voltmeter (in ohms), and R_T is the equivalent impedance of the circuit being measured (also in ohms). We can see that as R_T approaches or becomes greater than the value of R_M, the indicated voltage gets significantly smaller than the actual voltage being measured. Thus, the smaller the value of its input impedance, the greater is the likelihood of a serious loading effect being caused by a voltmeter.

Since the oscilloscope is basically a voltmeter, it is also prone to loading a circuit. Let us see how and when scope loading occurs and how its effects can be reduced.

The input impedance of the oscilloscope amplifiers can be represented by the circuit in Fig. 9-29(a). For a typical oscilloscope amplifier, R is about 1 MΩ and C is between 30 and 50 pF.

If a dc voltage is applied to the scope, the input impedance is just 1 MΩ because the capacitor appears as an open circuit to dc quantities. Hence, when measuring dc signals, an oscilloscope will begin to seriously load a cir-

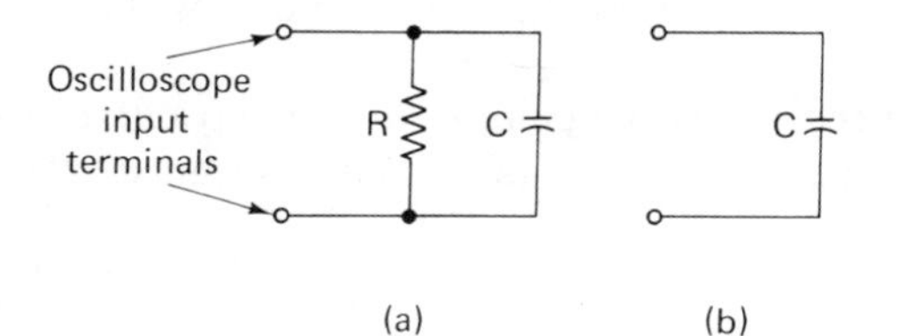

Figure 9-29

cuit (by causing a 10 percent or greater indication error) if the equivalent resistance of the test circuit is 100 kΩ or more.

EXAMPLE 9-5

If an oscilloscope is set to a sensitivity of 1 V/Div. and its inputs are connected to the circuit as shown in Fig. 9-30, find the vertical deflection of the spot.

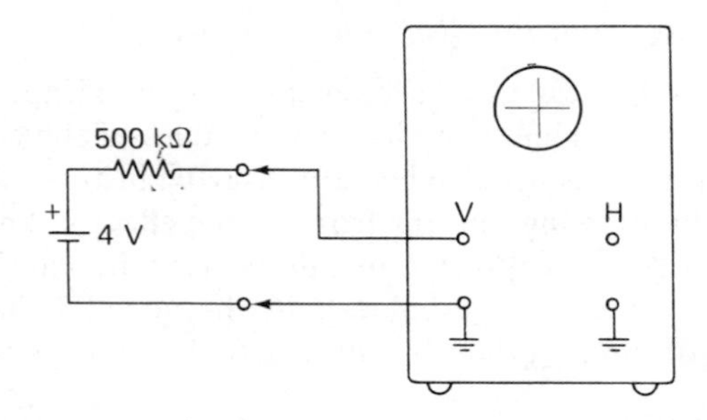

Figure 9-30

Solution: If the scope were an ideal voltmeter, the spot would be vertically deflected 4 Div. because the scope would be indicating the dc voltage of 4 V. However, due to loading effects, the deflection of the scope is

$$\text{Deflection} = \frac{\text{volts}}{\text{vertical sensitivity}} = V_o\left(\frac{R_{\text{scope}}}{R_T + R_{\text{scope}}}\right) \div 1 \text{ V/Div.}$$

$$= 4 \times \frac{1 \text{ M}\Omega}{0.5 \text{ M}\Omega + 1 \text{ M}\Omega} = 2.67 \text{ Div.}$$

Note that this represents an error in the indication of 33 percent from the actual voltage.

When an ac signal is applied to a scope, the capacitive reactance of C in Fig. 9-29 also begins to affect the input impedance. As the frequency of the input signal increases, the capacitor makes the effective input impedance of the scope decrease. This happens because the capacitive reactance X_C (in ohms) decreases with increasing frequency according to

$$X_C = \frac{1}{\omega C} \tag{9-9}$$

where ω is the radian frequency of the signal, and C is the capacitance in farads.[7]

At very high frequencies, the magnitude of the capacitive reactance becomes much smaller than the magnitude of the resistance. Then the input

[7]Example 6-4 works out the capacitive reactance of two different capacitors at different frequencies.

signal current virtually bypasses the higher-resistance path and flows through the lower-impedance path offered by the capacitor. At such high frequencies, the input circuit of the scope can be approximated by a capacitor along (Fig. 9-29(b)).

Thus we see how the overall loading effect of a scope will increase with increasing signal frequencies or with increasing impedance of the measured circuit.

To counteract the loading effect of an oscilloscope, voltage probes are used. *Voltage probes* are devices which increase the input impedance of a scope by inserting a high impedance in series with the scope inputs. Since the scope input impedance is increased, the error due to loading is decreased.

Because the capacitor of the input-impedance model shown in Fig. 9-29(a) has no effect on dc quantities, a tenfold increase in the input impedance for dc signals could be achieved by inserting a resistor into the input circuit as shown in Fig. 9-31. However, for progressively higher frequency signals, the decreasing capacitive reactance assumes a greater and greater role in establishing the scope input impedance.

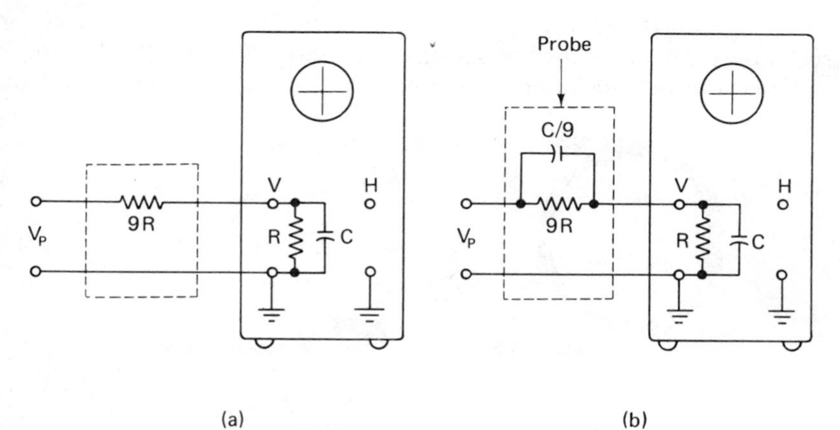

Figure 9-31. Connections for increasing the input impedance of an oscilloscope

Therefore, a high-valued series resistor alone will not be sufficient to increase the high-frequency input impedance of the scope. There must also be a decrease in the input capacitance in order for the probe to uniformly increase the input impedance for all frequencies.

To decrease C by a factor of 10, a capacitor of capacitance value $C/9$ is also connected in series with the scope capacitance. The circuit of Fig. 9-31(b) shows a suitable method for making such a connection.

With a probe such as that shown in Fig. 9-31(b), the overall input impedance of the scope as seen from the probe terminals can be represented by

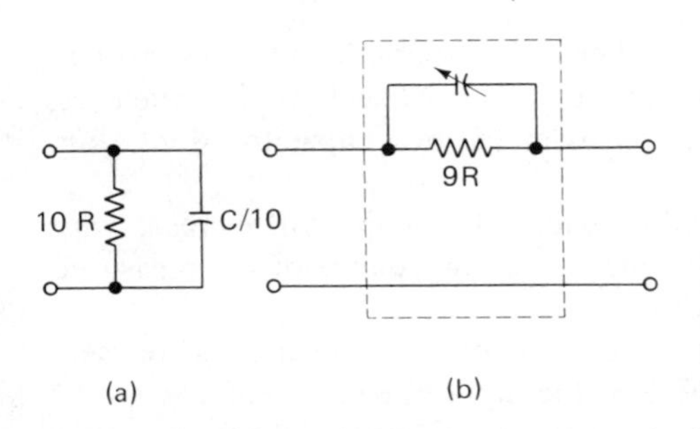

Figure 9-32. (a) Input impedance of an oscilloscope with a 10*x* voltage probe connected (b) Commonly used circuit model for representing a voltage probe

Fig. 9-32(a). The impedance of the scope with this probe inserted is increased by a factor of 10 for all frequencies.

In practice, the total input capacitance also contains a stray capacitance effect due to the cables carrying the test signal to the scope input ($\approx$50 pF). Since this capacitance is not precisely known, the capacitor within the voltage probe must be a variable capacitor to allow for an exact adjustment of its value to $C/9$. Such an adjustment is called *probe compensation.* A special square-wave signal generated within the scope can be displayed on the scope screen to check if the adjustment is correct (Fig. 9-33). If the probe capacitor is set to the right value, the test signal will look like Fig. 9-33(b). Improper adjustment will yield distorted waveform shapes such as those of Fig. 9-33(c)

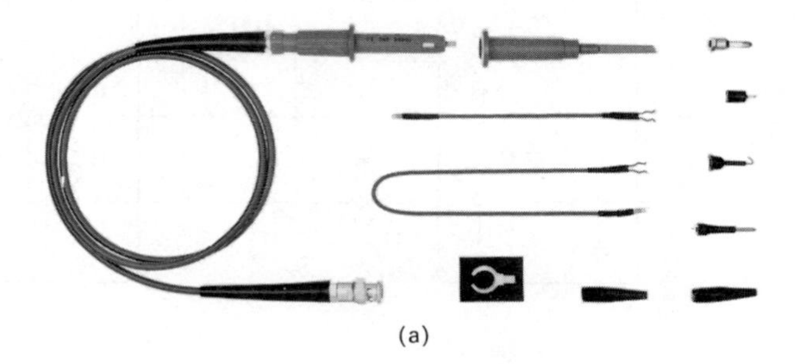

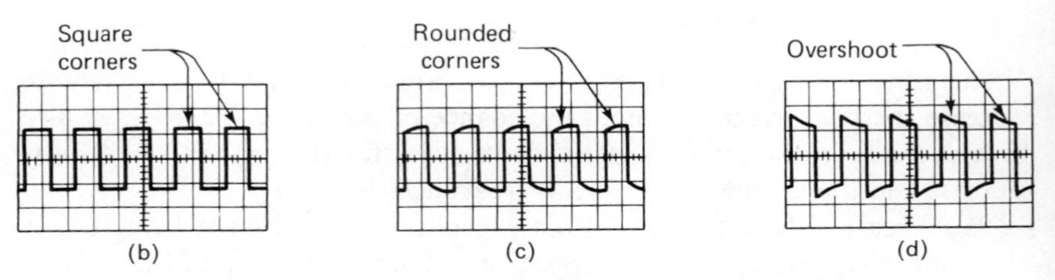

Figure 9-33. 1000-Hz calibrator waveform for adjusting probe compensation in 561B oscilloscope (a) One type of 10*x* probe (Reprinted with permission of Tektronix, Inc.) (b) Probe compensation properly adjusted

and (d). Probes must be compensated every time they are connected to a different scope amplifier. For this reason, the probe is represented by a circuit model such as that shown in Fig. 9-32(b). Figure 9-33(a) shows one type of 10*x* voltage probe. It allows a variation of its capacitance by rotating a sleeve on the probe body. Other types of probes may use a recessed screw on the probe handle to adjust *C*.

The high impedance of the voltage probe also reduces the amplitude of the signal received by the scope amplifier inputs. The 10*x* voltage probe described in this section increases the input impedance by a factor of ten, but it also attenuates (reduces) the magnitude of the signal tenfold. The voltage probe therefore sacrifices sensitivity for higher input impedance. When a 10*x* voltage probe is used with the oscilloscope, the sensitivity switch setting must be multiplied by ten to display the same V/Div. deflection as if no voltage probe is being used.

HUM AND NOISE PICKUP

The oscilloscope is designed to be capable of amplifying and displaying small input signals. This capability also makes it susceptible to amplifying small unwanted signals and noise, especially when set to its most sensitive V/Div. capability. To prevent the scope from picking up such interference signals, special precautions must be taken.

The most serious interfering signals in a scope come from the electric and magnetic fields generated by the power lines that carry 60-Hz electric power to the home and laboratory. Motors, fluorescent lights, and other electrical equipment being run from the power lines are also sources of such time-varying fields. These stray fields are always present in the lab and cause 60-Hz signals to be induced in the circuits being measured by the scope and in the scope input leads. This 60-Hz signal is referred to as *hum*.[8] The magnitude of hum picked up by an oscilloscope depends upon the impedance of the circuit being measured, the length of the scope leads, and the type of shielding used by these leads. Hum signals are largest when long, unshielded leads are connected to high-impedance circuits. The longer the lead length, the greater is the stray capacitance between the leads and nearby power-line cords. Since

$$i = C\frac{dv}{dt} \tag{9-10}$$

a larger unwanted signal will be induced in a test lead when a greater capacitive effect exists between it and some other conductor at a different potential. This equation also predicts that higher-frequency signals (larger value of dv/dt) will also cause a greater pickup current to flow in the leads. Thus

[8]See Chapter 3 for a discussion as to why 60-Hz noise is called "hum" and for a more detailed description of externally generated noise.

any wires carrying higher-frequency signals (from nearby oscillators, radios, etc.) can also be sources of interfering signals.

To verify the fact that a scope picks up and displays interfering signals, connect one end of an unshielded cable (or a shielded cable with an *ungrounded* shield) to the scope inputs. Leave the other end unconnected and lying on the work bench. The vertical amplifier should be set to the position of greatest sensitivity and the trigger level to the AUTO triggering position. The sweep time should be set to about 5 ms/Div. A 60-Hz sine wave should be displayed on the scope screen. If the resulting display is a distorted 60-Hz sine wave, the distortion may be caused by high-frequency harmonics or spikes arising from fluorescent lights, etc.

The best way to reduce the level of such interference signals is to use short cables which have a shield that can be connected to ground. Most scope probes are provided with a shield that can be connected to the scope ground. When the shield is grounded, it remains at approximately zero potential and thus protects the signal-carrying cable from being influenced by the external time-varying electric fields. This fact can also be verified by grounding the shield of the input cables in the experiment mentioned above. In practice, shielded cables are about 90 percent effective in guarding the leads from external interfering signals.

In addition to the use of shielded cables, obvious noise sources such as fluorescent lights, oscillators, and motors should be turned off or kept far from the scope during low-level measurements. Differential inputs are also designed to help reduce noise pickup in low-level signals.

OSCILLOSCOPE ERRORS

1. Reading Error—It is usually difficult to read the position of a scope trace to better than $\frac{1}{10}$ of a major division. Hence the reading error can be $\pm\frac{1}{20}$ of a division at best, and more if the observer is careless. This error, as a percentage of the reading, becomes larger if the deflection involves fewer divisions. To minimize the error, always use the V/Div. setting which yields the largest deflection while still displaying entire vertical excursion of the signal on the screen.

EXAMPLE 9-6

If the deflection of a signal displayed on a scope screen is five divisions, what is the minimum reading error? If the deflection is only one division, how large is this error?

Solution: (a) If the deflection is five divisions and the error is $\pm\frac{1}{20}$ or 0.05 divisions, the possible error is

$$\text{Error} = \frac{\pm 0.05}{5} \times 100\% = 1\%$$

(b) If the deflection is one division, the possible error is still ± 0.05 divisions or

$$\text{Error} = \frac{0.05}{1} \times 100\% = 5\%$$

2. Parallax Error—In some scopes, the graticule is positioned a short distance in front of the screen (Fig. 9-34). If the observer does not look at the screen properly, the spot will appear to be at a slightly different position on the graticule than it really is. The error due to this effect is called *parallax error.*

There are other scopes which are made with the graticule etched directly on the CRT screen. In these scopes, parallax error is eliminated. Check the operation manual of the particular scope being used to see if this feature exists. There is no parallax error in the HP 122A or Tektronix 561B.

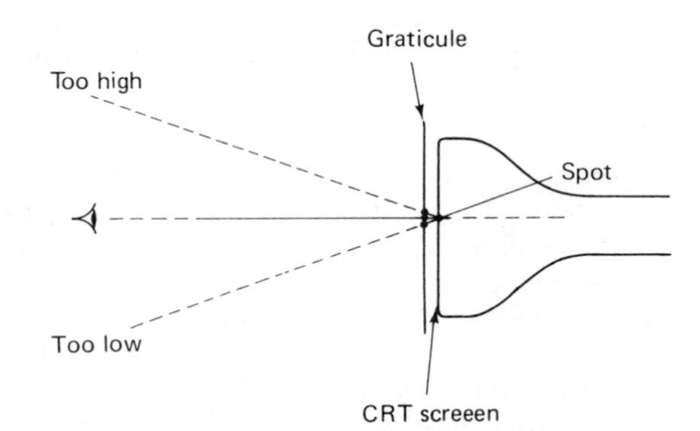

Figure 9-34. Parallax error in oscilloscopes

3. Scope Calibration Error—The sensitivity of a scope may be set to various levels. When the *calibration position* is employed, the deflection per division is set equal to the sensitivity step selected. But even when the scope is properly calibrated, there is still an error inherent in the calibration circuits. This calibration error varies with the instrument, but common values are from 1–3 percent. Thus a scope with a 3-percent calibration error, which is set to 1 V/Div. sensitivity, may actually be displaying 0.97 to 1.03 V/Div. Check the scope manual for the value of this error. (e.g., HP 122A, ± 3 percent; Tektronix 561B, ± 3 percent)

4. Frequency-Response Error—The amplifier of each oscilloscope is designed to provide constant amplification from dc up to some maximum frequency. Above this maximum frequency, the amount of amplification begins to drop off. Thus, if a signal is applied whose frequency exceeds the capability of the scope, the displayed trace will no longer yield an accurate vertical deflection of the signal. For the HP 122A, the maximum frequency is 200 kHz. For the Tektronix 561B equipped with a 2A63 plug-in, the maximum frequency is 300 kHz.

5. Loading Error—The general concept of loading was introduced in Chapter 3. The details of how loading causes an error in the readings of a scope were discussed in the section on *Probes and Scope Loading*.

High-Frequency and Other Special Oscilloscopes

In addition to the basic oscilloscope described in this chapter, there are other special purpose oscilloscopes and features which extend some of the basic scope capabilities. The most important types will be briefly described here, and a few facts about their special features and functions will be mentioned. However, the operating details and applications of such scopes are too specialized (and complex) to be fully explored by this text.

DELAYED SWEEP

The delayed sweep feature allows small, selected portions of the oscilloscope trace to be displayed. The selected portion can be shown in an expanded version at whatever magnification is desired. (The sweep magnifier also allows parts of the display to be magnified up to 10x.) For example, suppose a digital signal such as the one shown in Fig. 9-35(a) is displayed on a scope screen. If we wanted to examine the third pulse of this signal in more detail, we could use the delayed sweep feature to display the third pulse in expanded form.

If we tried to use a faster time/Div. setting to magnify the trace in order to display an expanded version of this pulse, we would end up displaying only the first pulse (or part of the first pulse) of the signal. This would occur because a shorter time/Div. setting would expand only the part of the signal at the beginning of the sweep. The part of the sweep that contains the third pulse would be out of the screen area (Fig. 9-35(b)).

The delayed sweep feature is made possible by the use of two time bases in the oscilloscope: a main time base (A) and a secondary time base (B). Time base A generates the same type of sweep waveform as in the basic oscilloscope. Time base B also generates a sweep waveform, but the starting time of this waveform is controlled by the sweep signal of time base A. The amount of time by which the start of sweep waveform B is delayed from the start of waveform A can also be varied.

When the oscilloscope is to be used in the delayed sweep mode, the entire trace is first displayed on the scope screen. Only sweep waveform A is used to trigger the display (Fig. 9-36(a)). The portion of the trace that is to be examined is then selected. Now time base B is activated so that it generates a sweep waveform to display the selected portion. The CRO adds the contri-

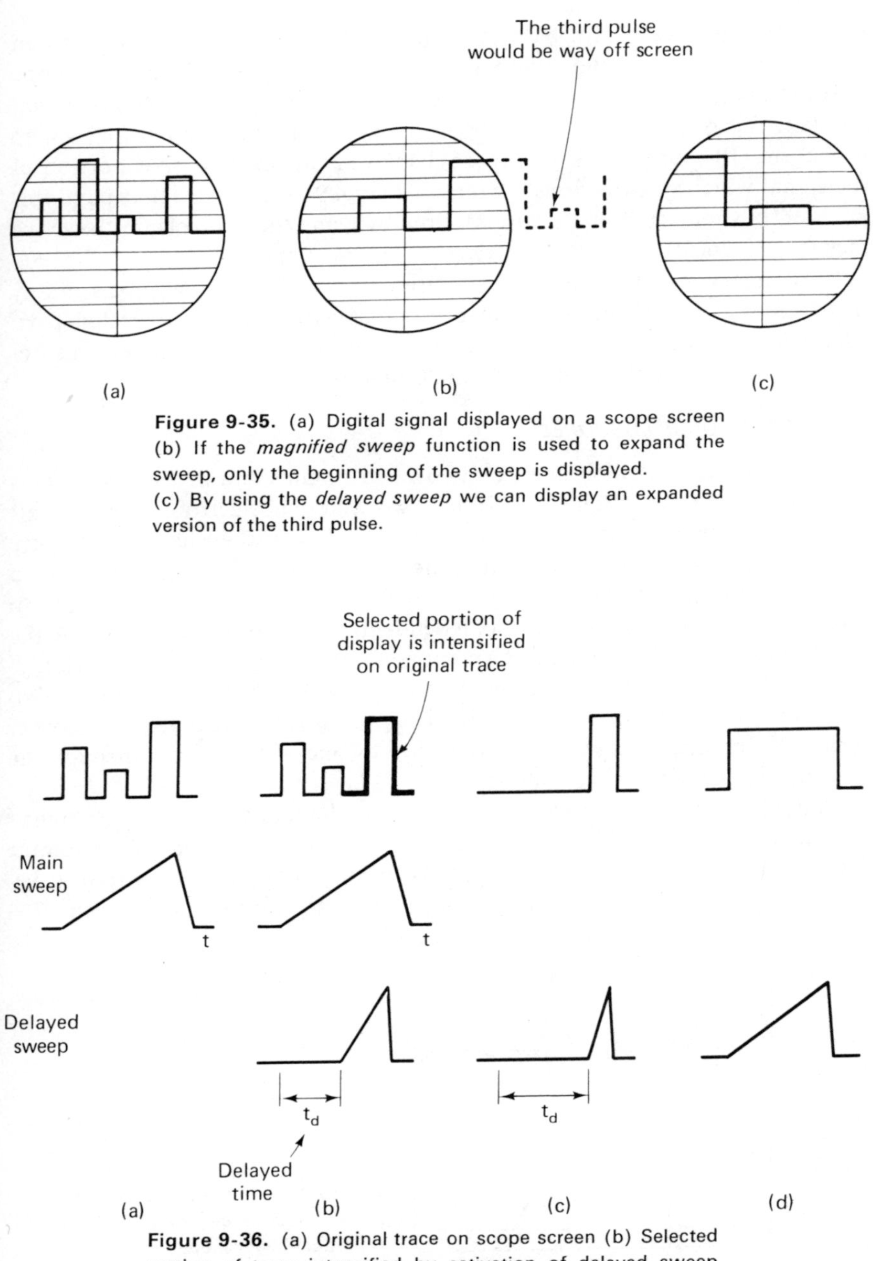

Figure 9-35. (a) Digital signal displayed on a scope screen (b) If the *magnified sweep* function is used to expand the sweep, only the beginning of the sweep is displayed.
(c) By using the *delayed sweep* we can display an expanded version of the third pulse.

Figure 9-36. (a) Original trace on scope screen (b) Selected portion of trace intensified by activation of delayed sweep waveform (c) Main sweep no longer causes a signal to be displayed—only delayed sweep waveform yields display (d) Selected portion of original trace in expanded display

bution of sweep waveform B so that the selected portion is presented as an intensified portion of the original display (Fig. 9-36(b)). Then the scope operation is changed so that sweep waveform A no longer causes the original signal to be displayed on the scope screen. Since only the delayed sweep waveform (B) causes the signal to be displayed, we see only that portion of the signal which was previously intensified (Fig. 9-36(c)). We can expand this portion to whatever magnification desired by using the controls on the scope face (Fig. 9-36(d)).

In summary, the delayed sweep feature allows selection of any part of any displayed signal with flexible control for displaying only the selected part. The delayed sweep may also be used to make more accurate time measurements than is possible by using a single time-base display.

DUAL-BEAM OSCILLOSCOPES

The dual-beam oscilloscope uses two electron guns and deflection systems mounted in the same CRT to display two different electron beams simultaneously. This feature should not be confused with the dual *trace* feature of other oscilloscopes (which alternately display two signals on one screen by time-sharing one electron gun and deflection system). By being able to display and trigger each beam independently, dual-beam scopes have the capability of displaying unusual combinations of signal characteristics.

For example, one beam may be used to show the phase shift between two sine-wave signals by displaying their Lissajous pattern. Meanwhile, the other beam can be displaying the time variation of one of the signals through the use of the triggered-sweep mode.

A dual-beam scope can also display two simultaneous, nonrecurrent signsls of short duration. This cannot be done on a dual-trace scope. An example of this application is shown in Fig. 9-37, where the input and output waveforms of a pulse-shaping circuit are both exhibited on the same screen.

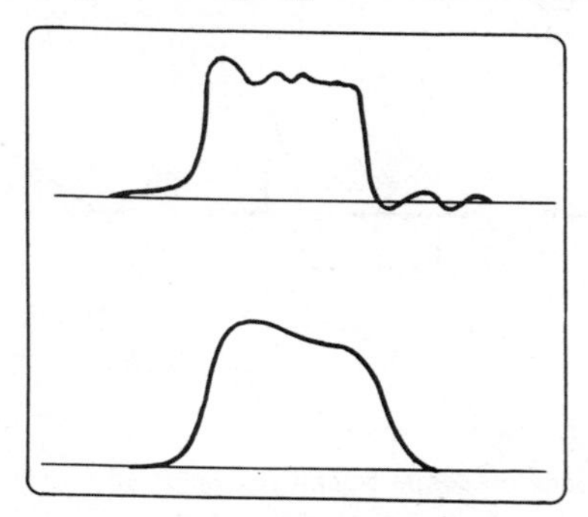

Figure 9-37. Input and output pulses of a circuit can be displayed simultaneously on a dual-beam scope

STORAGE OSCILLOSCOPE

If the vertical input signal being examined by a scope represents a single event rather than a repetitive waveform, or if a signal traces a very slowly occurring process, an ordinary scope image will fade before the signal can be completely observed. To preserve or *store* the image long enough to examine the entire sweep, a *storage oscilloscope* is used. This type of scope is equipped with a special CRT designed to maintain a visible trace of one sweep for extended periods of time. Some storage scopes also have the feature of *variable persistence* which can vary the trace storage time from seconds to hours. Storage scopes are finding increasing applications in mechanical and biomedical fields where events occur in nonrepetitive or long-duration cycles.

HIGH-FREQUENCY OSCILLOSCOPES

The simple oscilloscopes discussed earlier had upper frequency limits of about 300 kHz. However, there are many applications which involve the measurement or examination of signals whose frequencies are much higher than this upper limit. For such higher-frequency applications, conventional high-frequency oscilloscopes are built. These high-frequency scopes can display signals with frequencies up to 500 MHz. They differ in their internal construction from lower-frequency scopes by their special high-frequency CRTs and high-frequency amplifiers.

SAMPLING OSCILLOSCOPES

To display repetitive signals whose frequencies are higher than the limits of high-frequency scopes, another technique, called *sampling*, is used. In this technique, the signal is reconstructed from sequential samples of its waveform. These samples are obtained at slightly different points along successive cycles of the signal waveform (Fig. 9-38). This means that the waveform being displayed must be a repetitive one in order for sampling to be possible.

A sampling pulse generated within the scope turns on the scope measuring circuit for a brief instant. The vertical position of the electron beam is controlled by the resulting voltage observed at that point of the input waveform. The following pulse samples the waveform in its next cycle at a slightly different point. The horizontal position of the spot on the scope screen is

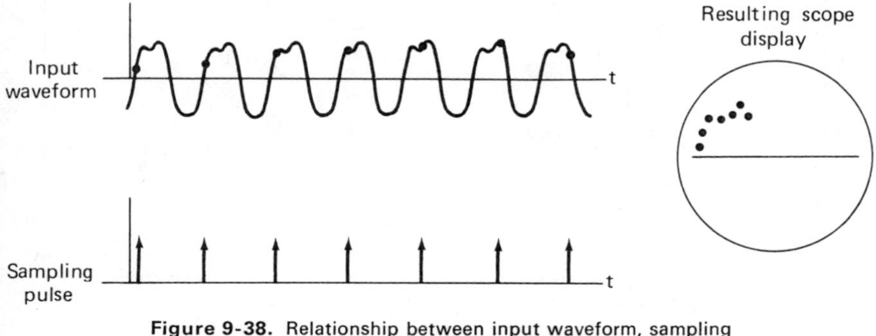

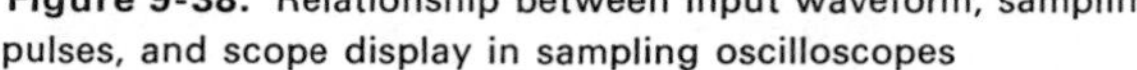
Figure 9-38. Relationship between input waveform, sampling pulses, and scope display in sampling oscilloscopes

meanwhile stepped forward very slightly, and the vertical position is determined by the new voltage value. As many as one thousand samples are taken to reconstruct one cycle.

By using the sampling technique, the bandwidth of an amplifier needed to handle high-frequency waveforms can be much lower than the frequency of the waveform being examined. As a result, sampling scopes are being built to display signals with frequencies of up to 14 GHz.

DIGITAL READOUT OSCILLOSCOPES

Digital readout oscilloscopes are high-speed oscilloscopes that also allow a digital readout of the signals being displayed by the scope. The digital readout feature consists of an electronic counter built into the same cabinet as the scope. The counter circuitry is connected to the scope outputs and allows values of signal amplitudes, time differences, and rise times to be displayed in digital form. The input signal of the scope is also sampled, and a reconstructed signal appears on the CRT.

PLUG-IN SCOPES

Many manufacturers of oscilloscopes are building so-called *mainframes* of oscilloscopes. The mainframe consists of a CRT, power supplies, and the final drive of the horizontal and vertical inputs. Various *plug-in* units can be inserted into this mainframe to provide various specialized oscilloscope functions. For example, plug-ins that perform dual-trace, high-gain differential, fast-rise high gain, or spectrum analysis operations are commercially available for vertical channels. Single-sweep, delayed-sweep, and sampling-sweep units are built for the horizontal plug-ins. Manufacturer's catalogs and operating manuals should be consulted for more details on each individual plug-in. The Tektronix 561B is a plug-in type scope, and Fig. 9-39 shows the plug-in units available for use with the 561B mainframe.

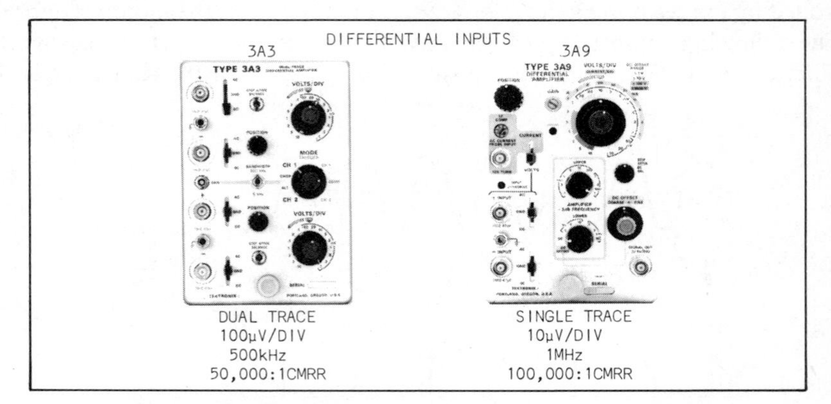

Figure 9-39. Plug-in units for 561B oscilloscope (Reprinted with permission of Tektronix, Inc.)

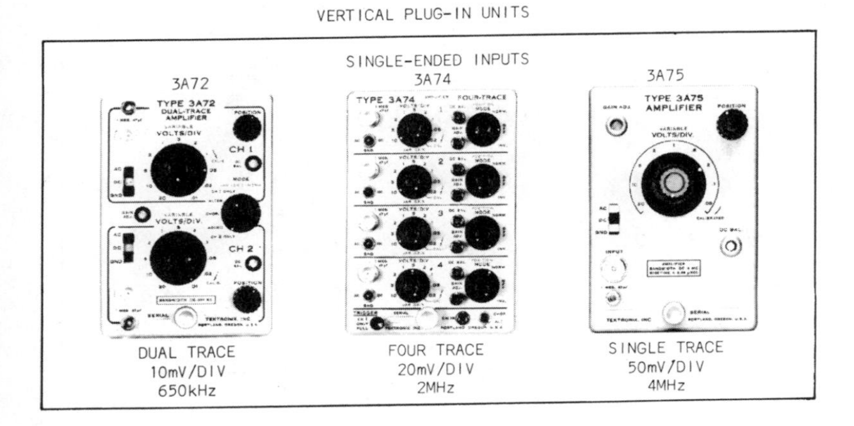

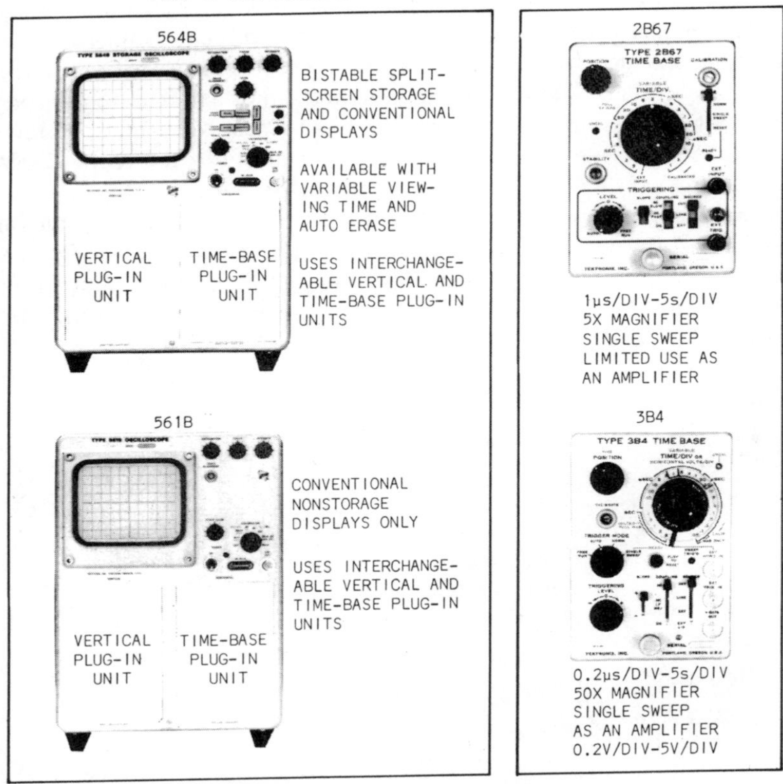

Figure 9-39. (continued)

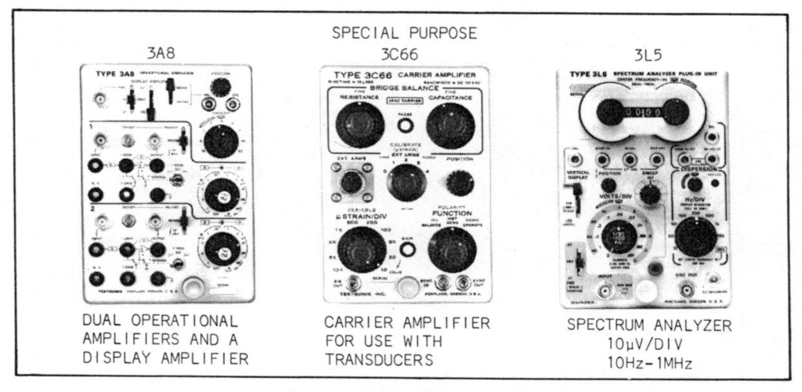

Figure 9-39. (continued)

Oscilloscope Photography

To obtain a permanent record of the signals displayed by an oscilloscope or transistor curve tracer, a hand trace or photograph can be made.

For hand tracing, special stick-on, transparent papers with inscribed graticules are available. This recording method is not as accurate as a photograph, but it is cheaper. When making hand traces, care must be taken to avoid scratching the screen or getting the intensity of the scope display too high, thus causing the screen phosphor to be burned.

For making photographs of scopes traces, special cameras employing Polaroid films are used (Fig. 9-40). These special cameras are designed to mount directly onto the scope face. The Polaroid film allows photographs to be made quickly, and it is unnecessary to expose the whole roll before developing the pictures.

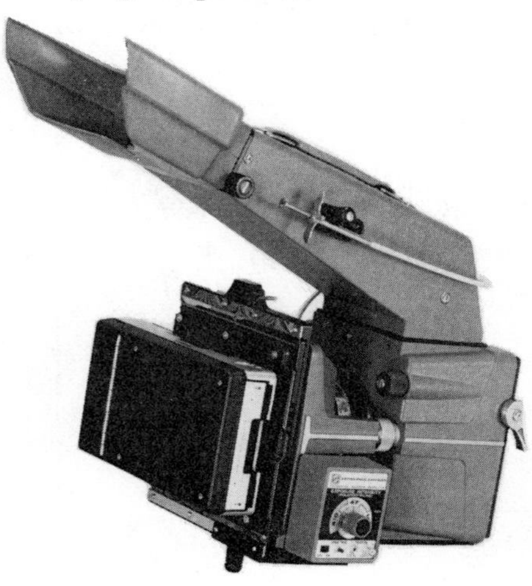

Figure 9-40. Typical oscilloscope camera (Reprinted by permission of Tektronix, Inc.)

Usually high-speed film (ASA rating of 3000 or 10,000) is used so that short exposure times are possible. After Polaroid film is developed, it must be carefully coated to protect the photograph from deteriorating. Suitable coating material is usually supplied with the film.

The following list of guidelines is provided to assist the reader in making oscilloscope photographs.

GUIDELINES FOR OSCILLOSCOPE PHOTOGRAPHY

1. Consult the instruction manual of the camera being used for details on loading the film, focusing the camera, and controlling its shutter speed.
2. When using ASA 3000 film, an exposure of f/5.6 and 1/5 second is good starting value.
3. Adjust the focus and astigmatism controls of the scope for a sharp trace. Make sure the camera is focused on the trace.
4. Keep the trace intensity low. Sharper pictures and less background lighting result by using low intensity traces.
5. Keep scale illumination (graticule) low.
6. If the photograph is not satisfactory when the above exposure is used, change the exposure by a factor of 4 or 8. A change by a factor of 2 will have only a mild effect on the outcome of the print.
7. When photographing multiple traces by making repetitive shots on one photograph, turn off the graticule illumination after the first shot.

Additional Oscilloscope Measurement Applications

The oscilloscope in the hands of a skilled operator can be used to make many more measurements than those discussed earlier in this chapter. Lack of space prevents us from presenting more details on any of these additional measurement procedures. Instead, the following list is provided to give the student an idea of the wide range of measurements possible with the oscilloscope. Further details on these and other measurements can be found in any number of specialty texts dealing specifically with oscilloscopes. For a list of some of their titles, see the list of references at the end of this chapter.

OSCILLOSCOPE MEASUREMENT APPLICATIONS

1. Checking individual electrical components such as vacuum tubes, diodes, transistors, magnetic components, relays, and choppers.
2. Checking amplifiers and amplifier circuits.
3. Displaying transducer outputs. Pressure, temperature, sound, velocity, and vibrations are all examples of quantities which can be measured with the oscilloscope-transducer combination.
4. Television, radio, and other communication equipment checking and repair.

5. Electrical diagnosis of automobile engines.
6. Special curve-tracer oscilloscopes are available for plotting the current-voltage (*I-V*) characteristics of transistors, diodes, and other elements. Figure 9-41 is a photograph of such a curve-tracer.

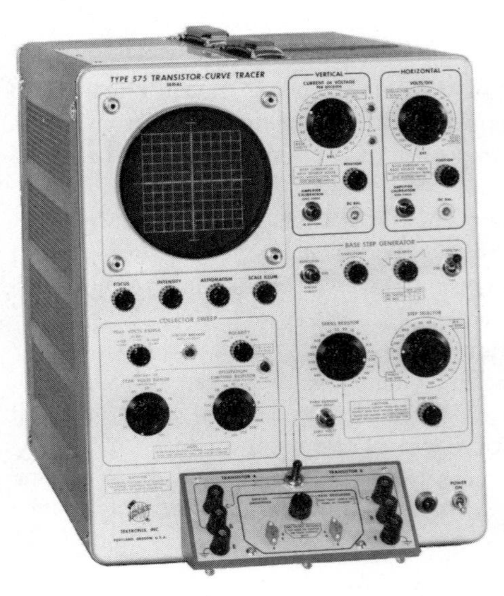

Figure 9-41. Transistor curve tracer Model 575 (Reprinted by permission of Tektronix, Inc.)

Problems

1. What are some of the advantages that oscilloscopes possess over other types of electronic measuring instruments?
2. Describe how the small spot of light seen on the oscilloscope screen is created by the CRT. Explain why permanent damage may be caused to the CRT if the spot is allowed to remain stationary on the screen.
3. Referring to Example 9-1 in the text, if the CRT shown in Fig. 9-2 is set to vertical sensitivity of 2 V/cm and a horizontal sensitivity of 5 V/cm, where will the spot appear on the scope screen (assuming it would appear at the center of the screen if no voltages were applied to the plates)?
4. What would be the position of the deflected spot on the scope screen if the polarity of the battery attached to the horizontal plates was reversed and the horizontal sensitivity was also 2 V/cm?
5. In Fig. P9-1 the horizontal and vertical sensitivities of the scope are set to 1 V/cm. With the circuit connected to the scope as shown, determine the position of the deflected spot.

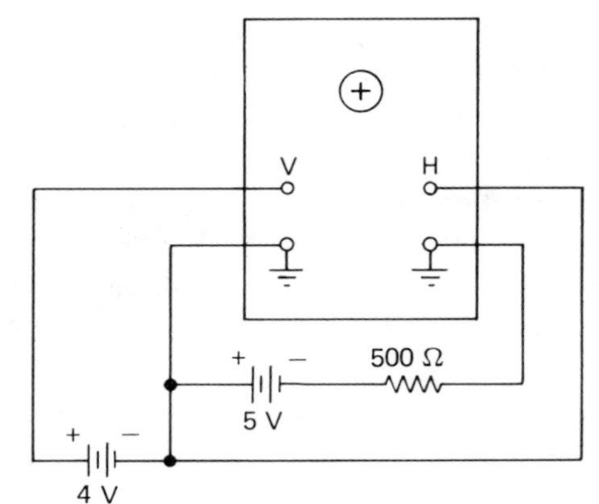

Figure P9-1

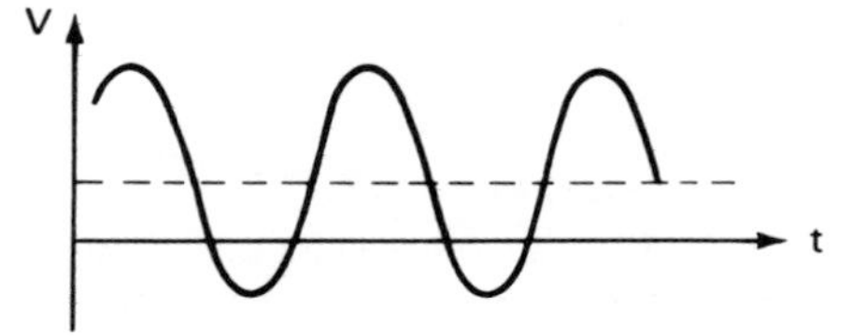

Figure P9-2

6. Explain the purpose and operation of the input coupling switch of the oscilloscope. What would be the waveform displayed on the scope screen if the waveform shown in Fig. P9-2 were applied to the scope and the input coupling switch was set to a) DC position, b) AC position?

7. What is the purpose of the *Ground* position of the Input Coupling Switch?

8. Describe the function of each of the following oscilloscope controls:
   a) focus,
   b) astigmatism,
   c) vertical position,
   d) external horizontal input,
   e) sweep vernier.

9. List one application in which each of the following connections that trigger the time base are used:
   a) Internal,
   b) External,
   c) Line.

10. If the vertical sensitivity switch of an oscilloscope is set to a) 50 mV/Div, b) 1 V/Div, what will be the peak-to-peak distance of the display shown on an oscilloscope of a sine wave whose amplitude is 0.25 V?

11. Describe the difference in the capabilities of measurement between an oscilloscope that has a single-ended input and one that has a differential input.

12. Describe the operation of the Time Base circuitry of an oscilloscope. How is the image on the oscilloscope screen stabilized by this circuitry?

13. Explain the functions of the *Trigger Slope* and *Trigger Level* controls of an oscilloscope.

14. Explain what the three positions (INTERNAL, LINE, EXTERNAL) of the Trigger Source switch correspond to on the oscilloscope.

15. Draw a block diagram of the major components of an oscilloscope which shows how the scope can be operated in either the X-Y or triggered sweep mode.

16. Describe the principles of operation of current probes. What are the advantages and disadvantages of using current probes rather than a test resistor to measure current?

17. Calculate the time elapsed between two points on a waveform separated by 3 cm if the sweep time of the scope is:
    a) 1 ms/cm
    b) 2 s/cm
    c) 50 μs/cm

18. If a linear sawtooth is applied to the vertical inputs of an oscilloscope and a sinewave is applied to the horizontal inputs, sketch the resulting pattern which will be displayed on the scope screen.

19. The sweep-time/cm of an oscilloscope is set to 0.2 msec/cm. The horizontal display switch is set in the "5x" position. If a sinusoid of unknown frequency is applied to the vertical inputs of the scope and it produces $2\frac{1}{2}$ cycles in a sweep of 10 cm, find the frequency of applied waveform.

20. If the sweep-time/cm of an oscilloscope is set to 0.1 ms/cm, and the width of the sweep is 10 cm, sketch the waveform patterns that the following signals would have if they were applied to the vertical amplifier input terminals:
    a) sinewave whose frequency is 5 kHz,
    b) cosinewave whose period is 0.5 ms,
    c) squarewave with a frequency of 10 kHz.

21. List some of the errors inherent in making measurements with an oscilloscope.

22. Explain how an oscilloscope loads the circuit being measured and how this loading effect varies with frequency.

23. List some of the capabilities of each of the following types of oscilloscopes:
    a) storage oscilloscope,
    b) dual-beam oscilloscope,
    c) sampling oscilloscope,
    d) digital readout oscilloscope.

References

1. ROTH, C. H., *Use of the Oscilloscope: A Programmed Text*. Englewood Cliffs, N.J., Prentice-Hall, Inc., 1971.
2. LENK, JOHN D., *Handbook of Oscilloscopes*. Englewood Cliffs, N.J., Prentice-Hall, Inc., 1968.
3. TURNER, R. P., *Practical Oscilloscope Handbook*, Vols. I and II. New York, Hayden Book Company, 1964.
4. CZECH, J., *The Cathode-Ray Oscilloscope*. New York, Interscience Publications, 1967.
5. RIDER, JOHN F., and USLAN, J. D., *Encyclopedia on Cathode Ray Oscilloscopes and Their Uses*, 2nd ed. New York, Hayden Book Company, Inc., n.d.

10

Low-Frequency Power and Energy Measurements

When voltage is applied to a conductor, current will flow in it. The amounts of current and voltage are indications of the power flowing in the conductor. In fact, the instantaneous power in the conductor is calculated from the product of v and i:

$$p = v \times i \tag{10-1}$$

The unit of power is the watt (W), and one watt is equal to one ampere of current which flows in a conducting path that has a voltage drop of one volt. The part of a circuit to which power is being delivered is called the *load.*

When measuring power, various types of instruments and methods are used. Their design depends upon the frequency range, the power dissipation level, and the type of load being powered. When the load is purely resistive and the frequency of the voltage and current applied is less than 10 MHz, the power can be measured quite easily. If the resistance value of such a load is known, an rms reading of the voltage across, or the current through, the load provides the added information needed to calculate the power dissipated by the load. However, since many loads are not purely resistive, the power dissipated by them cannot be measured as easily. Instead, it is necessary to use instruments that are especially designed for the particular frequency range.

To measure power in low-frequency (below 400 Hz) circuits where moderate to high power consumption levels exist, dynamometer instruments and basic ammeters and voltmeters can be used. For lower power dissipation levels and higher frequencies (below 100 kHz), electronic wattmeters may also be used. For still higher frequencies, the oscilloscope becomes the appropriate instrument (i.e., from dc to about 10 MHz). Above 10 MHz (at the so-called radio frequencies), rf wattmeters employing thermoelectric principles become a necessity.

The material of this chapter will be primarily concerned with low-frequency power measurements and instruments. Some of the ideas presented, however, will also be applicable to measuring power with the oscilloscope. (Oscilloscope operation is discussed in Chapter 9.) Rf power measurements involve concepts which lie beyond the scope of this text.

Power Measurements in dc Circuits

If dc currents and voltages are the only ones being applied to a load, the power dissipated by the load is quite easy to calculate. Since dc quantities are constant with time, the power can be found by taking the product of the magnitudes of V and I.

$$P = V \times I \tag{10-2}$$

In addition, since resistances are the only effective passive elements in dc circuits, the power in a load can also be determined from

$$P = I^2R \tag{10-3a}$$

or

$$P = \frac{V^2}{R} \tag{10-3b}$$

where R is the resistance of the load. If the resistance value of the load is already known, only one measurement is required in order for the power to be calculated. As long as dc quantities are involved, power measurements should be made with dc meters rather than with wattmeters. Dc meters are more sensitive than available wattmeters and thus their readings yield more accurate results.

If P is determined by measuring V and I, the dc voltmeter and ammeter are connected in the same way as in the voltmeter-ammeter method for measuring resistance (Fig. 10-1). Since the voltmeter and the ammeter consume some power themselves, their presence in the circuit will introduce an error into the value of the readings. To extract an accurate result, correction factors must be subtracted from the values indicated by the meters.

For example, when we use connection (a) of Fig. 10-1, the ammeter reads the current I that is flowing in *both* the voltmeter and the load. Thus we find the true power to the load by subtracting the power needed to operate the voltmeter from the product of V_L and I.

$$P = V_L I_L = V_L\left(I - \frac{V_L}{R_V}\right) = V_L I - \frac{V_L^2}{R_V} \tag{10-4}$$

where R_V is the resistance of the voltmeter, V_L is the voltage across the load (as indicated by the voltmeter reading), I_L is the current in the load, and I is the current being read by the ammeter.

For connection (b) of Fig. 10-1, the voltmeter is measuring the voltage being dropped across both the load and the ammeter. Thus, in calculating the true power, we subtract the voltage drop of the ammeter from the total voltage measured.

$$P = V_L I_L = (V - I_L R_A) I_L = V I_L - I_L^2 R_A \tag{10-5}$$

where R_A is the resistance of the ammeter and V is the voltage indicated by the voltmeter.

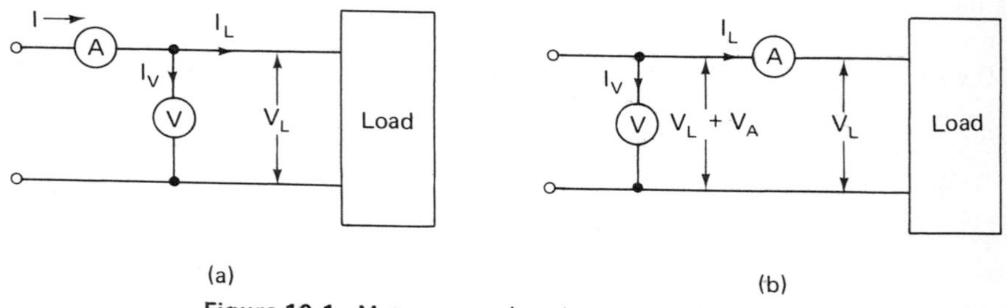

Figure 10-1. Meter connections for measuring dc power

The choice of connection ordinarily depends upon which meter resistance is known. It is usually easy to determine the voltmeter resistance from its ohms/volt rating. It is not always quite as easy to find the resistance of the ammeter. Therefore, the usual connection will be that of Fig. 10-1(a).

EXAMPLE 10-1

We are given a 50-V dc voltmeter with a 1000-Ω/V rating, and a 100-mA dc ammeter. They are connected to measure the power in a load as shown in Fig. 10-1(a). The voltmeter reads 40 V and the ammeter reads 50 mA. How much power is being dissipated by the load?

Solution: We find the resistance of the voltmeter from its ohms/volt rating.

$$R_V = 50\ \text{V} \cdot 1000\ \Omega/\text{V}$$
$$= 50\ \text{k}\Omega$$

From Eq. (10-4) we find that the power being dissipated by the load is

$$P = (2.00) - \frac{1600}{5 \times 10^4} = (2.00) - (0.03)$$

$$P = 1.97\ \text{W}$$

Power in ac Circuits

In all circuits (ac or dc), the instantaneous power, p, being delivered to a load can be found by taking the product of v and i in the load. However, in ac circuits, this instantaneous product varies from moment to moment. Hence a more useful quantity than the instantaneous power is the *average power*, $\bar{P}$, dissipated by a load. Because the ac quantities of voltage and current are not only sinusoidal but also tend to differ in phase, the calculation of this average power is more complex than for dc loads. These factors tend to complicate the *measurement* of ac power as well. For example, the procedure of measuring the values of V_{rms} and I_{rms} with ac meters, and then taking the product of the readings, is usually not a valid technique for measuring ac power. (Only if v and i are in phase in a load is the average power correctly found by using this method.)

To see how the average power in an ac load may be calculated, let us examine the relevant voltage and current waveforms. For a typical load, these waveforms would look like those shown in Fig. 10-2(a). Note first that the phase difference, θ, between v and i can be determined from their waveforms. Now the power at any instant is found by multiplying the magnitudes of v and i at the time of interest, and the resulting power waveform of such products is plotted as in Fig. 10-2(b). We see from this figure that the power being delivered to a load at any instant can be positive or negative. Positive power (the + portions of the curve) indicates that the load is *absorbing* power from the generator or source. Negative power (the − portions of the curve) means that the load is *returning* power to the source and is helping to run it (as in the charging of a battery or the supplying of power to turn the shaft of a generator).

The average power dissipated by the load is calculated by finding the average value of the power waveform of Fig. 10-2(b). It can be shown mathematically that this average value is given by the expression

$$\bar{P} = V_{rms} I_{rms} \cos\theta \tag{10-6}$$

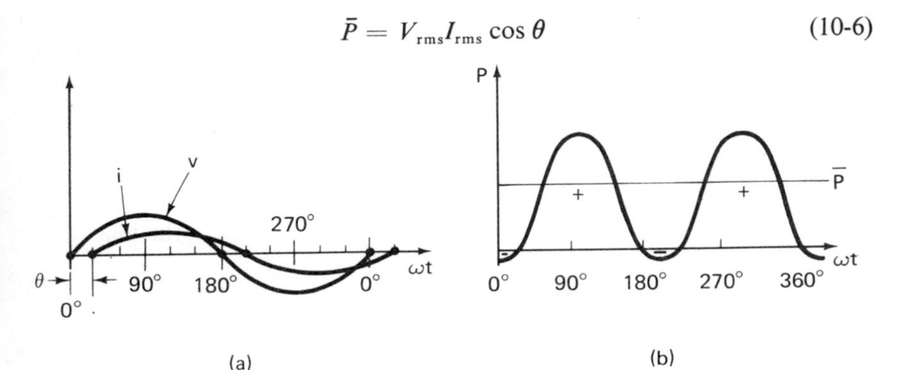

Figure 10-2. (a) Voltage and current waveforms in an ac circuit where $\theta = 30°$ (b) Power waveform in the same circuit

where θ is the phase angle that exists between v and i. We see from Eq. (10-6) why the product of the rms voltage and current readings would not ordinarily yield a correct value for $\bar{P}$. Their product would not take into account the factor due to the phase angle, θ. This factor, the $\cos\theta$ term in Eq. (10-6), is known as the *power factor* (pf).

If the load is purely resistive, the waveforms of v and i are in phase, and $\theta = 0$. Since $\cos(0) = 1$, $\bar{P}$ in this special case is equal to

$$P_{\theta=0} = V_{rms}I_{rms}$$

As noted in the introduction, if the resistance (R) of such a load is known, $\bar{P}$ can also be found from

$$\bar{P} = I^2_{rms}R$$

or

$$\bar{P} = \frac{V^2_{rms}}{R}$$

The instruments which are designed to sense the effect of the phase difference and yield correct values for the average power are called *wattmeters*. The name is derived from the unit used to describe power.

The energy which flows in an ac circuit but is not dissipated in the load is known as the *reactive power*, P_{var}. This reactive power is described by units called vars (volts-amperes-reactive) and is calculated from the expression:

$$P_{var} = V_{rms}I_{rms}\sin\theta \tag{10-7}$$

where θ again is the phase angle between the v and i waveforms. P_{var} alternatively flows into and out of the load. Even though this energy is not consumed by the load, there is some loss associated with its transmission back and forth in the circuit because transmission lines are not perfect conductors. Hence it is important to minimize this quantity, especially when large current flows are involved. P_{var} is measured with instruments known as *varmeters*.

The combination of $\bar{P}$ and P_{var} present in the load is called the *apparent power*, P_A, of the load. P_A is found from $\bar{P}$ and P_{var} by the expression:

$$P_A = \sqrt{\bar{P}^2 + P^2_{var}} \tag{10-8}$$

In addition, P_A is also equal to the product of V_{rms} and I_{rms} in the load.

$$P_A = V_{tms}I_{rms} \tag{10-9}$$

Therefore the product of the two ac meter readings (connected to a load as the dc meters were connected for dc power measurements) generally yields

P_A and not $\bar{P}$. The units of P_A are the volt-ampere (VA), but they are also often expressed in thousands of volt-amperes (kVA).

Finally, we see from Eqs. (10-6) and (10-9) that the ratio of $\bar{P}$ to P_A will also yield the power factor

$$\text{pf} = \cos\theta = \frac{\bar{P}}{P_A} \tag{10-10}$$

EXAMPLE 10-2

The voltage and current being applied to a load are sinusoidal waveforms whose amplitudes are 100 V and 5 A, respectively. The phase angle between them is 30°. Calculate the power, the reactive power, and the apparent power of the load.

Solution: (a) The rms values of any sinusoidal waveforms with amplitudes V_O and I_O are:

$$V_{\text{rms}} = \frac{V_O}{\sqrt{2}} = \frac{100}{1.41} = 70.7 \text{ V}$$

$$I_{\text{rms}} = \frac{I_O}{\sqrt{2}} = \frac{5}{1.41} = 3.53 \text{ A}$$

(b) The phase angle between v and i is 30°; so

$$\cos\theta = 0.866$$

$$\sin\theta = 0.50$$

(c) $\bar{P} = V_{\text{rms}}I_{\text{rms}}\cos\theta = 216$ W

(d) $P_{\text{var}} = V_{\text{rms}}I_{\text{rms}}\sin\theta = 124$ VAR

(e) $P_A = V_{\text{rms}}I_{\text{rms}} = 70.7 \times 3.53 = 249$ VA

or $P_A = \sqrt{\bar{P}^2 + P_{\text{var}}^2} = 249$ VA

POWER RATINGS OF AC EQUIPMENT

Ac electrical equipment (machines, transformers, etc.) is rated in terms of apparent power, P_A, as well as average power, $\bar{P}$. Thus an ac motor will have both a wattage and kVA rating. The dual ratings must be used because of the effect of the power factor. If the power factor of the circuit in which the equipment is connected is small, excessively high current through the equipment could result. (Since voltage from the power station remains relatively constant, only I can increase when P_A increases.) High currents could burn out the insulation or other parts of the equipment. Such an occurrence would be possible even if the average power dissipated by the actual operation of the equipment remained below its wattage rating. By specifying and observing a maximum kVA rating, such high current damage is avoided.

Single-Phase Power Measurements

WATTMETERS

The electrodynamometer movement is utilized as the sensing mechanism for the vast majority of low-frequency (below 400 Hz) power instruments. Dynamometer-type instruments can be built to measure the average power dissipated in a load, the power factor, or the reactive power in a circuit. They can determine these quantities even if the waveforms being measured are not sinusoidal. This allows them to be used to measure power in dc circuits as well as in other applications where the alternating waveform has a nonsinusoidal shape.[1]

In Chapter 4 we saw how the coils of the dynamometer movement were connected so that it could be used to measure ac currents and voltages. A series connection was employed, and the same current flowed in both the stationary and the rotating coils. In the dynamometer *wattmeter*, the stationary coils and the rotating coil of the movement are connected differently (Fig. 10-3). The current coming from the power source, i_c, is made to

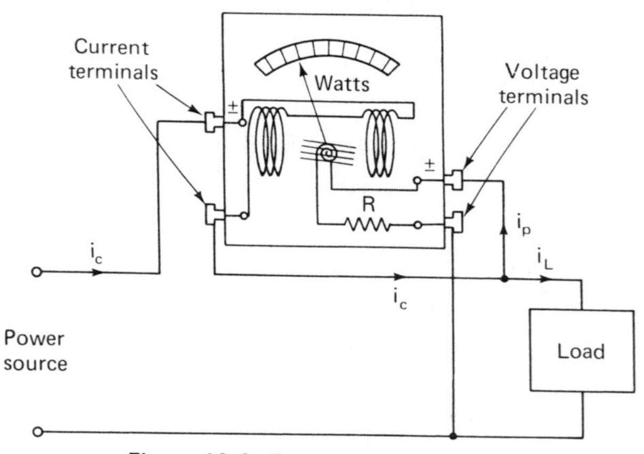

Figure 10-3. Dynamometer wattmeter

pass through the stationary coils, thus connecting them in series with the load. (These coils are also called the *current coils* or *field coils*.) The movable coil (rotating coil) has a large resistance, R, connected in series with it. This coil and resistance branch is connected *across* the load. The movable coil branch is also called the *voltage branch* and carries a small current, i_p (usually 10 to 50 mA).

The current in the stationary coils sets up a magnetic field in the space

[1]However, as we noted earlier, dc power measurements can be made more accurately be using dc meters.

between them, which is proportional to i_c. The current in the moving coil is proportional to the voltage across the load ($i_p \approx V_L/R$). Since the moving coil is located in the space between the field coils, i_p interacts with their magnetic field and causes the moving coil to rotate. The pointer attached to the rotating coil displays this rotation on a scale. Since the power at any instant is defined as

$$p = v \times i \tag{10-1}$$

the torque developed in the moving coil is proportional to the instantaneous power.

At frequencies above a few hertz, the inertia of the pointer is too great for it to follow the variations of p. Instead the pointer takes up a position proportional to the average of $v \times i$. This position is calibrated to indicate the average power, $\bar{P}$.

In the dynamometer ammeter, the current i was equal in both coils, so the deflection was proportional to the average of i^2. In the dynamometer wattmeter, the currents in the two coils are different, and the deflection is proportional to the average of $v \times i$. Figure 10-4 is a photograph of a typical dynamometer wattmeter.

The dynamometer wattmeter has four external terminals to which connections must be made in order to measure power. Two of them are designated

Figure 10-4. Dynamometer wattmeter (Courtesy of Weston Instruments, Inc.)

as the *voltage terminals* and the other two as the *current terminals*. The current terminals provide connections to the stationary coils, while the voltage terminals provide connections to the rotating-coil branch. One terminal of each type is marked $\pm$. It is necessary to connect the $\pm$ current terminal and the $\pm$ voltage terminal to the same wire of the incoming power line. In that way the stationary coils and the movable coil will be at about the same potential. (Because the value of the resistance of the series resistor (R) in the voltage branch is much greater than the resistance of the voltage coil, most of the voltage across the voltage branch is dropped by R.) Then no electric field will exist between the stationary and the moving coils. An electric field would arise between the voltage and current coils if they were at different potentials. The force of attraction due to the field could slightly restrict the rotation of the movable coil and produce an erroneous reading.

Dynamometer wattmeters are built with current ratings up to 20 A and voltage ranges up to 300 V. It is good practice, however, to limit the input current to the wattmeter to a maximum of 5 A. This can be done by using a current transformer to step down the value of the input current. When this practice is followed, the large magnetic fields associated with leads carrying heavy current are reduced. Such magnetic fields could sizably alter the relatively weak magnetic fields of the instrument coils. If the voltage being applied to the load exceeds 300 V, it is also wise to step it down to 115–125 V with a voltage transformer. In this way, damage to the voltage circuit of a wattmeter is avoided. Instrument transformers and their connections are discussed in Chapter 7.

The overall errors in commercially manufactured dynamometer instruments run between ± 0.1 and ± 0.5 percent when operated between their specified frequencies. The highest-accuracy meters are used as laboratory standards of power.

The wattmeter is rated in terms of its maximum current, voltage, and power. Each of these ratings must be observed to prevent damage to the wattmeter. Excess current could harm the current coils and their insulation. Excess voltage could cause the voltage-coil branch to suffer in a similar fashion. In low-power-factor circuits, either of these limits could be exceeded without exceeding the wattage rating.

ERRORS IN DYNAMOMETER WATTMETERS

Even if the connections to a dynamometer wattmeter are made properly, an error is still present in its readings. This error is caused by the power needed to maintain the magnetic field of the stationary coils and the power consumed by the voltage drop across the voltage branch. At 5 A, the power loss through the current coils is about 0.8 W. At 115 V, the power loss through the voltage coil circuit is about 2.9 W. For large power measurements this error is small, but it can become appreciable if the power levels being measured are small (5 watts or less).

A wattmeter can be connected so that the voltage branch is placed either before or after the current coils. When the wattmeter is connected in the manner shown in Fig. 10-5(a), the movable coil branch is connected at point *A*. In this connection, the voltage branch senses the true voltage across the load. However, the magnetic field of the current coils is too large because some of the current causing their field goes to the voltage branch of the movable coil instead of going to the load. The result is an average power reading that is higher than the actual power dissipated by the load. On the other hand, if the movable-coil branch is connected to point *B*, the current in the current coils is now correct. But the voltage across the voltage branch is too large, that is, the voltage branch is sensing the voltage drop of the load and the stationary coils in series. Thus a high reading again results.

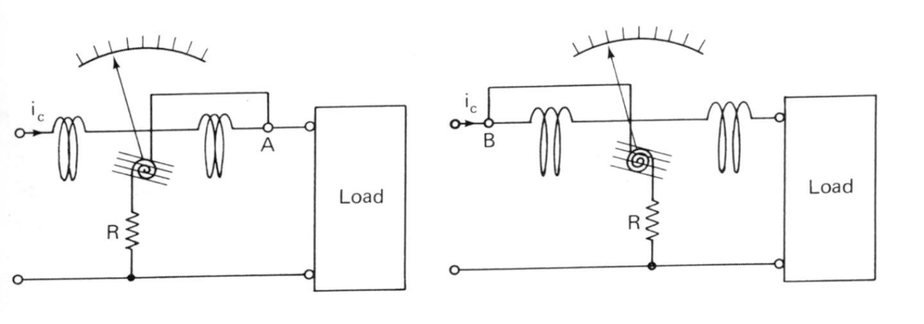

Figure 10-5. Connecting a wattmeter to measure power with the least error (a) Connection A—good for high-current, low-voltage loads (b) Connection B—good for high-voltage, low-current loads

There are two methods which can be used to reduce these errors. The first is to make use of the connection which will yield the lowest error and apply a correction factor to the resultant readings. When this method is used, the following guidelines may be helpful. The connection of Fig. 10-5(a) will be better for high-current, low-voltage loads, while connection (b) of the same figure is better for high-voltage, low-current loads.

The second solution is to use a *compensated wattmeter* (Fig. 10-6). The compensated wattmeter is constructed by winding the wire carrying the movable-coil current, i_p, back into the current coils which originally carried $i_c = i_L + i_p$. The extra winding is carried out in the opposite direction to the windings of the current coils. Then the magnetic flux due to the current in the extra winding will cançel that part of the magnetic flux due to i_p flowing in the original current coils. As a result, the compensated wattmeter eliminates the above errors and indicates low-power readings much more accurately.

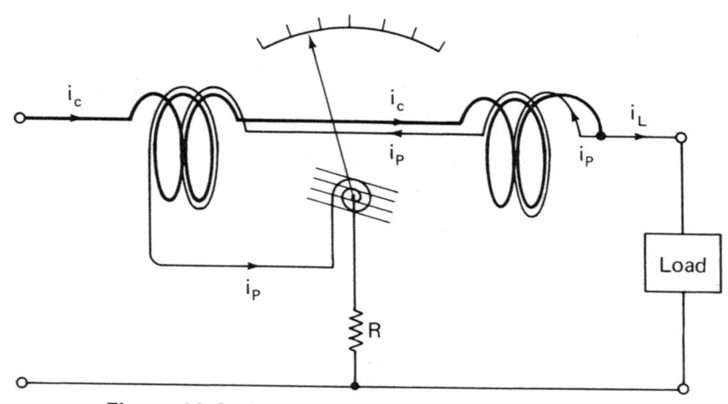

Figure 10-6. Schematic of a compensated wattmeter

MEASURING $\bar{P}$ AND P_A SIMULTANEOUSLY

Because the power dissipated in a load does not reveal all the power relationships of an ac circuit, the ac currents and voltages in the load are also sometimes monitored simultaneously. This allows calculations of P_A and the power factor to be made. Unfortunately, when three meters are connected into a circuit together, their total effect may markedly disturb the true currents and voltages. This is particularly true of ac meters, which are usually less sensitive than comparable dc meters. In general, the same effects arise as those described in the discussion on errors in wattmeters and in the discussion on the errors present in dc power measurements. If it is suspected that the errors caused by the three meters are large enough to be significant, the choice of connections that will minimize the errors should be used (along with a computation that corrects the resultant readings).

For example, if we are using a connection as shown in Fig. 10-7, we can calculate the true average power, the true voltage, and the true current in the load by making the following corrections. First we note that the voltmeter is reading the correct voltage across the load, and its value does not need correction. However, the reading of the wattmeter is too high because it is

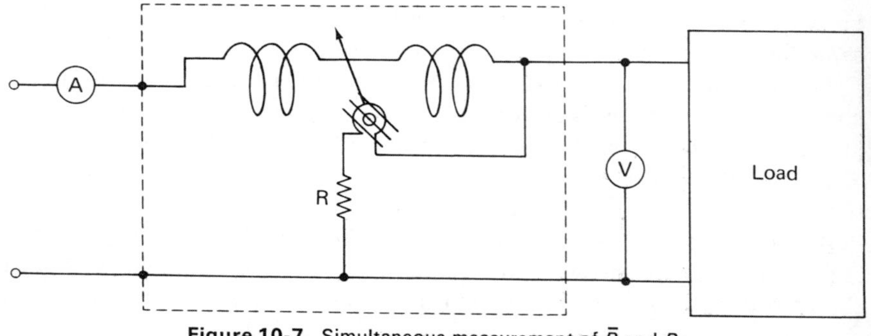

Figure 10-7. Simultaneous measurement of $\bar{P}$ and P_A

indicating the power of the load, the voltmeter, and its own voltage-sensing circuit. Thus the true value of $\bar{P}$ is found from

$$\bar{P} = W - \frac{V^2}{R_v} - \frac{V^2}{R_w} \qquad (10\text{-}11)$$

where W is the wattmeter reading in watts and R_w and R_v are the resistances (in ohms) of the wattmeter voltage branch and the voltmeter, respectively. Likewise, the true current in the load I_L is found from

$$I_L = \sqrt{I^2 + V^2\left(\frac{1}{R_w} + \frac{1}{R_v}\right)^2 - 2W\left(\frac{1}{R_w} + \frac{1}{R_v}\right)} \qquad (10\text{-}12)$$

where I is the reading taken from the ammeter and V is the voltage reading of the voltmeter. By applying the above corrections to readings obtained from this connection, more accurate data are obtained.

ELECTRONIC WATTMETERS

If it is desired to measure the power dissipated in a load at audio frequencies greater than 400 Hz (up to about 100 kHz) or at power levels below 5 watts, an electronic wattmeter can also be used. The electronic wattmeter measures power for frequencies up to about 100 kHz and for power ranges of 0.1 W to 100 kW full-scale. If the frequency is less than 40 kHz and if the pf of the load is greater than 0.1, its accuracy can be as good as ± 3 percent. For higher frequencies or for loads of smaller pf, the accuracy drops to ± 10 percent.

The basis of operation of such wattmeters relies on a circuit containing two amplifiers. One amplifier measures the current through the load and the other the voltage across the load. By properly combining and amplifying the current and voltage quantities, a reading proportional to the average power dissipated by the load is produced.

Although the accuracies quoted above are possible, the electronic wattmeter is fairly difficult to use properly and must be allowed a long warm-up time. The vacuum tubes that make up its amplifier circuits must also be properly matched in order to yield accurate readings. The overall inherent uncertainties associated with its operation and its relatively high cost have kept the electronic wattmeter from becoming widely accepted.

Electric Power Distribution

At this point, we will make a slight diversion to introduce the methods by which electric power is distributed to domestic and industrial consumers. Besides providing good background for measurement techniques, it will introduce power measurements in polyphase ac circuits.

Electric power-generating stations are often built near supplies of coal or water power (hydroelectric power). The power generated at the stations is then distributed to users by long-distance transmission lines—usually along wires suspended from steel towers. To transmit the power most efficiently, the losses due to voltage drops in the wire should be minimized. This is achieved by lowering the resistance of the transmission line wires and by transmitting alternating (ac) rather than direct (dc) currents[2]. Power in ac form is more efficiently transmitted because transformers can be used to step up (increase) the value of the voltage. This reduces the current flowing in the transmission-line wires. Since the power loss due to current flowing in the wires is found from

$$\text{Power loss} = I^2 R \tag{10-13}$$

where I is the rms current flowing in the wires and R is the wire's resistance, the voltage-drop loss can be diminished by decreasing I. To reduce losses even further, the power is delivered in *three-phase* rather than *single-phase* form.

The three-phase (3ϕ) designation is a description of the three sinusoidal waveforms by which the power is distributed. Three-phase generators are used to develop these three sinusoidal voltages (Fig. 10-8(a)). Each of the voltages is transmitted by a separate wire of a three-wire transmission system, and each is out of phase with the other two by 120°. The resulting waveforms are shown in Fig. 10-8(b).

This type of scheme is used to distribute electric power because more power can be delivered with smaller transmission losses than in the single-phase, two-wire system. Smaller losses mean that smaller copper transmission wires can be used. A more economical system is the result. In addition, the power delivered by a 3ϕ circuit is a constant rather than a pulsating power (unlike the power delivered by a single-phase circuit). As a consequence, single-phase power transmission is now used only in smaller or emergency-type power systems.

A generator at the power station may put out 1000 amperes of current at 240 volts ac (Fig. 10-9). In this example, the output is supplied to a high-voltage transformer which converts it to a voltage of 240,000 volts and 1 ampere. The high voltage is then fed to the transmission line which carries it to the population or industrial center requiring the power. At the consumption end, another transformer is used to step down or decrease the voltage to the final 115-volt level. Usually this is done in two steps, using two separate transformers. The first transformer steps down the voltage so

[2]The resistance of wires is greater when current is passed through them in ac rather than dc form. This phenomenon has led to some research into dc power transmission. In some countries, it is even being experimentally attempted.

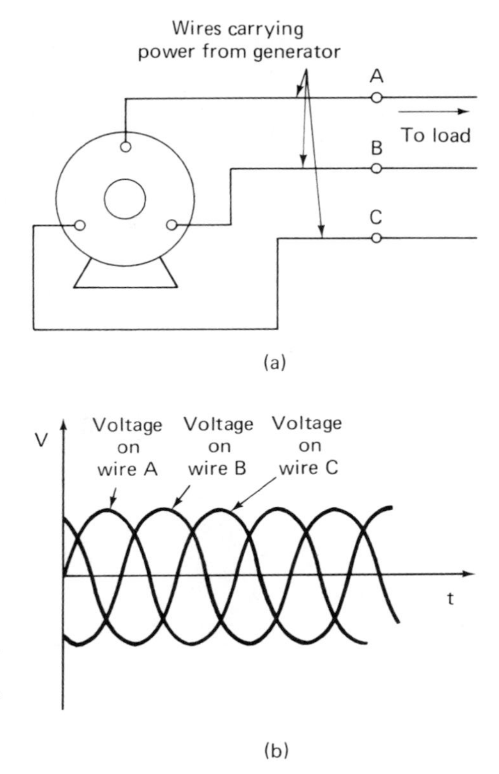

Figure 10-8. Three-phase power generator (a) Three-phase generator (b) Phase relation between the waveforms of the voltages on wires *A, B,* and *C* of a three-phase generator

that it can be carried on lower-voltage local transmission lines. The second transformer is located near the consumer's premises. The high-current-carrying wires from the local transformers to the consumer outlets are called *mains*. The final step also involves extracting a single-phase voltage from the transmitted 3ϕ voltage.

The voltage supplied by public utilities and other larger power systems is usually kept to within ± 5 percent of the stated value. The frequency is also usually closely regulated at 60 Hz[3] so that it can be used to drive precise instruments such as electric clocks.

Polyphase Power and Measurements

As indicated in the previous section, most electric power for commercial use is transmitted in the form of three-phase ac power. At the load end of a three-phase (3ϕ) system, a three-phase load connection is used. The two

[3] Power systems of aircraft and ships usually operate at 400 Hz. Smaller and lighter components of circuits can be built to operate at this higher frequency than at 60 Hz.

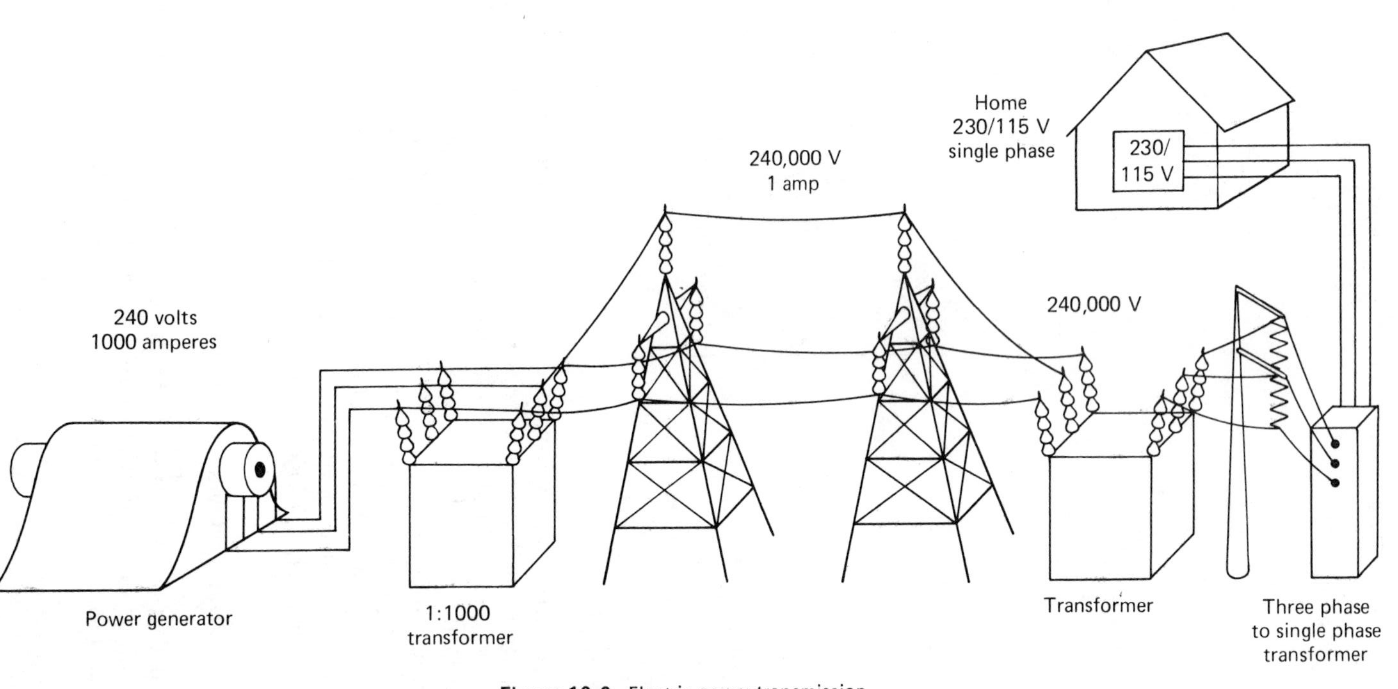

Figure 10-9. Electric power transmission

possible 3ϕ load connections are shown in Fig. 10-10. They are called the *wye* (Y) and *delta* (Δ) loads, respectively. In the Y load, a fourth wire connected to point N (called the neutral connection) may also be used. Usually N is connected to ground.

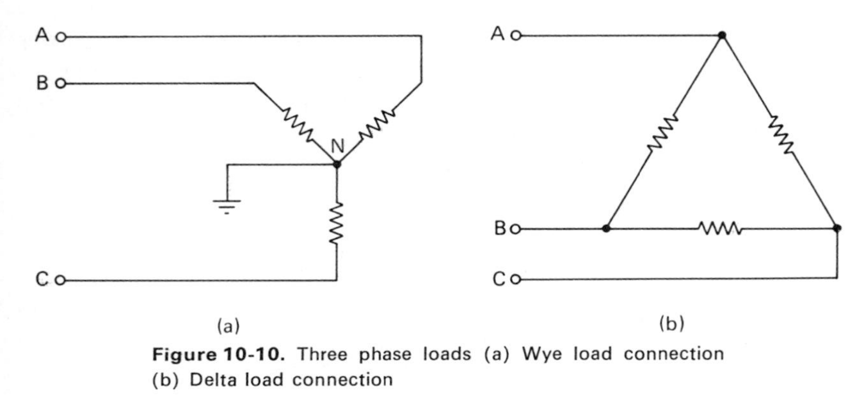

Figure 10-10. Three phase loads (a) Wye load connection (b) Delta load connection

In most urban centers, the common load connection is the Y load. The circuit shown in Fig. 10-11 is used for domestic power delivery. In this circuit, the voltage between the phases of the power mains is 4160 V. The voltage between any of the phase wires and the neutral point is 2400 V. For each household, a single-phase transformer is used to step down this 2400 V to 240 V. The single-phase transformer is center-tapped so that there are two 115-V, single-phase circuits available for each house. (A 5-V drop is allowed for the connection between the transformer and the house wall outlets.)

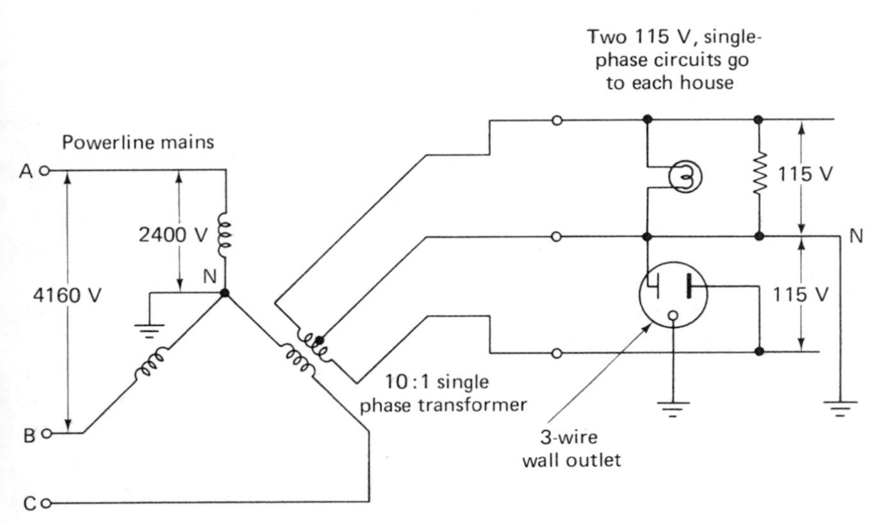

Figure 10-11. Common urban household power connection

Single-phase power is all that is required to run most common appliances and lighting.

For schools, factories, and farms, however, 3ϕ power is sometimes required to drive motors and other 3ϕ electrical equipment. The circuit used to deliver this 3ϕ power is a little different than the single-phase circuit. That is, as shown in Fig. 10-12, a 10: 1, 3ϕ transformer is used to step down the 2080 V between phases to 208 V between phases. This 208 V, 3ϕ voltage can then be applied to 3ϕ equipment. The voltage between the phase wires and the neutral wire is still 120 V, and this connection can be used to obtain the same single-phase power as in the previous circuit.

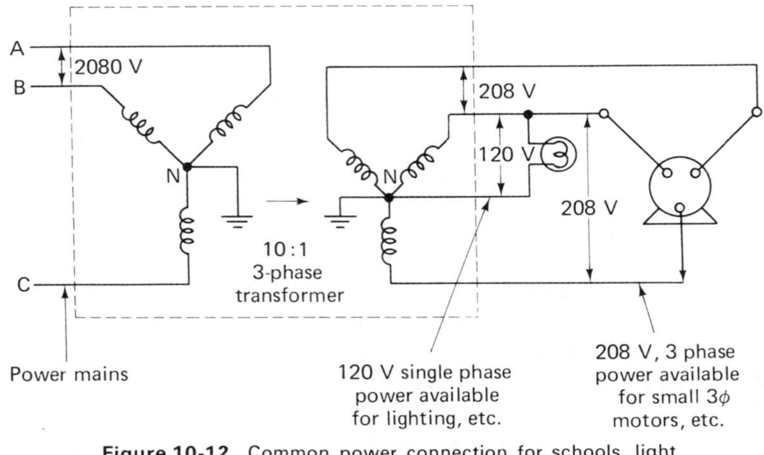

Figure 10-12. Common power connection for schools, light industry, etc.

POLYPHASE MEASUREMENTS

The power delivered to any one of the elements of the 3ϕ load can be measured by a wattmeter placed across the element of interest as shown in Fig. 10-13. However, if it is desired to measure the total power, P_T, being delivered to the entire Y or Δ load, the measurement is not as straightforward. The power delivered to a three-phase load, connected in either Y or Δ configuration, must be measured by means of a polyphase wattmeter or by a proper connection of two wattmeters.

One possible two-wattmeter connection is shown in Fig. 10-14. The same side of the voltage branch of each wattmeter is connected to the phase wire that does not have a wattmeter connected in series (in this case, wire B). We find the power dissipated by the entire load by taking the *algebraic sum* of the two wattmeter readings. (However, in this method, neither wattmeter is reading power by itself because there is no measurement being made of both v and i in a single arm of the load.) In making the initial connections, it

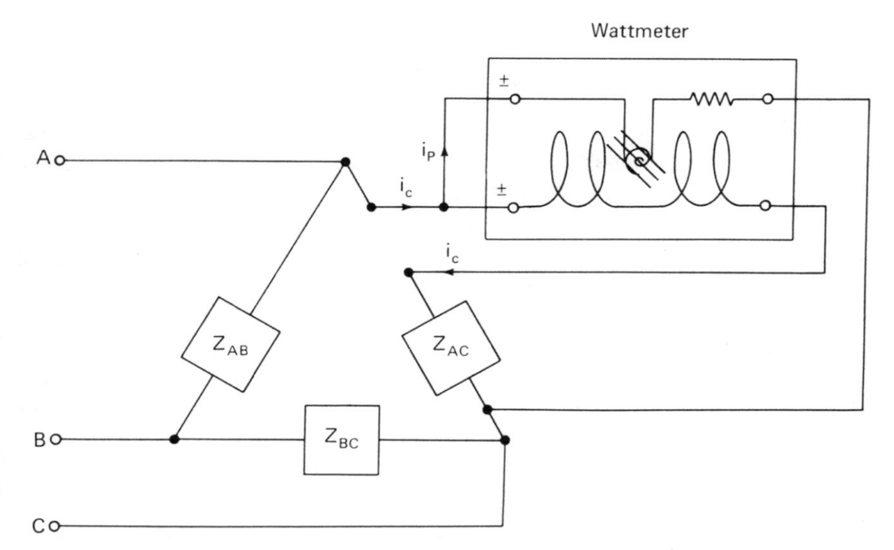

Figure 10-13. Using a wattmeter to measure power in one arm of a three-phase load

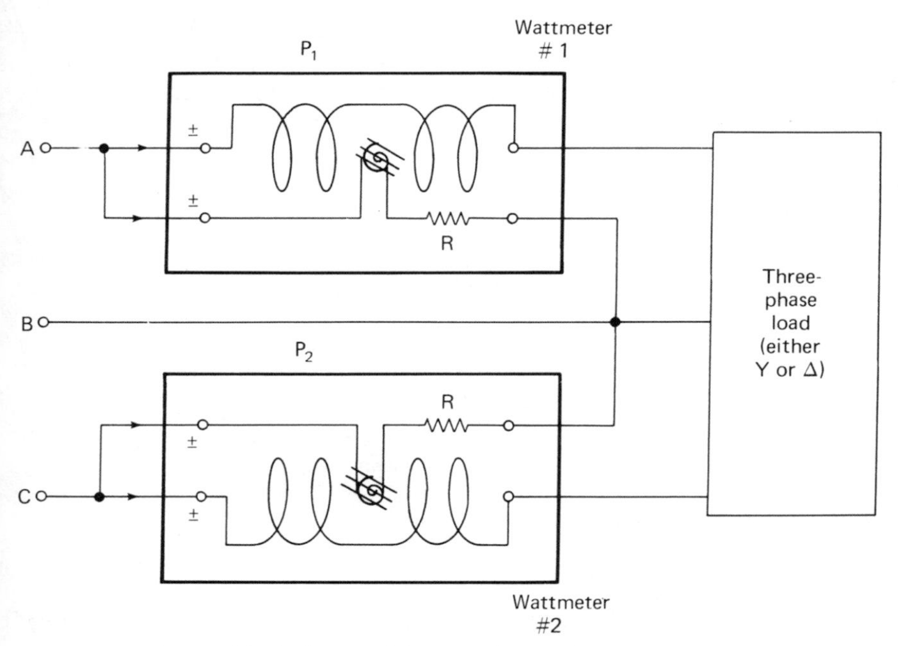

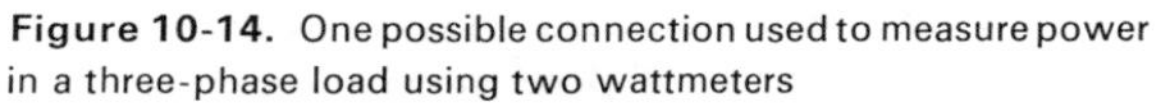
Figure 10-14. One possible connection used to measure power in a three-phase load using two wattmeters

is important to ensure that the connections to the phases are symmetrical (as shown in Fig. 10-14). A convenient guideline for ensuring symmetrical connections is to connect the wattmeters so that, as current flows from the source, it always enters the wattmeter through the $\pm$ terminals. This arrangement will establish a reference for determining the sign of the wattmeter readings. If the pf of the load is > 0.5, both wattmeters will read positive and $\bar{P}_T = P_1 + P_2$. If the pf of the load is $0.5 <$, one of the meters (e.g., meter 2) will give a negative reading. In this case, the current leads of the negative-reading meter should be reversed, and the resultant reading should then be subtracted from the positive wattmeter reading to yield the power dissipated in the load ($\bar{P}_T = P_1 - P_2$).

If only one wattmeter is available, this method can still be used if two separate readings are made. As long as the proper connection rules are followed, a correct result is obtained.

POLYPHASE WATTMETERS

Polyphase wattmeters are made by attaching two electrodynamometer movements onto one shaft (Fig. 10-15). When the proper connections are made to the meter (as in the two-meter method), the torque acting on the assembly will be the sum of the two torques acting on each of the moving coils separately. The total power is then summed automatically and indicated directly on one scale. The use of the polyphase wattmeter eliminates calculations and conserves space (since only one instrument performs the job of two). The instrument can also be used to measure single-phase power. In this case, connections are made to only one of the movements.

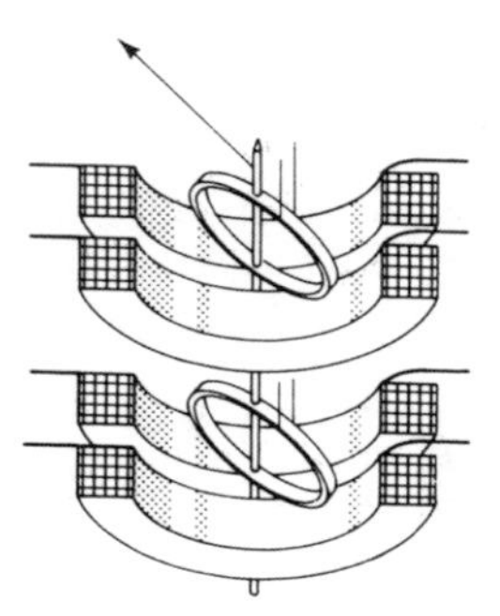

Figure 10-15. Polyphase wattmeter movement (Courtesy of Weston Instruments, Inc.)

Miscellaneous Meters

REACTIVE POWER MEASUREMENTS (Varmeters)

There are certain applications in which a direct measure of the reactive power provides desirable or even necessary information. Such a measurement can provide a direct check on other power measurements and calculations.

It can also provide information to the controller dispatching power at the generating station about the nature of particular loads.

The dynamometer movement can be adapted to allow measurements of reactive power by including reactive elements into the voltage branch. Then the impressed voltage sensed by the movement is 90° out of phase with the line voltage. Such an instrument is called a *varmeter*. Since $\cos(90° - \theta) = \sin\theta$, the dynamometer coils will be sensing $VI \sin\theta$, which is the expression for reactive power. The reactive elements will cause the proper phase shift only at the frequency at which the meter is calibrated, and only if a single sine-wave component is present.

For use in three-phase systems, ordinary wattmeters are used, except that their voltage coils are fed by voltages which are 90° out of phase with the actual line-voltage waveforms. This is accomplished by using two special transformers. As long as the waveforms of the two voltages are symmetrical in amplitude and phase, the three-phase varmeter will yield a correct value of the reactive power flow.

POWER FACTOR METERS

Sometimes it is desirable to measure the power factor without necessarily measuring all the other power characteristics of a load. We usually do this by measuring the phase difference of v and i. The phase difference between voltage and current in ac circuits can be measured in many different ways. For high frequencies, we saw in Chapter 9 how the oscilloscope could be used. In this section we will discuss the operation of low-frequency meters that measure the phase between the voltage and current. Such power-factor meters are designed to measure the phase angle and then display the result in terms of the pf.

One common type of power-factor meter is the crossed-coil meter (Fig. 10-16). When this type of meter is used for single-phase measurements, the movable coil of the dynamometer wattmeter is replaced by two coils

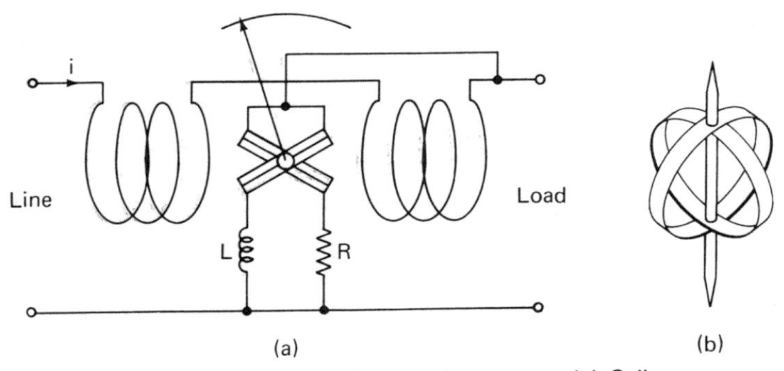

Figure 10-16. Crossed-coil power-factor meter (a) Coil connections in the meter (b) Crossed moving coils

mounted on the same shaft. The two coils are at right angles to each other, and the shaft itself can rotate. There are no springs to restrain the shaft's rotation. Instead, the torque felt by one coil is balanced by the torque exerted in the opposite direction on the other coil. The coils are subject to oppositely directed torques because of their orientation in the magnetic field of the fixed coils and because one coil is connected to a resistor and the other to an inductor. These elements divide the current i_p through the two coils, and this helps to determine the magnitude of the torques. In a balanced position, the resultant torque is zero. This balance is established by the mutual inductance of all the coils, the currents in the coils, and their interaction with the magnetic field of the current coils. The deflection at balance is proportional to the phase angle between the current and voltage in the circuit. This deflection is again indicated by a pointer which moves along a scale calibrated to give θ in terms of $\cos \theta$ (the power factor).

In a three-phase connection, the current coils are again connected in series with one of the three wires. However, the two crossed coils are both connected on one side to this same wire and on their other sides to the other two wires through resistors. This connection yields oppositely directed torques on the two crossed coils, and they take up a positions which depend on the power factor of the load.

Note that power-factor meters are accurate only at low frequencies (below 200 Hz). Thus they are quite suitable for use at power-line frequencies (60 Hz) but cannot be used for higher frequencies.

HOW TO USE THE DYNAMOMETER MOVEMENT METERS TO MEASURE POWER QUANTITIES

1. If the current exceeds 5 A, use an instrument transformer to step it down below 5 A before applying it to the wattmeter or other meters.
2. Try to avoid placing the wattmeter into regions where strong external magnetic fields are present.
3. Connect both $\pm$ terminals of the wattmeter to the same side of the load and line. This keeps the potential of the current and the voltage coils approximately equal.
4. Do not exceed the voltage, current, or power rating of the meter. Remember that the current or voltage ratings can be exceeded even if the power reading is not indicating a full-scale deflection.
5. When using a wattmeter, voltmeter, and ammeter together, corrections due to the meter disturbances of the circuit should be made (especially if it is suspected that the reading errors are significantly large).
6. When a high-current, low-voltage reading is taken, connect the voltage coil to position A of Fig. 10-5.
7. When a high-voltage, low-current reading is made, connect the voltage coil to position B as shown in Fig. 10-5.

8. If a negative reading occurs in one of the wattmeters during a two-wattmeter measurement of polyphase power, reverse the connections of the current leads of this wattmeter and treat the resultant reading as a negative number.

Electrical Energy Measurements

When the power being dissipated in a load is calculated in terms of time, the amount of energy consumed by the load can be found. If one watt is delivered for one second, the energy consumed in that time is equal to one *joule*. The joule is therefore also called the watt-second. In electric power calculations, the watt-hour or the kilowatt-hour are also used, since these are often more convenient units than the watt-second. One kilowatt-hour represents 1000 watts delivered for one hour.

Earlier in this chapter we saw how electric energy is generated and distributed to various consumers of electric power. At the distribution points there must be measuring devices to record the amount of energy used by each household or industrial consumer. In this way the supplier can bill the consumer for energy delivered during a given time span. The unit of energy that is sold by electric power companies is the kilowatt hour (kWh). The approximate cost per kWh is about 5 cents in the U.S. Table 10-1 gives a list of common electrical appliances and their typical wattage ratings.

Table 10-1 Typical Wattage Ratings of Electrical Equipment

Equipment	Wattage Rating
Air conditioner	2800
Clothes washer	400
Dishwasher	1400
Electric heater	1650
Hi-fi equipment	230
Electric iron	1000
Radio	30
Electric oven	10,000
Refrigerator	320
Color television	420
Oscilloscope	50–150

The most common energy-measuring device is the watt-hour meter. The type of watt-hour meter that is most widely used today was developed by Schallenberger in 1888. This meter is an inexpensive and accurate instrument which can operate properly for long periods of time with little maintenance.

Furthermore, it is not significantly affected by large changes in the load, power factor, or environmental conditions. It records the energy consumption of a load by counting the turns of a spinning aluminum disc. The spinning of the disc is caused by the power that passes through the meter.

A sketch of the important components of the meter is shown in Fig. 10-17. Its operation is similar to that of an induction-type motor. The current coil and the metal core on which it is wound set up a magnetic field. The voltage coil and its metallic core set up an additional magnetic field. In the aluminum disc (which is positioned to be influenced by both fields), eddy currents arise from the variation of the magnetic field of the current coil. (See Chapter 7 for a discussion of eddy currents.) These eddy currents interact with the magnetic field of the voltage coil, and a torque is exerted on the disc. Since there are no restraining springs, the disc continues to turn as long as power

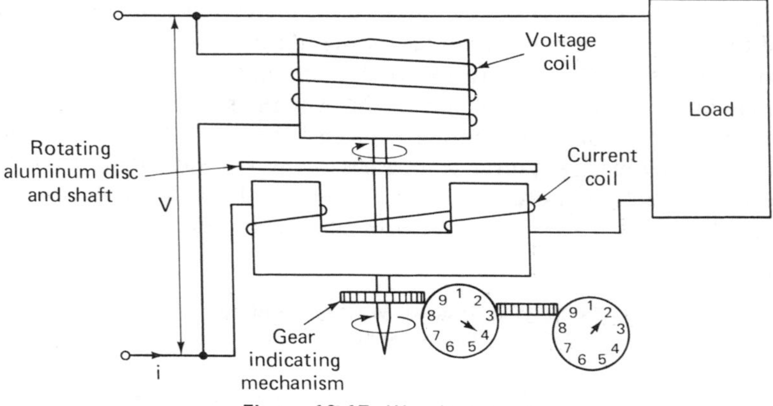

Figure 10-17. Watt-hour meter

Figure 10-18. A watt-hour meter for industrial or domestic application (Courtesy of Westinghouse Electric Corp.)

is fed through the meter. The torque on the disc is proportional to the product of $v \times i$. Thus, the greater the power passing through the meter, the faster the disc turns. The number of turns is a measure of the energy consumed by the load. The shaft on which the disc is mounted is geared to a group of indicators with clock-like faces. By reading the values on their faces at different times, one can determine how much energy passed through the meter during the interval between readings.

When the power to the meter is cut off, there are extra damping magnets which stop the disc from rotating further. Figure 10-18 shows a photograph of a typical watt-hour meter.

Problems

1. What is the power output of a 12 V battery when it is generating a current of 0.5 A?
2. A power supply can deliver a dc output of 250 mA at 150 V. What is the power output of this supply?
3. What is the power dissipated by a 10Ω resistor, if 4 mA of current are flowing through it?
4. What is the resistance of a resistor that dissipates 160 W when a voltage drop of 40 V exists across it?
5. A sinewave whose amplitude is 20 V and whose frequency is 60 Hz is applied across a 20 Ω resistor. Find the power dissipated by the resistor.
6. A 250 V voltmeter with a 20,000 Ω/V rating and a 5 A ammeter are used to measure the power being dissipated in a dc circuit. The voltmeter reads 180 V and the ammeter 4.5 A. What is the actual power being dissipated by the circuit?
7. The voltage and current being applied to a single-phase load are in the form of sinusoidal quantities whose amplitudes are 250 V and 3.5 A, respectively. The power factor of the load is 0.32. Calculate $\bar{P}$, P_A and P_{VAR} of the load.
8. Explain why one pair of terminals in a single-phase wattmeter is called the current terminals and the other pair of terminals is called the voltage terminals?
9. Explain why it is necessary that the "±" terminal of the current terminals and the "±" terminal of the voltage terminals of the dynamometer wattmeter must be connected to the same wire of the input power line?
10. Describe why connection A of Fig. 10-5 is better for measuring loads which have high current and low voltage levels.
11. If the cost of electric power is 4 cents per kilowatt-hour, what is the cost of using:
    a) a 1200 W toaster for 30 min,
    b) 12, 50 W light bulbs for 4 hours,
    c) a 205 W television for 3 hours,
    d) a 5600 W electric clothes dryer for 20 min?

References

1. BOYLESTAD, R., *Introductory Circuit Analysis*, Chaps. 4, 17, and 19. Columbus, Ohio, Charles E. Merrill Books, Inc., 1968.

2. STOUT, M. B., *Basic Electrical Measurements*, 2nd ed., Chap. 17. Englewood Cliffs, N.J., Prentice-Hall, Inc., 1960.

3. KINNARD, I. F., *Applied Electrical Measurements*, Chap. 6. New York, John Wiley & Sons, Inc., 1956.

11

Potentiometers and Recorders

The purpose of this chapter is to discuss the operation and use of potentiometers and recorders. *Potentiometers* may be used to make very accurate determinations of unknown voltages. There are many measurement applications where such accuracy is a necessity. For example, the calibration of voltmeters must be undertaken by comparing their readings to an accurately known *voltage standard.* Because the accuracy of the potentiometer is so high, it can be used in the laboratory in conjunction with such a standard to calibrate voltmeters and other instruments.

The accuracy of potentiometers arises from the fact that they measure voltage through the use of a comparison technique. The unknown voltage is compared to an accurately known but adjustable voltage within the potentiometer. When the known voltage is set equal to the value of the unknown voltage, the unknown voltage becomes identified. The existence of the equality between the two voltages is indicated as a null reading on a sensitive meter. The value of the unknown voltage is read from the dial settings of the potentiometer at this *null* or *balance* point. (Note that a similar comparison measuring technique was used in the Wheatstone bridge circuit for measuring resistance.)

There are two basic classes of potentiometers, the *manually operated* and the *self-balancing* types. The manually operated models require an observer to adjust the dials until the equality between the known and unknown voltages is located. The self-balancing models are automatic devices which seek out the condition of equality themselves and do not need a human manipulator. Also, the self-balancing models are usually equipped with a marking device and a moving chart system. This allows, their indications to be permanently and automatically recorded on a graph.

Recorders are devices which provide a permanent graphical record of the quantity being measured, as well as carrying out the measurement itself. We note that these functions are also performed by the self-balancing potentiometer which was just described. This coincidence is not accidental. Rather, it is due to the fact that self-balancing potentiometers actually belong to one class of recorders. Since this overlapping of functions between some recorders and potentiometers exists, it is logical to undertake the discussion of the other types of recorders in this chapter too. Consequently, the latter part of the chapter is devoted to an examination of *X*-*Y* recorders, galvanometer recorders, and other forms of null-balancing recorders.

Potentiometers

When dc voltages are measured with the D'Arsonval meter movement, a combination of factors restricts the overall accuracy of the measurement. Sources of error such as bearing friction, nonlinearity in the restraining springs, and human reading errors prevent the measured voltages from being known to a degree greater than 0.1 percent. If this accuracy limit is to be extended, a measurement procedure which utilizes highly accurate components and which eliminates undesirable mechanical aspects of meter movements must be devised. The potentiometer is an instrument which uses such a measurement procedure. As a result, even portable manual potentiometers are capable of measuring voltages to within ± 0.05 percent. Much higher accuracies are available in the more elaborate self-balancing types of instruments (± 0.01 percent or less). Such accuracy makes potentiometers ideal for the task of calibrating other voltmeters.

In addition to being used as calibration standards, potentiometers can also act as detectors of microvolt-level voltages. This capability is put to use in measuring the minute output voltages of such transducers as thermocouples, piezoelectric crystals, and photovoltaic devices.[1] Finally, current measurements in the nanoampere range (10^{-9} A) can be made with potentiometers.

MANUAL POTENTIOMETERS

Manual and self-balancing potentiometers both achieve a null or balance condition by the use of the same basic measuring principle. However, the manual potentiometer is the simpler of the two instruments. Since it is easier to understand the general principles of potentiometer measurements from an examination of a simple instrument, we will undertake a discussion of the manual potentiometer first.

[1]Chapter 15 describes the operation of these and other transducers.

The operation of the basic manual potentiometer can best be understood with the help of a block diagram which shows its major elements. (Fig. 11-1). The four most important of these elements are the precision slide-wire resistance,[2] the sensitive current detector, the working voltage source, and the highly accurate reference voltage source.

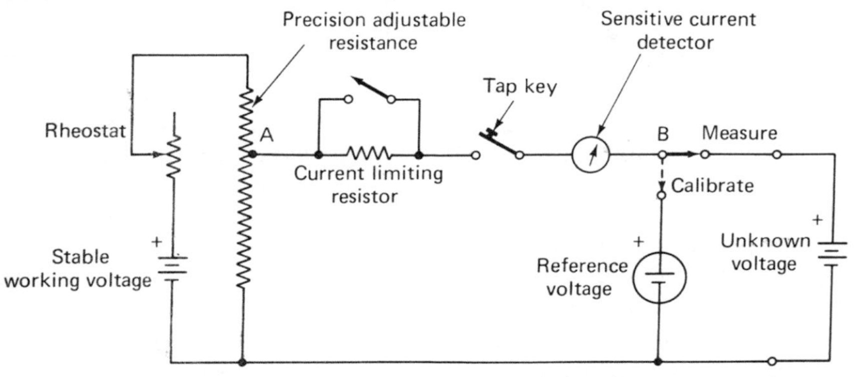

Figure 11-1. Block diagram of a manual potentiometer

The precision slide-wire resistance element is really the heart of the entire device. It basically consists of a length of wire whose total resistance is very accurately known (i.e., to within ± 0.02 percent or better). In addition, the cross section of the wire is kept extremely uniform so that any fraction of its total length will contain the same fraction of the total resistance. For example, if the wire has a total resistance of 150 Ω and is 150 cm in length, a portion of the wire whose length is 53.65 cm will have a resistance of 53.65 Ω (± 0.01 Ω). If a current of a 10.0 mA is caused to flow in such a 150-Ω resistance element, a voltage drop of 1.50 V will exist across its entire length.

A moving slider is attached to the precision resistor in such a way that an accurately known fraction of the total voltage across the resistor can be "picked off" by setting the slider to that fractional length of the resistor. The position of this slider along the length of the resistance element is indicated in voltage units by the dials on the potentiometer.

The purpose of the working voltage and rheostat is to provide an accurate amount of current flow in the precision resistor element. The value of this current is chosen so that it provides convenient numbers for calculating the voltages being measured by the instrument. To ensure that the working voltage and rheostat are actually providing the desired value of current to

[2]The slide-wire resistance is actually the same type of device as the *potentiometer* described in Chapter 5. In fact, this one element has given its name to the entire voltage-measuring instrument. To reduce confusion, in this chapter we will continue to call the variable-resistance element a slide-wire resistor and the complete voltage-measuring device a potentiometer.

the precision resistor, a very accurately known voltage source is used as a calibrating device. (If a secondary standard cell[3] is the calibrator, its voltage is known to be 1.019 V.)

In the process of calibrating the potentiometer, the slider on the precision resistance is set to a position which will ensure that some convenient current value (e.g., 10 mA) is flowing in the wire when it is properly calibrated. For example, if the 150-Ω, 150-cm resistor is being used, the slider is set to the 101.9-cm point on the wire. The resistance of the length of the wire from that point to the potentiometer circuit ground is 101.9 Ω. If a current of 10 mA were flowing in the wire, a 1.019-V drop would exist between that point and the ground point. To calibrate the potentiometer (that is, to make sure that 10 mA is actually flowing in the wire resistor), the positive terminal of the standard cell (which is also at +1.019 V) is connected to the same point on the wire as the slider. A current detector (in the form of a very sensitive D'Arsonval galvanometer) is connected between the slider on the precision resistor (point A in the circuit of Fig. 11-1) and the positive terminal of the standard cell (point B of the same figure). If the voltage drop across the precision resistor is actually equal to 1.019 V, no current will flow in the circuit branch containing the detector because there will be no voltage drop between points A and B. Since the value of the resistance of the potentiometer wire between point A and ground is known to be 101.9 Ω, we know that exactly 10.0 mA is flowing in the wire.

If, on the other hand 10.0 mA is not flowing in the resistor, point A will not be at a potential of 1.019 V. In that case, there will be a potential difference between points A and B, and a current will flow in the branch between them. This current flow will be sensed by the galvanometer detector. The rheostat connected in series with the working voltage battery and the precision resistance must then be adjusted so that the current flowing becomes 10 mA. At that point, the galvanometer detector will indicate that no current is flowing between points A and B of the circuit.

Once the magnitude of the current flowing in the precision resistance is calibrated in this way, the potentiometer is ready for making voltage measurements. From that time on, the rheostat (which was adjusted to allow this exact current flow) should no longer be touched. When an unknown voltage is connected to the UNKNOWN VOLTAGE terminals of the instrument, the slider of the precision resistor is moved until a position is reached where the galvanometer detector again shows no indication of current flow. At that point, the unknown voltage is known to be equal to the voltage drop across that fraction of the precision resistance which exists between the slider tap and ground. In the example of the potentiometer (which was calibrated to have 10.0 mA flowing in it), a position equal to 94.72 cm at

[3]See the section on *Standard Cells* in Chapter 8 for more details about its operation.

no-current (or *null*) would indicate that the unknown voltage was equal to 0.947 V.

A few additional features of the potentiometer and its components should also be mentioned at this point. First, when a null condition is reached, the unknown voltage is being measured without any current being drawn by the potentiometer. This means that the impedance of the potentiometer at balance is essentially infinite. An infinite impedance corresponds to the condition existing in an *ideal voltmeter*. Second, the mechanical aspects of the galvanometer operation (such as bearing friction and component nonlinearities) are bypassed because there is no deflection required of the detector at the time the final measurement is made. As a result of these features, two major sources of error which exist in other voltage measurement procedures are eliminated (the loading effect and the mechanical uncertainties).

The remaining sources of error in the potentiometer involve the accuracy of the resistance-to-position correlation along the precision resistor, the sensitivity of the detector to current flow, the stability of the working voltage, and the accuracy of the voltage reference. The importance of each of these errors is discussed in the following paragraphs.

The accuracy to which the resistance value along any point of the precision resistor is known is typically ± 0.02 percent or less. (The more elaborate the potentiometer, the greater the accuracy.) The galvanometers used as the null detectors (Fig. 11-2) have sensitivities which range from 2 mA/mm (portable type) to 0.0001 mA/mm (light beam, laboratory type). Since the galvanometers are such sensitive devices, they are provided with a clamp

Figure 11-2. Galvanometer detector used in potentiometers (Courtesy of Leeds & Northrup Co.)

for holding the pointer when it is not in use, as well as a protective current shunt. When making the initial adjustments in locating the null, one should set the shunt to its smallest resistance. This action protects the galvanometer from the large currents which might occur due to severe imbalance between the unknown voltage and that of the potentiometer.

The working voltage source is usually a *mercury battery*. As described in Chapter 8, this battery provides a constant voltage value over long periods of time and is therefore well-suited for its specified task. The standard cell (also discussed in Chapter 8) is a very accurate reference voltage source. If used and maintained properly, it produces a very stable voltage over long periods of time. (In fact, its output value drifts less than 30 μV per year!) To prevent more than 100 μA from ever flowing through the standard cell during calibration procedures, the galvanometer branch also contains a large protective resistor. As the point of calibration is reached, this resistor can be removed from the branch to increase the accuracy of the calibration. (A high-sensitivity button is provided on most commercially available potentiometers for this purpose.)

Since standard cells are relatively sensitive to damage by misuse, some potentiometers are using Zener diodes as their voltage calibration standards instead of standard cells. Under certain circumstances, the accuracies of Zener diodes rival those of secondary standard cells (and they are much more rugged).

COMMERCIAL MANUAL POTENTIOMETERS

Commercially available manual potentiometers are built with full-scale voltage ranges of 0.01 mV to 1.5 kV. However, the most commonly used instruments generally have ranges from 150 mV to 1.5 V. The accuracy of portable units (Fig. 11-3) is typically 0.05 percent of the *reading*.

The precision variable-resistance element of most commercial potentiometers consists of a group of precision decade resistors (for adjusting the

Figure 11-3. Manual potentiometer (Courtesy of Leeds & Northrup Co.)

coarser resistance values), plus a slide-wire for making fine resistance adjustments. The value of the measured voltage is then found from the dial settings by adding the values of the decade resistors to the value of the slide-wire resistance.

Some manually operated potentiometers are also equipped for measuring the voltage outputs of thermocouples. These models ordinarily have a control marked *Reference Junction*. When making ordinary voltage measurements, this control should be set to zero.

HOW TO USE A MANUAL POTENTIOMETER TO MEASURE VOLTAGE

1. The first step in the use of a manual potentiometer is to zero the galvanometer. Then we can begin calibrating the potentiometer. We start the calibration procedure by setting the magnitude of the resistance of the precision resistor equal to some convenient multiple of the magnitude of the standard-cell voltage being used in the potentiometer. The magnitude of the standard cell voltage is usually printed on the housing of the cell.
2. Set the switch of the potentiometer to the *Calibrate* position. Then, by tapping on the galvanometer key, open and close the circuit containing the galvanometer. This tapping action (rather than a holding down of the key) prevents large unwanted currents from flowing into or out of the standard cell in cases of great unbalance. While tapping the key, watch the galvanometer scale to observe the extent of the pointer deflection. Adjust the rheostat connected to the working battery until the galvanometer shows no deflection (null condition).
3. When the null condition is reached, set the galvanometer to its most sensitive setting. This is done by shorting out the protective resistor in the galvanometer branch. Make a final adjustment of the rheostat so that the most sensitive null condition is achieved. This indicates that the potentiometer is calibrated to the value of the standard cell. Record the value of the resistance of null from the dial settings.
4. Turn the function switch from the *Calibrate* to the *Measure* position.
5. Connect the unknown voltage to the appropriate terminals of the potentiometer. In doing so, it is very important to be sure that the polarities of the unknown voltage are connected to the proper terminals of the potentiometer.
6. Seek a null reading in the galvanometer again by first varying the decade resistor and then the position of the slider on the precision resistor. During this initial adjustment, leave the protective resistor in the circuit with the galvanometer.
7. As the null is approached, short out the protective resistor so that the most sensitive null setting can be located. Proceed to reach this null by adjusting the value of the slide-wire resistor.

8. When a final null is reached, the dial settings of the potentiometer will be equal to the value of the unknown voltage.
9. To ensure that the measured value is valid, reset the function switch to the *Calibrate* position, and return the dials of the slide wire to the setting that yielded a null during the initial calibration. If a null condition still exists, the measurement is valid. If a null no longer exists, this means the current flowing in the precision resistance has changed. In that case, recalibrate the potentiometer and make the measurement again. After the second measurement, a check of the calibration is again necessary to ensure the validity of the measurement.

SELF-BALANCING POTENTIOMETERS

When manual potentiometers were the only types of potentiometers available, potentiometric methods for determining voltages were limited to highly precise laboratory measurements. The major limitation which prevented potentiometers from being more widely used was that they required an operator to perform the careful manipulations necessary to bring them to a condition of balance. With the development of automatic or *self-balancing* methods for achieving the balance conditions, the advantages of potentiometric measurement techniques were applied to a much wider range of measurements.

In addition to merely providing a measurement indication, self-balancing potentiometers are also capable of *permanently recording* the values of the voltages as they are being measured. This is done by putting a marking device on the pointer and using a motor-driven chart as the recording surface.

The automatic balancing feature of the self-balancing potentiometer relies on a servomotor[4] working together with several other pieces of electronic equipment. Figure 11-4 is a block diagram of a self-balancing potentiometer which shows how these elements perform the balancing act.

In this figure we see that the voltage signal to be measured (V_i) is applied to one input terminal of a device called an *error detector*. Another voltage (V_s), obtained from an adjustable voltage reference source, is fed to the second input terminal of the *error detector*. The error detector electronically subtracts V_i from V_s and utilizes the result as its output signal. The output signal from the error detector is called the *error signal*, and it is used to drive the servomotor. However, before the error signal is applied to the servomotor, it must be amplified by an electronic amplifier. This amplification endows the signal with a magnitude that is sufficiently large to activate

[4]A *servomotor* is defined to be a motor which responds to a command. This response is in the form of a motion which corrects any difference between the actual and desired state of the system to which the motor is connected.

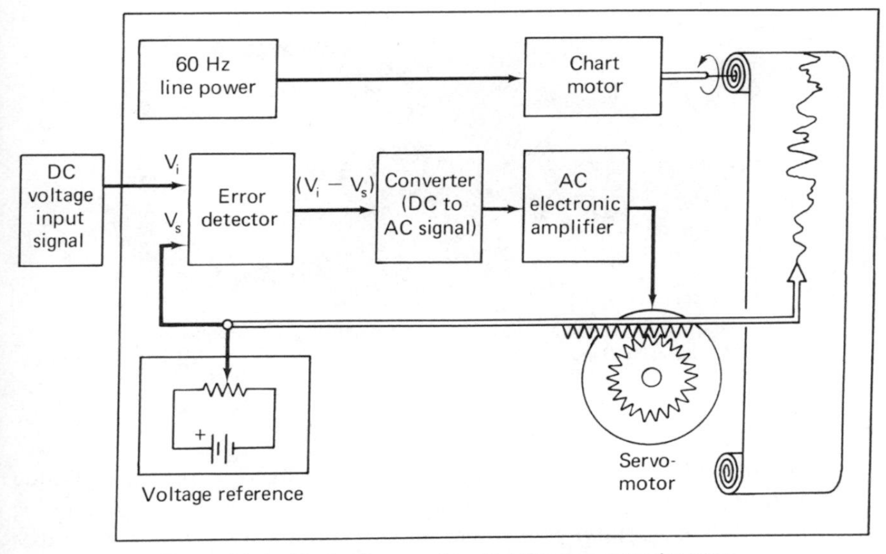

Figure 11-4. Block diagram of a self-balancing potentiometer

the servomotor.[5] The amplified error signal applied to the servomotor causes the shaft of the motor to rotate.

The servomotor shaft has two mechanical connections attached to it. The first couples the shaft to a voltage-divider circuit which is part of the variable dc reference source that provided V_s. As the shaft turns, it moves the slider of this voltage divider. The direction of the motion is such that V_s changes in value so that it approaches the value of V_i. When the values V_s and V_i are equalized, the error signal from the error detector becomes zero ($V_i - V_s = 0$) and the servomotor halts. (Note that the movement of the slider by the servomotor replaces the action that the observer performs in the manually operated models.) The second mechanical connection attached to the servomotor shaft is coupled to a pointer which indicates the degree of rotation of the shaft. When the servomotor comes to a halt, the reference voltage, V_s, is equal to the input voltage, V_i. Since the value of the reference voltage is known very accurately, the position of the pointer (as controlled by the motor shaft position) can be calibrated to indicate the value of the input voltage directly. If the pointer has a pen or other type of marking device connected to it, and if a separate motor is used to drive a ruled paper chart under the marker, a permanent record of the voltage measured versus time is produced. Figure 11-5 shows the insides of a self-balancing potentiometer.

[5]To avoid certain problems involved with dc amplifiers, the error signal is generally converted to an ac form before being fed to the amplifier.

Figure 11-5. Inside view of a self-balancing potentiometer (Courtesy of Leeds & Northrup Co.)

Recorders

A *recording instrument* is a device whose function is to record the value of a quantity as it is being measured. Such instruments may include graphic recording devices, computer printers, tape recorders, or cathode ray tubes (CRTs). Our present discussion will be concerned only with the operation and use of graphic recording instruments.[6]

There are three types of graphic recording devices in common use. They are the self-balancing recorder, the *X-Y* recorder, and the galvanometer recorder. All of these devices generally contain a built-in measuring apparatus as well as a recording mechanism. This gives *recorders* the capability of receiving electrical signals from detectors or sensors and converting their magnitudes (and variations in the magnitudes) into a permanent graphical record.

In the self-balancing and galvanometer recorders, a pen assembly is moved over a paper chart that is simultaneously being moved in an orthogonal direction by a separate motor. (These types of recorders are often called *oscillographs* as well.) In the *X-Y* recorders, the pen is simultaneously moved in two perpendicular directions while the paper chart is kept stationary.

The charts of the self-balancing and galvanometer recorders consist of either *round* or *strip-type* graphs (Fig. 11-6). The round graphs are circular pieces of paper which have concentric circles ruled on them to form their scales. In addition, there are printed arcs extending from the center of the chart to the paper's edge. As the pen of the recorder is moved, it swings along these arcs. The arcs are known as *time arcs*. (Since the chart is rotated at a uniform rate, the angular position between arcs indicates elapsed time.) The radial position of the pen at any time indicates the instantaneous value of the quantity being measured. The *strip-chart recorders* have graphs of

[6]CRTs were described in Chapter 9, and the other devices do not fall within the scope of the material being discussed in this text.

(a)

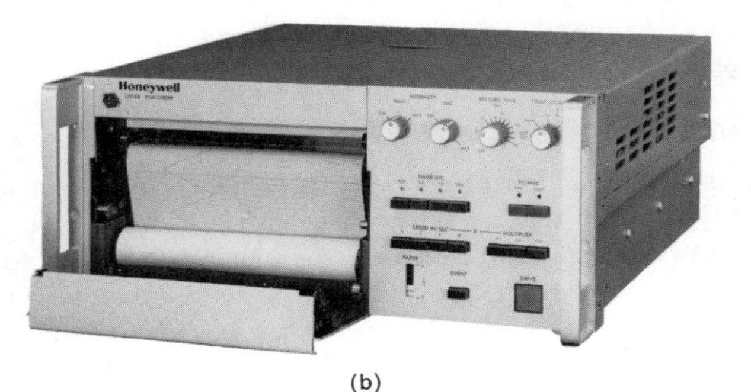

(b)

Figure 11-6. Recorders (a) Round-chart recorder (Courtesy of General Electric Co.) (b) Strip-chart, null-balance-type recorder (Courtesy of Honeywell Inc., Test Instruments Div.)

paper in the form of a long roll. As the graph is unrolled, it also moves along under the pen or marker at a uniform rate of speed. The lines on the paper are ruled parallel to the direction of motion and form the scales of the graph. The lines that are ruled perpendicular to the direction of the motion are the time lines. (It is also common to use this axis for other parameters which can be made proportional to time.) The pen assembly moves in a direction perpendicular to the motion of the paper. Thus the value of the measured variable versus time is traced on the chart as the paper rolls by the pen. The instantaneous value of the measured quantity is given by the position of the pen.

There are several common marking devices used in recorders. The pen-and-ink system is the most familiar type. It usually consists of an ink reservoir connected to a fiber-tipped pen through a narrow tube. The advantages of pen-and-ink systems are their simplicity and low operating costs. On the other hand, they also suffer several disadvantages. First, there is the hazard of ink spillage, which may spoil the graph. Second, the pen tip is easily damaged, and this can lead to an impairment of the flow of ink through it.

Finally, ink can dry in the tube connecting the pen and the reservoir. This can clog or ruin the entire pen assembly.

As a consequence of the disadvantages of pen-and-ink assemblies, several other marking devices have been developed. These inkless methods are particularly useful in those recorders that must remain unattended for long recording periods. One type of inkless marking device utilizes a heated pen on a heat-sensitive paper. In this type, an electric current is passed through the tip of the movable stylus. The current heats the stylus, and the heat causes a thin, clear line to appear on the special heat-sensitive paper. Other inkless graphing methods include light-sensitive and pressure-sensitive chart papers. These methods must also be used in conjunction with special styluses. The biggest disadvantages of inkless methods are that they are generally more complex than pen-and-ink assemblies, and they also require more expensive paper for their operation.

NULL-BALANCE RECORDERS

The self-balancing action of all *null-balance recorders* is provided by the same basic type of servomotor scheme as that used in self-balancing potentiometers. However, the source of the error signal which is fed to the servomotor may be derived from either a potentiometer circuit, a bridge circuit, or a linear, variable, differential transformer (LVDT) circuit. The choice of circuit used in a particular null-balance recorder depends on the types of signals the recorder is designed to monitor.

If the input signal to the recorder is in the form of a *voltage*, a potentiometric circuit is used to measure the signal. Such voltage quantities exist as the outputs of *active* transducers (e.g., thermocouples, pH meters, piezoelectric crystals, and photovoltaic cells). On the other hand, bridge types of circuits (Fig. 11-7(a)) are used by recorders when the signals they are monitoring are due to a change in some electrical parameter of a *passive* transducer. Examples of these types of signals include the resistance changes which occur in strain gauges, thermistors, and photoconductive cells. Finally, if the measured signal is manifested in the form of small displacements, a circuit which utilizes LVDTs is employed to generate a self-balancing capability[7] (see Fig. 11-7(b)).

All of the null-balancing recorders described above are very sensitive indicating instruments. They are capable of reading signal voltages from 1 mV to 100 V (dc) with a high degree of accuracy (i.e., to within ± 0.1 percent). However, since the null-balance mechanism is electromechanical in nature, such recorders are quite limited in their ability to respond to rapidly changing quantities. In the same vein, since the maximum speed of the pen assembly is limited, the frequency response depends upon the peak-

[7]The operational principles of the transducers mentioned in this paragraph (including the LVDT) are all described in Chapter 15.

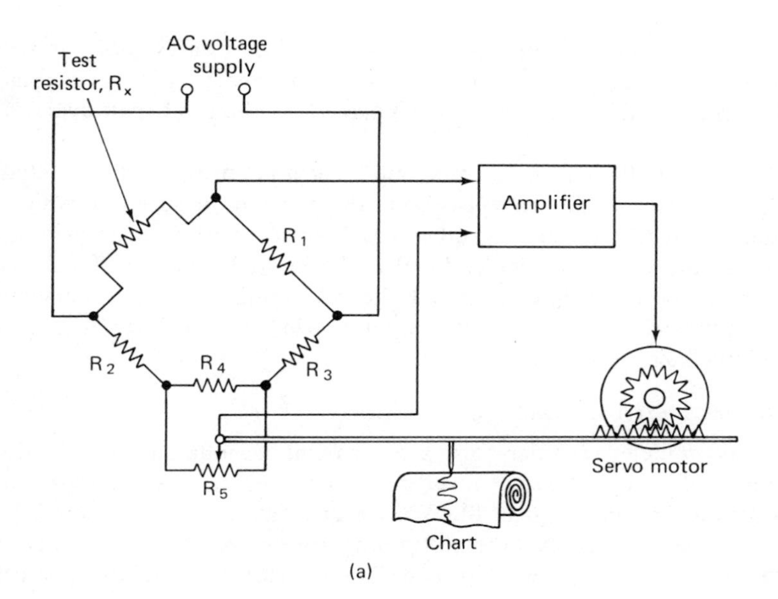

(a)

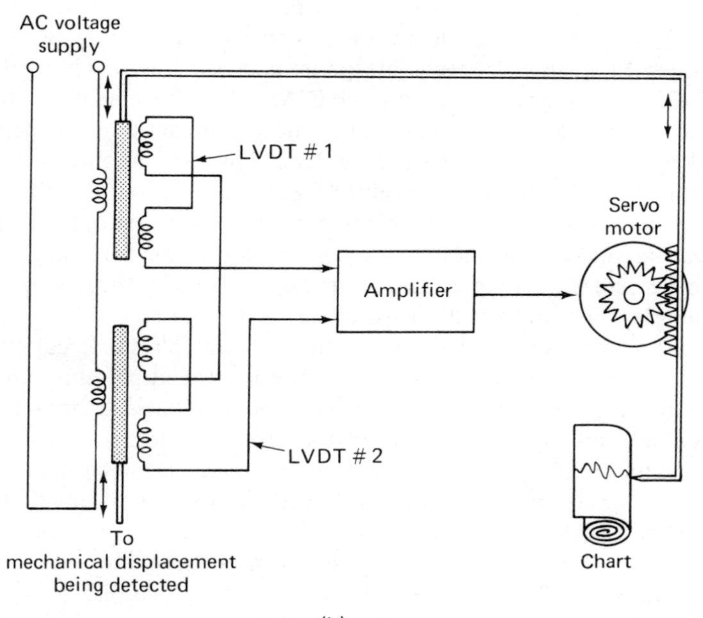

(b)

Figure 11-7. Bridge and linear differential transformer (LVDT) null balance recorders (a) Bridge-type recorder (b) LVDT-type recorder

to-peak excursion of the signal being recorded. For example, the typical maximum frequency response of a null-balance recorder is specified as 5 Hz when the excursion of the pen is 10 percent of the total span over which it can move.

To make null-balance recorders more versatile, provisions are included to allow variations in the *zero position* of the pen and the speeds at which chart paper is fed through the recorder. Furthermore, various chart widths from 4 to 12 inches are available. Finally, the input inpedance of null-balance recorders is made quite high. At the mV scales, the input impedance is potentiometric (i.e., essentially infinite), while at the less-sensitive scales, the impedance is 10 MΩ or more.

GALVANOMETER RECORDERS

Galvanometer recorders[8] use a pen assembly mounted on the end of the pointer of a rugged D'Arsonval movement similar to the movement used in basic dc meters (Fig. 11-8). When a quantity is being measured by such galvanometer recorders, the restraining springs of the movements (rather than a self-balancing signal) provide a counterforce which balances the force created by the quantity being measured. Strip charts are usually used with this type of recorder.

Although galvanometer recorders are not as sensitive as self-balancing models, they have some other advantages that self-balancing recorders do not possess. First, their frequency response can be made much greater than that of self-balancing types. Second, the D'Arsonval movement which directs their marking devices can be made much more compact than the pen-motor assemblies of the self-balancing recorders. Thus galvanometer recorders with multichannel outputs are feasible (Fig. 11-9). As a result, commercially available recorders are built which contain up to thirty-six output channels. This feature makes galvanometer recorders attractive for simultaneously monitoring and recording a large number of slowly varying quantities. For example, a multichannel recorder can be used to record the outputs of transducers which simultaneously monitor several physiological functions of a hospital patient (such as body temperature, blood pressure, and respiration rate). Another application involves the simultaneous monitoring of various conditions which exist aboard a satellite in space.

The maximum frequency of ordinary galvanometer recorders is about 100 Hz (for small pen excursions), while their maximum sensitivity is on the order of 25 mV/in. The input impedances of these instruments are typically 100 kΩ or more, and corresponding accuracies are ± 1.0 to ± 2.0 percent, full-scale. Special galvanometer recorders that use a light beam rather than a pointer-and-pen assembly are capable of recording signals whose frequency variations are as high as 13 kHz.

[8]Galvanometer recorders are sometimes also called *direct-writing* recorders.

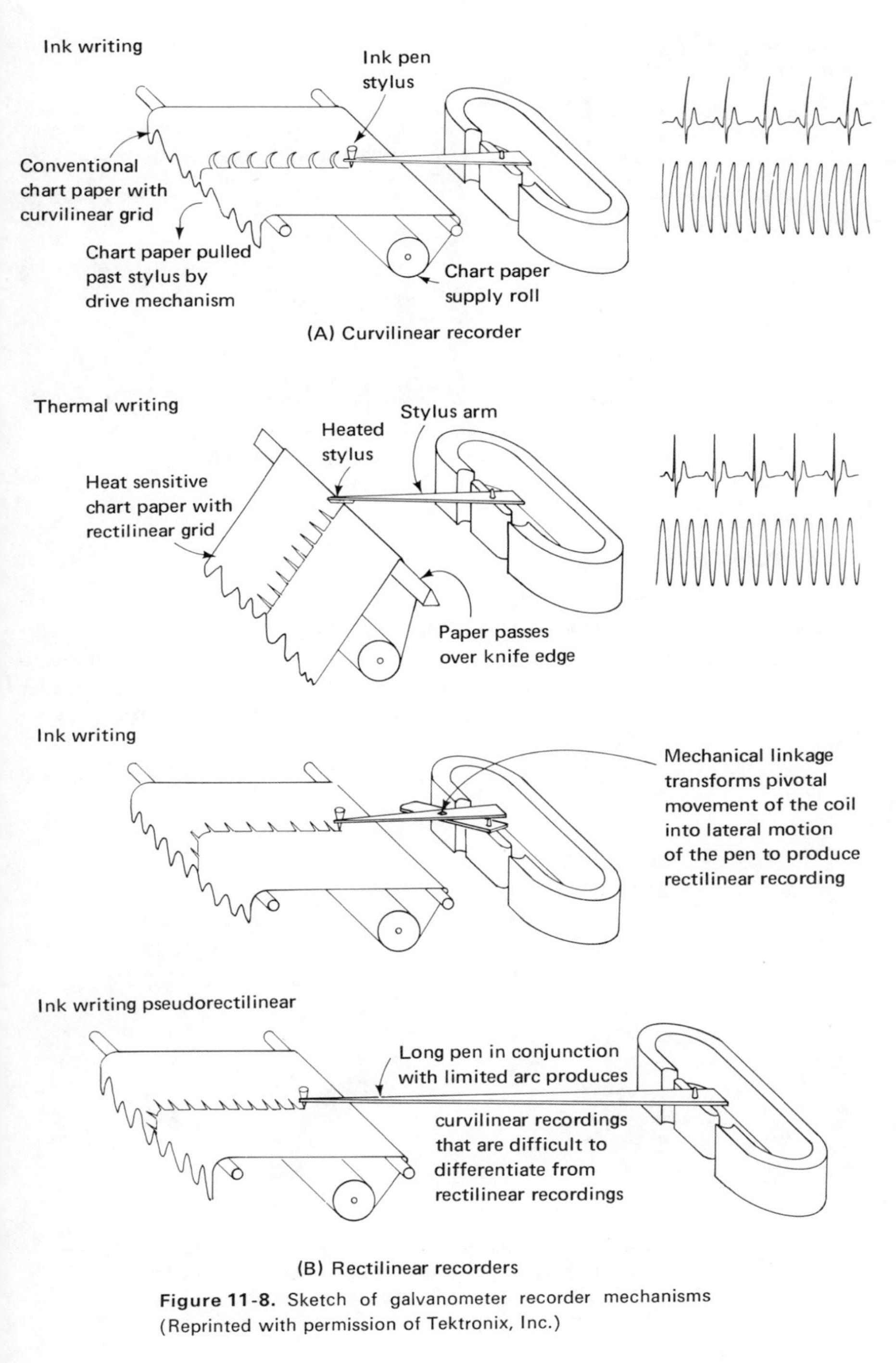

Figure 11-8. Sketch of galvanometer recorder mechanisms (Reprinted with permission of Tektronix, Inc.)

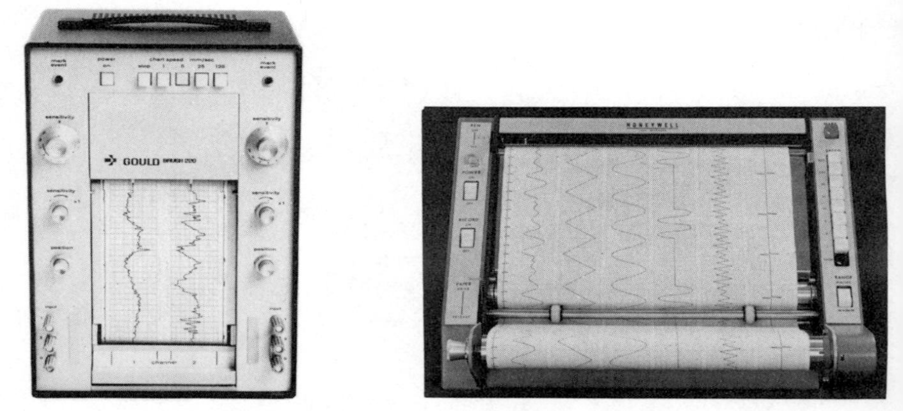

Figure 11-9. Galvanometer recorders (Courtesy of Gould, Inc., Instrument Systems Div. and Honeywell Test Instrument Div.)

X-Y RECORDERS

X-Y recorders are instruments which have the special capability of displaying two separately varying quantities on the *X* and *Y* axes of cartesian coordinates (Fig. 11-10). This means that one of the variables can be applied to the *X* input of the recorder, and the other to the *Y* input, and the recorder will plot their variations against one another. This feature exists because the *X-Y* recorder can simultaneously move a pen in both the *X* and *Y* directions across a fixed paper in response to electrical signals applied to its two

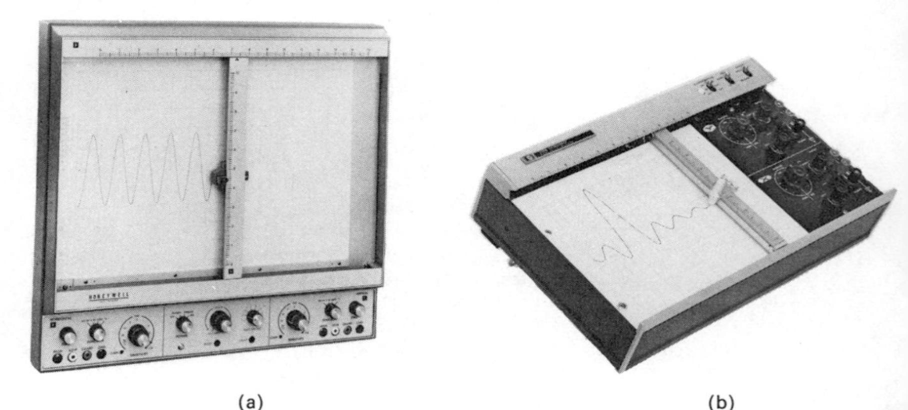

(a) (b)

Figure 11-10. X-Y recorders (a) Honeywell 55D *X-Y* recorder (Courtesy of Honeywell, Inc., Test Instrument Div.) (b) Hewlett-Packard 7035B X-Y recorder (Courtesy of Hewlett-Packard Co.)

input terminals. Since most *X-Y* recorders also contain a time base, they can be used to portray the variation of one variable versus time as well. This allows them to perform some measurements in the same manner as strip-chart recorders.

Another feature of *X-Y* recorders is that they are rather inexpensive to operate. Because most use a pen-and-ink marking system, ordinary low-cost paper can be used for the chart. In addition, they are rather easy to use.

On the negative side, *X-Y* recorders are slower than strip recorders and cannot be used for continuous monitoring applications. They are also considerably more expensive to buy than many simple strip recorders.

There are a wide variety of applications in which *X-Y* recorders are used. Some of these applications include plotting current vs voltage (*I-V*) curves of diodes and transistors, plotting *B-H* curves of magnetic materials, reproducing analog and digital (indirectly) computer readouts, and graphing voltage vs frequency plots from sweep-frequency oscillators.

The mechanism which locates the position of the pen along both the *X* and *Y* axes of the recorder is a closed-loop servosystem (very similar to the servomechanism used in the self-balancing potentiometer). The rotation of the servomotor moves the marking pen to its position by using a string-and-pulley arrangement. Because this mechanism is electromechanical, the frequency responses and speeds of *X-Y* recorders are inherently limited. Their typical frequency response is about 5 Hz for a one inch peak-to-peak signal. Likewise, their *slewing speed* (defined as the maximum velocity at which the pen assembly can be moved after its acceleration has stopped) is about 20 inches per second.

The other specifications of *X-Y* recorders include input and output characteristics, paper sizes, maximum sensitivities, and their accuracies. Typical values for these quantities are as follows. Input impedance—100 kΩ to 1 MΩ on the less-sensitive ranges and 50 MΩ or more on the most sensitive range; paper sizes—$8\frac{1}{2} \times 11$ inches or 11×17 inches; maximum sensitivity—100 mV/in to 1 mV/in, depending on the model; accuracy—±0.1 percent, full-scale.

Problems

1. Explain why measurements that involve the detection of a null to indicate the quantity being measured can be made more accurately than those involving the reading of a pointer deflection made along a scale.
2. What is the difference between the definitions of the term *potentiometer* as introduced in Chap. 5 and as used in Chap. 11?
3. Describe the differences between *manual* and *self-balancing* potentiometers.
4. Explain why the loading effects of potentiometers and conventional voltmeters differ greatly from one another in magnitude.

5. A manual potentiometer contains the following components: a working battery of 3.0 V and negligible internal resistance; a standard cell whose voltage is 1.0191 V and whose internal resistance is 200 Ω; a 200 cm adjustable precision resistance; a galvanometer with an internal resistance of 50 Ω. The rheostat of the instrument is set so that the potentiometer is calibrated at the 101.91 cm mark of the precision adjustable resistance.
   a) Calculate the current flowing in the rheostat as well as its resistance value.
   b) If a protective resistance is placed in series with the galvanometer, calculate the value of the resistance necessary to limit the current in the galvanometer to 10 μA.
6. List some of the factors which determine the accuracy of potentiometers.
7. When using the potentiometer it is recommended that the key should only be depressed momentarily to avoid damage to either the galvanometer or the standard cell. Explain in more detail why this recommendation should be followed.
8. What would be the effects of a sizable temperature change on the accuracy of a potentiometer?
9. Describe the functions of the following components of a self-balancing potentiometer:
   a) error detector,
   b) amplifier,
   c) servomotor,
   d) reference voltage source.
10. Draw the charts of both the round and strip chart recorders. Indicate how both time and the magnitudes of the quantities being measured are exhibited on these charts.
11. Explain the advantages that each of the following types of recorders have over the other types:
    a) self-balancing recorders,
    b) galvanometer recorders,
    c) X-Y recorders.
12. What causes the frequency response of recorders to be limited?
13. Describe two types of inkless recording methods.

References

1. STOUT, M. B., *Basic Electrical Measurements*, 2nd ed., Chap. 7. Englewood Cliffs, N.J., Prentice-Hall, Inc., 1960.
2. HARRIS, F. K., *Electrical Measurements*, Chap. 9. New York, John Wiley & Sons, Inc., 1952.
3. *Leeds & Northrup Catalog*, 1970. No. Wales, Pa., Leeds and Northrup Company.

4. *Honeywell Instrumentation Catalog*. Denver, Honeywell, Inc., 1968.

5. PRENSKY, S. D., *Electronic Instrumentation*, 2nd ed., Chap. 8. Englewood Cliffs, N.J., Prentice-Hall, Inc., 1970.

6. MALMSTADT, H. V., ENKE, C. G., and TOREN, E. C., *Electronics for Scientists*, Chaps. 6 and 7. New York, W. A. Benjamin, Inc., 1962.

12

Electronic Voltage and Current Meters

As we noted earlier, measurements of voltage and current are probably the most common electrical measurements made. In previous chapters we have discussed basic meters, oscilloscopes, and potentiometers as instruments which are used to make such measurements. In this chapter, we will consider the various types of electronically assisted meters which are also used to measure current and voltage. These instruments include such general purpose meters as vacuum-tube voltmeters (VTVMs), transistor voltmeters (TVMs), and digital voltmeters (DVMs). In addition, they include special meters such as electrometers, microvoltmeters, nanoammeters, and vector voltmeters. The approach of the discussion will be to describe the operation of these instruments along with the advantages and limitations associated with their use. Some guidelines to assist in the proper utilization of the more common instruments will also be presented.

Electronically assisted meters can be grouped into two general classes: *analog* meters and *digital* meters. Those types which use electromechanical movements and pointers to display the quantity being measured belong to the *analog* class. Those that display their indication as a specific numerical readout belong to the *digital* class of meters. The respective advantages of each type will be considered in the course of the discussion.

Analog Electronic Meters

Electronic meters belonging to the analog class basically consist of an amplifier connected to a D'Arsonval meter movement. In those meters designed to measure dc quantities, the amplifier receives the signal being measured, amplifies it and feeds it to the D'Arsonval movement. The D'Arsonval movement converts the amplified signal to an indication by causing a pointer to move along a calibrated scale. If the meters are designed to measure ac quantities, a rectifier is connected into the circuit along with the amplifier and D'Arsonval movement (Fig. 12-1).

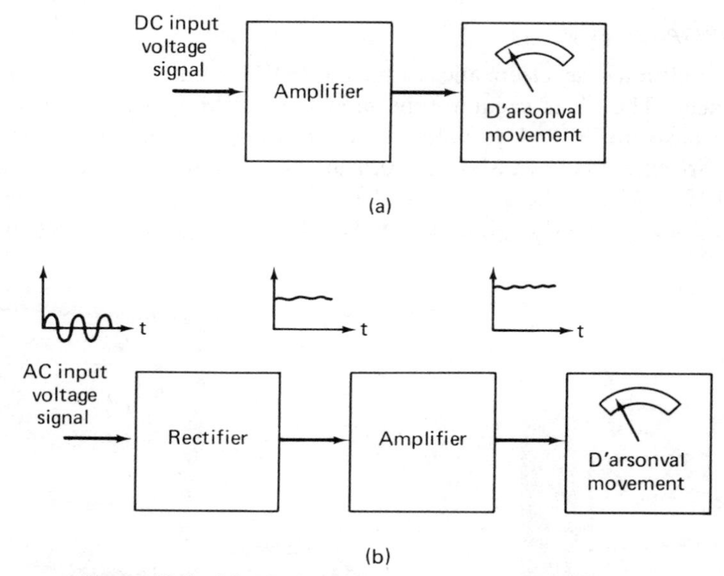

Figure 12-1. (a) Block diagram of a dc electronic analog meter (b) Block diagram of an ac electronic analog meter

The amplifiers used in analog electronic meters are usually of special design, but they can be either vacuum-tube or transistor powered. The older meters which incorporated vacuum-tubes in their circuits were called vacuum-tube voltmeters (VTVMs). The more modern meters use transistors in the amplifier circuits and are known as TVMs (transistor voltmeters) or FETVMs (field-effect transistor voltmeters).

Electronic analog meters are generally divided into three categories:

1. Multipurpose meters—capable of measuring dc voltages, ac voltages, resistance (and sometimes dc current).
2. High-sensitivity ac meters—for measuring ac voltages.
3. High-sensitivity dc meters—for measuring dc voltages and currents.

Since the *multipurpose meters* are the most common types encountered, we will discuss their operation in greatest detail. As a part of this discussion we will compare the capabilities of the multipurpose electronic meter to those of the multipurpose electromechanical meter (the VOM). Such a comparison will demonstrate the advantages of both types of instruments and indicate the applications where each might best be used.

We will then briefly introduce the high-sensitivity ac meters. The description of high-sensitivity dc meters will be undertaken when we consider the electrometer, a common high-sensitivity VTVM.

MULTIPURPOSE EVMs

The multipurpose electronic voltmeter (EVM) is a versatile measuring instrument. The older vacuum-tube models (VTVMs) were designed to be able to measure dc and ac voltages and resistances. Some of the modern transistorized meters (TVMs) can measure these quantities and dc currents as well (Fig. 12-2). The amplifiers of both types of EVMs are of the same general design, even though one is powered by tubes and the other by transistors.

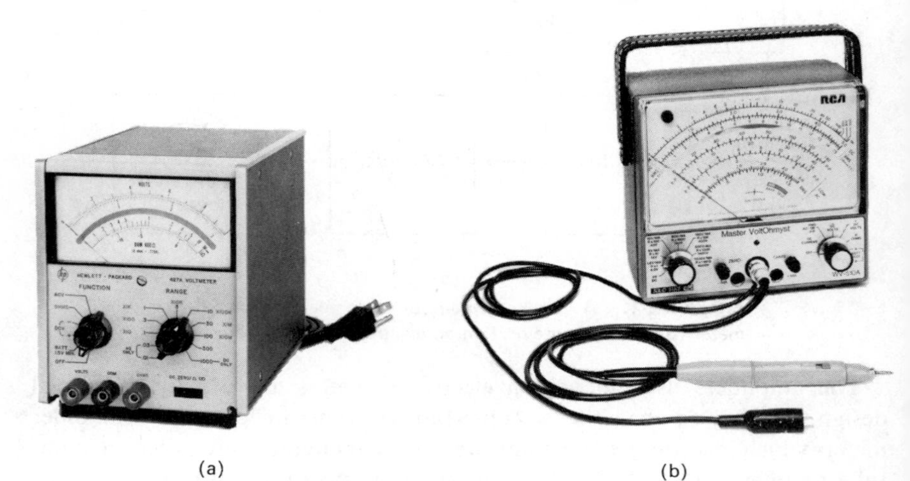

(a) (b)

Figure 12-2. Analog multipurpose EVMs (a) HP 427C (Courtesy of Hewlett-Packard Co.) (b) RCA Master VoltOhmyst (Courtesy of RCA Corp.)

sistors. The name given to this special type of amplifier is the *balanced-bridge, two-tube* (*or two-transistor*) *amplifier*. The circuit diagram of the tube version of this amplifier is shown in Fig. 12-3. When used alone (without a rectifier), it is very well suited for the task of serving as an amplifier for a dc voltmeter.

The two outstanding characteristics of the balanced-bridge amplifier which make it so attractive for use as an EVM amplifier are its high input impedance and stability.[1]

In order to make the EVMs equipped with balanced-bridge amplifiers

[1]High stability indicates that the amplifier is not apt to undergo changes in the value of its output due to inherent changes taking place in its components.

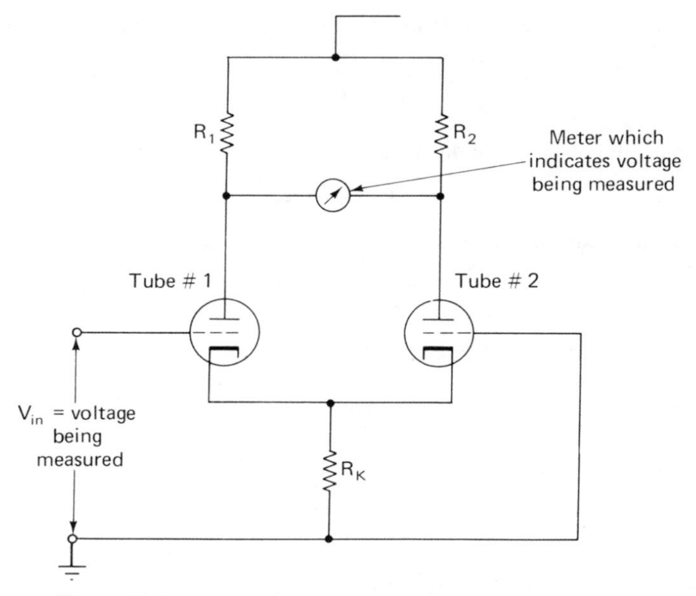

Figure 12-3. Circuit diagram of a balanced-bridge, two-tube amplifier

suitable for measuring ac voltages and resistance (as well as dc voltages), two additional circuits are built into their design. One is an ac rectifying circuit for converting the meter into an ac voltmeter. The other is a battery (or power supply) and voltage divider for allowing the meter to make resistance measurements. The manner in which these various components are connected together in an actual EVM is shown in Fig. 12-4.

The rectifier circuit of most multipurpose EVMs is of the type that causes the meter to respond to the peak-to-peak value of the signal being measured. Since the value of interest in ac signals is the rms value, a special calibration procedure is used so that the scale readings of the meters will yield an rms value. This procedure is similar to the one employed by the basic electromechanical rectifier meters described in Chapter 4, that is, the signal being measured is assumed to be a pure sine wave. Since it is known that the ratio of the peak-to-peak to rms values of a sine wave is 2.8 : 1, the actual voltage values read by the meter are divided by 2.8 to yield the scale markings of the meter. However, this means that if the signal being measured by the EVM is not a pure sine wave, an error will exist between the indicated and the true rms values of the signal being measured.

EXAMPLE 12-1

If a square wave with an amplitude of 5 V is measured by an EVM, what will be the error between the measured and the actual rms values of the signal?

Solution: The EVM responds to the peak-to-peak value of the square wave. This quantity ($V_{\text{peak-to-peak}} = V_{\text{p-p}}$) is

$$V_{\text{p-p}} = 2V_o = 2 \times 5\text{ V} = 10\text{ V}$$

The scale of the meter will indicate a voltage according to the following relation:

$$V = \frac{V_{\text{p-p}}}{2.8} = \frac{10}{2.8} = 3.6$$

The actual rms value of a square wave is found by using Eq. (1-16) (or by consulting Fig. 1-18):

$$V_{\text{rms}} = 5\text{ V}$$

Thus, the error of the meter reading for a square wave will be

$$\text{Error} = \frac{5 - 3.6}{5} \times 100\% = \frac{1.4}{5} \times 100\% = 28\%$$

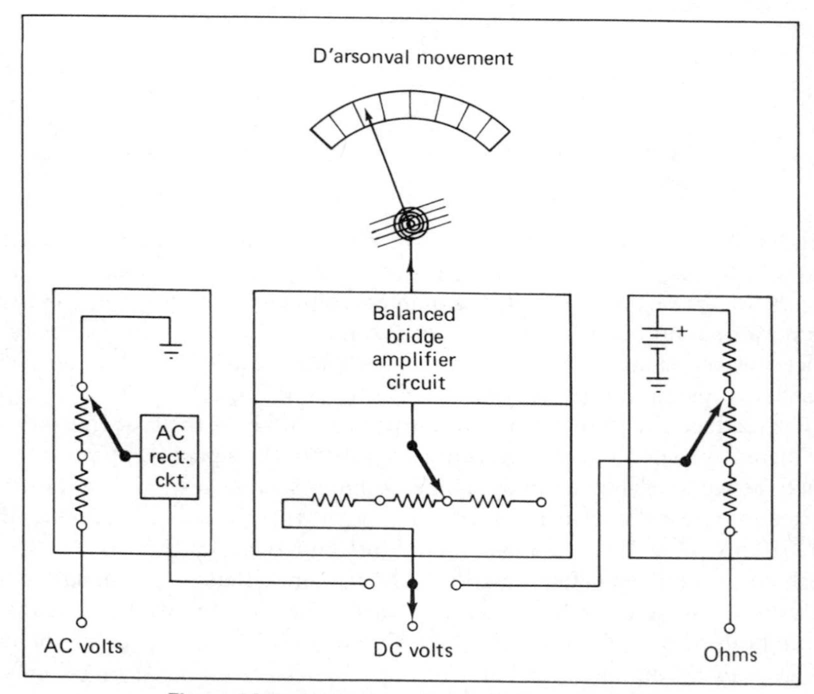

Figure 12-4. Major elements of a multipurpose EVM

EVM VERSUS VOM

The VOM (volt-ohm-milliammeter) was introduced in Chapter 4 as a multipurpose meter that had the capability of measuring dc voltages, ac voltages, dc currents, and resistances. In construction, VOMs are basically like the multipurpose EVMs, except that the VOMs lack an amplifier. As a means of emphasizing the advantages gained by the inclusion of the amplifier in the meter design, we will compare the capabilities of the VOM and the various types of multipurpose EVMs. Table 12-1 provides a source of information for making such a comparison. It lists many of the most important characteristics of typical VOMs, VTVMs, and two types of multipurpose TVMs. (The first of these TVMs is the service-type TVM. Such instruments are most commonly used in radio repair and other general electronic equipment servicing. The second type of TVM, the laboratory type, is used in a wide variety of laboratory measurements. The lab-type TVMs are somewhat more sensitive and expensive than the service-type models.)

The VOM is a smaller, simpler, and less expensive meter than any of the comparable multipurpose EVMs. It does not need to be plugged into the ac power line to be operated, and this makes it portable and free from hum and rf pickup (a feature also available in some, but not all, EVMs). Since it is simpler and is only in use during an actual measurement, the VOM gets less wear and is not as likely to break down as an EVM. Furthermore, if the VOM does break down, it is usually easier to repair than an EVM. Finally, VOMs have the capability of directly measuring dc currents. Not all EVMs are designed to measure current, and those types which can do so measure current by indirect methods. Thus, they are not able to measure dc currents as accurately as VOMs can.

On the other hand, EVMs as a group also possess some very definite advantages over VOMs. In general, we can say that the most important of these are:

1. Higher input impedance.
2. Higher upper-frequency limit.
3. Greater sensitivity.

Table 12-1 indicates that the input impedance of the various EVMs (when dc voltages are measured) can be 10–100 MΩ. For measuring ac voltages, VTVMs have input impedances from about 1.5 MΩ to 3 MΩ, whereas the input impedances of TVMs are up to 100 MΩ. In contrast, the input impedances of VOMs are generally 20,000 Ω/V for dc voltages and 5000 Ω/V for the ac voltages. Thus the input impedances of VOMs are generally much lower than those of EVMs. The lower input impedance makes the loading effect of VOMs greater than the loading effects of EVMs. For applications involving measurements of voltages across high-impedance elements, the larger loading effect of the VOM will yield more unreliable results. The higher input impedance of the EVM thus makes it applicable to a wider range of voltage measurements.

The frequency range of the EVM is also much greater than that of the

Table 12-1 Comparison of Specifications of VOMs and Multipurpose Analog EVMs

Characteristic	VOM	Multipurpose VTVM	Multipurpose TVM (Service Type)	Multipurpose TVM (Lab Type)
Dc volts	0–250 mV to 0–5000 V	0–3 V to 0–1200 V	0–0.3 V to 0–1000 V	0–50 mV to 0–1000 V
Ac volts	0–2.5 V to 0–5000 V	0–3 V to 0–1200 V	0–0.3 V to 0–1000 V	0–5 mV to 0–300 V
Dc current	0–50 μA to 0–10 A	—	0–1500 mA	0–5 μA to 0–1500 mA
Resistance	1 Ω to 10 MΩ	1 Ω to 1000 MΩ	1 Ω to 100 MΩ	0.1 Ω to 10 MΩ
Input impedance:				
Dc volts	20,000 Ω/V	10 MΩ	11 MΩ	11–100 MΩ
Ac volts	5000 Ω/V	1.5–3 MΩ	10 MΩ	11–100 MΩ
Accuracy:				
Dc volts	±2%	±3%	±3%	±2%
Ac volts	±3%	±5%	±3%	±3%
Frequency	0–20,000 Hz	0–1 MHz (w/0 rf probe)	0–100 kHz	0–1 MHz (w/0 rf probe)
	—	0–100 MHz (with rf probe)	0–100 MHz (w/rf probe)	≈ 500 MHz (with rf probe)
Power required	—	115 VAC/60 Hz	Battery or line	Battery or line

VOM. Table 12-1 shows that the upper frequency signal which can be accurately measured by a VOM is about 20,000 Hz. In contrast, most EVMs have a maximum frequency limit which is between 100 kHz and 1 MHz. In addition, special high-frequency probes are available for use with EVMs that extend their upper frequency limit to between 100 and 500 MHz (depending on the instrument design).[2]

The maximum sensitivities of multipurpose VTVMs are about the same as those for the VOMs. However, multipurpose TVMs tend to have greater sensitivities, particularly on the ac voltage scales. With the exception of dc voltage and current measurements, the accuracies of both VOMs and the EVMs are roughly comparable (if their respective operating limitations are not exceeded). For measuring dc quantities, the VOMs are somewhat more accurate.

The VOMs are instruments that do not require power of any kind (with the exception of a battery for resistance measurements). EVMs require either battery power or ac line power for general operation. (VTVMs need to be plugged into an ac power line. This restricts their portability and makes them susceptible to ground-loop interference.) TVMs are usually battery powered and are thus able to circumvent this limitation of the VTVM.

In summary, it can be said that the VOM is a simple, reliable, and comparatively inexpensive instrument. Because of these characteristics, it is probably best suited for measurements in low-impedance, low-frequency circuits where the voltage levels of interest are not too small. The multipurpose EVM is a higher-priced instrument, but it is more versatile than the VOM. In general, it is capable of measuring higher-frequency signals as well as voltages in higher-impedance circuits. Furthermore, special laboratory types of EVMs are available which are far more sensitive to low voltages than are the VOMs.

HOW TO USE THE MULTIPURPOSE EVMs

In following the directions below, refer to Fig. 12-5.

1. Before turning on the EVM, check to see that the pointer is set to the zero position. If the pointer is not set on the zero mark, use the mechanical zero setting screw to alter the pointer position (Fig. 12-5).
2. For line-powered meters, plug in the meter and allow a few minutes warmup time.

[2]A special rectifier designed to accurately convert high-frequency signals into dc form is placed within the high-frequency probe itself. In this way, the effect of the capacitance of the cables which carry the measured signal to the meter is minimized. At high frequencies this capacitance would shunt the signal to ground before it got to the meter for measurement. (The effect of such shunt capacitance is explored in more depth in the section on oscilloscope probe use in Chapter 9.)

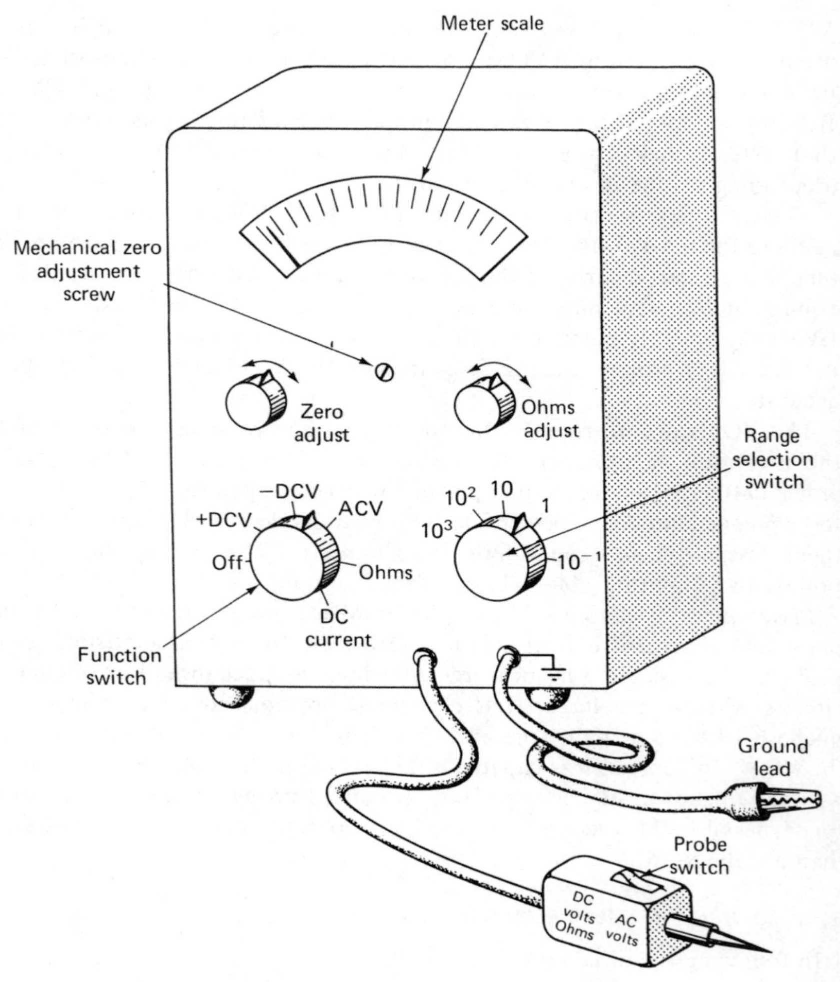

Figure 12-5. Controls of a multipurpose EVM

3. Set the *Function* switch to the desired setting. If using a probe, set the *Probe* switch to the same setting as the Function switch.
4. Select proper position of *Range* switch.
5. Short the *hot* lead to the ground lead (touch them together). The pointer should read zero when this is done. If the pointer does not remain at zero, use the *Zero Adjust* knob to bring it to the zero setting. If measuring resistance, the pointer should also read ∞ once the leads are disconnected from one another. If the pointer does not indicate an infinite reading with the leads disconnected, use the *Ohms Adjust* knob to adjust the pointer position until it does so.

6. Connect the probe leads across the element being measured. Read meter.
7. If a VTVM is being employed, check and (if necessary) read just the zero setting at half-hour intervals (or whenever a range change is made). VTVMs are subject to an electrical drift of the zero position with time.
8. If using a special rf probe for measuring high frequencies, set the meter to the *dc volts* function. The rectifier in the probe itself converts the rf signal to dc quantities.

COMMON SOURCES OF ERROR AND INCORRECT OPERATION OF EVMs

1. Incorrect initial zeroing procedure (i.e., setting the zero with meter *On*, but leads not shorted; this causes errors in readings of low-voltage measurements).
2. Failure to correct zero drifts in VTVMs after initial warmup of thirty minutes.
3. Failure to realize that erratic pointer deflections which occur when meter is set on low ranges (and leads are unconnected) are due to the sensitivity of the VTVM to capacitive pickup (and not due to a faulty instrument). From Chapter 3, we remember that a high-impedance instrument is quite subject to picking up voltages due to stray capacitive effects if its leads are not connected across a low-impedance element. The fact that the deflections of the VTVM pointer are due to such stray capacitance effects can be verified by connecting the leads of the meter across the components being measured. If the meter is operating properly, the fluctuations will usually cease when the connection is made.
4. Failure to observe proper grounding procedures when using the VTVM. In Fig. 3-4 we saw how the attempt to measure a nongrounded voltage with a VTVM (one of whose inputs is connected to ground through the power line) resulted in the shorting out of part of the circuit being tested. Failure to connect the leads of the VTVM in proper polarity when measuring grounded voltages is another possible mistake. (Fig. 12-6). In some cases this error can lead to a shorting out (and consequent burnout) of the power supply or VTVM.

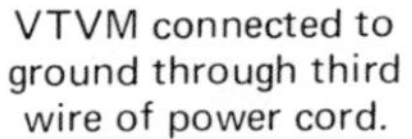

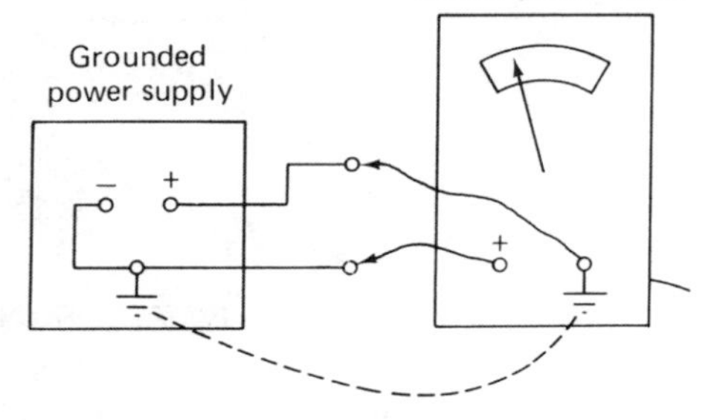

Figure 12-6. Inattention to proper polarity can cause a VTVM to short out a grounded power supply

5. Failure to correlate the *Function* switch and *Probe* switch settings. If a probe is being used with the VTVM, make certain that the probe switch is set to the appropriate position corresponding to the type of measurement being made.
6. Failure to correct the meter readings when nonsinusoidal waveforms are being measured (see Ex. 12-1).
7. Use of EVMs to measure voltage across very-high-impedance elements (e.g., dc voltage measurements across a capacitor). This causes unreliable readings due to loading effects.

AC ELECTRONIC VOLTMETERS

Although the multipurpose EVMs are versatile instruments, they are not sensitive enough for measuring very low voltages. For measuring very small ac voltage values, high-sensitivity ac electronic meters can be used. The configuration in which the components that make up the sensitive ac meters are connected is different than that found in the multipurpose EVM. In the ac meters, the rectifier circuit is connected in the circuit after, rather than before, the amplifier (Fig. 12-7).

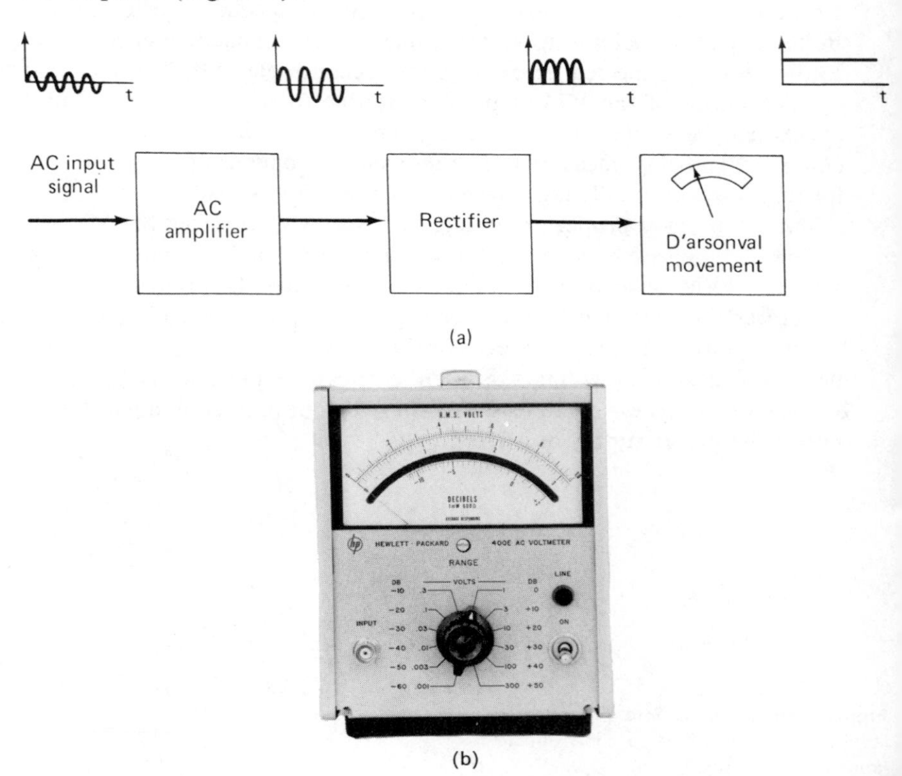

Figure 12-7. Components of high-sensitivity ac VTVM

The amplifier itself is an ac amplifier specially designed to provide high sensitivity and stability. In fact, the amplifier is so stable that Zero-Adjust knobs are not necessary on these meters.

The maximum sensitivity of typical ac meters is in the millivolt range, although special microvolt ac meters are also available. The frequency range is about 5 MHz. A full-wave rectifier is used in most ac meter instruments because it provides more accuracy than the peak-to-peak-type rectifiers. The accuracies of ac meters are typically 1% between 50 Hz and 500 kHz.

Digital Electronic Meters

Digital meters (abbreviated DVM, for digital voltmeter) indicate the quantity being measured by a numerical display rather than by the the pointer and scale used in analog meters (Fig. 12-8). The numerical readout gives the DVM special advantages over analog instruments in many applications. First, the accuracies of DVMs are much higher than those of analog meters. For example, the best accuracy of analog meters is about 0.5 percent, while the accuracy of DVMs can be 0.005 percent or better. Even on simple DVMs, the accuracy is at least ±0.1 percent. Second, a definite number is provided for each reading made by the DVM. This means that any two

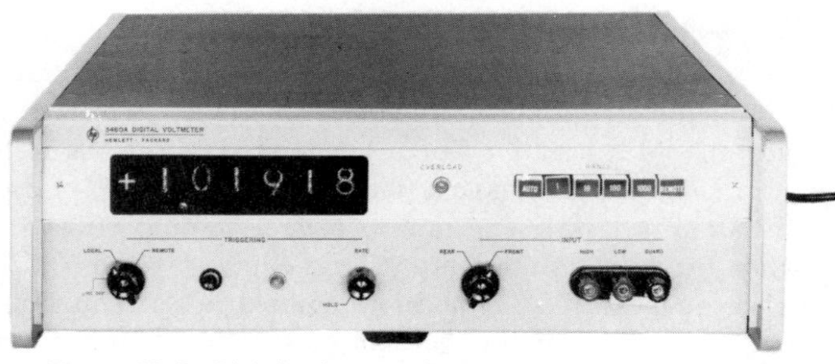

Figure 12-8. Digital voltmeter (Courtesy of Hewlett-Packard Co.)

observers always see the same value. As a result, such human errors as misreading the scale and parallax are eliminated. Third, the numerical readout increases the speed of reading and makes the task of taking measurements less tedious. In situations where a great number of readings must be made, this might be an important consideration. Fourth, the repeatability (precision) of DVMs becomes greater as the number of digits displayed is increased. The DVM also contains automatic ranging and polarity features which protect it from overloads or reversed polarity. Finally, the DVM output can be fed directly to recorders (printers or tape punches) where a permanent record of the readings is made. These recorded data are in a form suitable for processing by digital computers. The biggest disadvantage of DVMs is their relatively high cost compared to ordinary analog instruments. However, with the advent of integrated circuits (ICs), the cost of DVMs has been reduced to the point where some simple models are now competitive in price with conventional analog electronic meters.

OPERATING PRINCIPLES OF DIGITAL VOLTMETERS

The main feature of the DVM is the circuitry which converts the dc voltage to a digital form. There are various methods for achieving this conversion, including:

1. Ramp method (voltage-to-time conversion).
2. Comparison method (potentiometric principle).
3. Integrating method no. 1 (voltage-to-frequency conversion).
4. Integrating method no. 2 (dual-slope integration).

The voltage value is displayed by means of indicating tubes or a solid-state light-emitting display.

To describe the operation of the various types of DVMs, we will have to make reference to several devices which will not actually be introduced until the later chapters of this text. Therefore, if the reader should want more information on these devices, he should consult the relevant chapter [i.e., oscillators (Chapter 13); frequency counters, timers, and digital displays (Chapter 14); integrators (Chapter 16)].

1. Voltage-to-Time Conversion Method (Ramp Type)—The dc voltage being measured is applied to the input terminals of the meter. At the same time, a ramp waveform with a negative slope is generated within the meter. This ramp waveform ideally should have a linear slope whose rate of decrease is accurately known (Fig. 12-9). The ramp starts at some positive value and decreases. When the value of the ramp voltage equals the value of the input voltage, an electronic *coincidence detector*[3] emits a pulse. This pulse opens a gate. [A gate is simply a circuit that can be switched between On (open) and Off (closed) states.] When the voltage reaches zero, another pulse from the coincidence detector closes the gate.

The length of time the gate is open is measured by a counter that counts the number of wavelengths emitted by a very accurate, fixed-frequency oscillator. Since the slope of the ramp voltage is known, as well as the time between the opening and closing of of the gate, this yields the value of voltage applied to the input. If the voltage input is negative, the gate is not opened until the ramp reaches zero, and it is shut off as the negative value of the ramp becomes coincident with the negative voltage value.

[3]Coincidence detectors are designed to produce a pulse only when the voltage values being applied to each of its two terminals are equal.

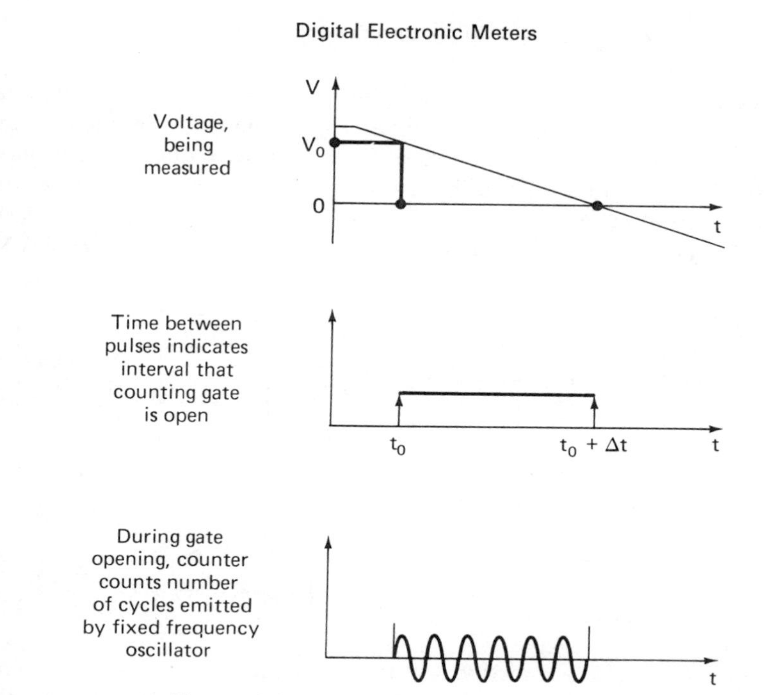

Figure 12-9. Ramp method of measuring voltage

The ramp type of digital conversion is the simplest technique being used in DVMs and has the advantage of not being affected by slowly varying voltages. However, it suffers from the disadvantages of having nonlinearities in the shape of the ramp waveform and a lack of noise rejection. A typical four-digit ramp-type meter is accurate only to 0.05 percent, while comparable integrating types are accurate to 0.01 to 0.02 percent.

2. Servo-Balance Potentiometer Types—The self-balancing potentiometer DVMs operate on a principle similar to the self-balancing potentiometer described in Chapter 11, where an unknown voltage and a known voltage are compared in value. The difference between their values is used to create a signal that leads to the adjustment of the known voltage. The adjustment continues until the values of the unknown and known voltages are equal. At the point where the two voltages are equal, the value is displayed by a digital readout.

The conversion of the input signal to a digital form is performed by a servomotor which responds to the magnitude of the difference between the unknown and known voltage values. The servomotor drives a mechanical drum-type digital indicator. At the point corresponding to a difference signal of zero, the position of the servomotor shaft is indicated by the drum-type indicator and identifies the unknown voltage. The potentiometer-type DVM is a low-cost instrument providing excellent performance for its price.

3. Integrating Type No. 1 (Voltage-to-Frequency Conversion)—In this type of DVM, the input dc voltage is converted (by a voltage-to-frequency converter) into a set of pulses whose repetition rate (or frequency) is proportional to the magnitude of the input voltage (Fig. 12-10). The pulses are counted by an electronic counter similar to the way the number of wavelengths was counted by the time-interval counter in the ramp-type DVM.

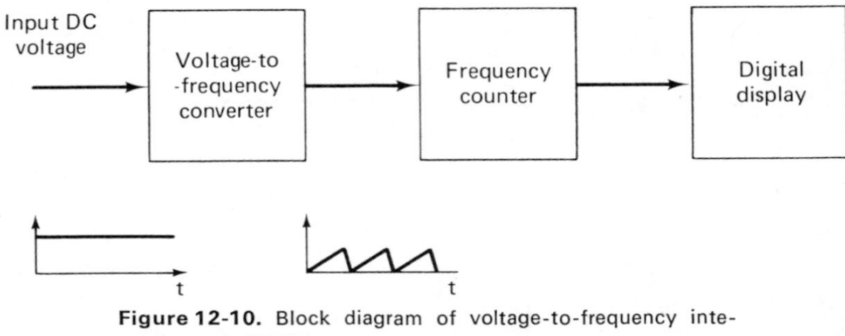

Figure 12-10. Block diagram of voltage-to-frequency integrating-type DVM

Therefore the count is proportional to the magnitude of input voltage. Since random (normal-mode) noise tends to have an average value of zero, this type of DVM is able to reject ac noise. That is, its displayed value is equal to the average value measured during some specific time interval. This noise-rejection ability is the main advantage of the voltage-to-frequency type of DVM.

4. Integrating Type No. 2 (Dual-Slope Integration)—The dual-slope-integration DVMs are relatively simple, yet accurate instruments. In addition, they retain their high noise-rejection capability because they use an integrating measuring method. Figure 12-11(a) shows a block diagram of the major components of the dual-slope integrating-type DVM.

The dc voltage to be measured is fed to an integrator which produces a ramp waveform output whose slope is proportional to the magnitude of the dc level of the signal (Fig. 12-11(b)). The ramp signal starts at zero and increases for a fixed time interval, T. At the end of the interval, switch S_1 is automatically moved to the V_{ref} position. In this position, the voltage fed to the meter is no longer the input voltage, but an internal reference voltage. The magnitude of the reference voltage is such that the voltage output of the integrator decreases linearly with time until a zero value is reached.

The time, t, required for the voltage output of the integrator to reach zero is measured. Since T is known, V_{ref} is known, and t_1 is measured, we can find V_x from the geometric proportions of the triangle.

$$\frac{V_x}{T} = \frac{V_{ref}}{t_1}$$

or

$$V_x = \frac{V_{ref} T}{t_1}$$

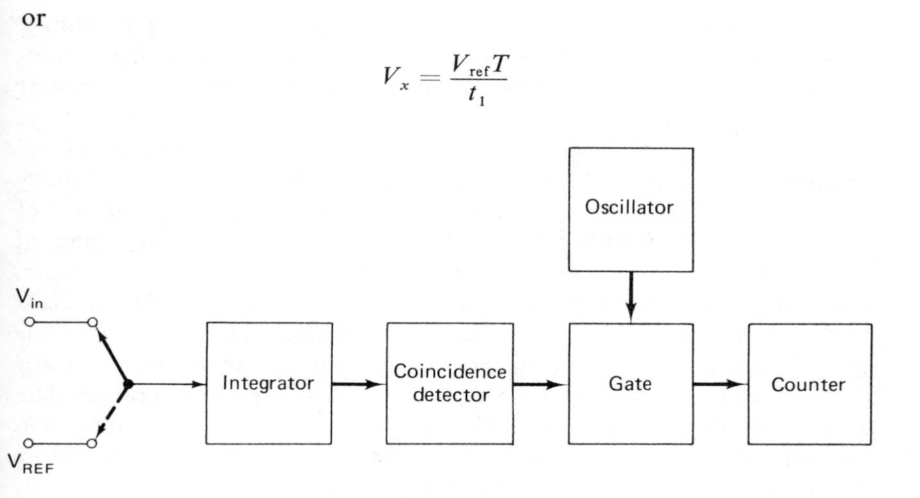

(a)

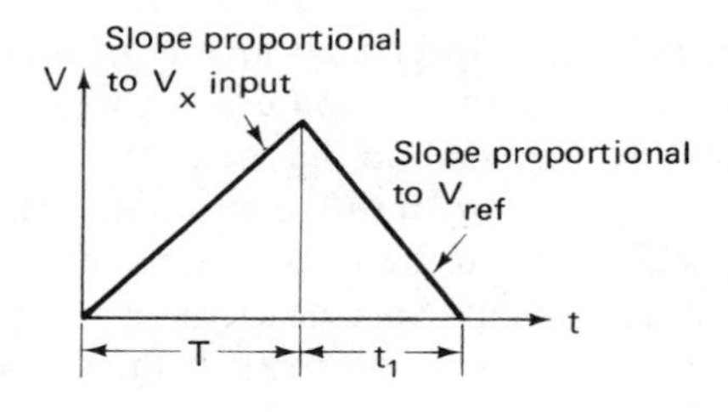

(b)

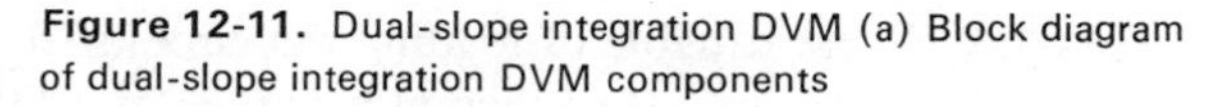

Figure 12-11. Dual-slope integration DVM (a) Block diagram of dual-slope integration DVM components

INTERPRETING THE ACCURACY SPECIFICATIONS OF DVMs

The accuracies of DVMs are usually greater than those of analog meters, but the accuracy specifications listed by a manufacturer should be clearly understood. There are three key concepts involved in being able to comprehend the accuracy specifications of DVMs: *resolution*, *constant error*, and *proportional error*.

The *resolution* of a DVM indicates the number of digits in the display. *Constant errors* are any errors which remain constant over the full range of the instrument. Such errors are expressed in terms of the number of digits or the percentage of full-scale reading (or range). *Proportional errors* are errors which are proportional to the magnitude of the digital indication.

Thus, proportional errors are expressed in terms of a percentage of reading. Most manufacturers specify the accuracy of a DVM in terms of a combination of constant and proportional errors. For example, the accuracy of a DVM may be expressed by such combinations as "±0.01 percent of reading ±0.01 percent of range"; or as "±0.05 percent of reading ±1 digit." As an example, if 5.000 V is measured with a four-digit DVM whose accuracy is "0.01 percent of reading +1 digit," the maximum error is 0.01 percent of 5 V + 0.001 V, or 0.0015 V total. The resolution of a DVM is important because the resolution should be greater than the accuracy of the meter. For example, it requires an instrument which has a resolution of five digits to enable measurements to be made to 0.01 percent over 90 percent of the total dynamic range of the meter. However, it cannot be automatically assumed that a DVM with a six-digit display has a greater accuracy than a meter with only five digits in the display (even though its resolution is greater). The specifications of both instruments must be examined before the accuracy of either meter is known with certainty.

For measurements of dc voltages, the accuracy of DVMs ranges from 0.1 to 0.001 percent of reading ±1 digit. If the DVM is also capable of measuring ac voltages, resistance, and currents, the accuracy with which the DVM measures each of these quantities is usually different (and less accurate) than the dc voltage accuracy. Specific accuracies of some typical multimeter-type DVMs are presented in the section dealing with these types of DVMs.

A final note should be included on the use of the terms *reference* and *rated* conditions. *Reference* (or short-term) conditions are ideal or laboratory conditions, and specifications at these conditions represent the best accuracy obtainable with the instrument. Specifications listed under *rated* conditions include degradation of accuracy due to such factors as temperature, component aging, and humidity.

ADDITIONAL FEATURES AND SPECIFICATIONS OF DVMs

Input Impedance—DVMs are capable of loading the circuits they measure, just like any other voltmeter. Since the inherent accuracy of a DVM can be made so very high, it is important that such loading effects do not cause an error greater than the uncertainties due to the meter alone. Usually the input impedances of DVMs are quite high (10 MΩ to 10 GΩ) and should not introduce serious loading. However, the following guideline is presented to enable one to determine if a DVM will cause loading errors that are in excess of the errors caused by its inherent inaccuracies: *The DVM must have an input impedance which exceeds the measured source impedance by at least a factor of* 10^n, *where n is the number of digits in the display*. Thus if the number of digits in a DVM's display is 5 and it has an input impedance of 1 GΩ (10^9 Ω), the maximum impedance across which the DVM can measure voltage without causing excessively large loading errors is 10 kΩ.

Speed of Reading—In most laboratory applications, a speed of one reading per second is usually satisfactory. However, there are some cases where faster speeds are necessary. Some DVMs are capable of making up to 100 readings per second (if an external recording device is used with the DVM).

Range Selection—DVMs can be made to have manual or automatic range selection. Where a large number of measurements having a wide range of random voltages must be made, automatic range selection can be a useful feature.

Overranging—Overranging allows a DVM to measure voltage values above the normal decade transfer points without the necessity of having to change ranges. This allows the meter to keep the same resolution for values which are near the decade transfer points (1 V, 10 V, 100 V). The extent to which overranging is possible is expressed in terms of the percentage of the range or full scale. Overranging from 5 to 300 percent is available, depending on the model of the DVM.

Normal-Mode Noise Rejection—Normal-mode noise refers to the type of noise which appears superimposed on the high side of the input signal (in the form of ripple or spikes). To eliminate this noise, which prevents the DVM from correctly determining the true dc level, a method of removal or averaging of the noise must be employed. In the integrating types of DVMs, this noise is rejected by averaging. In all other types of DVMs a filter is used to remove as much of the noise as is feasible. Filtering need not degrade the DVMs accuracy, but it does reduce the speed of measurement. The ability of a DVM to reject normal-mode noise is specified by a quantity called the NMR and is expressed in terms of decibels at a specific frequency (e.g., 30 dB at 60 Hz).

Common-Mode Rejection—Common-mode signals are those that appear at both high and low terminals simultaneously. They arise from ground-loop currents in the circuit to which the DVM is connected. These common-mode signals can be a severe problem in some measurements. So-called *guarding* techniques are used to reject the common-mode signals. These techniques involve completely surrounding the measurement circuitry and the input leads of the DVM with a metal shield that is insulated from the measurement circuitry. This shield is connected to an additional voltage source which provides a voltage whose value is equivalent to that of the voltage at the input lead being measured. Since the guard and the input leads are connected to points of equivalent potential, no potential difference exists between the input terminals and the guard shield. This prevents the ground-loop currents from flowing into the measuring circuitry; instead these ground-loop currents flow harmlessly via the guard shield to ground.

The rejection of common-mode signals by a DVM is specified by the quantity called CMR and is also expressed in decibels at a specific frequency. A typical CMR for a DVM might be 120 dB at 60 Hz.

MULTIPURPOSE DVMs

Although the DVMs we have been describing up to this point are only designed to measure dc voltages, other quantities can also be measured if additional circuitry is included within the meter. Some such multipurpose DVMs are designed to be able to measure all of the following quantities: dc voltage, ac voltage, dc and ac currents, and resistance (Fig. 12-12). The

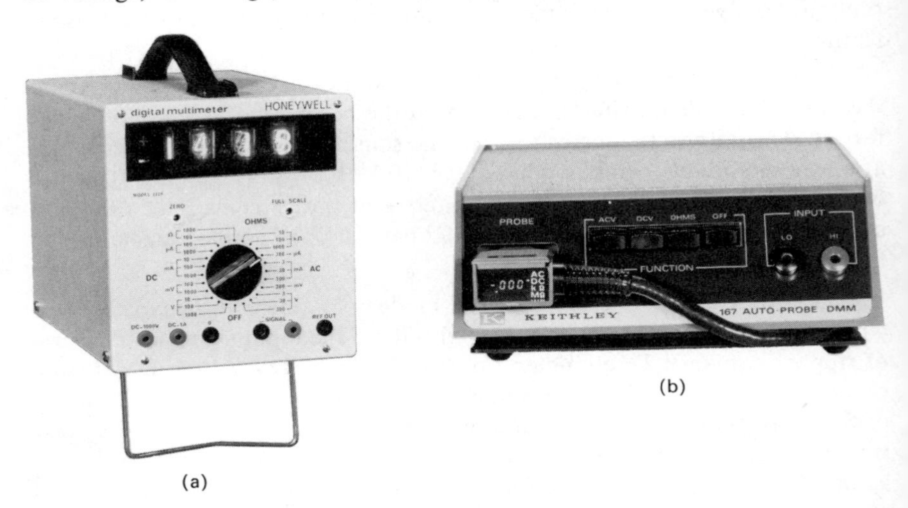

(a)

(b)

(c)

Figure 12-12. Photographs of multipurpose DVMs (a) Honeywell Model 3335 (Courtesy of Honeywell Inc., Test Instrument Div) (b) Keithley Model 167 (Courtesy of Keithley Instruments, Inc.) (c) NLS Model Series X-I (Courtesy of Non-Linear Systems, Inc.)

upper-frequency limit of such DVMs ranges between about 10 kHz and 1 MHz, depending on the instrument design.

To measure ac voltages, a rectifier is included in the meter design. Since the accuracies of rectifiers are not as high as the accuracy of the dc-voltage-measuring circuitry of the DVM, the overall accuracy of ac voltage measurements is lower than when measuring dc voltages (ac voltage accuracies range from ±0.1 to ±1 percent ±1 digit). Currents are measured by having the DVM determine the voltage drop across an accurately known resistance value. Although the value of a resistor can be very closely specified, there is some additional error due to the resistance change as a function of the heating effect of the current passing through the resistor. In addition, caution must be exercised when using the current-measuring function—care must be taken not to allow excessive current to be passed through the resistor. Typical accuracies for dc current measurements range from ±0.2 to ±2 percent of reading ±1 digit, while ac current accuracies are ±0.5 to ±2 percent ±1 digit.

The DVM becomes an ohmmeter when a very accurate current source is included within the meter. This source circulates the current through the resistance being measured, and the remainder of the DVM circuitry monitors the resulting voltage drop across the element. The current source is accurate only for voltages below the full-scale voltage range of the DVM. If the resistance being measured is too large, the test current from the current source will decrease. The accuracies of multipurpose DVMs which are used to measure resistance vary from ±0.1% of reading ±1 digit to ±1 percent of reading ±1 digit.

The advent of light-emitting diodes has greatly expanded the usefulness of DVMs. For example, in one DVM multimeter, a readout made of light-emitting diodes is built into the probe of the DVM (see Fig. 12-12(b)). This allows the operator to make readings without taking his eyes off the circuit under test.

Special Purpose Electronic Meters

ELECTROMETERS

Electrometers are dc voltmeters which have extremely high input impedances (up to 10^{16} Ω). Such high input impedances are a necessary property of voltmeters that are used to measure voltages in very-high-impedance circuits. (Recall from Chapter 4 that a voltmeter must have an input impedance at least 100 times as great as the impedance of the circuit being measured in order to keep loading errors below 1%.) One application in which such high-impedance voltmeters are a necessity is in the measurement of the pH (hydrogen ion concentration) of a chemical solution. In such pH meters, voltages of about 50 mV must be measured across the walls of a glass tube. Such glass walls have resistances of 500 MΩ or more. Therefore, a voltmeter with an input impedance of 5×10^{10} Ω or more is needed to accurately measure these voltages. (The standard 10-MΩ input impedance of the multipurpose VTVM or TVM would be far too small to make this measurement accurately.) In another application, the meager output current of phototubes is fed through very-large-valued resistors so that a voltage drop of measurable magnitude can be generated. The voltages across these resistors must be measured by very-high-input-impedance meters. A third application involves the measurement of leakage current in capacitors.

Electrometers are designed to fill these and other applications calling for high-input-impedance instruments. They can also act as current and electric charge detectors of the highest sensitivity. Some electrometers can detect currents of 10^{-16} A or less, and measure charge as low as 5×10^{-16} C. Figure 12-13 is a photograph of a commercially available electrometer.

Figure 12-13. Electrometer (Courtesy of Keithley Instruments, Inc.)

The electrometer possesses these unique characteristics because it contains specially designed circuits and materials. There are several types of devices used to build electrometers, depending on the particular application of the electrometer. These devices include MOSFETs (metal-oxide-semiconductor field-effect transistors), electrometer tubes, and vibrating capacitors. Table 12-2 summarizes the characteristics of each of these (and a few additional) input devices.

Electrometers do not suffer from drift problems as badly as dc microvoltmeters do because their smallest scales are generally in the millivolt range. A typical electrometer may be specified as drifting less than 2 mV/hr. However, the insulation required by electrometers must be extremely good so that leakage or stray pickup do not contaminate its readings.

MICROVOLTMETERS

Microvoltmeters are basically sensitive D'Arsonval movements connected to stable, high-gain amplifiers. They are capable of measuring microvolt dc levels (Fig. 12-14(a)). Similarly, nanovoltmeters can measure dc voltage

Table 12-2 Comparison of Various Type Input Devices Used In Electrometers

Characteristic	MOSFET	Electrometer Tube	Vibrating Capacitor	Varactor Bridge	Junction FET
Input resistance	$>10^{14}\ \Omega$	$>10^{14}\ \Omega$	$>10^{16}\ \Omega$	$>10^{12}$	$>10^{12}$
Input offset current (rms)	$<5 \times 10^{-15}$ A	$<2 \times 10^{-14}$ A	$<2 \times 10^{-17}$ A	$<10^{-14}$ A	$<10^{-12}$
Voltage stability					
1) time	1 mV/24 hrs	4 mV/24 hrs	30 μV/24 hrs	100 μV/24 hrs	50 μV/24 hrs
2) temp	150 μV/°C	500 μV/°C	30 μV/°C	30 μV/°C	10 μV/°C
Current stability					
1) time	$<10^{-15}$ A/24 hrs	$<10^{-15}$ A/24 hrs	5×10^{-17} A/24 hrs	10^{-13} A/24 hrs	
2) temp	$<10^{-15}$ A/°C	10^{-15} A/°C	—	X2/8°C	X2/10°C
Noise (rms)					
1) voltage					
0.1–10 Hz	10 μV	5 μV	<1 μV	10 μH	5 μV
10–500 Hz	100 μV	30 μV	1 μV	(not usable)	12 μV
2) current					
0.1–10 Hz	5×10^{-15} A	5×10^{-15} A	$<2 \times 10^{-17}$ A	10^{-14}	2×10^{-14} A
Minimum input C	0.5 pf	5 pf	2 pf	30 pf	2 pf

(Courtesy of Keithley Instruments, Inc.)

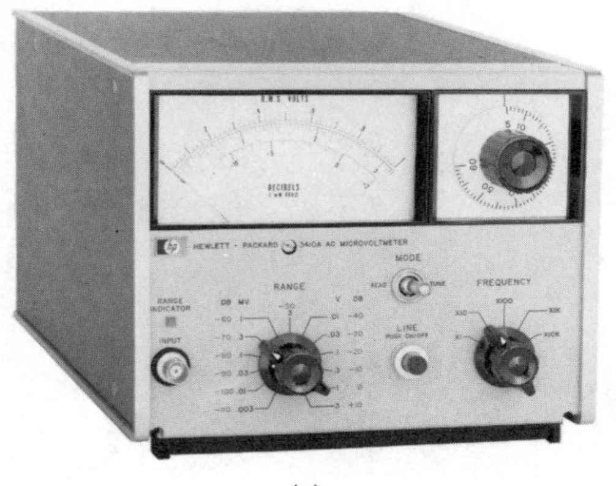

(a)

(b)

Figure 12-14. (a) Microvoltmeter (Courtesy of Hewlett-Packard Co.) (b) Keithley Model 410A picoammeter (Courtesy of Keithley Instruments, Inc.)

which are even smaller. One of the main uses of these types of meters is as detectors for potentiometers.

The most sensitive voltmeters have full-scale ranges as small as 10 nV at $\pm 2\%$ accuracy. To overcome problems of drift, the dc voltage is usually converted to ac, then amplified, and finally reconverted again to dc. The amplified dc voltage is used to drive the movement of the meter.

NANOAMMETERS AND PICOAMMETERS

Nanoammeters and picoammeters are specially designed meters for measuring very small current levels (Fig. 12-14(b)). A microvoltmeter measures the voltage drop across a shunt resistor, and the voltmeter reading is calibrated to indicate current rather than voltage. Meters which can measure currents as small as 0.3 pA (3×10^{-13} A) full-scale and with accuracies of ± 2 to $\pm 4\%$ (depending on the range) are available.

VECTOR VOLTMETERS

Vector voltmeters are used to measure the phase difference between two points in a circuit at the same time the voltage between these two points is measured. This phase-difference measurement is useful as a method for determining such quantities as amplifier phase shifts and filter transfer functions. The operating frequencies over which vector voltmeters provide phase determinations are typically 1 MHz to 1 GHz. Voltage sensitivities usually range from 100 mV to 10 V.

VECTOR IMPEDANCE METERS

The impedance of even simple electrical components becomes almost impossible to calculate theoretically at high rf frequencies (i.e., 10 MHz or more). Rather than attempting to make such calculations, a measurement of the quantity is usually made. The vector impedance meter is an instrument designed to measure the magnitude and phase of the impedance at the frequency of interest. The element or circuit of interest is merely connected to the inputs of the meter, the desired frequency is dialed in, and the magnitude and phase angle of the impedance are displayed by the two front-panel meters. Vector impedance meters are made to cover either af or rf frequencies. Af meters have ranges of 5 Hz to 500 kHz and 1 Ω to 10 MΩ. The ranges of rf vector impedance meters are 5 MHz to 100 MHz and 1 Ω to 100 kΩ.

Problems

1. State the advantages of electronic voltmeters (EVMs) over electromechanical meters. Where does each type find its principle usage?
2. If two EVMs measure the same voltage quantity and one of them reads 6.1 V while the other reads 6.5 V, explain the most probable reason for the difference in the values of the readings.
3. What are the characteristics of the amplifier circuit used in electronic multimeters that make this type of amplifier so well suited for its task?
4. List some of the advantages of digital electronic meters over comparable analog meters. Why are many analog meters still in use?
5. List some of the performance differences between the VTVM and the TVM.
6. Which instrument has the higher input impedance: an EVM with an 11 MΩ input impedance or a 20,000 Ω/V VOM which is set on its 1500 V scale? What is the answer if the same VOM is set on a 150 V scale or less?
7. Explain why the VTVM or TVM exhibits non-zero readings when the meter is set to its most sensitive scale and the meter probes are not connected to the circuit (and are not short-circuited to one another).

8. If a triangular waveform (Fig. P12-1a) with an amplitude of 10 V is applied to a general purpose EVM, what will be the error between the measured and the actual rms values of the applied signal?

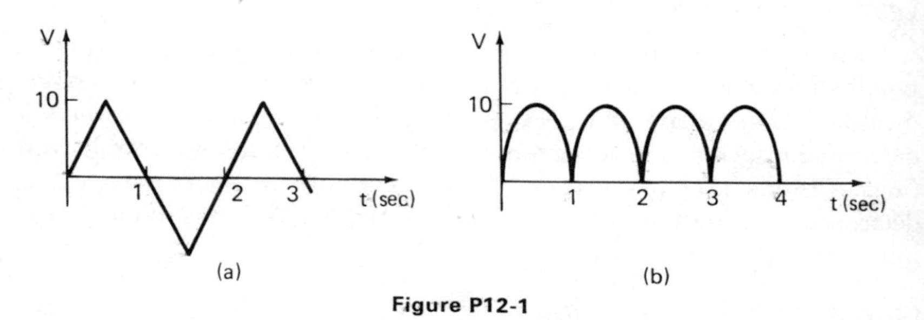

Figure P12-1

9. Repeat Prob. 8 for the waveform of Fig. P12-1b.
10. Refering to Fig. 12-5 in the text, explain when the ZERO ADJUST control knob on the EVM is used. When and how is the OHMS ADJUST control knob used? What determines the position of the switch which is found on the probe of the meter?
11. Describe the potential hazards of not paying careful attention to the importance of proper grounding techniques when making measurements with a grounded VTVM.
12. Why are special scales generally provided for low ranges of ac voltages on multipurpose EVMs?
13. Define the terms *resolution*, *proportional error*, and *constant error* as used when referring to DVMs.
14. Given a DVM with accuracy specifications as follows, what will be the maximum error of each of these instruments if 3.50 V are indicated by each of the meters?
    a) resolution = 3 digits, accuracy = ±0.1% of reading and ±0.1% of range.
    b) resolution = 5 digits, accuracy = 0.005% of reading and ±0.005% of range.
    c) resolution = 4 digits, accuracy = ±0.01% of reading and ±2 digits.
15. Explain what is meant by the terms *normal-mode rejection*, *common-mode rejection*, and *guarding*.
16. If a DVM has a 100% overranging feature and is set to a 100 V full-scale setting, what is the maximum voltage which can be measured by the meter before the range of the meter has to be changed?
17. Why are the accuracies of a DVM lower when the instrument is used to measure ac quantities and resistance than when it is used to measure dc voltages?
18. What are the advantages and disadvantages of electrometers compared to general-purpose electronic voltmeters?

References

1. PRENSKY, S. D., *Modern Electronic Voltmeters*. New York, John F. Rider, 1964.
2. *Hewlett-Packard Instrumentation Catalog*. Palo Alto, Calif., Hewlett-Packard Co., 1971.
3. LENK, JOHN D., *Handbook of Electronic Meters*. Englewood Cliffs, N.J., Prentice-Hall, Inc., 1969.
4. MILLMAN, JACOB, and TAUB, HERBERT, *Pulse, Digital, and Switching Waveforms*. New York, McGraw-Hill Book Co., 1965.
5. Keithley Instruments, *Electrometer Measurements*. Cleveland, Ohio, Keithley Instruments, 1972.
6. *Non-Linear Systems Catalog*—Guide to Digital Voltmeters. Del Mar, Calif., Non-Linear Systems, Inc., 1970–71.

13

AC Signal Sources

For the process of providing power for electronic instruments and measurements, both dc and ac power sources are needed. In Chapter 8, we discussed some devices that are used as sources of dc power. This chapter will examine the operation and use of instruments that produce time-varying signals. Although the term *ac signal source* can be used to describe the entire family of such instruments, more specific classifications are used for identifying particular subcategories. The classifications are based on the type of output signals which can be generated by the particular signal source.

The subcategories of ac signal sources which we will examine in this chapter are the following:

1. Oscillators
2. Signal generators
3. Sweep-frequency generators
4. Pulse generators
5. Function generators.

Oscillators

Oscillators are instruments which generate sinusoidal output signals. Although there are also other ac signal sources capable of producing sinusoidal outputs, we usually reserve the term *oscillator* for those instruments which are designed to produce sine-wave signals exclusively. Most oscillators are capable of generating sinusoidal signals whose frequencies and amplitudes are adjustable over specific ranges. However, there are also a few fixed-frequency oscillators available.

Since sinusoidal waveforms are encountered so frequently in electronic measurement work, oscillators find uses in a large variety of applications. Some of the more common ones are listed in Table 13-1.

Table 13-1 Common Oscillator Applications in Electrical Measurement

1. Energy source for driving impedance bridges. Usually an *RC* oscillator at a fixed frequency (i.e., 1000 Hz) is built into impedance bridges which are designed for measuring *R*, *L*, and *C*. On some special bridges, radio-frequency (rf) oscillators are also available. In addition to such built-in oscillators, most bridges are also designed to be driven by external oscillators.
2. Energy source for driving *Q*-meters. A wide-range osicllator is usually included as an internal source for driving *Q*-meters.
3. Source for measurement of amplifier frequency response. In this capacity oscillators are used to determine the frequency responses of amplifiers in hi-fi's, audio transmitters, televisions, and oscilloscopes.
4. Ac test source for assisting in amplifier design. The bandwidths of amplifiers are calibrated by comparison with the the frequencies of standard oscillators set to the frequencies of interest.
5. Use as a source of ac signals in frequency meters and timer-counters.

The basic principles involved in the operation of oscillators will be introduced with the help of the analogy of the child's swing. Although not all oscillators operate on a principle exactly analogous to the child's swing, the analogy remains quite instructive. In addition to helping the student visualize the operation of some oscillator circuits, the analogy develops a qualitative feeling for the conditions which must be satisfied in all cases where an oscillatory signal is to be generated. In the later discussion on the various type of oscillators, we will identify those oscillators which closely resemble the child's swing and those that do not.

If a child's swing is initially at rest, a substantial amount of energy must be expended to start it oscillating. However, once the desired amplitude of oscillation is reached, a much smaller amount of energy must be added per cycle to sustain a constant amplitude of oscillation. This energy must be supplied to the swing at appropriate times during the oscillation cycle in order for it to be effective in sustaining the oscillation. If we stop adding energy altogether, the oscillatory motion of the swing will continue for some time. But the frictional and air-resistance losses during each cycle will gradually diminish the amplitude of the oscillations until the swing finally comes to rest. Therefore, to keep the swing oscillating, energy must be continually added to overcome the losses of the system.

An oscillator also produces an output voltage whose magnitude oscillates back and forth between positive and negative values—much like the positions of the moving swing. In those oscillators whose operation resembles that of a swing, a natural sinusoidal voltage variation takes place within a portion of the oscillator circuitry (when that portion is initially excited). The frequency of the oscillation is determined by the values of the electrical components comprising the oscillatory portion of the circuit. The natural voltage oscillation is then amplified by an electronic amplifier, and the resulting amplified signal is used as the output of the oscillator. However, a small portion of the amplified output is diverted and returned to the oscillatory part of the circuit. (This procedure of returning a portion of the signal to its source is called *feedback*.) By feeding back a fraction of the output signal so that it is in phase with the natural voltage oscillations, the losses which would lead to a diminishing voltage amplitude can be counteracted. If the energy contained in the feedback signal is just equal to the energy lost during each oscillatory cycle, the amplitude of the voltage oscillations in the oscillatory circuit is kept constant. The frequency of the output of the oscillator can be varied by changing the values of the electrical components which make up the oscillatory portion of the circuit.

From this discussion it can be inferred that an oscillator must contain the following three elements (Fig. 13-1):

1. Oscillatory element or circuit.
2. Amplifier.
3. Feedback path to provide the energy to *regenerate* the oscillating signal within the oscillatory element.

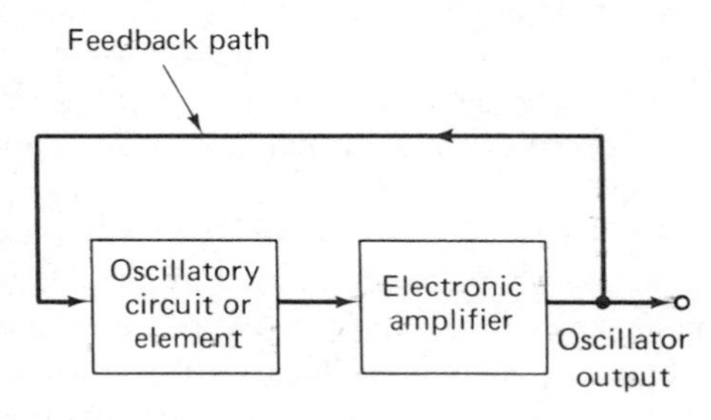

Figure 13-1. Oscillator block diagram

We should not think, however, that the oscillator is a perpetual motion machine; neither does it violate the law of conservation of energy. The amplifier portion of the oscillator is actually a device which converts dc power into the ac power contained in its output signals. This dc power must be supplied by some external energy source such as a battery or power supply.

Figure 13-2 shows two models of commercially available oscillators.

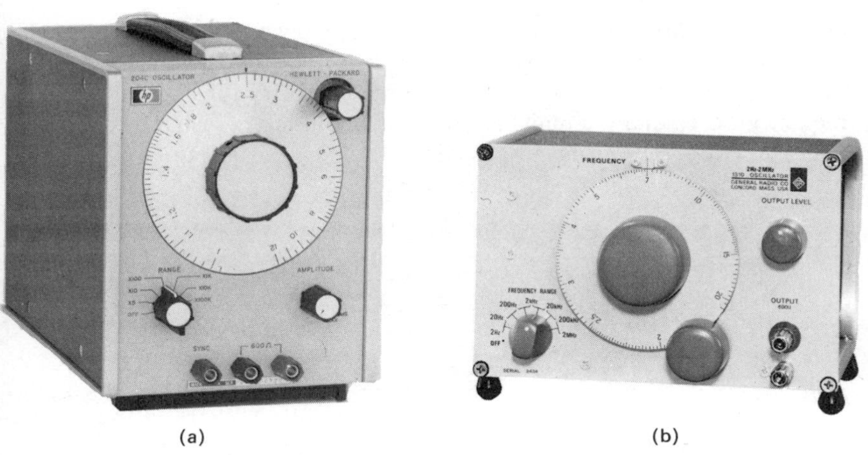

(a) (b)

Figure 13-2. Oscillators (a) HP model 204C (Courtesy of Hewlett-Packard Co.) (b) General Radio Model 1310 (Courtesy of General Radio Co.)

OSCILLATOR TYPES

The frequency spectrum over which oscillators are used to produce sine-wave signals is extremely wide (from less than 1 Hz to many hundreds of gigahertz). However, no single oscillator design is practical for producing signals over this entire range. Instead, a variety of designs are used, each of which generates sine-wave outputs most advantageously over various portions of the frequency spectrum. In this section we will discuss some of these oscillator types.

The details of their electronic circuitry will not be covered. Instead, the emphasis of the discussion will be to identify the most commonly used designs and provide a description of their chief characteristics. By being exposed to such a catalog of the major types of oscillators, the reader should be able to recognize and select appropriate instruments for various measurement applications. On the other hand, for those readers interested in the electronic circuit aspects of specific oscillators, a list of references containing this information is provided at the end of the chapter.

Oscillators which use *inductance-capacitance* (*LC*) circuits as their oscillatory elements resemble the child's swing in their operation rather closely. There is a particular frequency at which such circuits are resonant (that is, produce a natural sinusoidal variation). Such *LC* oscillators are very popular for producing high-frequency (rf) outputs (e.g., 10 kHz to 100 MHz). The most widely used *LC* oscillator designs are the *Hartley* and the *Colpitts* oscillators. Although they are slightly different from one another in their electronic circuitry, these types of oscillators have virtually identical frequency ranges and frequency-stability characteristics. However, *LC* oscillators as a whole are not well suited for producing low-frequency sine-wave outputs. This is due to the fact that the components necessary for constructing low-frequency *LC* resonant circuits are too bulky and heavy. Therefore, *resistor-capacitor* (*RC*) oscillators are usually used for generating low-frequency sine waves (from 1 Hz to about 1 MHz).

Table 13-2 Common Oscillator Types

Type	Approximate Frequency Ranges
Wien bridge (*RC*)	1 Hz–1 MHz
Phase-shift (*RC*)	1 Hz–10 MHz
Hartley (*LC*)	10 kHz–100 MHz
Colpitts (*LC*)	10 kHz–100 MHz
Negative-resistance	>100 MHz
Crystal	Fixed frequency

The two most common *RC* oscillators are the *Wien bridge* and *phase-shift* types. (The basis of their operation is somewhat different than that described by the child's swing analogy.) The Wien bridge design is used in almost all oscillators which produce signals in the audio-frequency range (20–20,000 Hz). This type of oscillator is simple in design, compact in size, and remarkably stable in its frequency output. Furthermore, its output is relatively free from distortion. However, the maximum frequency output of typical Wien bridge oscillators is only about 1 MHz.

Phase-shift oscillators also employ a simple circuit and produce sine-wave outputs that are quite distortion free. Their principle advantage over Wien bridge oscillators is that they have a wider frequency range (up to several thousand kHz). But they also have the disadvantage of not being as frequency stable as Wien bridge oscillators.

Other less frequently used oscillators include the *crystal oscillator* and the *negative-resistance oscillator*. Crystal oscillators use a piezoelectric crystal[1] to generate a sinusoidal signal of constant frequency. The output frequency is extremely stable with time but cannot be tuned (adjusted). Therefore crystal oscillators are used only in applications requiring a fixed frequency of high stability. Negative-resistance oscillators are primarily used to produce very-high-frequency signals. Table 13-2 summarizes the frequency ranges of the most common oscillator types.

[1]See the section entitled *Sound Transducers* in Chapter 15 for a brief discussion on the properties of piezoelectric crystals.

OUTPUT IMPEDANCE OF OSCILLATORS

When we described the operation of the battery in Chapter 8, we noted that it did not possess the same electrical characteristics as the ideal dc voltage source. The major difference between the two devices was that the battery contained some internal resistance, while the resistance of the ideal voltage source was zero. The internal resistance of the battery was represented in its circuit model by a resistor in series with an ideal dc voltage source.

Similarly, the circuit model of the oscillator can be represented by an ideal ac voltage source in series with a resistor, R_G (Fig. 13-3(a)). The origin of this resistance value arises in connection with the electronic circuitry of the instrument; in the absence of a reactive component, it is the *output impedance* of the oscillator. This impedance is designed to remain constant as the frequency of the output is varied. In af oscillators, R_G usually equals 600 Ω, while in rf oscillators (and in signal generators and pulse generators), R_G usually equals 50 Ω. The 600-Ω value is used in af oscillators because the characteristic impedance of audio-frequency communications systems (e.g., telephone circuits) have been standardized at 600 Ω. Likewise, rf oscillators and pulse generators have 50-Ω output impedances because rf signals and pulses are transmitted along coaxial cables (when not being propagated through space). Such cables typically have a characteristic impedance of 50 Ω.

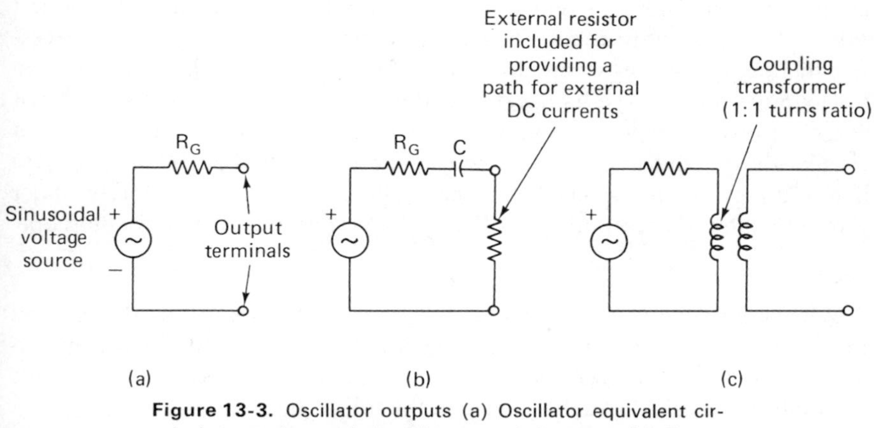

Figure 13-3. Oscillator outputs (a) Oscillator equivalent circuit (output dc-coupled) (b) ac-coupled output (c) Transformer-coupled output

Why is it so important that the output impedance of an oscillator be equal to the characteristic impedance of the system to which it is connected? The answer lies in the concept of impedance matching which was discussed in Chapter 3. In order to transfer the maximum power from the generator to the load, the impedances of the generator and the load must be equal (*matched*). The rated power output of an oscillator is therefore conditional upon its being connected to its rated load impedance. This factor is an important consideration in many measurement applications where the full rated power of the oscillator is required by the test circuit.

If the oscillator is connected to a load whose value does not equal (match) its output impedance, the value of the maximum output voltage (as well as the maximum power output) will change.

To see the effect of the load impedance on the voltage, we use the circuit model of the oscillator shown in Fig. 13-3(a). From this model we see that, if a load whose impedance value is equal to R_x is connected to the oscillator, the voltage, V_o, across R_x can be found from

$$V_o = \frac{R_x}{(R_G + R_x)} V \tag{13-1}$$

We note from Eq. (13-1) that as R_x approaches infinity (open circuit), V_o approaches V in magnitude. On the other hand, if R_x becomes very small, V_o approaches zero. Finally, if $R_x = R_G$ (matched load), V_o is one-half the maximum voltage output.

When a load is connected to the oscillator, current flows through the load. As defined in Chapter 3, the drawing of current from a circuit is called *loading*. Since a progressively smaller impedance connected across an oscillator will cause more and more current to be drawn from it, we say that this action causes the oscillator to become more *loaded down*. In some oscillators, excessive loading introduces severe distortion of the output signal. In others, even a maximum loading (i.e., a short-circuiting of the output leads) should not cause serious distortion. However, if the oscillator is to be used to drive a circuit whose input impedance is considerably smaller than R_G, and if there is any doubt about the ability of the oscillator to operate properly under severe loading, the output signal should be monitored with an oscilloscope.

SELECTION OF AN OSCILLATOR

When an oscillator is being chosen for a particular application, the requirements of the task at hand should be compared with the performance capabilities of the oscillator. Examples of some of the most common requirements an oscillator is asked to satisfy include the following.

1. *Frequency Range*—The oscillator should be able to supply an output signal whose upper and lower frequency limits exceed those required by the measurement.

2. *Power and/or Voltage Requirements*—The measurement may have a specific voltage or a specific power requirement. The oscillator should be able to produce the pertinent quantity with a magnitude large enough to fulfill the requirement.

3. *Accuracy and Dial Resolution*—The accuracy of an oscillator specifies how closely the output frequency corresponds to the frequency indicated on the instrument dial. Dial resolution indicates to what percentage of the output frequency value the dial setting can be read.

4. *Amplitude and Frequency Stability*—The *amplitude stability* is a measure of an oscillator's ability to maintain a constant voltage amplitude with changes in the frequency of the output signal. *Frequency stability* determines how closely the oscillator maintains a constant frequency over a given time period. Sometimes the frequency stability is included in the accuracy specifications of the oscillator.

5. *Waveform Distortion*—This quantity is the measure of how closely the output waveform of the oscillator resembles a pure sinewave signal. Sometimes the oscillator is used as a source in a test which measures the tendency of a circuit to distort a sine-wave signal. In such tests, the distortion produced by the oscillator should be much less than the anticipated distortion due to the circuit under test.

6. *Output Impedance*—The aspects of the output impedance of oscillators were discussed in the previous section.

GUIDELINES FOR OSCILLATOR USE

There are some practical suggestions involving the use of oscillators which will be discussed in this section. A number of these suggestions apply to the use of other signal sources as well.

1. *Warmup Time*—When an oscillator is first turned on, it requires some time for its circuits to warm up and stabilize. In transistor-powered oscillators, the warmup time is usually so short that any inaccuracies due to warmup effects are often included in the overall accuracy figure of the instrument. However, in older, tube-powered oscillators, the minimum recommended warmup time is usually five minutes. If the oscillator is used during this warmup period, the frequency of the output will not be reliable. If such oscillators are to be used for a measurement in which the frequency accuracy is critical, a longer warmup time is usually recommended (i.e., one-half hour to one hour).

2. *Impedance Matching*—As discussed earlier, the output impedance of an oscillator specifies the impedance value of the load which must be connected to it for maximum power transfer. On the other hand, the maximum voltage signal, V_m, that an oscillator can produce exists when the load is an open circuit. The maximum voltage that the oscillator can supply to a matched load is approximately $V_m/2$.

3. *Reducing Output Impedance*—If the specified output impedance of the oscillator is too high for a particular application, its value can be reduced by

one of two methods. (An instance where this requirement might occur is when the oscillator is to serve as a voltage source with a very low source impedance. The oscillator of the Q-meter discussed in Chapter 7 needs to have such a characteristic.) The first method merely involves the connection of a low-impedance resistor across the output terminals of the oscillator (Fig. 13-4(a)). It is easier to see why the impedance is reduced by this method if we redraw Fig. 13-4(a) into the form of Fig. 13-4(b). From this diagram we clearly see that the impedance is given by

$$R_{eq} = \frac{R_G R_o}{(R_G + R_o)} \tag{13-2}$$

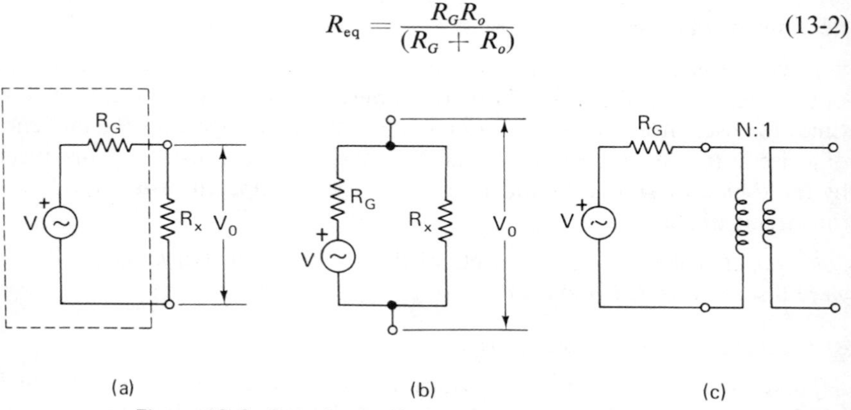

Figure 13-4. Reducing oscillator output impedance (a) Use of external resistor ($R_X < R_G$) (b) Figure (a) redrawn to show how $V_o = V(R_X R_G/R_X + R_G)$ (c) Use of an external transformer to lower output impedance

This method is simple, but it has the disadvantage of reducing the maximum power that the oscillator can supply to the circuit under test. The second method uses a transformer to convert the output impedance to a lower value. (See Chapter 7 for details on how transformers are used in this capacity.) This method does not reduce the power output of the oscillator, but it is limited by the frequency characteristics of the transformer.

4. Reducing Voltage Output—There are some applications where the voltage or power output from the oscillator is too large, even when the amplitude control knob is turned to its lowest position. Even the small output voltage in this situation could be too large if the oscillator is to be used to drive some amplifiers. In such cases, the minimum voltage output of the oscillator may be too great for the transistors of the circuit to amplify in a linear manner. To reduce the power or voltage from the oscillator below this minimum value, an attenuator must be connected externally to the oscillator. A simple voltage attenuator is a voltage divider circuit like the one shown in Fig. 13-5. If $R_1 \ll R_G$, the voltage drop across R_1 of this figure will be approximately

$$V_o = V\frac{R_1}{9R_1 + R_1} = \frac{V}{10}$$

Thus, the output voltage from the oscillator as seen by looking across R_1 is approximately one-tenth the voltage across the oscillator's output terminals.

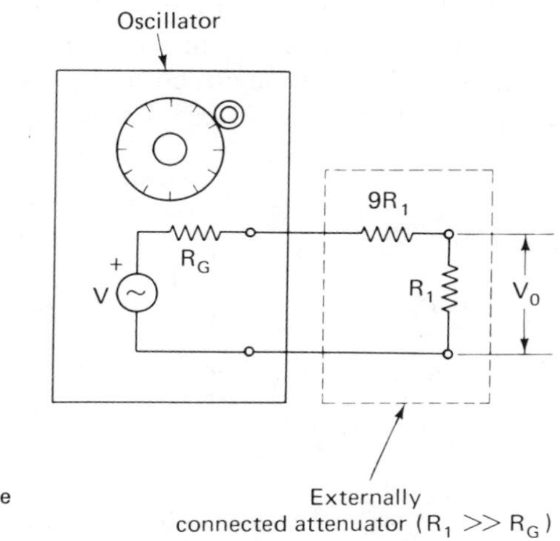

Figure 13-5. Method for reducing the output voltage from an oscillator

5. *Proper Grounding Techniques*—The output terminals of most oscillators look like those shown in either Fig. 13-6(a) or 13-6(b).

If there are three terminals (as in the oscillator of Fig. 13-6(a)), the voltage of the output signal from the oscillator can be either referenced to ground or floating. If the shorting bar between the minus and ground terminals is connected, the voltage will be grounded. If the shorting bar is disconnected, the voltage between the plus and minus terminals will be floating.

If the oscillator has only two terminals, the voltage signal output is always grounded. It is with the grounded voltage outputs that grounding problems

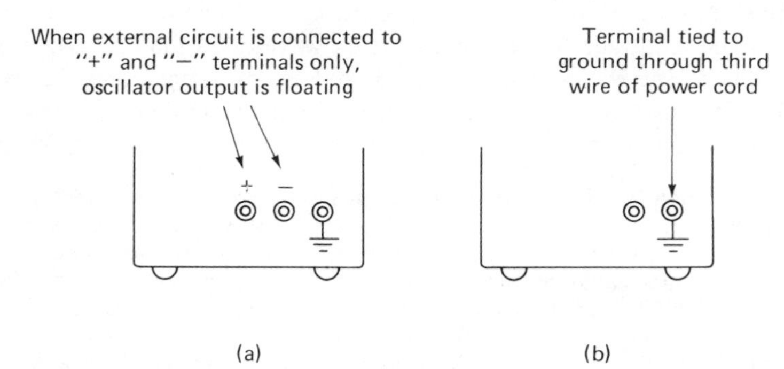

Figure 13-6. Output terminals of various oscillators (a) Floating output (from three-terminal output) (b) Grounded output

may sometimes arise. An example of such a problem involves the series connection of a power supply and oscillator. These two instruments are often connected in series when it is desired to produce a sine-wave output superimposed on some dc voltage level (Fig. 13-7(a)). The diagrams of Fig. 13-7(b) and (c) show incorrect and proper ways of connecting the two instruments. As a result of the incorrect connection of Fig. 13-7(b), the power supply is shorted to ground. (Can you see why connecting the plus terminal of the power supply to the nonground terminal of the oscillator in the connection of Fig. 13-7(b) would still be incorrect?)

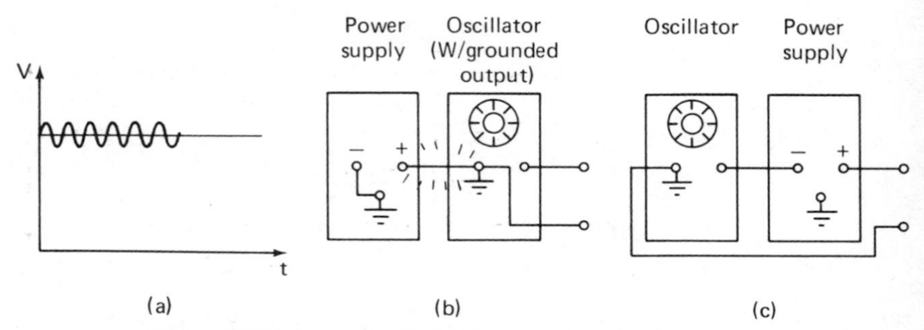

Figure 13-7. Example of the importance of using proper grounding procedures with oscillators (a) Desired signal (sinusoid superimposed on dc level) (b) Incorrect connection, power supply shorted to ground (c) Correct connection

Signal Generators

The signal generator, like its close cousin the oscillator, is designed to generate sinusoidal output signals. However, signal generators are also designed to be capable of modulating[2] their sinusoidal output signals with other signals, and this is the major difference between the two instruments. When signal generators are used to produce an unmodulated sine-wave output they are said to be producing a CW (or *continuous* (height) *wave*). When the output signal is modulated, the modulating waveforms may be either externally applied sine-waves, square waves, pulses, or more complex signals, as well as internally generated af sine waves. Most signal generators employ amplitude modulation (AM), but a few types are designed for frequency

[2]Communication by radio waves is based on the phenomenon that electromagnetic waves in the radio-frequency range (500 kHz to 100 MHz) can propagate over long distances of free space and can be detected by reasonably sized antennas. Electromagnetic waves in the audio-frequency (af) range do not lend themselves as well to detection or to fulfilling various communication requirements if used alone. Therefore a scheme called *modulation* is employed to propagate af signals. In this scheme, af waves are superimposed onto rf waves. The modulated rf wave then *carries* the af pattern to a receiving instrument where the af variation is separated from the rf *carrier*.

modulation (FM) instead. (Fig. 13-8 shows the principles of AM and FM.) Our discussion will center on the more common AM types.

Signal generators are primarily used to provide appropriate signals for calibrating, testing, and troubleshooting the amplifier circuits used in communications electronics (e.g., radio and television amplifiers). They are also employed for measuring the characteristics of antennas and transmission lines.

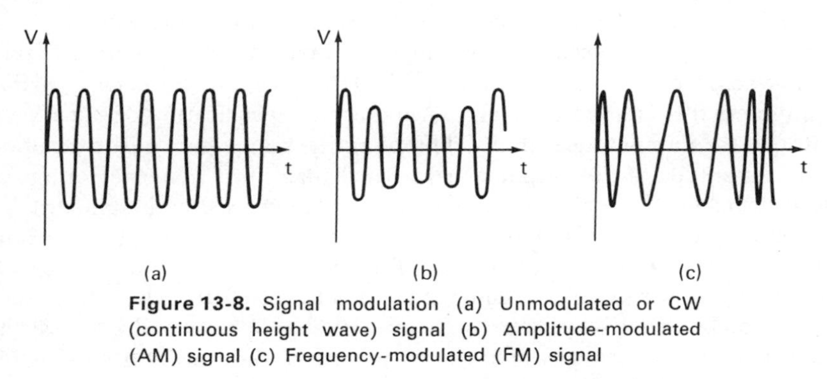

Figure 13-8. Signal modulation (a) Unmodulated or CW (continuous height wave) signal (b) Amplitude-modulated (AM) signal (c) Frequency-modulated (FM) signal

The most important components of laboratory-type signal generators are shown in Fig. 13-9. An rf oscillator is used to produce a carrier waveform whose frequency can be adjusted (typically) from about 100 kHz to 30 MHz. This oscillator is designed to be very stable and accurate. A second very stable oscillator, called a *modulation oscillator*, generates the af modulation frequency. The frequency of the modulation oscillator output is usually a fixed frequency of 1000 Hz (in some models, 400 Hz is also available). The rf and the modulation-frequency signals are fed to a wide-band amplifier

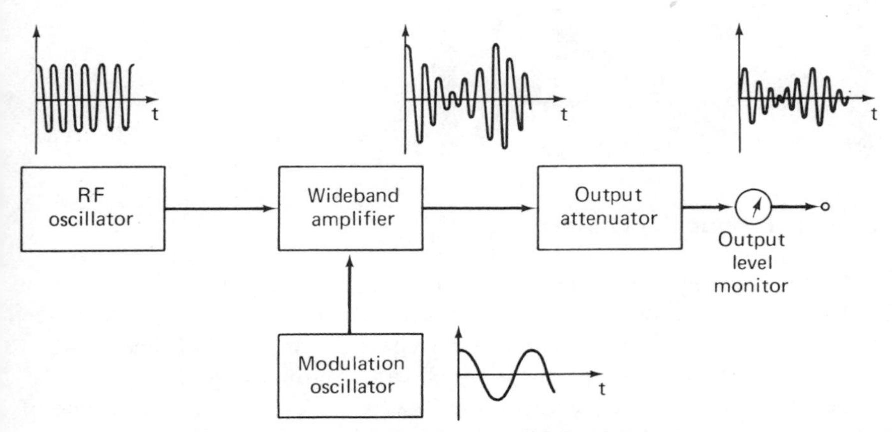

Figure 13-9. Block diagram of an AM signal generator

which combines them. The result of the signal combination is an rf signal which is modulated by the af signal. This modulated signal is then fed to an *attenuator* which allows the maximum amplitudes of the output signal to be adjusted.

The accuracy to which the frequency of the rf oscillator is known is an important specification of the performance of the signal generator. Most laboratory-type models are usually calibrated to be within 0.5–1.0 percent of the dial setting. This accuracy is usually sufficient for most measurements. However, if a greater accuracy is required for a particular test, a crystal oscillator (whose frequency is known to within 0.01 percent or better) is available as an internal rf calibration source. Another key specification of signal generators is their amplitude stability. This quantity is a measure of the variation of the amplitude of the output signal with changes in rf frequency or age of the instrument. Since the signal generator is often used as a standard for determining the gain of amplifiers, it is important that the amplitude of the output signal remain constant as the rf frequency is varied.

Some signal generators are equipped with an rf oscillator whose frequency output can be changed by a motor-driven capacitor. When one dials a selected frequency and pushes a switch, the motor drives the capacitor so that the frequency of the rf oscillator output is varied by about 7 percent per second. This feature allows motor-driven signal generators to be automatically controlled at remote locations. It further allows the instruments to produce an output signal which is automatically varied between different frequency limits. Figure 13-10 shows a photograph of a commercially available signal generator.

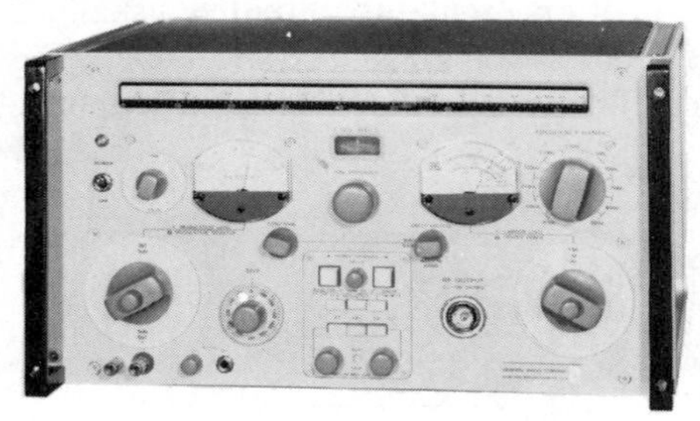

Figure 13-10. Signal generator (Courtesy of General Radio Co.)

Sweep-Frequency Generators

Sweep-frequency generators are instruments which produce a sine-wave output whose frequency is automatically varied (swept) between two chosen frequencies. One complete cycle of the frequency variation is called a *sweep.* The rate at which the frequency is varied can be either linear or logarithmic, depending on the particular instrument design. However, the amplitude of the signal output is designed to stay constant over the entire frequency range of the sweep. Figure 13-11 shows a block diagram of the major components of a sweep-frequency generator.

Sweep-frequency generators are primarily used to measure the responses of amplifiers, filters, and electrical components over various frequency bands. For example, when amplifier circuits or filters are built, they must be tested to ensure that their responses over a specific band of frequencies (bandwidth) meet the design requirements. Performing the measurement of the bandwidth over a wide frequency range with a manually tuned oscillator would be a time-consuming task. In addition, since every change in design requires that the measurement be repeated, many hours could be spent performing this measurement by the end of the project. By using a sweep-frequency generator, a sine-wave signal that is automatically swept between

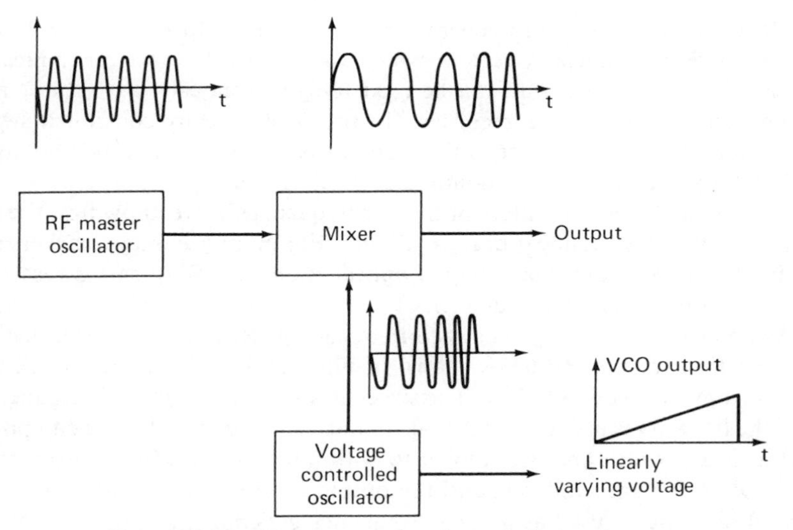

Figure 13-11. Block diagram of an electronically tuned sweep-frequency generator

two selected frequencies can be applied to the circuit under test, and its response versus frequency can be displayed on an oscilloscope or *X-Y* recorder. Therefore, the measurement time and effort is drastically reduced. Sweep generators can also be used in the same capacity to check and repair amplifiers used in televisions and radar receivers.

The major component of the sweep generator is a master oscillator (usually an rf type) with several operating ranges which are selected by a range switch. The frequency of the signal generator's output signal can be varied by either a mechanical or electronic process. In the mechanically varied models, the frequency of the master oscillator output signal is changed (tuned) by a motor-driven capacitor.

In the electronically tuned types, the frequency of the master oscillator is kept fixed, and a second oscillator is used to provide a varying frequency signal. The second oscillator contains an element whose capacitance value depends on the voltage applied across it. This element is used to vary the frequency of the sine-wave output of the second oscillator. As a result, this second oscillator is called a *voltage-controlled oscillator* or VCO. The output of the VCO is then combined with the output of the master oscillator in a special electronic device called a *mixer*. The output of the mixer is a sine wave whose frequency depends on the difference between the frequencies of the two applied signals. Thus, if the master oscillator frequency is fixed at 10.00 MHz and the variable frequency is varied from 10.01 MHz to 42 MHz, the output of the mixer will be a sine wave whose frequency is swept from 10 kHz to 32 MHz.

The sweep rates of sweep generators can be adjusted to vary from 100 to 0.01 seconds per sweep. A voltage which varies linearly (or logarithmically) according to the sweep rate can be used to synchronously drive the X axis of an oscilloscope or X-Y recorder. (In the electronically controlled sweep generators, the same voltage which drives the VCO serves as this voltage.)

The frequency of various points along the frequency-response curve can be interpolated from the values of the end frequencies if we know how the frequency is varied (i.e., linear or logarithmic). For more accuracy, markers can be used. (Markers are pulses which appear along the frequency-swept output at accurately known frequencies.)

As an example of a typical sweep-frequency generator, let us examine the specifications of the HP 675A shown in Fig. 13-12. This instrument can be swept from 10 kHz to 32 MHz. The sweep has a linear variation in frequency, and the linearity is accurate to 0.5 percent of the sweep width. The end points of the sweep width are also known to within 0.5 percent. Sweep rates from 0.01 to 100 Hz are available, and the output level is kept constant to about 1 V. Besides being used as a sweep-frequency generator, this instrument can

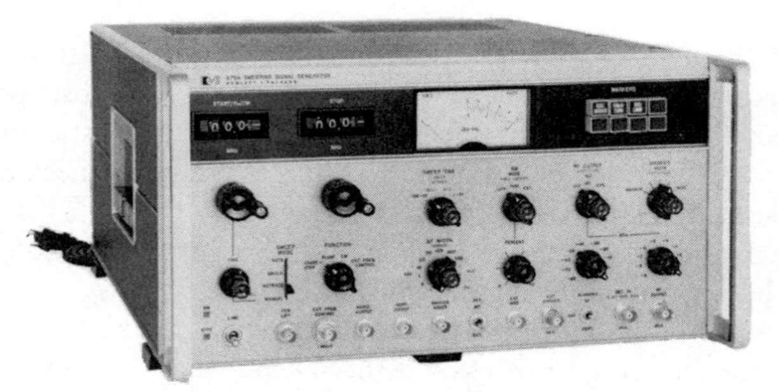

Figure 13-12. Sweep-frequency generator (Courtesy of Hewlett-Packard Co.)

also be used to produce sine-wave outputs with fixed frequencies in the 10-kHz to 32-MHz range.

Pulse Generators

Pulse generators are instruments which are designed to produce a periodic train of equal-amplitude pulses (Fig. 13-13(a)). In pulse generators, the duration of the *on* time of a pulse may be independent of the time between pulses. However, if the pulse train has the property of being *on* 50 percent of the time and *off* 50 percent of the time, the waveform is called a *square wave* (Fig. 13-13(b)). Square-wave generators can be considered to be a special class of pulse generators.

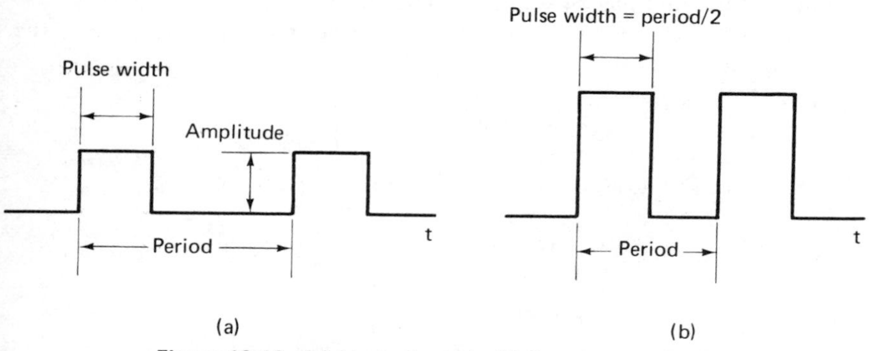

Figure 13-13. (a) Ideal pulse train (b) Square-wave signal

To describe the output of pulse generators and the applications in which they are used, the terminology associated with pulses must be introduced. The first group of such terms denotes the characteristics of ideal rectangular pulses. The second group of terms provides measures of the deviation from the shape and periodicity of ideal pulses.

The terms which characterize a train of ideal periodic pulses include the following:

1. *Period*—The time (in seconds) between the start of one pulse and the start of the next. The frequency (or *pulse repetition frequency*) is inversely related to the period.
2. *Amplitude*—The peak voltage value and polarity of the pulse.
3. *Pulse width*—The time duration of the pulse (in seconds).
4. *Duty cycle*—The ratio of the pulse width to the period (expressed in percent of the period). Square waves have a duty cycle of 50 percent.

However, real pulses and pulse trains only approximate the characteristics of their ideal counterparts. As a consequence of this fact, some additional terms are used to describe the nonideal aspects of real pulses. These terms and their definitions are listed below and are illustrated in Fig. 13-14.

1. *Rise time*—(t_r) the time (in seconds) it takes for the pulse to increase from 10 percent to 90 percent of its amplitude.
2. *Fall time*—(t_f) the time (in seconds) it takes for the pulse to decrease from 90 percent to 10 percent of its amplitude.
3. *Overshoot*—The extent (in percent of amplitude) to which the pulse surpasses its correct value during the initial rise.
4. *Ringing*—Oscillation occurring (in percent of pulse amplitude) as a result of overshoot.
5. *Sag* or *droop*—Any decrease (in percent of pulse amplitude) in the pulse amplitude taking place during the pulse width.
6. *Jitter*—Specifies the maximum variation in period from one cycle to the next (in terms of the percentage of the period).

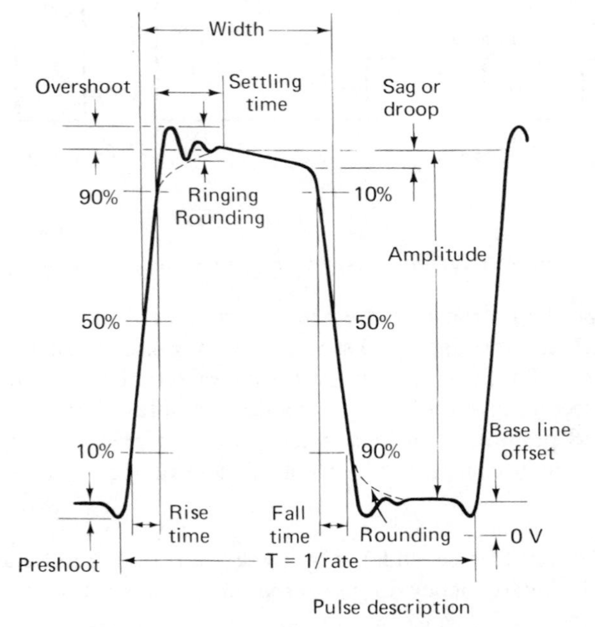

Figure 13-14. Actual pulse (Courtesy of Hewlett-Packard Co.)

Pulse generators are designed to produce pulses which approximate ideal pulses as closely as possible. High-quality pulses ensure that any distortion in the output pulse from a test circuit is due to the test circuit alone. The amplitude, pulse width, and period of the generated pulses are usually adjustable over various ranges. The duty cycle is also made adjustable; but if the power contained by each pulse is large, the maximum duty cycle must be kept small. As the maximum duty cycle of a pulse generator is reached, the pulse waveform becomes irregular or the width of the pulse no longer increases.

The source or generator impedance of pulse generators is 50 Ω. This value is chosen so that pulse generators will be matched to the cables which transmit the output pulses of the generator (the cables are coaxial cables which have a 50-Ω characteristic impedance). Such matching is necessary because loads connected to pulse generators (at the other end of the coaxial cables)

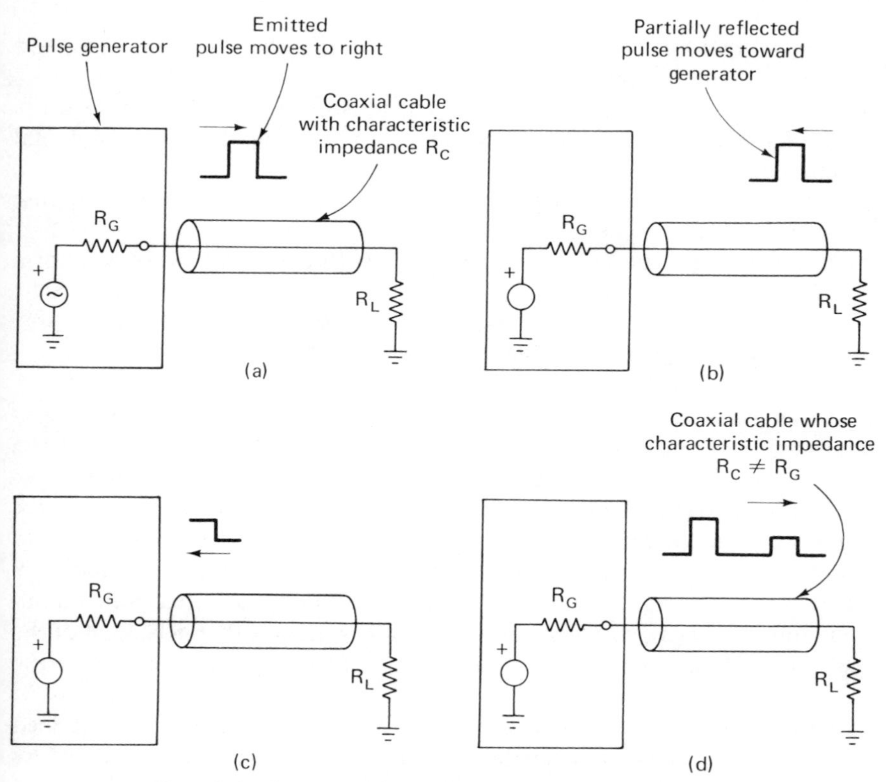

Figure 13-15. The importance of matching the impedance of a pulse generator and the pulse-transmission cable. (a) Pulse emitted by pulse generator is transmitted to the load along a coaxial cable. Coaxial cable has characteristic impedance $R_C = R_G$. Load has impedance $R_L \neq R_G$. (b) Pulse is partially reflected by the load R_L, because $R_L \neq R_C$. Reflected pulse travels toward the left on coaxial cable. (c) Since $R_C = R_G$, reflected pulse is absorbed when it returns to the generator. (d) If $R_G \neq R_C$, reflected pulse would be re-reflected and would propagate to the right along the cable again. This pulse would appear at the load in addition to the desired pulses.

are not always matched to the source impedance or cable impedance. Any such *mismatching* causes a part of the pulse to be reflected back to the pulse generator along the coaxial cable. Since the cable and the generator are matched, the reflected signal is completely absorbed upon returning to the generator. If this total absorption did not take place, a portion of the pulse would be re-reflected, and spurious pulses would appear to be generated from the pulse generator (Fig. 13-15).

Pulse generators find application in a wide variety of measurements. Some of these are:

1. Test sources for measuring the rise times of amplifiers, oscilloscopes, and electrical components.
2. Digital circuitry testing—measuring the rise times and switching times of digital circuits and components (e.g., diodes, transistors) used in digital circuits.
3. Impulse testing—providing very narrow pulses to circuits whoses response to impulse excitation needs to be determined.
4. External trigger pulse source—triggering such devices as sweep waveform generators.
5. Providing pulses of varying widths for determining the minimum energy required to trigger such devices as silicon-controlled rectifiers.

As an example of the capabilities of a commercially available pulse generator, let us consider the Hewlett-Packard 8002A (shown in Fig. 13-16). This instrument has a repetition rate which is continuously variable from 0.3 Hz to 10 MHz. The pulse width can be varied from 30 ns to 3 s, and the duty cycle can be made as high as 90 percent (50 percent from 1 MHz to 10 MHz). The maximum voltage output is 5 *V* into a 50-Ω load and 10 V into an open circuit. The rise time of the pulses can be adjusted from a minimum of 10 ns to 2 s. The overshoot and ringing are less than 5 percent of the pulse amplitude. The maximum jitter is specified to be 0.1 percent of the repetition rate.

Besides the features included in the ordinary pulse generators described above, other options are available on special instruments. One such feature is the ability of the generator to generate two very closely spaced pulses, followed by a relatively long period before the next two pulses appear. This capability is useful for measuring the resolving power of special nuclear measuring instruments which must be able to discriminate between pulses that occur in rapid succession. Another feature is found in so-called *digital word generators*. These instruments allow a pattern of several pulses to be produced and then a length of time before this same pattern is repeated. As the price of such instruments is increased, a greater flexibility in adjusting the pulse amplitudes and widths within a given word is made available.

(a)

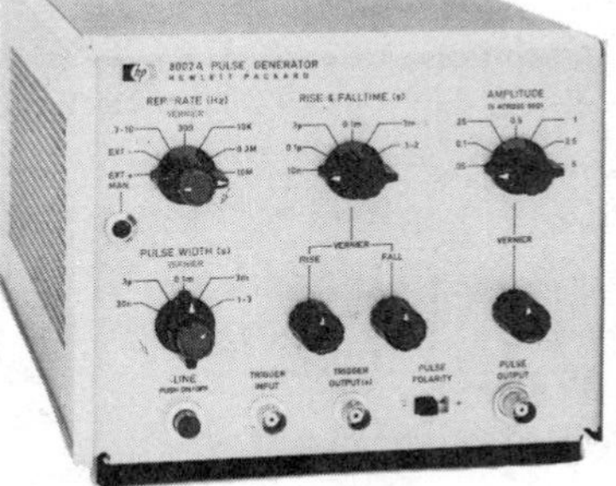

(b)

Figure 13-16. Pulse generator (a) Tektronix Model 114 (Reprinted by permission of Tektronix, Inc.) (b) HP 8002A (Courtesy of Hewlett-Packard Co.)

Function Generators

A function generator is a signal source that has the capability of producing several different types of waveforms as its output signal. Most function generators can generate sine waves, square waves, and triangular waves over a wide range of frequencies. Other models are capable of generating pulses or sawtooth waves as well as the three types of waveforms mentioned. The frequency range of a function generator is generally 0.01 Hz to about 1–2 MHz.

Because they can produce a wide variety of waveforms and frequencies, function generators are very versatile instruments. In fact, each of the wave-shapes they produce are particularly suited for a different group of applications. The uses of the *sine-wave output* were already described in the earlier section on oscillators. The *square-wave* signal can be employed for testing electronic amplifiers and the transient responses of other circuits. Since low-frequency square waves are composed of a wide frequency range of sinusoidal components, they provide a unique measurement capability for the testing of amplifier circuits.[3] In other words, the response of an amplifier circuit to a square-wave input signal yields the same type of data about its electrical characteristics as if it were tested sequentially with sine-wave inputs of many different frequencies. To use the square wave for this type of measurement, it must be applied to the input of the circuit under test. The output of the circuit is displayed on an oscilloscope. The signal on the oscilloscope

[3]For a more elaborate discussion of this point, see any mathematics or circuits text which covers Fourier analysis. The section in Chapter 14 entitled *Harmonic Analysis* also has a brief discussion on some of the introductory aspects of Fourier analysis.

will look like one of those shown in Fig. 13-17. From this figure, we see how the departure from the input square-wave shape can be interpreted in terms of the amplifier's frequency response.

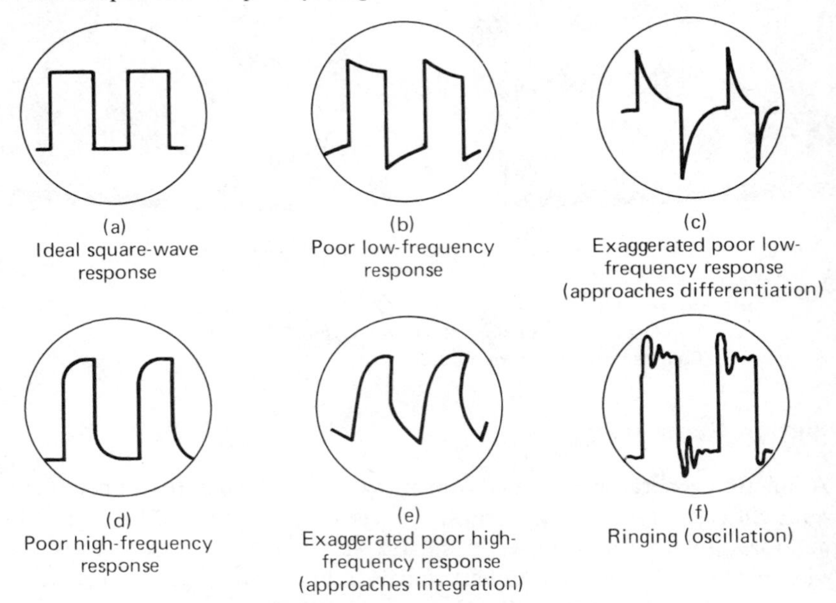

Figure 13-17. Amplifier response in square-wave testing (a) Ideal square-wave response (b) Poor low-frequency response (c) Exaggerated poor low-frequency response (approaches differentiation) (d) Poor high-frequency response (e) Exaggerated poor high-frequency response (approaches integration) (f) Ringing (oscillation) (Courtesy of S. Prensky, *Electronic Instrumentation,* 2nd ed., Englewood Cliffs, N.J., Prentice-Hall, Inc., 1971, p. 279.)

The triangular-wave and sawtooth-wave outputs of function generators are commonly used for those applications which require a signal that increases (or decreases) at a specific linear rate. They are also useful for driving sweep oscillators in oscilloscopes and the *X* axis of *X*-*Y* recorders. Many function generators are also able to generate two different waveforms simultaneously (from different output terminals, of course). This can be a useful feature when two generated signals are required by a particular application. For example, a triangular wave and a sine wave of equal frequencies can be produced at the same time. If the zero crossings of both waves are made to occur at the same time, a linearly varying waveform is available which can be started at the point of zero phase of a sine wave.

Another important feature of some function generators is their ability to *phase-lock* to an external source. This means that when the *phase-lock* feature is employed, each cycle of the waveform from a function generator will bear a fixed phase relationship to an applied external signal. (This phase relationship is also adjustable.) One example of how this feature is used is when a sine-wave output of a function generator is phase-locked to another sine wave from a second function generator. If this other sine wave is the second harmonic of the first, the summation of the two sine waves at different amplitudes and phase shifts can produce a wide variety of unusual waveforms. In addition, if the function generator is connected to an accurate frequency standard (and the phase-lock feature is being employed), all the output waveforms from the function generator will have the same frequency, stability, and accuracy as the standard. Figure 13-18 shows some common function generator models.

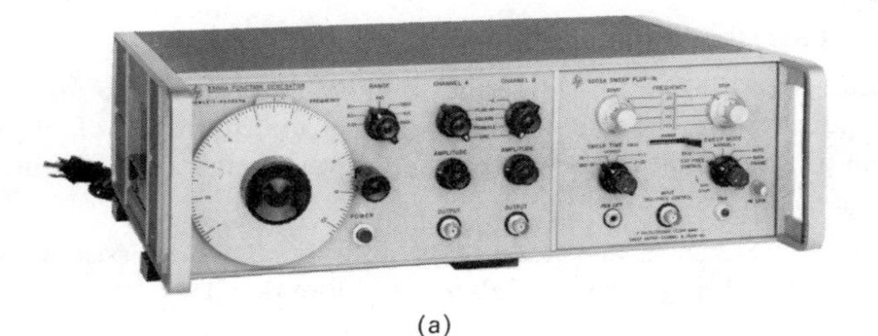

(a)

(b)

Figure 13-18. Function generators (a) HP 3300A (Courtesy of Hewlett-Packard Co.) (b) Wavetek Model III (Courtesy of Wavetek)

Problems

1. Briefly describe the various types of output signals that are generated by each of the following ac signal sources:
   a) oscillators,
   b) signal generators,
   c) sweep frequency generators,
   d) function generators.

2. Qualitatively describe the operation of the oscillator. From where does the oscillator draw its power to overcome the losses present in its oscillating circuitry?
3. Explain why af and rf oscillators usually possess output impedances of 600 Ω and 50 Ω, respectively.
4. Describe what effect the value of the load impedance connected to the output of an oscillator will have on the maximum voltage which can be put out by the oscillator.
5. Define the following terms used to specify the performance of an oscillator:
   a) dial resolution,
   b) frequency stability,
   c) range,
   d) amplitude stability.
6. In reference to Fig. 13-7(b), describe what would occur if the minus terminal of the power supply was connected to the ground terminal of the oscillator and the plus terminal of the power supply was connected to the ungrounded terminal of the oscillator.
7. By making references to other texts describe the processes of amplitude modulation and frequency modulation in more detail.
8. If a pulse has the following period and width, calculate its duty cycle:
   a) pulse width = 1 μs, period = 10 μs
   b) pulse width = 3 ms, period = 1 s.
9. Explain why the output impedance of a pulse generator and the impedance of the cable along which the output pulses of the generator are transmitted should have the same values.
10. Describe what is meant by the capability of a function generator to phase lock to an external source.
11. Define the terms a) rise time, b) fall time, c) overshoot, d) ringing, e) jitter, in reference to the description of a pulse.
12. Using a block diagram, design a test setup which utilizes an oscilloscope to calibrate the frequency of an oscillator.

References

1. HEWLETT-PACKARD, *Instrumentation Catalog*. Palo Alto, Calif., Hewlett-Packard Co., 1971.
2. COOPER, W. D., *Electronic Instrumentation and Measurement Techniques*. Englewood Cliffs, N.J., Prentice-Hall, Inc., 1970.
3. MANDL, M., *Fundamentals of Electronics*, Chap. 6. Englewood Cliffs, N.J., Prentice-Hall, Inc., 1960.
4. MILLMAN, JACOB, and TAUB, HERBERT, *Pulse, Digital, and Switching Waveforms*. New York, McGraw-Hill Book Co., 1965.

5. MALMSTADT, H. V., ENKE, C. G., and TOREN, E. C., *Electronics for Scientists*, Chap. 5. New York, Benjamin, 1962.

6. TERMAN, F. E., and PETTIT, J. M., *Electronic Measurements*, 2nd ed. New York, McGraw-Hill Book Co., 1952.

7. RYDER, John D., *Electronic Fundamentals and Applications*. Englewood Cliffs, N.J., Prentice-Hall, Inc., 1964.

14

Time and Frequency Measurements

The accurate measurement of time is an important task in all fields of scientific investigation. Since time is one of the fundamental units from which all other units are derived, it is necessary that time measurements be very accurately made. Although time intervals are measured by many different techniques and devices, the most accurate measurements involve electronic measuring devices (i.e., for accuracies of up to one part in 10^{10}). We will discuss such accurate time-interval measuring instruments, as well as some simpler electrical timing devices.

Frequency is defined as the number of recurring events taking place in a unit interval of time. When we refer to electrical signals, frequency (f) is usually meant to be the number of cycles of a periodic waveform that occur per second. The length of time (in seconds) of one complete cycle of such a waveform is called the period (T) of the waveform. As pointed out earlier in the text, the frequency and the period of a periodic waveform are related by

$$f = \frac{1}{T} \tag{14-1}$$

The unit of frequency is the hertz (Hz—formerly referred to as cycles per second). Equation (14-1) indicates that if we can measure either time or frequency, we can determine the other of these two quantities.

As a final topic in this chapter, we will discuss the instruments used for harmonic analysis of waveforms. Harmonic analysis involves the determination of the various frequency components of which signals are composed. Harmonic-analysis instruments are also used to detect the extent of distortion of various waveforms.

TIME STANDARDS

The earliest time standards were based on the rotation of the Earth about its axis. One *day* is the time it takes for the Earth to make one complete revolution about its axis. When clocks were first developed, the *second* was chosen to be approximately the time of one complete oscillation of a pendulum. Eventually, the second was made into the unit which was 1/86,400 of a day (24 hours $\times$ 60 minutes $\times$ 60 seconds).

As more precise astronomical measurements of the Earth's rotation were made, however, it was learned that slight variations occur in the time of the axial rotation of the Earth. Thus, it was realized that a time standard which was tied to the Earth's rotation would never be completely constant. Since modern measurements require very accurately known time intervals, various standards were developed which are not dependent on astronomical observations.

The modern standard of time is defined in terms of an *atomic standard.* In this standard, the resonant frequency of cesium atoms is measured and converted to time. Since the frequency is dependent only on the internal structure of the cesium atoms (as long as the atoms are not disturbed by external conditions such as magnetic fields), the accuracy of this standard is guaranteed to within approximately one part in 10^{10}. One second is identified as 9,192,631,770 periods of oscillation of such cesium atoms.

An atomic clock based on this principle of operation (and whose precision exceeds 1 μs a day) is now operated by the National Bureau of Standards as the primary time standard. The time signals from this clock are broadcast on WWVB (60 kHz) and are accurate to five parts in 10^{10}.

In the near future, it is likely that the time standard will be based on the number of wavelengths in a beam of laser light. The accuracy of such a standard promises to be several orders of magnitude higher than the atomic standard.

Time Measurements

Time can be measured by various mechanical, electrical, and astronomical methods. Our discussion will be concerned with only the electrical and electronic methods. The complexity and expense of time-measuring equipment increases as the length of the time interval being measured decreases (and as the accuracy with which the measurement must be made increases). We will discuss time-measuring instruments in an ascending order of complexity and accuracy. The most accurate time-measuring instruments (i.e., digital timers) will be discussed in the section on digital frequency counters.

ELECTRIC (CLOCK) TIMERS

Electric timers are relatively simple time-measuring instruments. As such, they are used in those applications where the time intervals being measured are one second or longer and the accuracy of the measurement need not be extremely high. They are also used in some industrial applications for automatically initiating and stopping processes at prescribed times. The ordinary home electric clock is also a form of electric timer. The smallest time interval electric timers can measure is about one-hundredth of a second.

Electric timers are driven by synchronous motors which spin in exact synchronization with the 60-Hz frequency of delivered electric power. Since the electric companies maintain the frequency of the power they distribute very close to 60 Hz, the synchronous motors provide a time standard which is adequate for many general timing needs.

Electric timers can be built with switches to open and close circuits at desired intervals and with counters that record time intervals.

ELECTRONIC TIMERS

One method for measuring shorter time intervals than is possible with electric clock timers is by using *electronic* timers. Such electronic timers can measure time intervals whose durations are as small as one microsecond.

The most common principle used in electronic timers involves the charging of a capacitor in an *RC* circuit. Electronic timers make use of the fact that it takes a finite amount of time to charge a capacitor when a dc voltage source is suddenly applied across the series resistor-capacitor connection. The variation of the voltage versus time across the capacitor will have a curve whose shape is shown in Fig. 14-1. The rapidity with which the capacitor

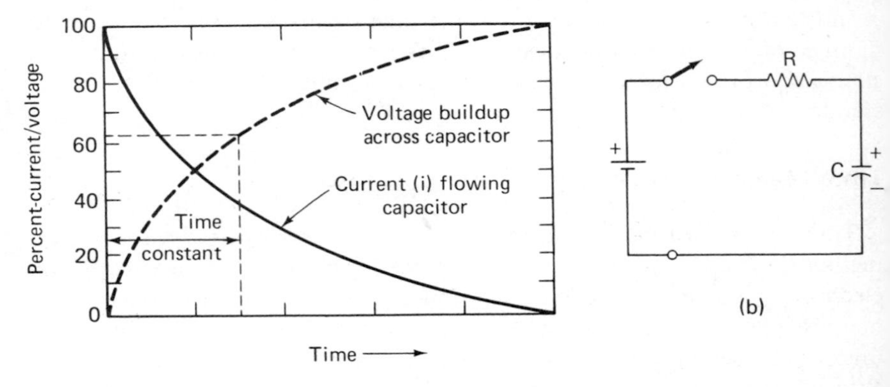

Figure 14-1. (a) Current and voltage of a capacitor when a dc voltage is applied to it at $t = 0$ (b) RC circuit of the electronic timer

charges depends upon the values of the resistance and capacitance in the connection. Specifically, a quantity known as the time constant (τ) of the resistance-capacitance combination is defined as the time it takes for the voltage across the capacitor to reach 63 percent of its full-charged value. The time constant is equal to

$$\tau \text{ (seconds)} = \text{R(ohms)} \times \text{C(farads)} \quad (14\text{-}2)$$

For example, if we had a 500-pF capacitor and a 4-MΩ resistor in series, it would take

$$\tau = R \times C = (4 \times 10^{6}) \times (5 \times 10^{-10})$$

$$2 \times 10^{-3} = 2 \text{ milliseconds}$$

to charge the capacitor to 63 percent of the value of any dc voltage applied across the series connection.

If we apply a known voltage across the series RC connection at time $t = 0$ and then interrupt the voltage at the end of the measured time interval, the voltage across the partially charged capacitor will be an indication of the time interval.

COINCIDENCE CIRCUITS

A coincidence circuit produces an output signal only when both of its two input terminals receive a pulse within a short time of each other. This short time interval is known as the *resolving time* of the circuit. The shorter the resolving time, the less likely the coincidence circuit will be to produce a pulse due to a random-chance coincidence. The ability to respond to pulses only when they are coincident in time is a useful feature for distinguishing applied pulses from noise pulses.

A typical application of coincidence circuits is in detecting radioactivity. If two photomultiplier tubes (devices designed to detect weak flashes of light)[1] are used to monitor the same fluorescent crystal, they will both simultaneously detect a flash of light when a radioactive particle strikes the crystal. Thus they will each produce an electric pulse in response to the detection of the light flash. If a coincidence circuit is connected to the outputs of both tubes, it will put out a pulse signal as a result of this event. However, if either one of the photomultiplier tubes emits a pulse which is a result of noise within either of the tube's circuits, the coincidence circuit will receive only one pulse and will not produce an output. In this fashion, bonafide radioactive events can be separated from random noise pulses.

[1] See the section on *Light Radiation Transducers* in Chapter 15 for details on photomultiplier tube operation.

Frequency Measurement

Various types of instruments are used to measure frequency, depending on the frequency range and accuracy desired. The following is a list of some of the instruments commonly used for measuring various frequency ranges.

1. Oscilloscopes.
2. Dynamometer frequency meters (10–200 Hz).
3. Wien bridge frequency meters (audio-frequency range).
4. Zero-beat frequency meters (rf ranges).
5. Grid-dip meters (coarse, rapid measurements in rf range).
6. Digital frequency counters (wide frequency range, very accurate).

The oscilloscope can measure frequencies over a wide range by using the triggered-sweep and Lissajous pattern methods described in Chapter 9. However, the accuracy of the frequency measurements made with an oscilloscope is somewhat limited. Thus, for more accurate measurements, other frequency-measuring instruments are more likely to be used.

DYNAMOMETER FREQUENCY METERS

Dynamometer frequency meters are used in a limited number of applications where low frequencies ($\approx$60 Hz) need to be measured. Their most common use is in the monitoring of the frequency of 60-Hz electric power (where only a small frequency variation is ever expected).

These meters employ the same kind of movement as the electrodynamometer ammeters, voltmeters, and wattmeters. However, the field coils of the movement used in the frequency meter each form part of a separate *LC* circuit (Fig. 14-2). Each of the circuits is tuned to a different frequency. For

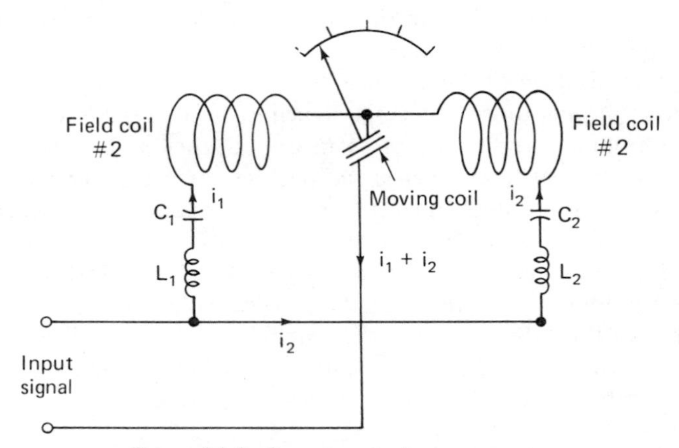

Figure 14-2. Dynamometer frequency meter

example, if the meter was to be used to monitor the power-line frequency (60 Hz), the circuit of field coil 1 in Fig. 14-2 would be tuned to 50 Hz. The circuit of field coil 2 would be tuned to 70 Hz. The current through the moving coil is the sum of the currents passing through each of the field coils. The torque on the moving coil is proportional to the current moving through it.

Since the frequency being measured falls between 50 and 70 Hz, the torque on the moving coil due to the current in field coil 1 will be attempting to rotate the moving coil in a different direction than the torque due to coil 2 at the same instant. As a result of the opposing torques, the resulting torque will be dependent on the frequency of the applied voltage. The meter can be calibrated to indicate this frequency directly.

The countertorque of the moving coil of this meter is not supplied by control springs. Instead, an iron vane is mounted in the center of the rotating coil. The tendency of the iron vane to remain aligned with the magnetic field produced by the field coils provides the restraining torque of this movement.

WIEN BRIDGE FREQUENCY METERS

In Chapter 13 we mentioned that the Wien bridge oscillator is a device which is designed to produce accurately known audio-frequency sine-wave signals. The bridge portion of the oscillator circuit can also be used as an instrument for measuring frequencies in the audio-frequency range. In this type of meter, the signal of interest is applied to the bridge (Fig. 14-3). The

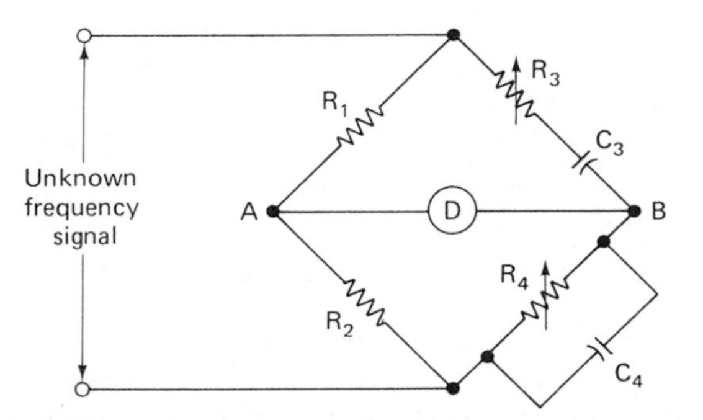

Figure 14-3. The basic circuit of the Wien bridge used to measure audio frequencies

arms of the bridge contain passive adjustable electrical components. For each combination of values of the components, there is a specific frequency at which the bridge is balanced (i.e., the value of the voltage difference between points A and B is zero). A set of headphones, a CRT, or an electronic ac voltmeter can be used to detect this balance condition. In most Wien bridges, the values of the components are chosen such that $R_1/R_2 = 2$,

$R_4 = R_3$, and $C_4 = C_3$. Then, at balance, the unknown frequency is found from

$$f = \frac{1}{2\pi R_3 C_3} \tag{14-3}$$

ZERO-BEAT FREQUENCY METER

Zero-beat frequency meters are used to make accurate measurements of the frequencies of rf signals. The operation of such meters is based on the zero-beat (or heterodyne) principle. This principle says that if two signals of different frequencies are combined in a nonlinear circuit[2] (in this case, a *mixer* circuit), the output of the circuit will contain a signal whose frequency is equal to the difference between the two original signals. This frequency is called the *difference heterodyne.*

In the zero-beat frequency meter, the signal whose frequency is being measured is one of the two signals applied to the mixer (Fig. 14-4). The other signal is generated within the meter from a variable-frequency oscillator (VFO). As the frequency of the VFO is adjusted so that it approaches the value of the unknown frequency, the difference between the two frequencies gets smaller. If a pair of headphones is used to detect the difference heterodyne signal, a sound will be produced in the headphones when the frequency

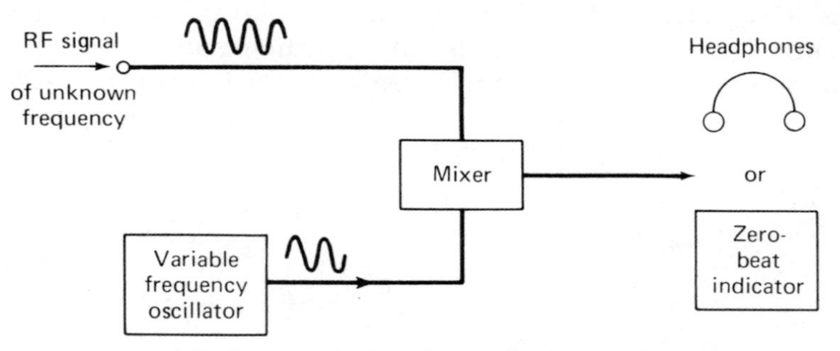

Figure 14-4. Block diagram of a zero-beat frequency meter

difference becomes smaller than 15,000 Hz. As the frequency of the VFO continues to be varied so that it gets closer and closer to the unknown frequency, the pitch of the sound produced by the headphones will become lower and lower. When the two signals are of equal frequencies, the sound will disappear. If the frequency of the VFO is raised above that of the unknown frequency, the sound in the headphones will again reappear because the dif-

[2]In a nonlinear circuit there are elements, such as vacuum tubes or transistors, whose current-vs-voltage characteristics are not straight lines. The fact that these characteristics are curved in some fashion, and not straight, implies that their operation is nonlinear. The heterodyning takes place in the region represented by the curved portion of these characteristics.

ference in the two frequencies will produce a nonzero difference signal. Thus, if the frequency of the VFO is accurately known at the point when the headphone sound disappears, the unknown frequency is identified. A recorder can be used instead of headphones to determine the point at which the difference heterodyne is zero.

The accuracy of the VFO can be checked by means of a crystal oscillator which is built into the instrument. The crystal oscillator produces signals whose specific output frequencies are known with a very high accuracy. By comparing the frequency of the VFO to the frequency of the crystal oscillator, we can calibrate the VFO so that the zero-beat frequency meter will yield values of the measured frequencies to within 0.01 percent of their true values.

The range of the zero-beat frequency meter can be extended beyond the maximum frequency range of the VFO by virtue of the nonlinear characteristics of the mixer. In other words, because it is a nonlinear circuit, the mixer also acts on each individual signal applied to it and produces components whose frequencies are double, triple, etc., those of each input signal. For example, if a 100-Hz signal is applied to the mixer, the mixer will produce an output signal with components whose frequencies are 100 Hz, 200 Hz, 300 Hz, etc. These higher-frequency components can also be used to produce a difference heterodyne signal with the unknown frequency if the unknown frequency is higher than the maximum frequency of the VFO. To get a rough idea of the frequency of the unknown signal, a grid-dip meter can be used.

GRID-DIP METERS

The grid-dip meter is a device which can be used for rapid (but not very accurate) determinations of the frequency of rf signals, the resonant frequency of LC circuits, and the Q of inductors.

When it is used as an rf frequency meter, it provides a VFO, a mixer, and a connection to which an external pair of headphones can be attached. When the grid-dip meter is held near a source of rf signals, an inductor within the meter couples the rf signal and applies it to the mixer. Then the grid-dip meter determines the zero-beat frequency in a manner similar to the meter described in the previous section. However, the grid-dip meter does not contain a crystal oscillator for calibration and hence cannot provide the accuracy of a bonafide zero-beat frequency meter. However, since it is a small and relatively simple instrument, it is useful for providing a quick check of the frequency of an rf signal source.

When used for determining the resonant frequency of LC circuits or the Q of inductors, grid-dip meters become oscillators whose signals are loosely coupled to the circuit or inductor under test. An ac voltmeter within the grid-dip meter monitors the magnitude of the oscillator output voltage as the frequency is varied. When the grid-dip meter is placed near the LC circuit of interest and the frequency of the grid-dip oscillator equals the resonant frequency of the circuit, energy is absorbed by the circuit from the oscillator. This means that the output voltage of the oscillator (as shown on the ac voltmeter) decreases or *dips*.

DIGITAL FREQUENCY COUNTERS

Digital frequency counters are the most accurate and flexible instruments available for measuring unknown frequencies (Fig. 14-5). The highest accuracy digital frequency counters can approach the accuracy of the atomic time standards described earlier. Frequencies from dc to the GHz range can be measured with digital frequency counters. In addition, since most events can be converted into an electrical signal consisting of a train of electrical pulses, digital frequency counters can also be used as counters of almost all types of quantities. Thus, events such as heartbeats, the passing of radioactive particles, motor or shaft revolutions, light flashes, and meteorites can all be counted with digital counters. However, digital frequency counters are usually considerably more expensive than the other methods used for measuring frequency. If the very high accuracy of a digital counter is not required by a frequency measurement, another device may be an adequate choice.

The major components which make up digital frequency meters are shown in Fig. 14-6. They are the digital counting and display assembly, the time-

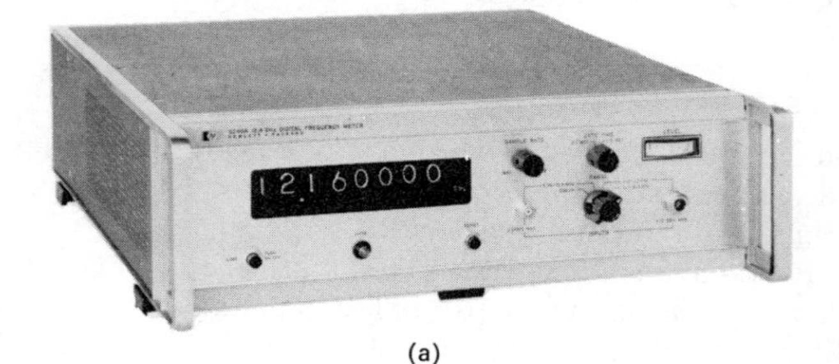

(a)

(b)

Figure 14-5. Digital frequency meters (a) HP model 5245M (Courtesy of Hewlett-Packard Co.) (b) General Radio model 1192 (Courtesy of General Radio Co.)

base generator, the pulse-forming circuit (usually a Schmitt-trigger type of circuit), and the gate circuit. (If a quantity other than the frequency of an electrical signal is to be counted, a transducer must also be used to convert the events into electrical pulses.)

In Fig. 14-6 we see that the signal of unknown frequency is fed into the counter and enters a pulse former. This circuit creates a pulse for every cycle of the input signal. These pulses are then applied to a gate. If the gate is open, the pulses can pass through it and be counted and displayed by a digital counting and display assembly. If the gate is open only for a fixed and accurately known time period, the number of pulses counted during the time period will yield the frequency (i.e., cycles per second) of the unknown signal.

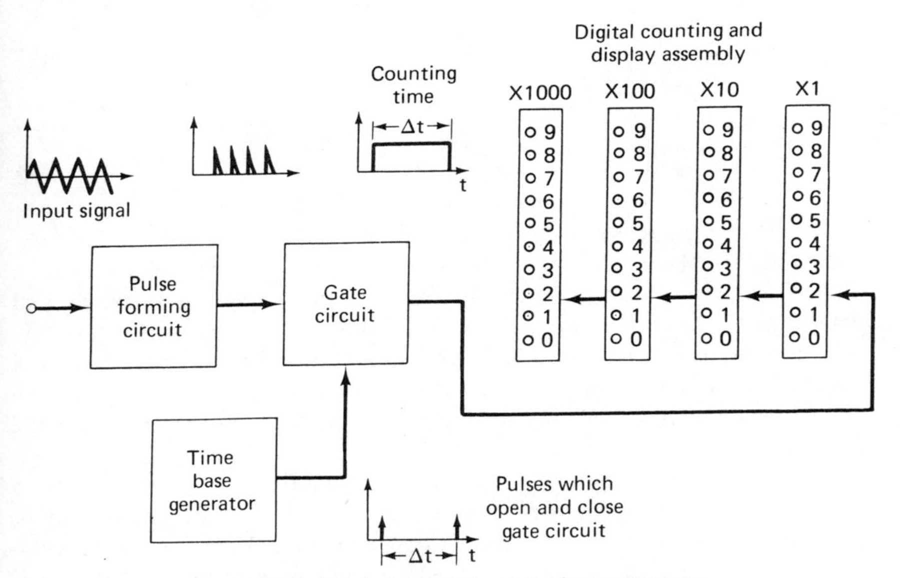

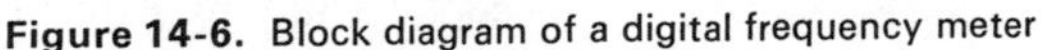
Figure 14-6. Block diagram of a digital frequency meter

The method by which the gate is allowed to be open for an accurately known interval is shown in Fig. 14-7. A crystal oscillator produces a signal at 1 MHz or 100 kHz (depending on the particular instrument design). To keep the output of this oscillator at exactly the desired frequency, the crystal is enclosed in a temperature-controlled oven. The oscillator output signal is then fed to another pulse-forming circuit. If the oscillator has a frequency output of 1 MHz, the pulse-forming circuit will produce a train of narrow pulses at 1 μs intervals. These pulses are applied to a number of decade frequency dividers which are designed to increase the time between pulses by factors of ten. As a result, we can choose to have pulses separated

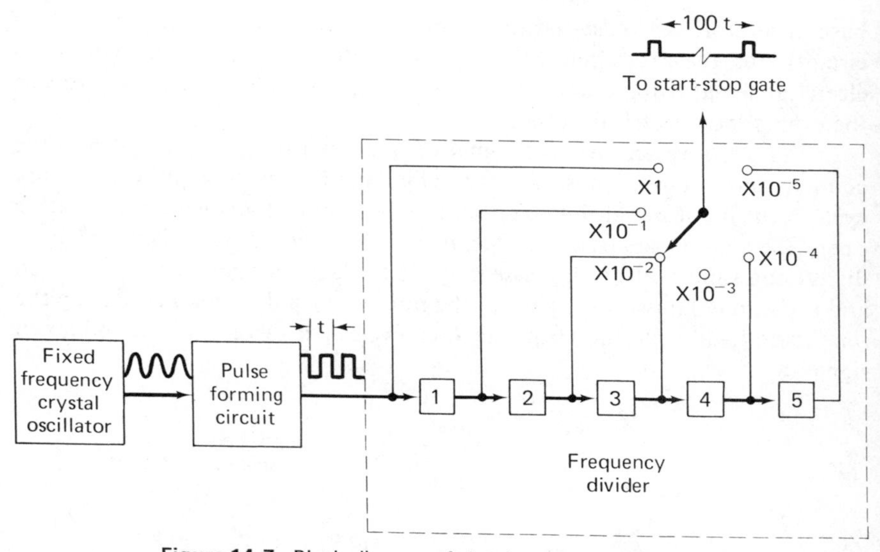

Figure 14-7. Block diagram of the time-base generator of a digital frequency counter

in time from 1 μs to 1 s. Any two of these pulses will act as the pulses that open and close the gate circuit of the counter. The accuracy to which the time interval between pulses produced by the time-base generator is known will determine the accuracy to which the frequency can be measured. Typically, this accuracy is one part in 10^8. After a short time, the counter is reset and another two pulses from the time-base generator again open and close the gate circuit. The frequency of this reset cycle is called the *sampling rate* of the counter.

If longer intervals than one second are required for counting quantities other than frequency, the opening and closing of the gate circuit can be achieved by means of an external time-base generator. Likewise, even higher-accuracy time intervals can be achieved by applying external, high-accuracy time-base generators.

The digital counting and display assembly uses one of three types of electronic display devices—indicating tubes, in-line readout indicators, or light-emitting diodes (LEDs). The indicating tubes each have ten filaments which are shaped in the form of the digits 0 through 9. As a voltage corresponding to the value of a digit is applied to the tube, the appropriate filament is made to glow and indicate the quantity. Figure 14-8 shows a photograph of one type of indicating tube and its schematic representation. The in-line readout indicator is made up of banks of ten lamps, each bank being numbered from 0 to 9. A switch controls the lamp which becomes lit and indicates the quantity being measured. The block diagram of the frequency counter shown in Fig. 14-6 shows an in-line readout indicator.

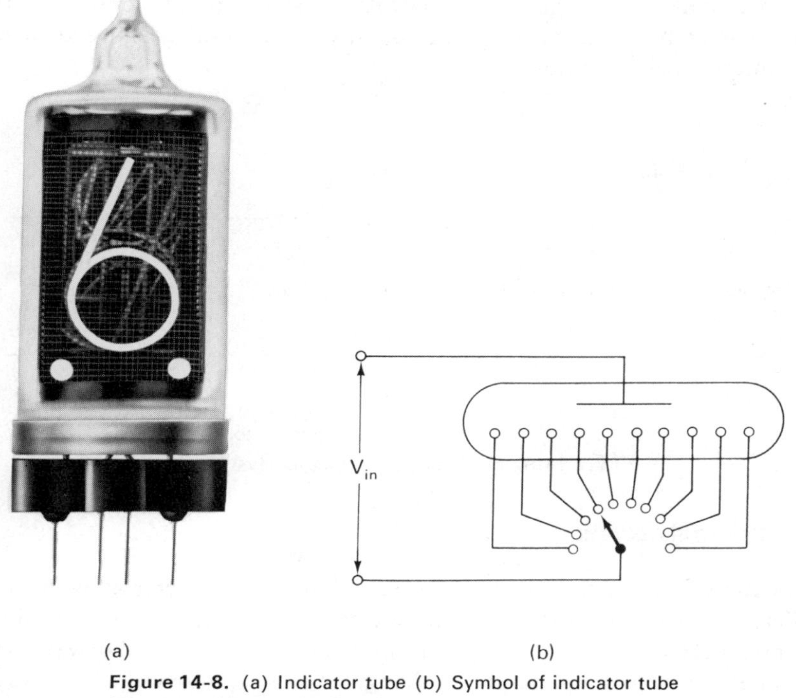

(a) (b)

Figure 14-8. (a) Indicator tube (b) Symbol of indicator tube (Courtesy of Burroughs Corp.)

Light-emitting diodes are semiconductor devices which give off light when excited by an electric current. When such diodes are placed in particular arrays, all ten digits can be constructed by applying electric current to the appropriate diodes in the array. LEDs are particularly attractive because of their ruggedness, long life, small power consumption, and tiny size.

TIME-INTERVAL COUNTERS

Time intervals can also be determined with an instrument that operates very much like the digital frequency counter. In fact, since the two instruments are so similar, *universal timer-counters* are built which perform the functions of measuring both frequency and time intervals. The time-interval counter is also used instead of the digital frequency counter to accurately measure low frequencies. This is done because measuring the period of a low-frequency signal allows more counts to accumulate during one period. Thus the resolution and accuracy of the measurement are both improved.

The time-interval counters employ an oscillator whose output frequency (typically 1 MHz) is constant and very accurately known (Fig. 14-9). They count the number of cycles emitted by this oscillator during the time interval of interest. Therefore, since the time of each cycle is known, the time interval can be calculated. The gate that controls the starting and stopping of the counter is activated by pulses which signify the beginning and end of the time interval being measured.

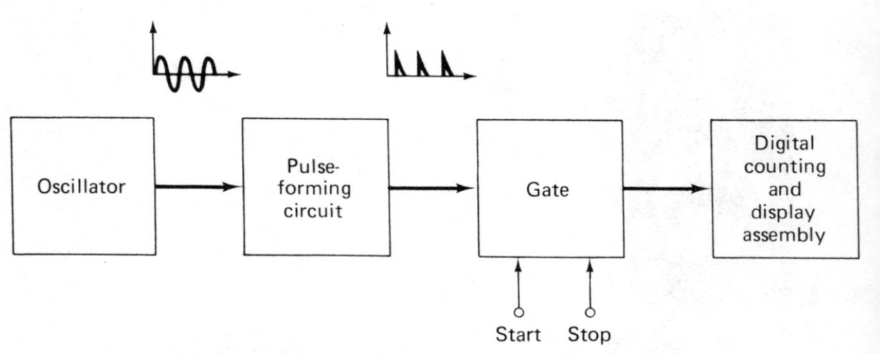

Figure 14-9. Block diagram of time-interval counter

Harmonic Analysis

In the course of this book we have encountered waveforms with many different waveshapes. Some of them were periodic waveforms (i.e., waveforms that repeat their shapes at regular intervals). These periodic waveforms included sine waves, square waves, triangular waves, half-rectified sine waves, and full-rectified sine waves. There are also many other irregularly shaped, but still periodic, waveforms. Although each of these waveforms can be plotted on a graph vs time as one signal waveform (or displayed as such on an oscilloscope screen), the same signal waveform can be decomposed into a number of sinusoidal waveform components. Conversely, any periodic waveform can also be constructed by adding together the waveforms of a particular group of sine-wave components. (These components are known as the *harmonics* of the waveform.) Of course the amplitudes and frequencies of each of the components must be of the proper value for the group of components to mesh together to form the waveform which is to be reconstructed. The generalized mathematical procedure for determining the amplitudes and frequencies of these harmonics was developed by the French mathematician, Fourier. We will not present the mathematics of this procedure (called *Fourier analysis*), but we will content ourselves with a description of its qualitative ideas.

As an example that shows that an actual waveform can be decomposed, or analyzed, into a number of harmonic components, let us consider the square wave shown in Fig. 14-10. The first four harmonic components of the waveform are shown in the column on the left. The column on the right indi-

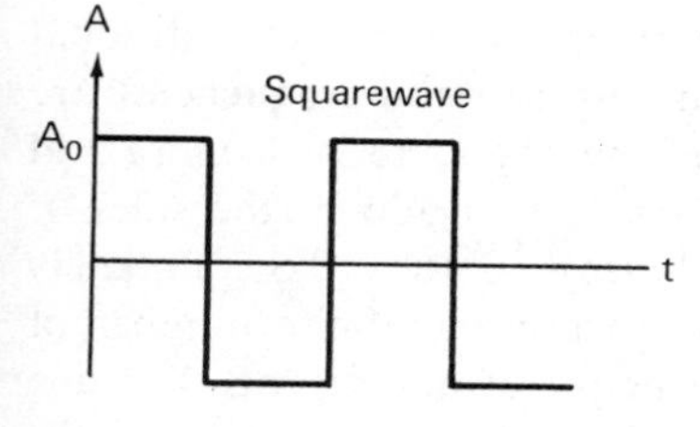

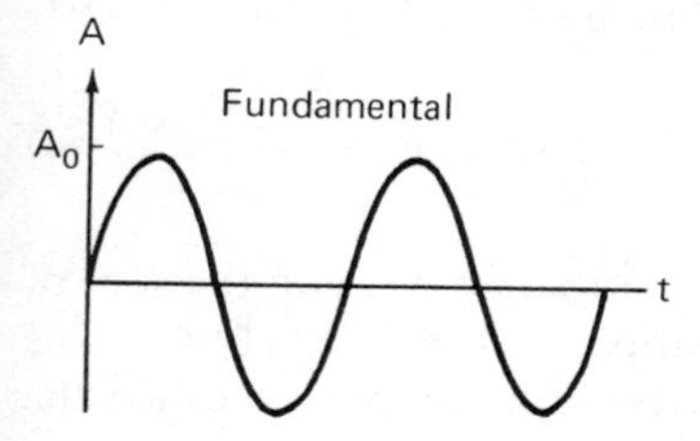

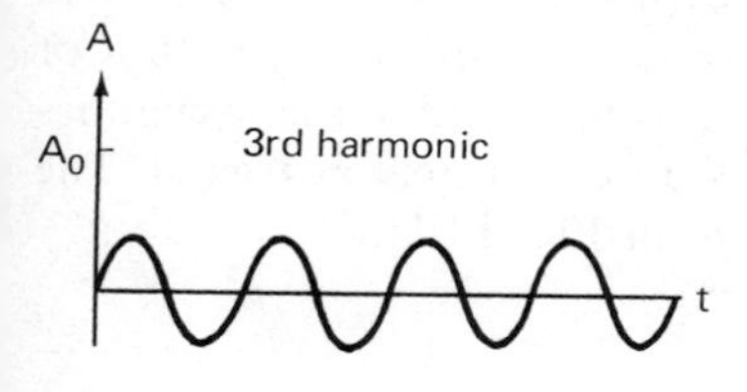

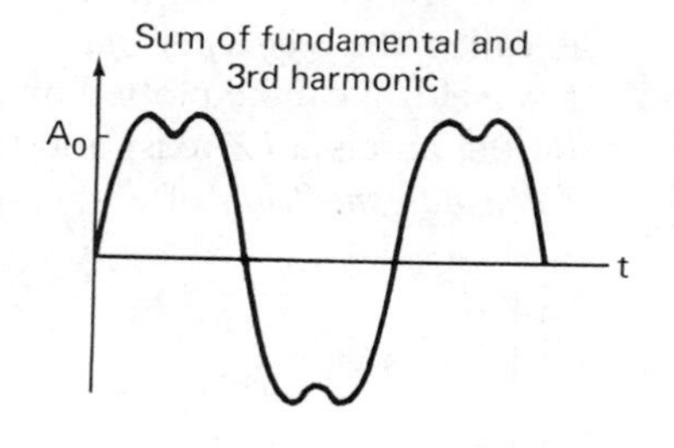

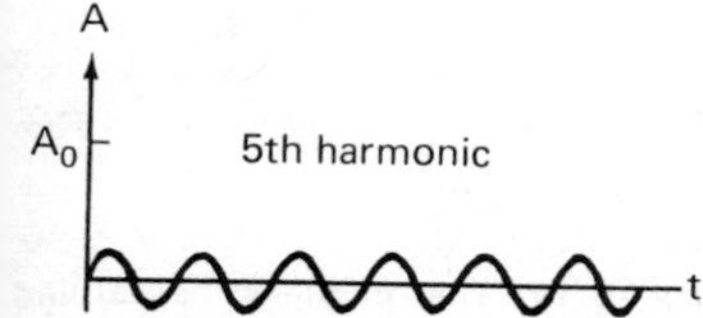

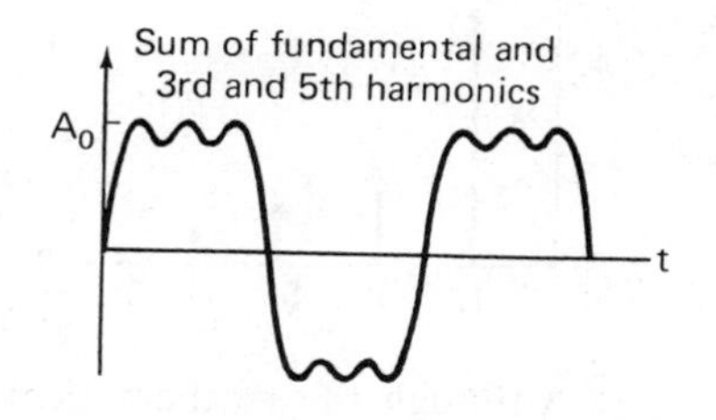

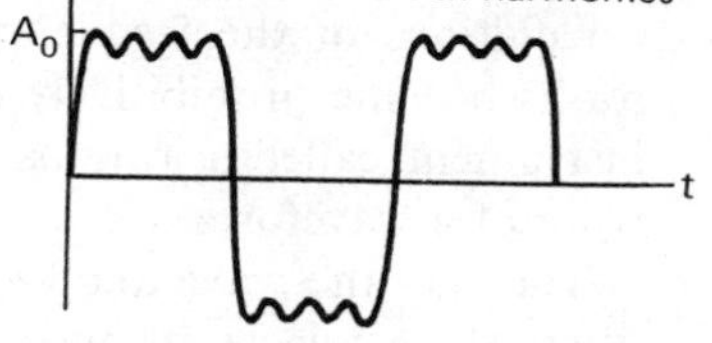

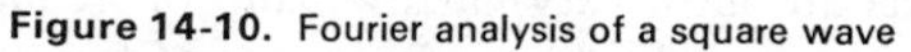

Figure 14-10. Fourier analysis of a square wave

cates how the shape of the *fundamental* component changes as additional harmonic components of the appropriate amplitudes and frequencies are added to it. We see that the composite waveform begins to look more and more like a square wave as each higher harmonic is added (the sides of waveform become steeper and the top and bottom, flatter). To completely reconstruct the square wave, we would have to use an infinite number of harmonics. However, we can see that the principle of adding together sine waves to get a square wave is not at all far fetched. The equation of the harmonics of a square wave having an amplitude of +1 is given by

$$y = \frac{4}{\pi}\left(\sin \omega t + \frac{1}{3}\sin 3\omega t + \frac{1}{5}\sin 5\omega t + \frac{1}{7}\sin 7\omega t + \cdots\right) \tag{14-4}$$

The same procedure can be applied to any periodic wave, and the sum of the waveforms can be expressed in an equation such as Eq. (14-4). This equation, made up of an infinite sum (or series) of harmonics, is called the *Fourier series* of the waveform.

The frequency composition of a signal as expressed by the Fourier series is called the *frequency spectrum* of a signal. Such a frequency spectrum of a waveform can be plotted on a graph, with the frequencies of the harmonics on the abscissa (*X* axis) and their amplitudes on the ordinate (*Y* axis). The *frequency spectrum* of a square wave is shown in Fig. 14-11.

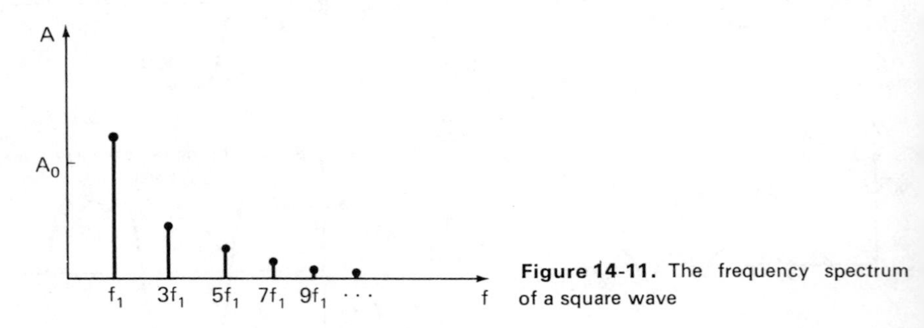

Figure 14-11. The frequency spectrum of a square wave

Although the mathematical calculation for a regularly shaped waveform (such as a square wave) can be performed rather easily, actual waveforms encountered in systems rarely have such mathematical simplicity. The calculations of the frequency spectrums of such irregularly shaped waveforms become prohibitively complex and thus are rarely made. Instead, an instrument called a *wave analyzer* is used to determine the frequency spectrum of a waveform.

The way the wave analyzer is able to perform this analysis is by using a filter which rejects all but a very narrow band of frequencies (Fig. 14-12). The filter is tunable (i.e., the center frequency of the narrow band can be varied, while the frequency interval is maintained). By sweeping the frequency

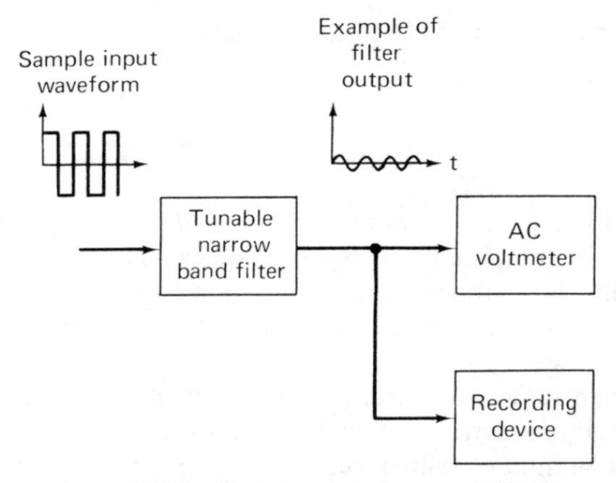

Figure 14-12. Block diagram of a wave analyzer

range of the instrument, the frequencies of the harmonics of the waveform applied to the wave analyzer, and their amplitudes, can be determined.

An ac voltmeter is used to monitor the signal at the output of the filter. Thus the voltmeter responds only to the rms voltage of the narrow band of frequencies coming through the filter at a given setting. The amplitudes of the various harmonic components of which the test signal is composed are thus experimentally determined, yielding the frequency spectrum. A 50-kHz wave analyzer allows frequencies to pass which are within about 3 to 50 Hz of the selected frequency. Higher-frequency wave analyzers use filters which generally sample 1 percent of the analyzer's range at a given setting. The output of the wave analyzer can be a meter, a recorder, or a frequency counter.

Wave analyzers are used in sound and vibration analysis as well as in electrical measurements. For example, the sounds or vibrations produced by a machine or aircraft are analyzed to help determine the origin of each sound component. By understanding the cause of the vibration, it becomes easier to develop methods for reducing the overall vibrational or sound level. Wave analyzers are also utilized to decompose such complex waveforms as those produced by the electrical activity of the human brain. The analysis of the purity of the response of an amplifier to sine waves of specific frequencies can likewise be performed by a wave analyzer. If a sine wave of one frequency is fed to the amplifier, the extent of higher harmonics at the output of the amplifier can be determined by the wave analyzer. If no output voltages are measured at any frequency other than the frequency under test, there would be no *harmonic distortion.*

DISTORTION ANALYZERS

The harmonic distortion caused by a circuit or electronic device is defined to be the ratio of the total portion of the output signal produced by the harmonics to the portion of the output signal at the fundamental frequency

$$\text{Harmonic distortion } (\%) = \frac{\left[\sum_{n=2}^{\infty} A_n^2\right]^{1/2}}{A_1} \tag{14-5}$$

In this expression A_1 is the amplitude of the *fundamental* and A_n (from $n = 2$ to $n = \infty$) are the amplitudes of the harmonics. As an example, if the rms value of the signal due to the harmonics is 3 mV while the amplitude of the output due to the fundamental was 100 mV, the harmonic distortion would be 3 percent.

The instrument used to measure the total harmonic distortion caused by amplifiers or other electronic equipment is called a *distortion analyzer*. It consists of an amplifier which suppresses the signal at the fundamental frequency and amplifies all the others. A Wien bridge circuit is used in the instrument as the *rejection filter*, that is, the Wien bridge circuit allows all the harmonics to be passed and amplified and fed to a voltmeter. The voltmeter indicates the rms value of the total signal due to the harmonics.

Distortion analyzers are used for fast, quantitative determinations of the *total distortion* in a waveform. Wave analyzers provide detailed information concerning each harmonic component of a test waveform.

Problems

1. Find the period of the periodic waveform whose frequencies are:
   a) 60 Hz
   b) 1000 Hz
   c) 50 MHz
2. Find the frequency of a repeating waveform with a period of
   a) 20 s
   b) 40 μs
3. Find the angular frequency of a sinewave whose frequencies are
   a 60 Hz
   b) 100 kHz
4. Find the frequency of a sinewave whose angular frequency is 314 radians/s.
5. List the applications and advantages of each of the following time-measuring devices:
   a) electric timers,
   b) electronic timers,
   c) universal timer-counters.
6. An electronic timer consists of a resistor whose value is 100 kΩ, a capacitor of 25 pF, and a 10 V source which is applied across the RC connection at time t = 0. How much time will have elapsed if the voltage across the capacitor is 8.6 V at the instant the applied voltage is interrupted?

7. Describe the role of the mixer in the zero-beat frequency meter.
8. In what type of circumstances might a grid-dip meter be used to measure frequency?
9. Describe the three most commonly used display methods used in digital frequency meters.
10. Explain the operation of a simple timer-counter with the help of a block diagram. Explain the function of each block in the diagram.
11. Calculate or look up the Fourier series of a half-wave rectified sinewave. Draw the fundamental and the first four harmonics on a graph using the same scale on each graph. Finally, graphically add the waveforms of each of the harmonics together.
12. Draw the frequency spectrum of the first five harmonic components of the half-wave rectified sinewave.
13. Define the terms a) distortion analyser, b) rejection filter.
14. Under what circumstances are harmonic analysers and distortion analysers used, respectively.

References

1. MILLMAN, JACOB, and TAUB, HERBERT, *Pulse, Digital, and Switching Waveforms*. New York, McGraw-Hill Book Co., 1965.

15

Electrical Transducers

Transducers are broadly defined as being devices which convert energy or information from one form to another. They are widely used in measurement work because not all quantities that need to be measured can be displayed as easily as others. A better measurement of a quantity can usually be made if it can be converted to another form which is more easily or accurately displayed. For example, the common mercury thermometer converts changes in temperature to changes in the length of a column of mercury. Since the change in the length of the mercury column is rather simple to measure, the mercury thermometer becomes a convenient device for sensing changes in temperature. On the other hand, the actual temperature change is not as easy to display directly.

The purpose of this chapter will be to introduce some of the transducers that convert physical quantities into electrical signals (as well as a few transducers that convert them to nonelectrical forms). Electrical transducers (i.e., microphones, loudspeakers, etc.) make up the vast majority of transducers in use today. The reason for their popularity is that electrical signals possess many desirable measurement traits. The first of these is that electrical instrumentation is so highly developed that there are usually several different methods for converting most physical quantities into electrical signals. Next, if weak signals are converted to an electrical form, they can be faithfully amplified until their amplitudes become large enough to be easily displayed. Finally, there are display and recording devices that can follow very-high-frequency variations in electrical signals. Thus a nonelectrical quantity which also has a high-frequency variation (e.g., the vibrations of a solid) can be converted to an electrical form and accurately monitored.

When we discuss the various types of transducers, a means for classifying them must be selected. For the most part, our method will be to classify transducers according to the type of quantity to which they are designed to respond.[1] The two exceptions to this procedure will involve the discussion of the *strain gauge* and the *linear variable differential transformer*. These two transducers are used for such a wide variety of measurements that we will treat them as multipurpose devices (rather than as single-purpose transducers which are capable of converting only one or two physical quantities to an electrical signal). However, the discussion of this chapter is not meant to be an examination of *all* possible electrical transducers. Instead, its aim is to introduce several of the most commonly used varieties and thereby alert the reader to some of the ways that measurement possibilities can be expanded through the use of transducers.

ROLE OF TRANSDUCERS IN MEASUREMENT SYSTEMS

There are, in general, three major elements which are common to most measuring systems (Fig. 15-1). The first of these is the *detecting element* (or *sensor*). The purpose of the detecting element is to respond to the magnitude (or changes in the magnitude) of the quantity being measured. The response of the sensor takes the form of an output signal whose magnitude is proportional to the magnitude of the quantity being measured.

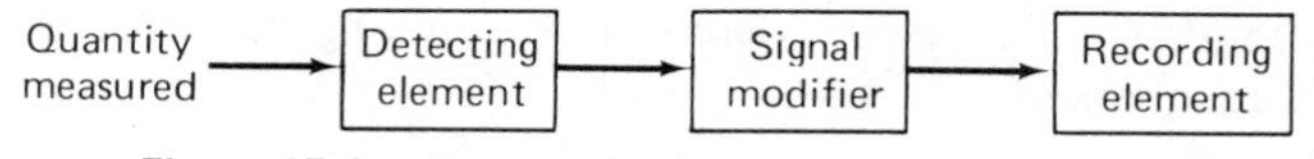

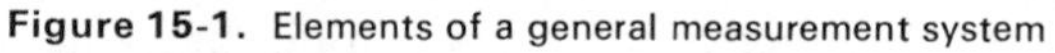

Figure 15-1. Elements of a general measurement system

The second element is the *signal modifier*. This element receives the output signal of the detecting element and modifies it by amplification or by suitable shaping of its waveform. When the signal emerges from the signal modifier, it should be in a form which is appropriate for display or recording. The third element of measurement systems is the *display* or *recording device*. In electrical systems, display or recording devices include such instruments as meters, cathode ray tubes, chart recorders, tape recorders, X-Y recorders, and digital computers.

If the measurement system is one in which a nonelectrical quantity is to be measured by converting it into an electrical form, an electrical transducer

[1] Another method groups transducers according to the type of electrical parameter that undergoes changes in the process of producing the transducer output signal.

is used as the detecting element.[2] If the electrical transducer produces a signal without requiring an electrical excitation, it is called an *active* transducer. If the transducer is capable of producing an output signal only when it is used in connection with an excitation source, the transducer is called a *passive* transducer. A complete transducer system includes the transducer and the voltage excitation source if one is required.

GUIDELINES FOR SELECTING AND USING TRANSDUCERS

When a measurement of a nonelectrical quantity is to be undertaken by converting the quantity to an electrical form, an appropriate transducer (or combination of transducers) for carrying out this conversion must be selected. The first step in the selection procedure is to clearly define the nature of the quantity that is to be measured. This awareness should include a knowledge of the range of magnitudes and frequencies that the quantity may be expected to exhibit. When the problem has thus been stated, the available transducer principles for measuring the desired quantity must be examined.[3] If one or more transducer principles are capable of producing a satisfactory signal, we must decide whether to use a commercially available transducer or undertake to build the transducer. If commercially manufactured transducers are available at a suitable price, the choice will probably be to purchase one of them. On the other hand, if no transducers are made which can perform the desired measurement, one may have to design, build, and calibrate his own device.

When the specifications of a particular transducer are examined, the following points should be considered in determining its suitability for a particular measurement:

1. *Range*. The range of the transducer should be great enough to encompass all the expected magnitudes of the quantity to be measured.
2. *Sensitivity*. To yield meaningful data, the transducer should produce a sufficient output signal per unit of measured input.
3. *Electrical output characteristics*. The electrical characteristics (such as

[2] Sometimes the initial transducing element of an electrical measurement system converts the measured quantity into a nonelectrical form. In such cases, the nonelectrical output signal of the first transducer can be fed to a second transducer. This second transducer is an electrical transducer which converts the signal to an electrical form. One example of such a two-transducer arrangement is the Bourdon tube and linear variable differential transformer (LVDT) which is often used to measure fluid pressure. The Bourdon tube converts fluid pressure to a mechanical displacement. An LVDT converts mechanical displacements to an ac electrical signal. Thus, if the LVDT is connected to the Bourdon tube, pressure can be converted to an ac electrical signal.

[3] A fine reference which presents a more complete exposition of the subject than is possible in the present chapter is a text by H. N. Norton, *The Handbook of Transducers for Electronic Measuring Systems* (Englewood Cliffs, N.J.: Prentice-Hall, Inc., 1969).

the output impedance, the frequency response, and the response time) of the transducer output signal should be compatible with the recording device and the rest of the measuring system equipment.

4. *Physical environment*. The transducer selected should be able to withstand the environmental conditions to which it may be subjected while making the test. Such parameters as temperature, moisture, and corrosive chemicals might damage some transducers but not others.
5. *Errors*. The errors inherent in the operation of the transducer itself, or those errors caused by environmental conditions of the measurement, should be small enough or controllable enough that they allow meaningful data to be taken.

Once the transducer has been selected and incorporated into the measurement system design, the following guidelines should be observed to increase the accuracy of the measurements:

1. Transducer Calibration—The transducer output should be calibrated against some known standards while it is being used under actual test conditions. This calibration should be performed regularly as the measurement procedes.

2. Changes in the environmental conditions of the transducer should be monitored continuously. If this procedure is followed, the measured data can later be corrected to account for any changes in environmental conditions.

3. By artificially controlling the measurement environment, we can reduce possible transducer errors. Examples of artificial environmental control include the enclosing of the transducer in a temperature-controlled housing and isolating the device from external shocks and vibrations.

Strain Gauges

The *strain gauge* is one of the most commonly used electrical transducers. Its popularity stems from the fact that it can detect and convert *force* or *small mechanical displacements* into electrical signals. Since many other quantities such as torque, pressure, weight, and tension also involve force or displacement effects, they can also be measured by strain gauges. Furthermore, if the mechanical displacements to be measured have a time-varying form (such as vibrational motion), signals with frequencies of up to 100 kHz can be detected.

Strain gauges are so named because when they undergo a *strain* (defined to be a fractional change in linear dimension caused by an applied force), they also undergo a change in electrical resistance. (The strain takes the form of a lengthening of the special wire from which the gauge is constructed.) If the wire is bonded to a thin paper or plastic base, the gauge is called a *bonded strain gauge* (Fig. 15-2(a)). This type of strain gauge is used to detect displacements caused by large forces. The bonded strain gauge is cemented with a special adhesive to the structure being measured. The adhesive must hold the gauge firmly onto the structure and yet be able to "give" under the strain without cracking. For some applications, the adhesive must also be resistant to humidity, temperature, and other extreme environmental conditions.

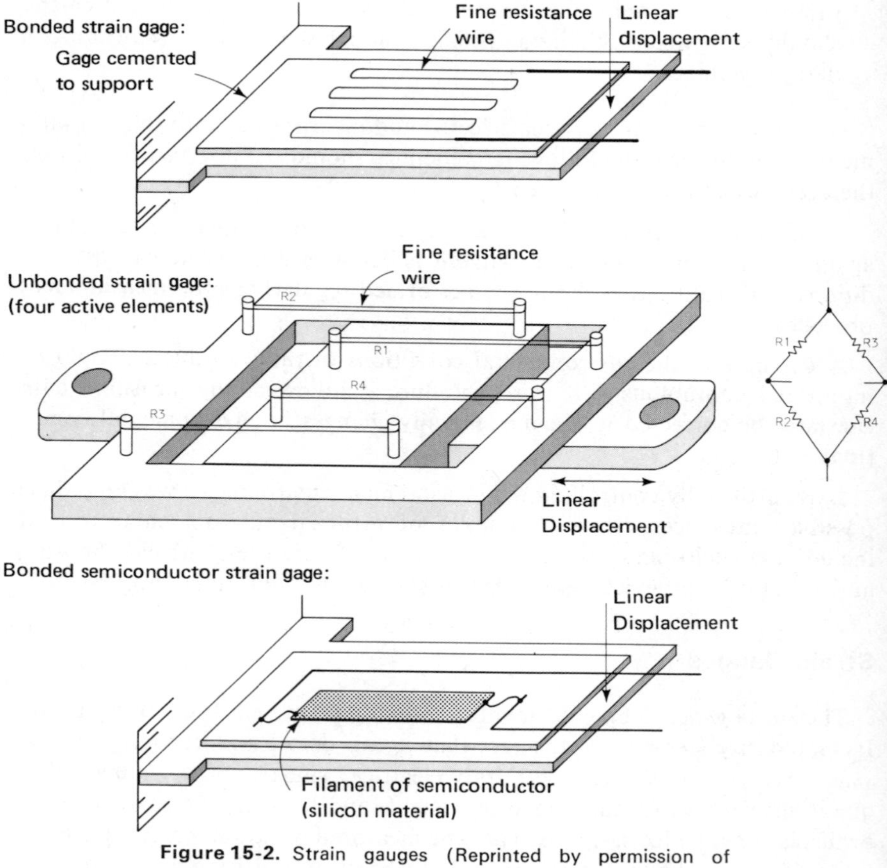

Figure 15-2. Strain gauges (Reprinted by permission of Tektronix, Inc.)

When a force is applied to the structure to which the gauge is attached, the entire base of the gauge is stretched. If the direction of the force is applied along the appropriate direction, the wire bonded to the base is stretched by a length Δl in each of its *fingers*. The magnitude of the overall extension is proportional to the force exerted on the gauge. Since it is known that the resistance of a wire will increase when it is stretched in length,[4] the change in the resistance of the strain gauge can provide an indication of the force causing the extension. However, the magnitude of the resistance change of bonded gauges is only on the order of 0.1 percent of the unstrained resistance value. (The initial resistance of metal-wire strain gauges usually lies between 120 Ω and 400 Ω.) Therefore the output signal from a strain gauge must be monitored by a carefully designed Wheatstone bridge circuit. Figure 15-3 is a photograph of an instrument used to detect strain gauge signals. Bonded strain gauges are manufactured in sizes from about $\frac{1}{8}$ in × $\frac{1}{8}$ in to a maximum of 1 in × 1 in.

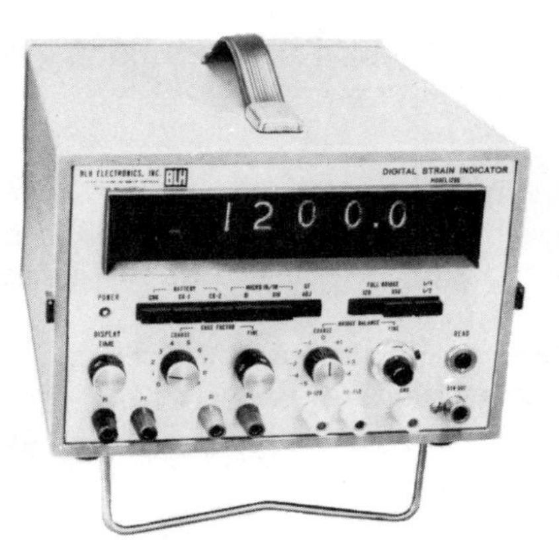

Figure 15-3. Strain gauge indicator (Courtesy of BLH Electronics)

One of the difficulties associated with the use of strain gauges is that their resistance also depends on temperature. Therefore a temperature change in the surrounding environment (or induced by the stress) can cause the gauge to produce an erroneous output signal. To counteract this temperature effect, a second identical gauge can be mounted on the same structure in a direction perpendicular to the direction of the force. (Figure 15-4 shows how this is done in a typical load cell used to measure heavy loads.) Each gauge can then be used as the resistance element of one arm of a resistance bridge. Any resistance changes due to temperature will be equal in both gauges, and the bridge will remain balanced. (Any strain effects will take place only in gauge 1 and not in gauge 2.) In this manner, variations due to temperature can be eliminated as a source of error.

Since bonded strain gauges require a great deal of force to measurably change their dimensions, smaller forces must be measured with strain gauges of a different design. One design which produces more sensitive devices is used in the so-called *unbonded* strain gauges. Their name is derived from the

[4]The increase in resistance is due to both the increased length and the stress exerted on the material.

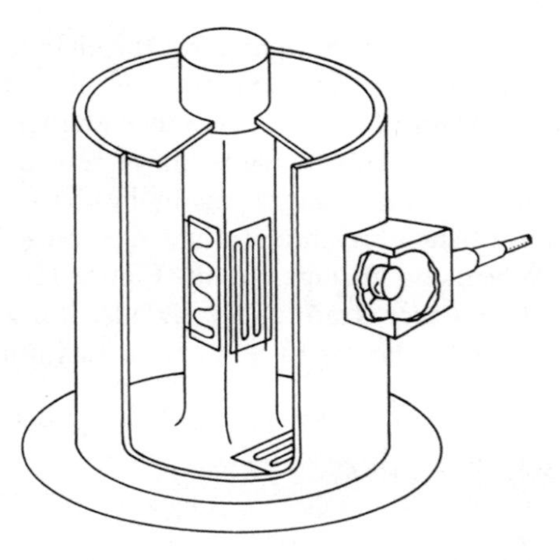

Figure 15-4. Load cell showing how strain gauges can be mounted to compensate for temperature changes (Courtesy of BLH Electronics)

fact that the wires of the gauge are attached between a moving and a fixed support, and thus a smaller force is needed to change the wire's length (Fig. 15-2(b)). Unbonded strain gauges are capable of detecting forces between 0.15 oz and 5 lb.

The ratio of the fractional change of resistance ($\Delta R/R$) to the fractional change of length of the wire ($\Delta l/l$) provides a measure of the sensitivity of strain gauges. This ratio is called the *gauge factor* (GF).

$$\text{GF} = \frac{(\Delta R/R)}{(\Delta l/l)} \tag{15-1}$$

A typical GF is about 2 to 3 for metal-wire strain gauges. However, if the strain gauge is built from special semiconducting materials, the associated GF can be as high as 130 (Fig. 15-2(c)). Thus the semiconductor strain gauge is suitable for measuring smaller strains than those which can be measured with metal wire gauges. Unfortunately, semiconductor strain gauges are much more sensitive to temperature fluctuations than metal-wire gauges (even though they are specially designed to be less temperature-sensitive than ordinary semiconductors). As a result, they must be used in systems with effective temperature-compensation designs.

EXAMPLE 15-1

The wire in a strain gauge is 10 cm in length and has an initial resistance of 150 Ω. If an applied force causes a change in its length of 0.1 mm and a change in resistance of 0.3 Ω, find the gauge factor of the device and the strain caused by this force (assume that the increase in resistivity with length is linear).

Solution: (a) The strain in found from $S = \Delta l/l$ or

$$S = \frac{0.1\text{ mm}}{10\text{ cm}} = \frac{10^{-4}\text{m}}{10^{-1}\text{m}} = 0.001$$

(b) The gauge factor is found from Eq. (15-1):

$$\mathrm{GF} = \frac{(\Delta R/R)}{(\Delta l/l)} = \frac{.002}{.001} = 2$$

Linear Variable Differential Transformers

The second general purpose transducer is the linear variable differential transformer (abbreviated LVDT). Like the strain gauge, it produces an electrical signal which is linearly proportional to mechanical displacements.[5] The displacements detectable by LVDTs are relatively large compared to those detectable by strain gauges. Thus LVDTs are suitable for use in applications where the displacements are too large for strain gauges to handle. (Conversely, strain gauges are generally more suitable for the smaller displacements.) For example, LVDTs can detect displacements that range from microinches to inches. Since the LVDT can also be connected to other transducers whose outputs are a mechanical displacement, they are often used together with other transducers (as well as alone).

The LVDT senses displacements by the motion of the ferromagnetic core within a special transformer (see Fig. 15-5(a)). The transformer has one primary winding and two secondary windings. All three of these windings are wound on the same hollow insulating tube. The primary winding is wound at the center of the tube, and the two secondary windings (which have an equal number of turns) are connected in series-opposition. This means that if the mutual coupling between each secondary winding and the primary winding is equal, the voltage, V_o, across the secondary winding wires will be zero, even when the primary is excited by an ac signal.

If the ferromagnetic core is centered (with respect to the length of the transformer), the mutual coupling between each secondary winding and the primary will be equal. As long as this condition exists, $V_o = 0$. However, if the core is moved from this centered position, the mutual coupling between each secondary winding and the primary will no longer be equal. For example, if the core in Fig. 15-5(b) is moved to the right, the mutual coupling between secondary winding No. 2 and the primary will increase, while the mutual coupling between secondary winding No. 1 and the primary will decrease. A shift of the core position to the left will have the opposite effect. As a result of the changes in the mutual coupling, the voltage, V_o, across the output wires connected to the secondary windings will no longer be zero. Instead (for small displacements of the core), this output voltage will be linearly proportional to the magnitude of the displacement.

The *sensitivity* of an LVDT is rated in mV/0.001 in. Its actual output voltage is thereby found by multiplying the sensitivity, the displacement,

[5]Angular displacements can be measured by a similar device called a rotary variable differential transformer (RVDT).

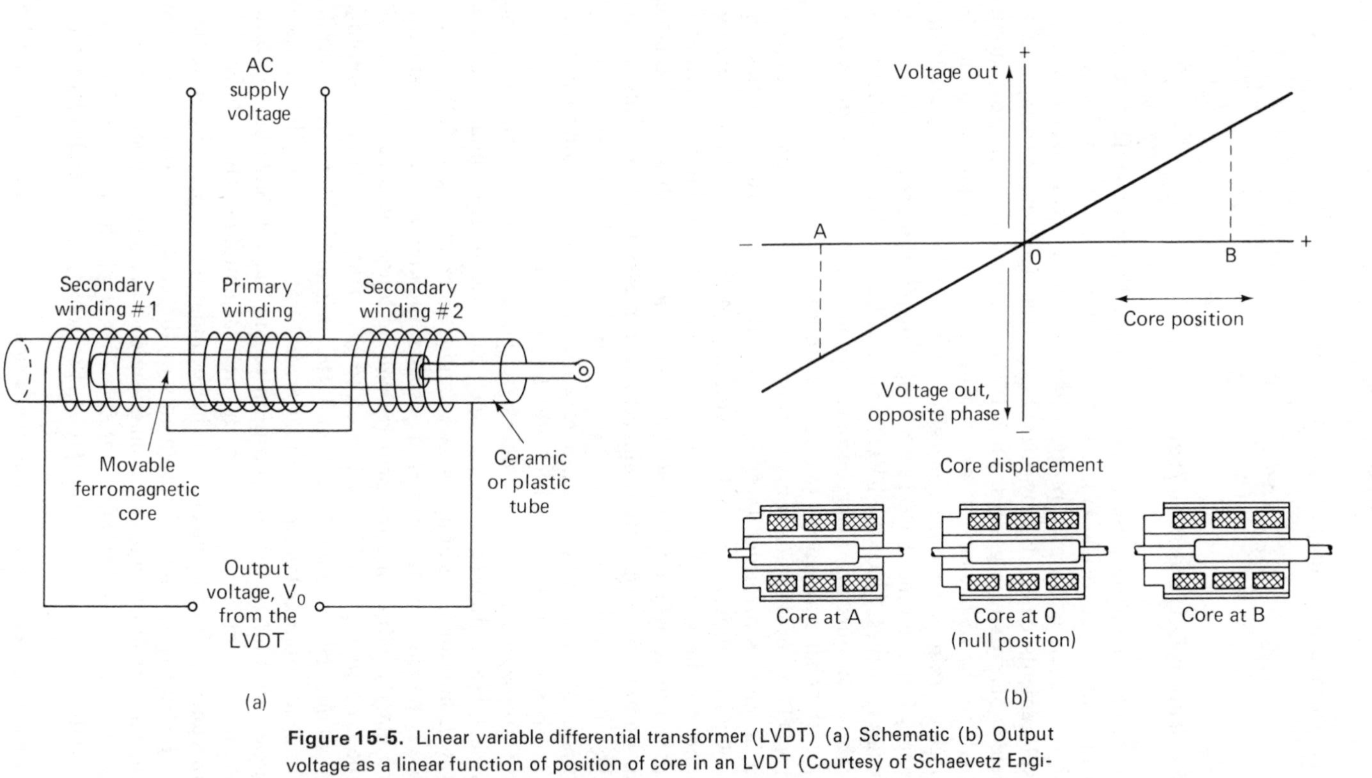

Figure 15-5. Linear variable differential transformer (LVDT) (a) Schematic (b) Output voltage as a linear function of position of core in an LVDT (Courtesy of Schaevetz Engineering)

and the rms value of the input voltage. As a result, the output voltage can be several volts or more. This gives the LVDT a large voltage signal output compared to many other transducers. In addition, LVDTs provide an output whose resolution is continuous.

LVDT ACCELEROMETER

As an example of the wide variety of uses in which the LVDT is utilized, we will examine the LVDT accelerometer. Such accelerometers are used to detect earthquakes and to measure missile accelerations. (Other accelerometers which measure shock and vibration primarily use piezoelectric crystal transducers. Piezoelectric crystals are discussed in a later section.)

A form of the LVDT accelerometer is shown in Fig. 15-6. We see that the magnetic core of the LVDT is connected by two cantilever springs to a larger external piece of equipment. If this piece of equipment is accelerated in the direction shown, the core undergoes a force proportional to the acceleration and therefore bends the cantilever springs. The change in the position of the core is thereby proportional to the acceleration, and this shift yields a voltage signal which is also proportional to the acceleration.

A typical missile accelerometer has a rated output voltage of 2 V_{rms}/g with a 115-V, 400-Hz input voltage.

OTHER POSITION AND VELOCITY TRANSDUCERS

Although the strain gauge and the LVDT produce accurate indications of position and velocity, there are other devices which also find use as transducers of these quantities. The two most common ones are the linear-motion potentiometer and the linear-motion variable inductor. These two transducers are simpler, cheaper, and easier to use than the strain gauge and the LVDT. However, their accuracies and sensitivities are not as high, and this limits their use in many cases.

The linear-motion potentiometer is shown in Fig. 15-7. We see that it is a variable-resistance device whose resistance value is changed by the motion of the slider along the resistance element. The slider is connected to an arm which couples the motion being measured to the transducer. If the centered position of the slider corresponds to the zero-value position, a resistance change will accompany any change of position in either the positive or negative direction. A Wheatstone bridge can be used to measure these resistance changes.

The linear-motion inductor works on a similar principle as the linear-motion potentiometer. However, the moving element in this device is a magnetic core which is placed inside the coil of an inductor. As the core is moved in relation to the coil, the inductance value of the inductor changes. The change in inductance can be monitored by an impedance bridge to indicate changes in position.

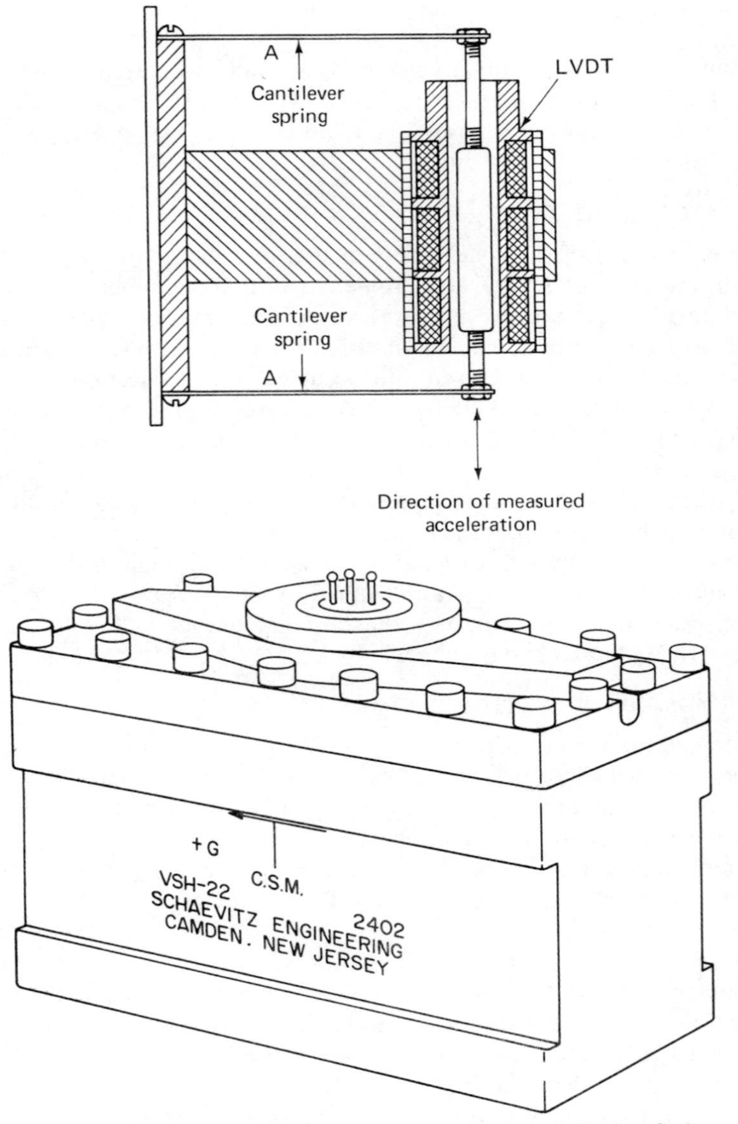

Figure 15-6. LVDT accelerometer (Courtesy of Schaevetz Engineering)

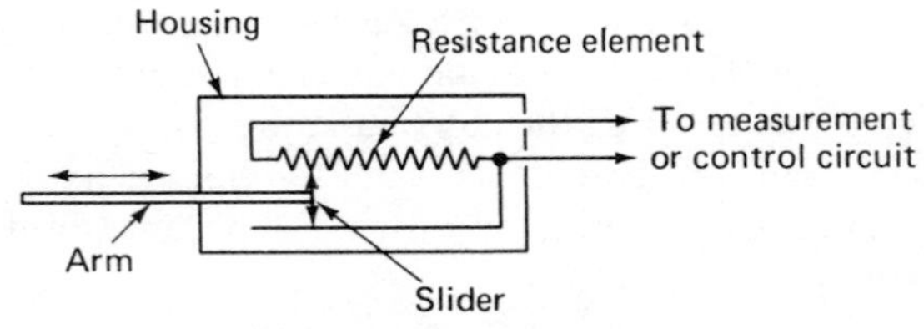

Figure 15-7. Linear motion potentiometer

Fluid-Property Transducers (Pressure and Flow Rate)

Since they both act like fluids in many respects, the most common properties of liquids and gases can usually be monitored by using the same type of transducers. This is especially true of the properties of pressure and flow rate.

FLUID PRESSURE

The quantity, *pressure*, can be described in many ways. If the value of pressure is being described relative to vacuum, this type of pressure is called *absolute pressure*. When the value of pressure is compared to the absolute pressure of air at sea level, the type of pressure is referred to as *relative pressure*. If the pressure of interest is the difference of the pressures of two fluids (or the difference of the pressures of the same fluid in different parts of a system), what is being described is the *differential pressure*.

Pressure can be electrically measured if it directly changes an electrical parameter (such as capacitance). It may also be measured if it produces a mechanical displacement. The mechanical displacement can then be made to activate a linear-displacement transducer, thereby causing an electrical signal.

A variable-capacitor transducer is shown in Fig. 15-8(a). It is very similar in operation to the capacitor microphone (to be described later). The reference pressure of the transducer of this example can be atmospheric pressure

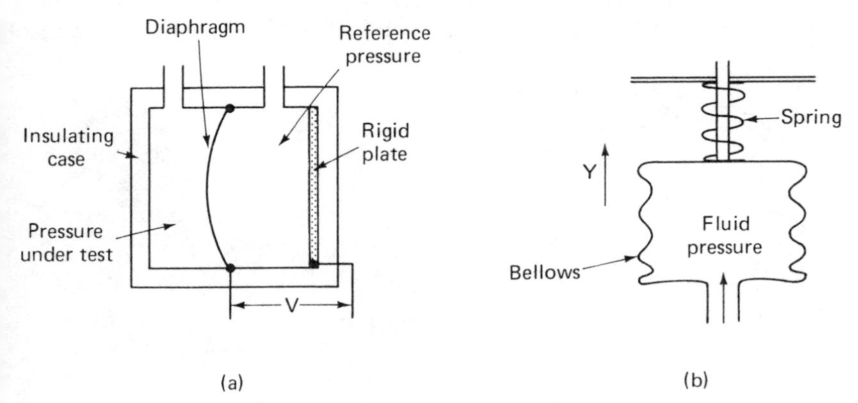

Figure 15-8. Fluid-pressure transducers (a) Capacitor pressure transducer (b) Bellows pressure transducer

(for relative pressure measurement), a vacuum (for absolute measurements), or a fluid from the second pressure of interest (for differential pressure measurements).

A metal diaphragm within the capacitor transducer moves closer or farther away from a rigid plate and thereby causes a change in the capacitance of the structure. If the capacitance value is made part of an oscillator circuit, the frequency of the oscillator will change as the capacitance value changes. The frequency changes can be monitored to indicate the pressure change.

The capacitor pressure transducer is one of the most rugged and accurate transducers available for measuring pressure. It can be built to respond to a wide range of pressure values as well as to high-frequency pressure changes.

The devices which are used to convert pressure into a mechanical displacement are made in many ways. We will mention only a few of the most common ones. The first is a flexible bellows like the one shown in Fig. 15-8(b). The fluid is allowed to enter the bellows, and its pressure extends them in the Y-direction. In low-pressure bellows, the external spring shown in Fig. 15-8(b) is not present. Then the springiness of the bellows alone is used to resist the pressure. For higher pressures, an external spring is used to add its restraining force to the force against the pressure. The extension of the bellows due to the force moves a rod which is connected to a position transducer. The position transducer converts the displacement to an electrical signal. Depending on the design of the bellows and springs, relative and absolute pressures can be measured with this device.

Another common pressure-to-displacement transducer is the Bourdon tube (shown in some of its various forms in Fig. 15-9). The Bourdon tube is a flat, hollow tube that is curled into a spiral or helical shape. When a fluid under pressure is introduced into it, the tube tries to straighten out. The extent of the straightening is proportional to the pressure. For low pressures, a simple shape like that shown in Fig. 15-9(a) is used. For higher pressures, the tube is wound into a helical shape (Fig. 15-9(b)). Since the output of

Figure 15-9. Bourdon tubes (a) Simple bourdon tube (b) Helical-spiral bourdon tube (Courtesy of the Foxboro Company)

the Bourdon tube is a mechanical displacement, the end of the tube must be connected to an additional electrical transducer which will convert the displacement to an electrical signal.

The *pressure cell* and *pressure transducer* shown in Fig. 15-10 use strain

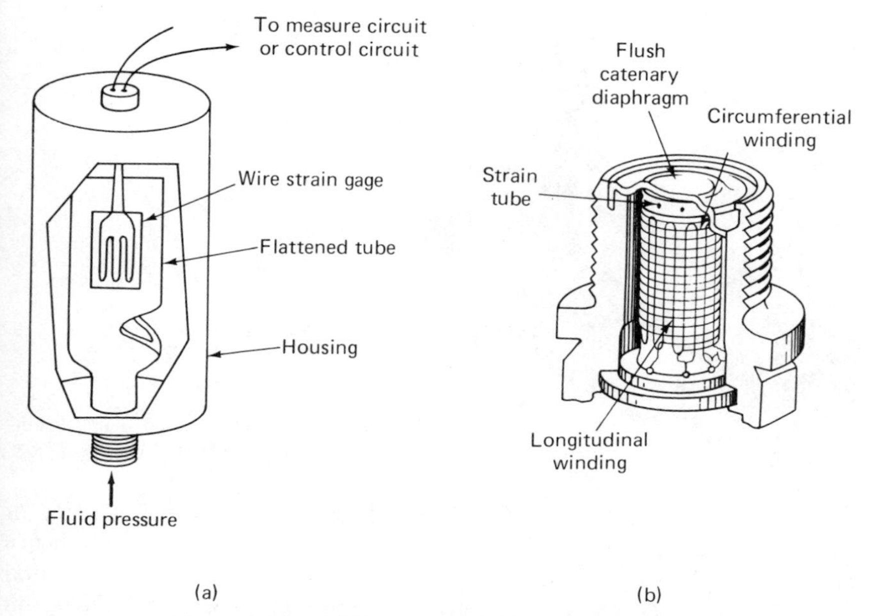

Figure 15-10. (a) Pressure cell (b) Pressure transducer

gauges to measure the effects of pressure on a sealed metal tube. In the pressure cell, the fluid causes a sealed tube to expand. A strain gauge mounted on the surface of the tube senses the extent of the expansion by changing its resistance value. The *pressure transducer* is a cylindrical tube with strain gauges attached to its circumference. The pressure is applied to a diaphragm at one end of the cylinder. This pressure causes the tube to contract lengthwise while increasing in diameter. The increased diameter causes the circumferentially attached gauges to change their resistances. This type of transducer is used to measure the compression capability of the cylinders in automobile and other internal-combustion engines.

FLUID FLOW TRANSDUCERS

There are many types of instruments used to measure the rate of flow of fluids. The three we will discuss in this section are the turbine flowmeter, the magnetic flowmeter, and the hot-wire anemometer.

Turbine flowmeters are probably the most commonly used type of flowmeter (Fig. 15-11). They provide a direct method for measuring both liquid and gas flow rates. They are also particularly useful for remote monitoring and aircraft applications. The turbine flowmeter consists of a rotor mounted in a pipe through which the fluid flows. The flowing liquid causes the rotor to turn. The greater the rate of flow, the faster is the speed of rotation. The vanes of the rotor are metal, and a *magnetic pickup* mounted on the wall of the pipe

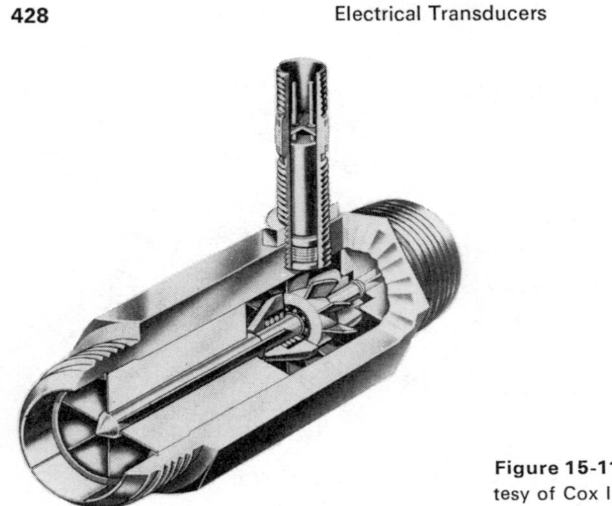

Figure 15-11. Turbine flowmeter (Courtesy of Cox Instruments Div.)

senses the passing of each vane as an electrical pulse. The frequency of the pulses is proportional to the rate at which the rotor is spinning and hence to the rate of flow of the liquid. Turbine flowmeters are available for liquid flow rates from less than 0.01 gallons per minute (gpm) to over 35,000 gpm.

The *magnetic flowmeter* is a transducer which is used to measure the flow of electrically conductive fluids. It has the advantage of not presenting any sort of obstruction to the fluid flow during the measurement. This type of flowmeter operates on the principle that a voltage is induced in a conductor when the conductor moves in a magnetic field. Since the voltage depends on the rate at which the conductor moves through the magnetic field, the strength of the voltage can be used as an indication of the rate of flow of the liquid.

The *hot-wire anemometer* is a fine resistance wire which is heated by a current passing through it. If a cooler fluid flows past the wire, heat is removed from it by the fluid. The rate of heat transfer varies with the type of fluid, but it also tends to vary as the square root of the velocity at which the fluid flows past the wire. If the current in the wire is kept constant, the change in resistance due to the cooling will yield a voltage signal. Because the diameter of the wire element can be made very small, the device can be made very sensitive and responsive to high-frequency changes in the flow rate. One of its chief uses is in aerodynamic research.

Temperature Transducers

A wide variety of transducers are used to measure temperature. Some of them convert temperature directly to an electrical signal, and others must be used in combination with an electrical transducer to convert the temperature indication into an electrical form. The most common temperature transducers include:

1. Bimetallic strips
2. Thermistors
3. Resistance thermometers
4. Thermocouples
5. Radiation pyrometers

Each is best suited for a particular application or range of temperatures.

BIMETALLIC STRIP

The *bimetallic strip* is made up of two strips of different metals welded together. Because of the difference in the coefficients of thermal expansion of the two metals, a heating of the entire strip will cause one of the metals to expand in length more than the other. Since the strips are welded together along one entire edge, the complete strip will bend in the direction of the metal which expands the least. The extent of the bending is directly proportional to the degree of temperature change. If one end of the strip is firmly clamped while the other end remains free, the extent of the bending can be used to indicate temperature change. This is done by attaching a position transducer (such as an LVDT) to the free end of the strip and calibrating its displacement due to the temperature change.

Bimetallic strips are actually used more frequently as controlling devices than as temperature-indicating devices. In this role they are used most commonly as the thermostats which control the on/off switches of heating furnaces and automotive automatic chokes. As pointed out in Chapter 3, bimetallic strips are also used in some forms of circuit breakers. Overload currents cause the strips to bend and break the circuit connections. (See Fig. 3-8.)

THERMOCOUPLES

Thermocouple operation is based on the physical principle that, if two dissimilar metal wires are joined together and the point of joining is heated (or cooled), a voltage difference appears across the two unheated, unjoined ends. The magnitude of the resultant voltage difference is only on the order of millivolts. Nevertheless, the voltage difference is directly proportional to the temperature difference that exists between the heated junction and the cooler ends. If a sufficiently sensitive detector is used, temperature difference can be measured with the thermocouple. The combinations of metals most commonly used for constructing thermocouples are the following: iron and constantan; chromel (alloy of nickel and chromium) and alumel (alloy of aluminum and nickel); and platinum and rhodium-platinum. The first combination is suitable for temperatures up to 900°C; the second can be used for temperatures up to 1150°C; and the third is for temperatures up to 1600°C.

The thermocouple is usually placed *in* the region whose temperature is to be determined. The temperature of both cool[6] ends is kept fixed, and the voltage between them is measured. From this data, the temperature of the junction is calculated. Figure 15-12 shows the relation between the voltage output of the two thermocouples vs temperature for each of the three metal combinations listed above.

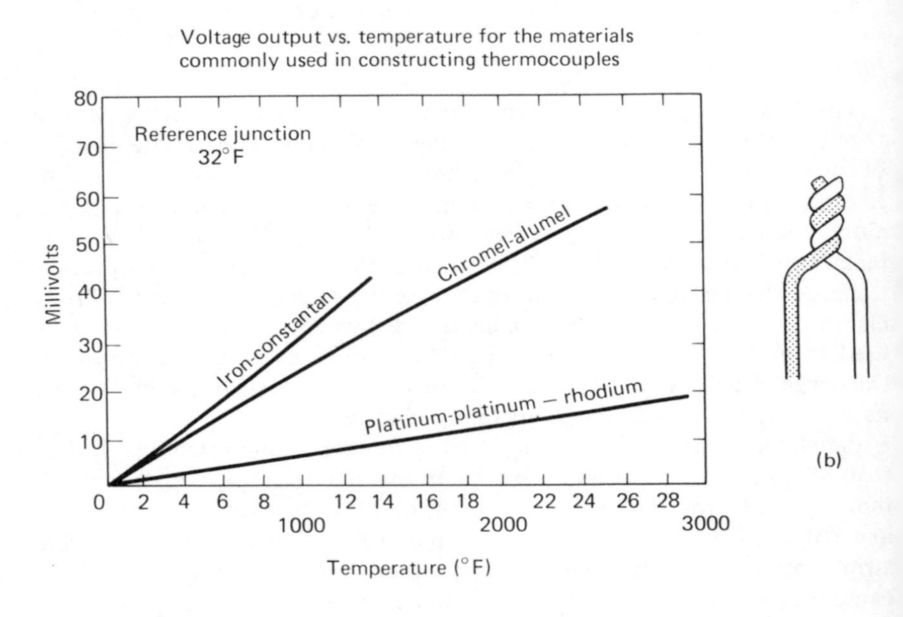

Figure 15-12. Thermocouples (b) Twisted and fused thermocouple

Since the output across the cooler ends of the thermocouple is only a few millivolts, a sensitive voltmeter is needed to measure this voltage. Therefore, potentiometers or highly sensitive digital voltmeters are generally used as the measuring devices. Figure 15-13 shows a practical method for connecting a thermocouple to an indicating instrument to measure temperature.

Thermocouples are rugged and accurate temperature-measuring devices. However, they do not respond quickly to changes in temperature, and this prevents them from monitoring temperatures that change rapidly with time.

[6]In some measurement applications (e.g., cryogenic measurements), the reference junction may be warmer than the measuring junction.

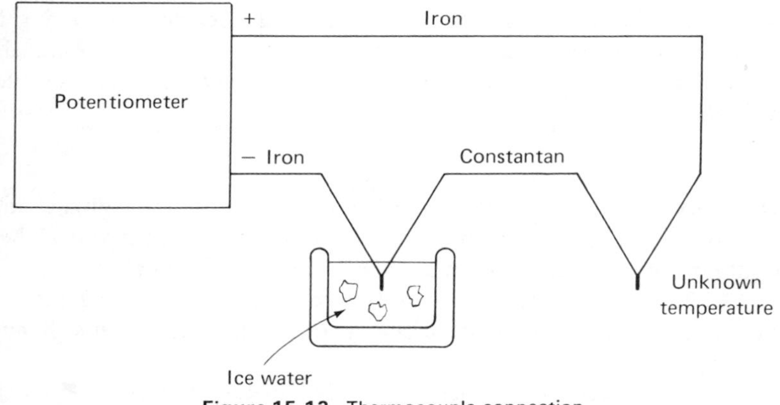

Figure 15-13. Thermocouple connection

RESISTANCE THERMOMETERS

As we noted in Chapter 5, materials change their resistance value with temperature. This effect makes it possible to measure temperature by measuring the change in resistance of a current-carrying device. The *resistance thermometer* is a device whose operation is based on this effect. The materials used to construct resistance thermometers are usually coils of fine copper, nickel, or platinum wire. For very low temperature work (i.e., less than 50°K), carbon resistors are used.

The coil of the resistance thermometer is encased in a metal tube to protect it from damage when inserted into the regions where it measures temperature. The tube is made of a high-thermal-conductivity material to allow fast response to temperature changes (Fig. 15-14). The resistance of platinum vs temperature is almost linear, while the resistance of copper and nickel are progressively less linear. Thus, platinum-wire resistance thermometers are the most accurate types and can be used to indicate temperatures between −250°C and +1200°C. In fact, platinum-wire thermometers are so accurate that they are used as interpolation standards for temperatures between −183°C (boiling point of liquid oxygen) and +630.5°C (freezing point of antimony).

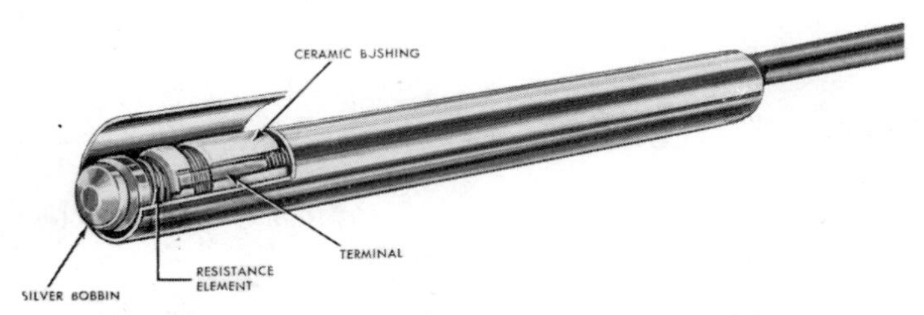

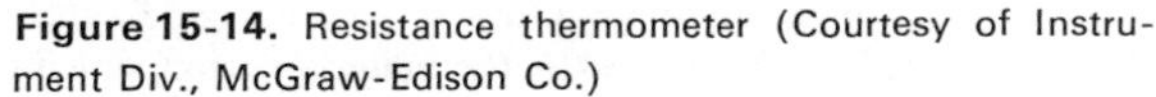
Figure 15-14. Resistance thermometer (Courtesy of Instrument Div., McGraw-Edison Co.)

Wheatstone bridges are used to sense the resistance changes arising in resistance thermometers. The bridges are usually calibrated to indicate the temperature that caused the resistance change rather than the resistance change itself.

THERMISTORS

Thermistors are devices which also measure temperature through a changing resistance effect. However, the resistance of materials from which thermistors are made[7] decreases with increasing temperature from about −100°C to +300°C. In some thermistors, the decrease in resistance is as great as 6 percent for each 1°C of temperature change (although one percent changes are more typical).

The decrease in resistance which takes place in thermistors involves the chemical bonding properties of electrons in semiconductor materials. In these materials the valence electrons are locked in covalent bonds with their neighbors. As the temperature of the thermistor is increased, the thermal vibrations of its atoms break up some of these bonds and release electrons. Since the electrons are no longer bound to the specific atoms in the lattice, they are able to respond to applied electric fields by moving through the material. These moving electrons add to the current in the semiconductor, and the material appears to have a smaller resistance.

Because the change of resistance per degree of temperature change in thermistors is so large, they can provide good accuracy and resolution when used to measure temperatures between −100°C and +300°C. If an ammeter is utilized to monitor the current through a thermistor, temperature changes as small as ±0.1°C can be detected. If the thermistor is instead put into a Wheatstone bridge, the measuring system can detect temperature changes as small as ±0.005°C.

Thermistors are most commonly made in the form of very small beads. This shape and others are shown in Fig. 15-15. Because of their small size, they can be inserted into regions where other larger temperature-sensing devices might not fit.

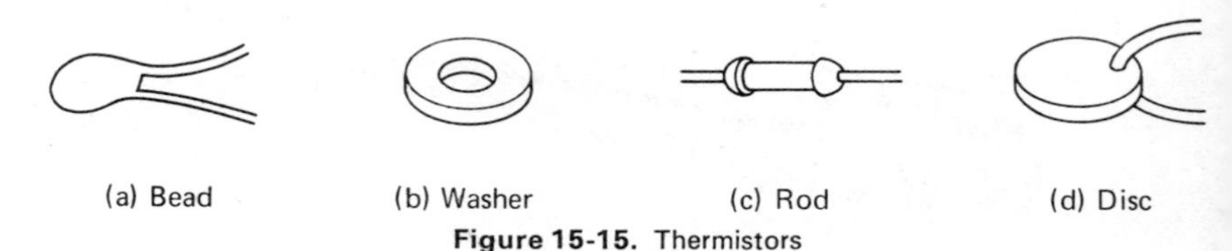

Figure 15-15. Thermistors

[7]Semiconductor ceramics consisting of a mixture of metallic oxides (such as manganese, nickel, cobalt, copper, and iron) are used as the materials from which thermistors are made.

RADIATION PYROMETERS

Radiation pyrometers are devices which sense temperature by measuring the optical radiation emitted by hot bodies. The higher the temperature to which a body is heated, the higher is the dominant frequency of radiation it emits. This means that as the temperature of a body increases to a point where it begins to emit visible light, the heated surface will first have a dull red color. As the body gets hotter and more incandescent, its surface becomes progressively less red and more white.

In order to detect the emitted radiation, the radiation pyrometer does not have to be placed in the furnace or region it is measuring. It is merely necessary to point it at the heated surface of interest to make the measurement.

The *disappearing-filament* type of pyrometer uses a heated-wire filament to provide a radiant temperature standard. An electric current passing through the filament provides an accurate heating method. When the filament is heated to the same temperature that exists at the surface being examined, the filament image is no longer visible because it has the same color as the surface (Fig. 15-16). Since the current through the filament is a known quantity, the pyrometer can be calibrated to yield the temperature of the surface from this current value. Because a body begins to emit visible light when heated to about 775°C, a disappearing-filament pyrometer can measure temperature from this point to about 4200°C.

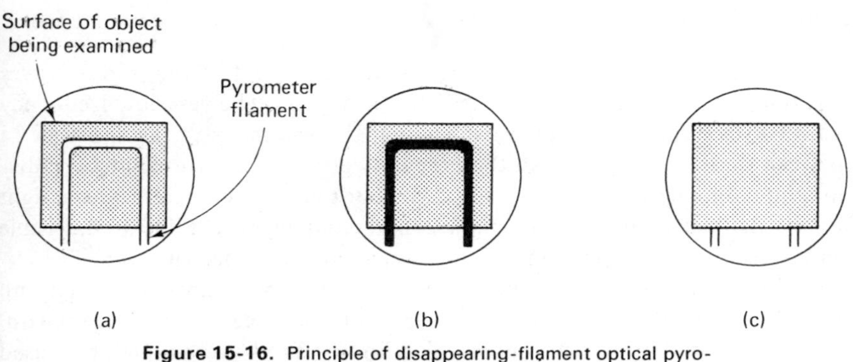

Figure 15-16. Principle of disappearing-filament optical pyrometer (a) Too much current through filament raises it to a higher temperature than the surface (b) Filament of wire too cold. Not enough current is being passed through it (c) Filament "disappears" when it reaches same temperature as surface being examined.

In the *brightness* type of pyrometer, the radiation from the heated surface being examined is collected by a lens and focused onto a thermistor or thermocouple. Since all surfaces emit radiation in proportion to T^4 (where T is the temperature of the surface, in degrees Kelvin), the radiation measured by this instrument can be calibrated to yield the value of surface temperature directly. However, the brightness type of pyrometer is calibrated by comparison to near-ideal *blackbody* radiators. Therefore, temperature readings from surfaces which do not possess such ideal surface emissivities must be corrected to account for this discrepancy. Figure 15-17 shows two commercially available radiation pyrometers.

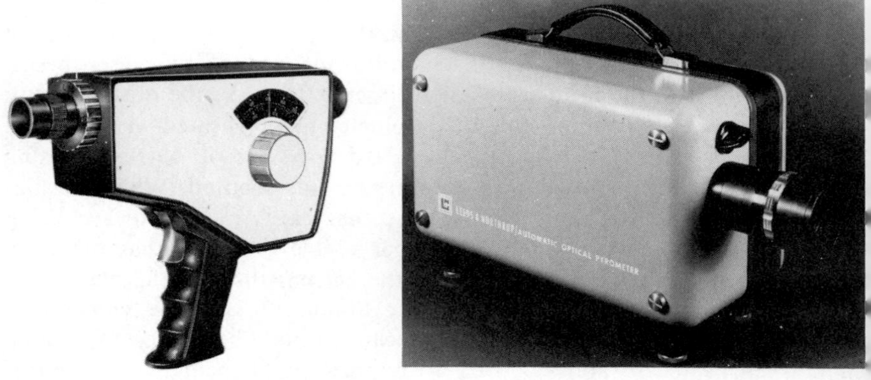

Figure 15-17. Optical pyrometers (Courtesy of Leeds & Northrup Company)

Light and Radiation Transducers

The spectrum of electromagnetic radiation extends from radio waves (less than 10 Hz) to gamma rays (10^{20} Hz or higher). The very-low-frequency radio waves have the longest wavelengths; the gamma rays the shortest. In between these extremes are all the other categories of electromagnetic radiation, including light. (See Fig. 15-18.) For our purposes of classification, light will be defined to include the radiations belonging to the infrared, visible light, and ultraviolet portions of the electromagnetic spectrum.

In this section we will be most interested in those transducers that can sense light radiation and convert it into an electrical form. The general class of light-radiation transducers, also known as *phototransducers*, is used to detect the presence and intensity of light under various circumstances. In fact, many phototransducers can be made much more sensitive to light radiation than the human eye is. The three primary types of light-to-electrical energy transducers are:

1. Photoemissive devices
2. Photoconductive devices
3. Photovoltaic devices

Each of these types possesses special advantages over the others.

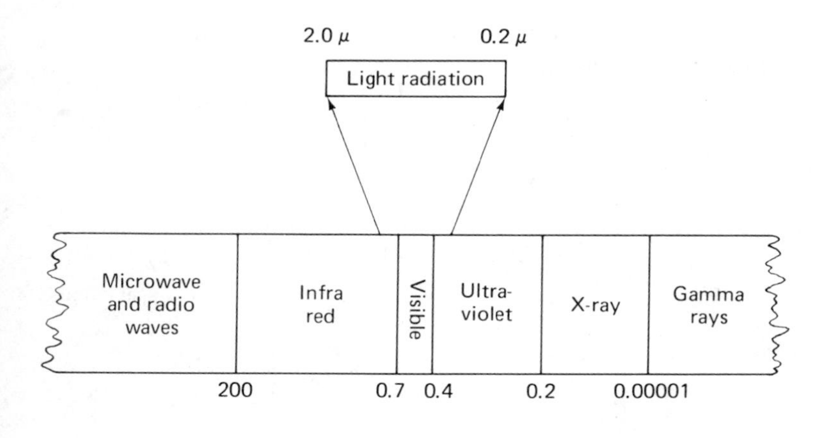

Figure 15-18. The electromagnetic spectrum

PHOTOEMISSIVE LIGHT SENSORS

Photoemissive light sensors are so named because they contain materials whose surfaces emit electrons when struck by light radiation. The electrons are emitted when the photons of the incident light are able to transfer enough energy to the electrons to break them free from both their atomic bonds and the forces of the entire material lattice. Materials in which this phenomenon takes place easily enough to produce many electrons when struck by visible light are called *photoemissive* materials. Because the photoemissive material is usually housed in a glass tube, photoemissive devices are also often known as *phototubes*. Figure 15-19 shows the basic principle behind the operation of phototubes.

The surface of a specially shaped cathode is coated with a photoemissive material (such as cesium-antimony). The cathode (now called a *photocathode*) is housed in a sealed glass tube along with another electrode called the *anode*. A voltage is created between the photocathode and the anode (with the anode having the positive voltage level). When light strikes the photocathode, the electrons emitted from the surface are attracted and collected by the positive anode. The stronger the intensity of the light incident on the photocathode, the more electrons it emits. Therefore, the magnitude of the current flowing in the circuitry connected to the electrodes of the tube is directly proportional to the intensity of the light incident on the photocathode.

Before going on with the description of the various types of photoemissive tubes, one precautionary note on their use should be sounded here. Because phototubes suffer from a condition known as *phototube fatique*, the data obtained with the use of phototubes are often misinterpreted. Phototube fatique is the loss of sensitivity of a photoemissive surface when it is subject to constant illumination by an intense source of light. The time constant of

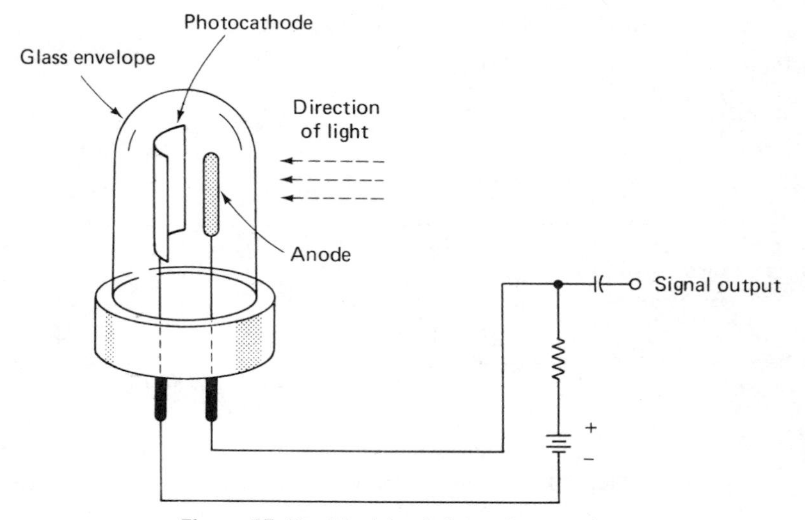

Figure 15-19. Principle of phototube operation

this effect varies from tube to tube, but it is on the order of one-half hour. The mistakes are made in interpreting the cause of the decreased intensity of the output signal of the phototube. Instead of attributing the resultant decrease to the characteristics of the tube, the cause for the decrease is sought as a change in the magnitude of the quantity being measured. One way to avoid having this effect occur is to use a mechanical chopper ahead of the tube. By chopping the light beam, the chopper converts the constant light intensity to a pulsating form. This greatly reduces the effect of phototube fatigue.

There are three commonly used types of phototubes. The first is the *vacuum phototube* shown in Fig. 15-19. In this type a vacuum exists within the glass tube. When light strikes the photocathode of the tube, electrons are emitted from its surface. If a sufficient voltage exists between the anode and the photocathode, the resultant current is almost linearly dependent on the intensity of the light. In fact, the response of vacuum phototubes is linear over such a wide range of light levels that they are used as standards in light-comparison measurements. In addition, the response time of phototubes to the incident light is so rapid that they are suitable for applications where very-short-duration light pulses must be observed.

The second type of phototube is the *gas-filled phototube.* In this device, the tube that houses the photocathode and anode is filled with an inert gas (such as argon) at a very low pressure. As electrons are emitted from the photocathode, they are also accelerated toward the anode by a voltage difference. While in transit between the electrodes, these electrons collide with the argon gas atoms. If the energy of an electron is high enough, the collision ionizes (i.e., tears away electrons) the argon atoms and therefore creates a positive ion and additional free electrons. The electrons are attracted by the anode and ions by the cathode. As a result, a greater current appears to flow between the anode and photocathode. Because of the multiplication effect of charge carriers arising from the collisions, the resulting current collected by the electrodes is often large enough that it does not require amplification. This makes gas phototubes very simple and inexpensive devices.

The relatively slow motion of the positive ions towards the cathode makes the response of gas phototubes to variations in the intensity of the incident light relatively slow. Therefore, gas phototubes are only suitable for applications where such a slow response time is not a hindrance. They are chiefly used to reproduce the sound tracks for movie films because their response-times are sufficiently rapid for this task.

The last type of phototube is the *photomultiplier tube* (PM tube). These devices are probably the most widely used types of light detectors. Their outstanding characteristic is that they can detect very-low-level light intensities.

The ability to detect very small intensities of light is due to the fact that photomultiplier tubes are actually amplifying devices. In Fig. 15-20, we see

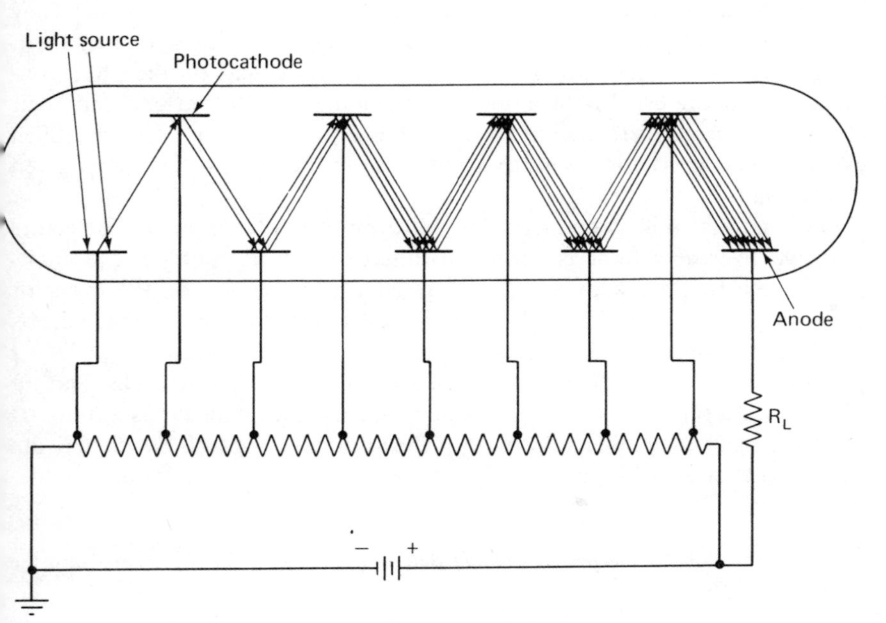

Figure 15-20. Principle of photomultiplier tube operation

that the incident light beam is first made to strike a photoemissive surface in the same manner as in the ordinary vacuum phototube. However, the emitted electrons are not drawn immediately to an anode. Instead, they are attracted (by a voltage difference) to another electrode called a *dynode*. The dynode emits *secondary electrons* when struck by an electron beam. Thus each original photoelectron is accelerated by an electric field and knocks several (i.e., 3 to 6) secondary electrons out of the dynode. There are usually ten dynodes in a PM tube, and each one is designed to form electric field lines which guide the secondary electrons emitted by the previous dynode to itself. [Each dynode is at a higher potential ($\approx$100 V) than the preceding one.] Thus the secondary electrons are *multiplied* in number at each dynode, and the final burst is collected by the anode. In this way, a multiplication factor of over 10^6 is achieved in commercially available tubes. Such amplifications allow PM tubes to detect the event of even a single electron being emitted from the photocathode. The response time of PM tubes is also very rapid, and frequencies of up to hundreds of megacycles can be followed by them. However PM tubes are generally not suitable for detecting infrared radiation because materials are not photoemissive in response to infrared radiation.

PHOTOCONDUCTIVE LIGHT DETECTORS

Photoconductive light detectors are basically light-sensitive resistors. They are made of materials whose resistance decreases when illuminated by light. The resistance decrease occurs because the light incident on the photoconductive material breaks the bonds between the material atoms and their electrons. These electrons are then available as free charge carriers. Such extra free charge carriers lead to an increase in the current flow for a given voltage value.

One material that is commonly used to build photoconductive detectors sensitive to *visible* light is cadmium-sulfide. If such resulting cadmium-sulfide detectors are kept dark, their resistance values are on the order of many megohms. When the cell is illuminated with light, the resistance drops to a much lower value (i.e., several kilohms).

Photoconductive cells are often used as the photosensitive elements in photoelectric relays,[8] as devices which measure light intensity, as automatic iris controls in cameras, and as part of the automatic controls which activate street lights at dusk. Figure 15-21 shows a cutaway view of a photoconductive device.

[8]See the section on *Electromagnets and Relays* in Chapter 7 for more details on the photoelectric relay.

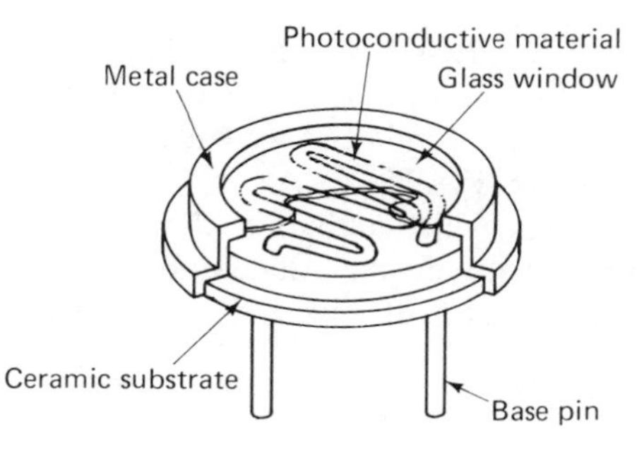

Figure 15-21. Cutaway view of a photoconductive cell

PHOTOVOLTAIC LIGHT SENSORS

Photovoltaic sensors are devices across which a voltage appears when they are illuminated by light. This voltage arises in connection with the electronic properties of the material from which the device is made.[9] The *solar cell* described in Chapter 8 is one type of photovoltaic sensor. In fact, solar cells convert the light energy to electrical energy well enough that the cell can act as an electrical power source. Silicon solar cells are used in this way to provide electrical power to space satellites. Selenium solar cells are used in photographic exposure meters.

Photovoltaic sensors are the only photodetectors which do not require an external power source. However, they also possess the disadvantage that their outputs are extremely nonlinear in relation to the intensity of the incident light.

X-RAY AND NUCLEAR RADIATION TRANSDUCERS

X-ray and nuclear radiation sensors utilize some of the same principles of operation as light-radiation sensors. For example, one type of X-ray sensor is like a photoconductive device, while another is made of a material that emits visible light when struck by X rays. The intensity of the emitted visible light in this latter type is proportional to the intensity of the incident X rays. The strength of the glow is then measured by a phototube. X-ray sensors are used to monitor the thickness of materials which are manufactured in sheet form. They are also used for locating flaws in metal structures and other construction materials, and in determining liquid levels in sealed tanks.

The most common nuclear radiation sensors are the *Geiger-Müller* tube and

[9]In most cases the voltage appears across a *PN junction* which exists in the material. Such junctions mark the boundaries between two regions of the semiconductor material that have different electrical properties.

the *scintillation counter*.[10] The Geiger-Müller tube is a sealed tube filled with an inert gas (such as argon). Within the tube is a cathode in the shape of a long cylinder, as well as a long wire anode placed along the axis of this cylinder (Fig. 15-22). At one end of the glass tube is an extremely thin window through which radiation can enter the tube. If a high voltage ($\approx$900 V) exists between the anode and the cathode, each burst of nuclear radiation (in the form of beta particles or gamma rays) which enters the tube will

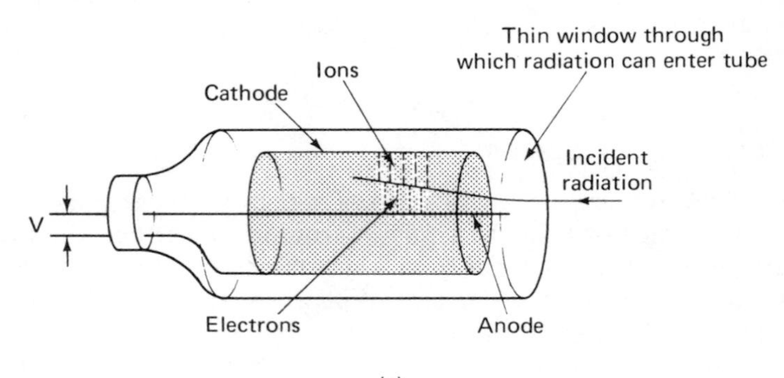

(a)

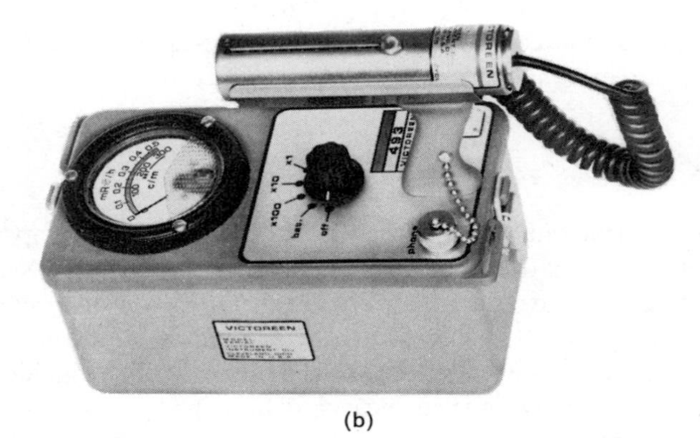

(b)

Figure 15-22. Geiger-Müller radiation detector (a) Schematic (b) Victoreen model 493 (Courtesy of Victoreen Instrument Co.)

[10]There are many other types of nuclear radiation detectors which find application in special radiation-measurement systems. They include the ionization chamber (similar to the Geiger-Müller tube), the cloud chamber, the Cerenkov counter, neutron detectors, and radiation dosimeters. Rather than discuss them further in this book the reader is asked to refer to other texts (listed as references at the end of the chapter) which have been written specifically on radiation detection. These books cover the operation and uses of the devices in extensive detail.

ionize some of the argon atoms. The ionized particles will rush toward the electrodes of the tube. During their voyage to the electrodes, the electrons and ions are accelerated and collide with other argon atoms, which are in turn ionized. When the burst of ions is collected by the electrodes, a pulse of current is caused in the circuit to which the tube is connected.

As the pulse passes through the connecting circuitry, the voltage between the anode and cathode drops below the value which causes further ionization by collision. The gas in the tube returns to its deionized state, and current in the connecting circuit ceases. Then the voltage between the cathode and anode rises again, and the tube is ready to sense the next radiation burst. The "dead time" between bursts is about 100–200 μs.

The number of pulses occurring in a given time is a measure of the strength of the radiation in the region near the tube. The pulses can be used to cause audible clicks from a loudspeaker or they can be counted by an electronic counter to yield a cumulative value over a specific time. Typical GM meters can detect up to 15,000 events (or counts) per minute.

The *scintillation counter* is a device which uses a photomultiplier tube to count light flashes which are produced by certain crystals when struck by nuclear radiation. These crystals (such as zinc sulfide or sodium iodide) produce a brief *scintillation* of light each time such an event occurs. These emitted flashes are reflected by mirrors and are carried to a photomultiplier tube through fiber-optic *light pipes*. The photomultiplier tubes convert and amplify the feeble light flashes to electrical pulses which have sufficient magnitude to be measured. The number of pulses the photomultiplier tube puts out is a measure of the intensity of the nuclear radiation.

Because of the amplifying ability of the photomultiplier tube, *scintillation counters* are much more sensitive than Geiger-Müller tubes in measuring nuclear radiation.

Sound Transducers

Because sound can exist as the vibration of solids or liquids as well as air, transducers that convert sound energy to electrical energy must be available for operation in many different media. The air-pressure-to-voltage transducer used to convert sounds in air is called a *microphone*. Those transducers that can transform the sound of solid vibrations to electrical signals are called *vibration pickups*. The most common microphone types are the following:

1. Capacitor microphone
2. Dynamic microphone
3. Carbon microphone

Vibration pickup transducers are mostly of the piezoelectric type.

MICROPHONES

Capacitor microphones such as the one shown in Fig. 15-23(a) perform the most accurate conversions of sound vibrations in air to electrical signals. As such, they are used as the standard for precise acoustical measurements. The principle of their operation is based on the fact that the capacitance between two conductors changes if the distance separating them is varied. In the capacitor microphone, a metal diaphragm is used as one of the plates of the capacitor. A rigid metal plate makes up the other capacitor plate. The variation of the air pressure in a sound wave which strikes the diaphragm makes the diaphragm move in and out. The changing separation between the two capacitor plates changes the capacitance between them. A fixed dc voltage is applied between the plates, and the changing voltage due to the incident sound exists as an ac component superimposed on the dc level. The ac component is then amplified and delivered to a measuring instrument.

Because a capacitor appears as a high-impedance source, the output of this device must be coupled to an instrument which also has a very high input impedance (to reduce loading errors). A special circuit which has such a high impedance (and is called a *cathode follower*) is often used to couple the output to an amplifier.

The *dynamic microphone* reacts to the sound vibrations of air by moving a coil of wire in a magnetic field (Fig. 15-23(b)). The motion of the coil in the field sets up a changing voltage across the coil, and the signal can be amplified and measured. The dynamic microphone can also act in reverse, causing a sound to be produced by applying a changing voltage to the coil. This is the principle on which a loudspeaker is based.

The *carbon microphone* (Fig. 15-23(c)) detects sounds by variations in the resistance of carbon granules. The resistance variations occur when

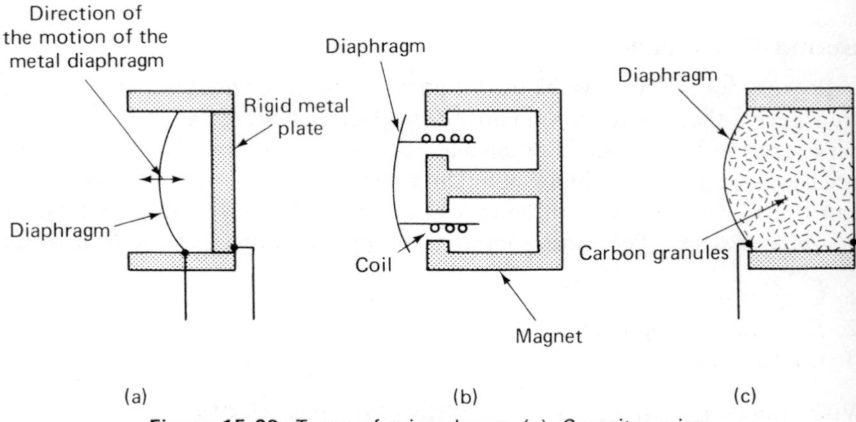

Figure 15-23. Types of microphones (a) Capacitor microphone (b) Dynamic microphone (c) Carbon microphone

the density of the carbon granules is changed by the pressure of incident sound waves. A diaphragm on the front of the carbon microphone moves in response to the sound pressure and compresses the carbon granules within the housing. A dc current is passed through the microphone, and the modulation of the current due to changes in resistance of the carbon granules superimposes an ac signal on this dc level. As in the capacitor microphone, the ac signal is separated from the dc level and amplified before being delivered to a display device.

The carbon microphone is almost never used to make accurate measurements of sound. The nonlinearities of the resistance variation of the carbon prevent the signal from being a faithful replica of the sound vibrations. Nevertheless, the device is perfectly adequate for voice transmission, and it is used as the microphone in almost all telephones. The fact that they are inexpensive, highly reliable, and very rugged also makes carbon microphones well suited for this type of application.

VIBRATION PICKUPS

Piezoelectric crystals are crystalline materials which develop a voltage across themselves when they are deformed from their natural shapes. (The deformation need only be on the order of micrometers to produce this effect.) If the applied force which deforms the crystal has a time-varying magnitude (i.e., as a force due to the vibration of a piece of material to which the piezoelectric crystal is attached), the output voltage from the crystal will also have time-varying form which is patterned after the time-varying form of the force.

To detect acoustic vibrations in solids, crystal transducers that employ the piezoelectric effect are used (Fig. 15-24). The crystals used for constructing ac pickups are usually made of quartz or Rochelle salt crystals. The former are more accurate but are also more expensive than the latter; however, Rochelle salt crystals are hampered by the fact that they melt at 65°C and are limited by high humidity conditions. Crystal microphones with a diaphragm make this type of transducer sensitive to sound vibrations.

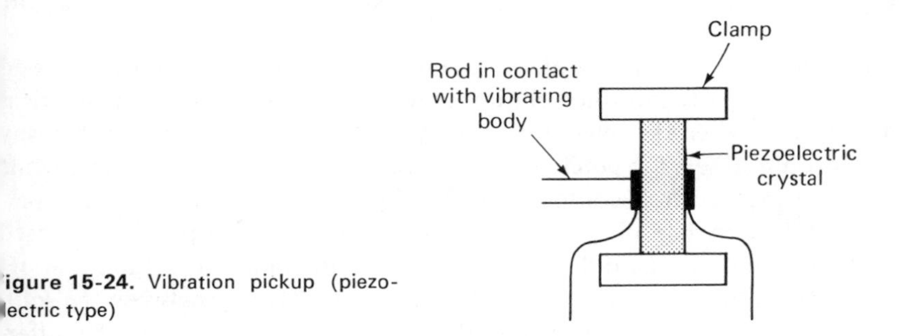

Figure 15-24. Vibration pickup (piezoelectric type)

Chemical Property Transducers

Electrical transducers are available for converting some chemical properties into the form of electrical signals. This can be a useful ability when it is necessary to control such properties in order to allow a chemical process to proceed efficiently. The conversion to electrical form allows the value of the essential chemical quantities to be continuously monitored. In addition, if the chemical concentrations of various solutions or gases can be measured electrically, such quantities as pollutant concentrations in the atmosphere can be quickly and accurately detected.

One chemical property which must often be monitored closely is the pH level of a liquid solution. The pH of a solution is an indication of its acidity or alkalinity. The value of pH can vary from 0 (purely acidic) to 14 (purely alkaline). When the pH is 7, the solution is neutral (as in the case of pure water). The pH of a solution can be accurately measured by means of an electronic pH meter (Fig. 15-25). This meter consists in part of two

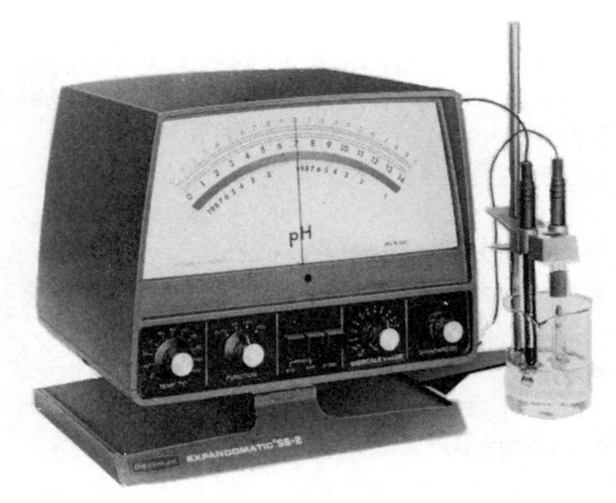

Figure 15-25. A pH meter (Courtesy of Beckman Instruments, Inc.)

special electrodes which are inserted into the test solution (Fig. 15-26). One electrode is kept at a constant voltage level, and the other electrode is allowed to develop a potential due to the pH of the solution being measured. Then the potential difference between the electrodes depends on the pH of the solution.

The reference electrode contains a solution of electrolyte that is allowed to leak very slightly into the solution under test. This provides an electrical connection between the reference electrode and the test solution. In this way the test solution sets the potential level of the reference electrode. Meanwhile the measuring electrode contains a buffer liquid of known and constant pH. It is contained in a special glass tube with a very thin wall that is immersed in the test solution. The difference in the pH of the buffer liquid and the pH of the test solution causes a potential difference to develop between the solution in the measuring electrode and the test solution. As a result, the differ-

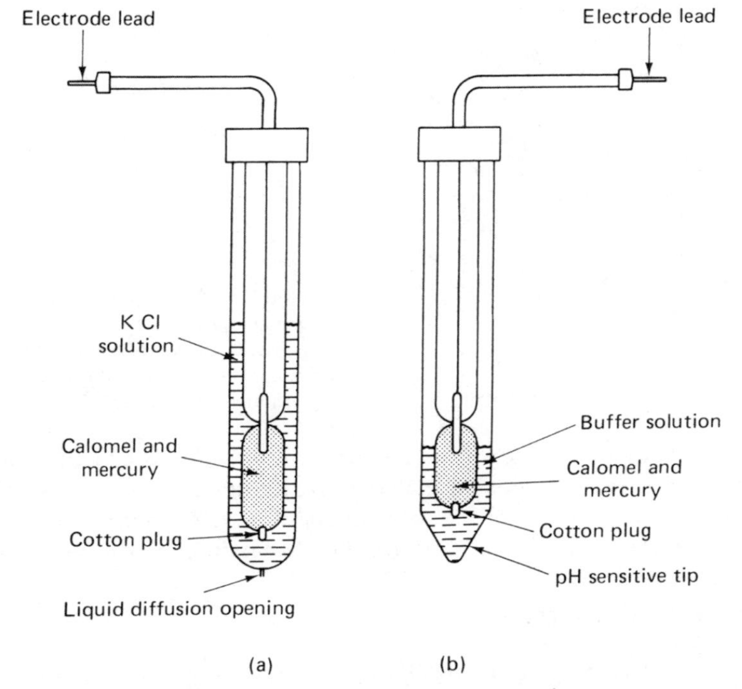

Figure 15-26. Principle of pH meter operation

ence in potential between the reference electrode and the measuring electrode is proportional to the difference in pH of the known buffer liquid and the unknown test solution.

Because the potential difference is measured across the glass tube wall of the test electrode and the glass wall has a resistance of about 500 MΩ, a special very-high-impedance voltmeter must be used to measure the potential difference accurately. The *electrometer* described in Chapter 12 is the type of voltmeter ordinarily used to measure this voltage.

The second type of chemical-property transducer is the detector element of the gas chromatograph. This detector measures the chemical purity of a gas by detecting changes in its thermal conductivity. The measurement is possible because various gases have very different values of thermal conductivities. Advantage is taken of this phenomenon by allowing the gas to cool heated resistors in a Wheatstone bridge circuit (Fig. 15-27).

The bridge is first calibrated by allowing the same gas to surround two heated test resistors. The values of the resistors are adjusted to yield a balanced bridge condition. Then one of the gases is replaced by the test gas, flowing under the same pressure and temperature conditions. If the test gas contains elements which have different thermal conductivities than the reference gas, the resistor will be cooled at a different rate. A different cooling

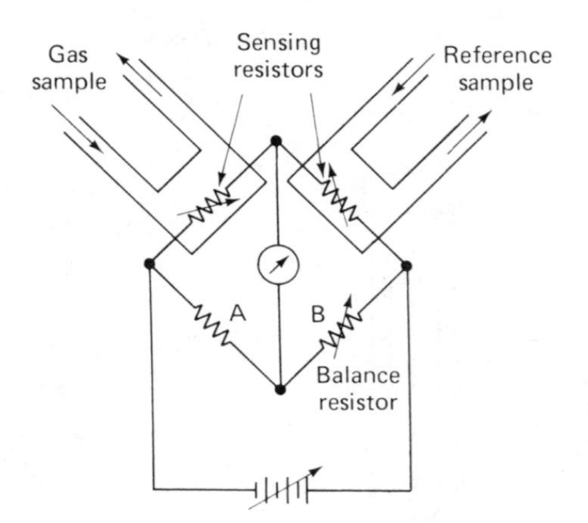

Figure 15-27. Basic thermal-conductivity gas-analyzer circuit (Courtesy of A. Marcus and J. D. Lenk, *Measurement for Technicians,* Englewood Cliffs, N. J.: Prentice-Hall, Inc., 1971, p. 230.)

rate results in a variation of the temperature of the resistor and a variation in its resistance. The resistance change creates an unbalanced condition in the bridge. If the purity of the gas is in question, the condition of unbalance can actuate a meter which indicates the change in composition of the gas.

Magnetic Measurements

In this section we will briefly describe how measurements of magnetic effects are made with gaussmeters and with two types of magnetic transducers.

Gaussmeters are simple meters which use the principle that a permanent magnet placed in another external magnetic field will try to align itself with the field. In the gaussmeter there is a small permanent magnet attached to a shaft that is connected to a spiral restraining spring (Fig. 15-28). When the gaussmeter is placed in a magnetic field and is oriented properly, a maximum amount of torque will be exerted on the permanent magnet. This torque is opposed by the restraining spring, and the resulting deflection is proportional to the strength of the magnetic field (in Wb/m^2).

The first magnetic transducer we will discuss involves the so-called *Hall effect.* (It was discovered by E. H. Hall in 1879.) The effect can be explained with the help of Fig. 15-29. We see that a slab of conducting material is connected to a battery so that a current, I_C, will flow through the slab in the manner shown. (The electrons which make up the current actually flow in the opposite direction to the direction of the current, since we are using the conventional current description.) If the magnetic field is applied so that the magnetic lines of force are at right angles to the slab of material the electrons of the current are acted on by a force due to the magnetic field. The force acts in a vertical direction, and the electrons are forced toward the top of the slab. This results in an excess of electrons near the top of the slab and a deficiency of electrons near the bottom. A potential difference between the top and bottom of the slab is thereby created. The magnitude

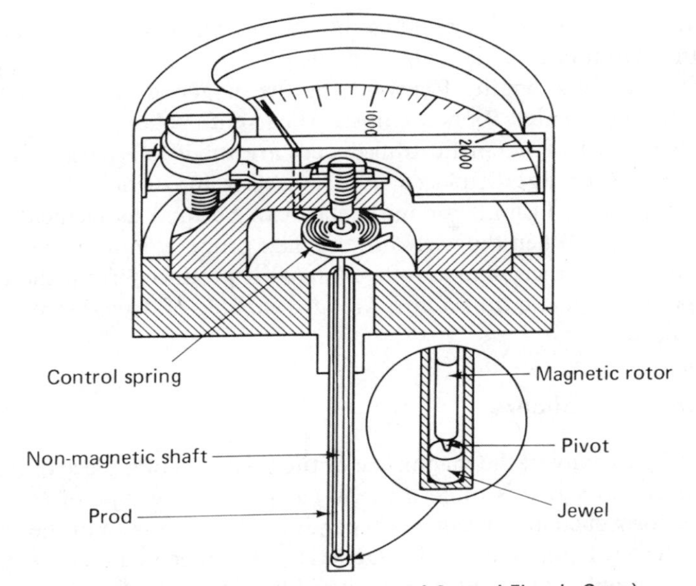

Figure 15-28. Gaussmeter (Courtesy of General Electric Corp.)

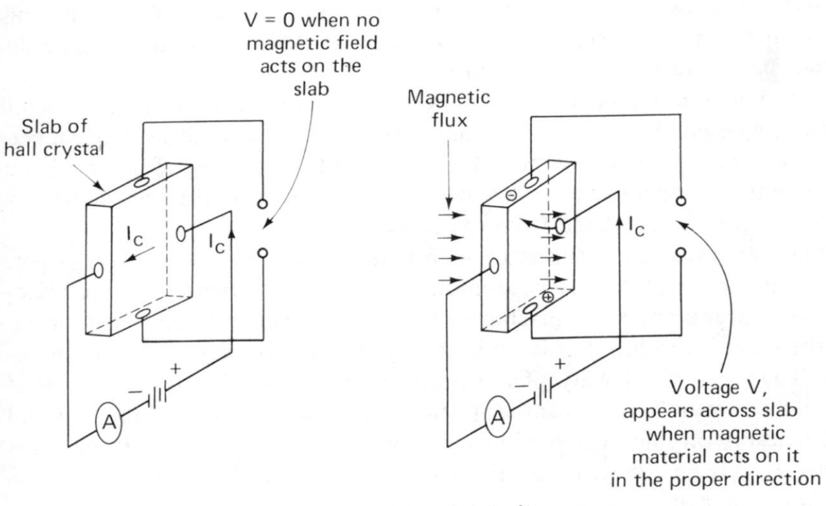

Figure 15-29. Principles of Hall-effect meters

of the voltage is proportional to the strength of the magnetic field (in Wb/m^2). Therefore, when a known value of current is passed through the conducting slab, the voltage reading across the device can be calibrated to yield the magnetic strength directly. Hall-effect transducers can be built to be sensitive enough to detect very small magnetic fields.

Commercial Hall-effect transducers are made from germanium or other semiconductor materials. They find use in instruments that measure magnetic fields with small flux densities. They are also found in some current probes, which measure the strength of currents by the magnetic field produced by the current. (See Chapter 9 on oscilloscope current probes.)

The other types of magnetic transducers are made of special materials whose resistance changes if placed in a transverse magnetic field. Wires of such materials (i.e., bismuth or mu-metal) can be used as elements in an impedance bridge. When the wires are oriented properly in a magnetic field, their resistances change, and a voltage signal is produced from the bridge. These types of magnetic transducers are also capable of detecting very weak magnetic fields.

Thickness Transducers

The ability to convert the magnitude of the thickness of a piece of material into an electrical signal is often a useful capability. The type of transducer used to perform such measurements depends on the properties of the material being produced. If the material is magnetic, a U-shaped coil can be pressed with its open end against the sheet. The inductance of the coil will change slightly as the thickness of the material with which it is in contact changes (provided the composition of the material remains constant). The change of inductance can be measured with a bridge calibrated to yield the magnitude of the changes in the sheet's thickness.

If the material whose thickness is being determined is an insulator, metal plates can be placed on either side of the sheet, and the resulting capacitance value of the structure can be noted. If the dielectric constant of the insulator is known, the spacing between the metal plates due to the test material's thickness can be calculated from the capacitance.

When manufacturing sheets of such materials as plastics, rubber, paper, etc., it may be important to continuously monitor the thickness of the material being produced. X-ray machines can be used to perform such monitoring. As the sheet is produced, it is subjected to an X-ray beam of constant intensity. The level of intensity of the beam which passes through the sheet is monitored by an X-ray sensor. Any changes in the thickness of the material will cause a varying degree of absorption of the beam to take place. This variation can be converted by the sensor to indicate the magnitude of the thickness change.

Problems

1. List five devices which act as nonelectrical transducers [i.e. the mercury thermometer (temperature to dimensional length transducer)].
2. List the series of steps you would take if you were required to select a transducer for a particular measurement application.

3. Are the following transducers active or passive transducers and why:
   a) strain gauge,
   b) LVDT,
   c) thermocouple,
   d) phototube,
   e) solar cell,
   f) carbon microphone,
   g) piezoelectric crystal.
4. Comparing the three types of strain gauges (bonded metal, unbonded metal, and semiconductor), list the advantages and disadvantages of each type.
5. A semiconductor strain gauge has a length of 5 cm and an initial resistance of 2 kΩ. An applied force causes a 0.5 percent change in its length and a change in resistance of 400 Ω. What is the GF of the strain gauge?
6. How does temperature affect the operating characteristics of strain gauges?
7. Describe the operation of the linear-variable-differential-transformer (LVDT).
8. An LVDT has an output of 5 V when the displacement is 8×10^{-2} in. Determine the sensitivity of the device in mV/0.001 in.
9. Choose the most suitable temperature transducer for measuring the temperature in each of the following applications:
   a) rapidly changing temperatures,
   b) very small temperature variations about 40°C,
   c) very high temperature (>1500°C),
   d) highly accurate temperature measurements,
   e) wide temperature variation,
   f) application calling for a rugged, accurate temperature sensing device.
10. Describe the differences in the principles of operation of photoemissive, photoconductive, and photovoltaic transducers.
11. What is phototube fatigue and how can it be avoided?
12. Explain why an electrometer is used to measure the voltages produced in the measurement of pH values?
13. Describe the principles of operation of a gaussmeter.
14. A capacitive type pressure transducer has two metal plates that are 3 cm in diameter and 0.4 cm apart. A 200 psi pressure on one plate will cause the separation between the plates to decrease by 0.04 cm. If no pressure is applied to the capacitor, it has a capacitance of 350 pF. Determine the capacitance value if 200 psi is applied to the transducer.

References

1. TEKTRONIX, *Use of Transducers*. Beaverton, Ore., Tektronix, Inc., 1971.
2. KINNARD, I. F., *Applied Electrical Measurements*, Chaps. 9, 12, 13, 14, 15, 16, 17, and 18. New York, John Wiley & Sons, Inc., 1956.
3. TUVE G. L., and DOMHOLDT, L. C., *Engineering Experimentation*. New York, McGraw-Hill Book Co., 1966.
4. NORTON, H. N., *Handbook of Transducers for Electronic Measurement Systems*. Englewood Cliffs, N.J., Prentice-Hall, Inc., 1969.

5. Partridge, G. R., *Principles of Electronic Instruments*, Chaps. 13, 14, 15, 16, 17, 18, and 19. Englewood Cliffs, N.J., Prentice-Hall, Inc., 1958.

6. Marcus, A., and Lenk, J. D., *Measurements for Technicians*. Englewood Cliffs, N.J., Prentice-Hall, Inc., 1971.

7 Lenk, J. D., *Handbook of Electronic Meters*. Englewood Cliffs, N.J., Prentice-Hall, Inc., 1969.

8. Kowalski, E., *Nuclear Electronics*. New York, Springer-Verlag, 1969.

9. Tektronix, *Biophysical Measurements*. Beaverton, Ore., Tektronix, Inc.

10. Price, W. J. *Nuclear Radiation Detection*. New York, McGraw-Hill, 1958.

16

Electronic Amplifiers

One of the primary reasons why electrical methods are so widely used in scientific instrumentation is that very weak electrical signals can be *amplified* to the point where they can directly activate indicating or recording devices and be thereby measured. The *electronic amplifier* is the component of electrical systems which provides the necessary power to the signals to make such measurements feasible. The purpose of this chapter is to describe the general characteristics and terminology which are associated with electronic amplifiers—especially those amplifiers that are most commonly used in electrical measuring instruments.

The level of the discussion will be such as to introduce the qualitative aspects of the most common amplifier types and list some of their important applications. The quantitative electronic design principles of amplifier circuits will not be stressed. Even a basic coverage of such circuit design aspects would require far more space than this text can devote to the subject.[1] As a result, we will treat the amplifier as a black box that somehow amplifies the signals that are fed into it.

One important amplifier type which will be covered more fully (but still in a very introductory fashion) is the *operational amplifier*. These amplifiers are extremely versatile and are finding use as key building blocks in a wide variety of modern measurement instrumentation systems.

[1] Readers desiring more information on the electronic design aspects of amplifiers should consult any of the standard texts on electronics, such as H. V. Malmstadt, C. G. Enke, and E. C. Toren, *Electronics for Scientists* (New York, Benjamin, 1962).

General Properties of Amplifiers

The primary function of an amplifier is to increase, or amplify, the signals that are fed to its inputs. Although amplifiers are also used in some cases to isolate one part of an electrical system from another, our overall concern will be with the amplifying properties. Thus, we will view amplifiers as instruments which are designed to receive a signal (a_{in}) and produce a replica of this signal with an amplitude that has been multiplied by some factor, K. The value of the output signal of the amplifier (a_{out}) may therefore be expressed as

$$a_{out} = Ka_{in} \tag{16-1}$$

From this relation we see that K (which is known as the *gain* of the amplifier) is equal to the ratio of a_{out} to a_{in}. K is a function of the amplifying device, the bias level, and the circuit configuration. If $K > 1$, it is evident that the amplifier increases the value of signals fed to its inputs. The circuit symbol for an amplifier is —▷K .

If a_{in} and a_{out} are expressed in terms of the voltage value of the input and output signals, respectively, the ratio of their magnitudes yields the *voltage gain* of the amplifier. On the other hand, if a_{in} and a_{out} are expressed in terms of the power contained in the input and output signals, their ratio yields the *power gain* of the amplifier. These two gain values are often quite different from one another.

The frequency range over which an amplifier is designed to amplify input signals with a constant gain is known as the *bandwidth* of the amplifier. This quantity is formally defined as the interval between those frequencies where the *power gain* of the amplifier has dropped to one-half of its midfrequency value (Fig. 16-1).

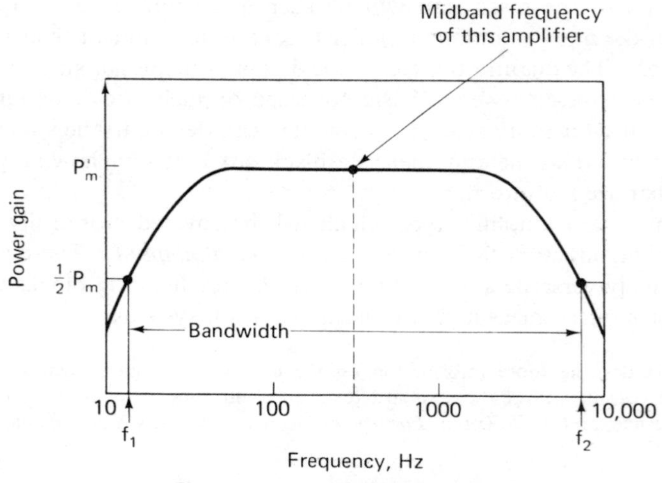

Figure 16-1. Amplifier bandwidth

In many cases, the voltage gain of an amplifier is specified rather than the power gain. However, the bandwidth is not the frequency interval between the two frequencies at which the *voltage* gain falls off to one-half its midfrequency value. Instead, since the voltage value of the half-power points is $0.707 V_m$ (where V_m is voltage value at midfrequency), the bandwidth is defined as the frequency interval between the points where the voltage gain drops to 0.707 of its midband value.

One broad distinction between types of amplifiers is based on their low-frequency response capabilities. If an amplifier is able to produce a constant dc output signal in response to a constant dc input, it is classified as a *dc amplifier*. The bandwidth of dc amplifiers extends from dc to the frequency at which the voltage gain of the amplifier drops to 0.707 of the voltage gain at dc. All other types of amplifiers are known as *ac amplifiers* (Fig. 16-2(a)).

Ac amplifiers are sometimes further divided into categories according to the band of frequencies over which they can amplify signals. If the amplifier is designed to be able to amplify a wide band of frequencies, it is called an *untuned* amplifier (Fig. 16-2(b)). On the other hand, if an amplifier is designed to be able to amplify frequencies within a very narrow band, the amplifier is known as a *tuned* type (Fig. 16-2(c)).

Untuned amplifiers are subdivided even further into the following classes: *audio-frequency* (af) amplifiers, designed for amplifying signals from 30 Hz

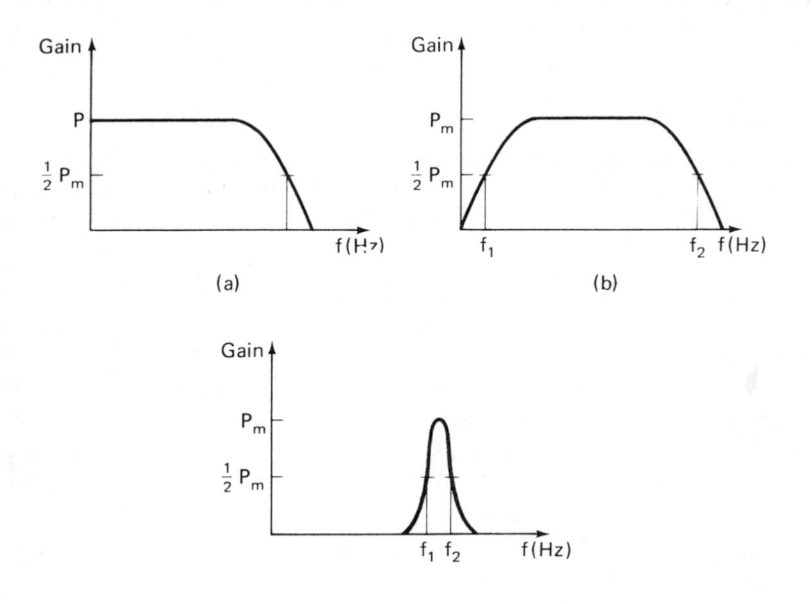

Figure 16-2. Frequency characteristics of various amplifiers (a) dc amplifier (b) Untuned ac amplifier (c) Tuned ac amplifier

to 15,000 Hz; *radio-frequency* (rf) amplifiers, capable of amplifying high-frequency waves—typically 500 kHz or higher—whether these signals are found in radios, radar, or other applications; *video* amplifiers, designed for amplifying the wide band of signals used in television receivers—typically from 30 Hz to 4 MHz.

Dc and ac amplifiers both have their advantages. Ac amplifiers tend to be less expensive than dc amplifiers, particularly if high gain is required. Thus, if dc amplification is not needed, ac amplifiers are usually a more sensible choice.

Dc amplifiers are required when dc and very-low-frequency signals must be amplified. Weak dc voltage signals are produced by such sources as thermocouples, photovoltaic devices, and strain gauges. Many bioelectrical phenomena also generate weak dc and very-low-frequency signals. A dc amplifier must to used to amplify such signals. In addition, the oscilloscope becomes a much more versatile instrument when equipped with dc amplifiers.

Dc amplifiers have a tendency, however, to suffer from the problems of *offset voltages* and *drifts*. The *offset voltage* is a small dc voltage which appears at the output of the amplifier, even when no voltage is applied to its inputs. To correct this voltage error (which would be zero in an ideal dc amplifier), a small compensating voltage must be applied to the input of the amplifier. *Drift* is the rate at which the output voltage of an amplifier changes due to temperature variations or aging of the amplifier components. Because of drift, the zero level of dc amplifiers must be checked and adjusted from time to time in order to get accurate readings.

MISCELLANEOUS SPECIFICATIONS OF AMPLIFIER OPERATION

In addition to the characteristics of amplifiers described in the previous section, there are other specifications which also provide a measure of how well amplifiers perform certain functions. Some of the more common of these are listed below.

Linearity—The output of an *ideal* amplifier is directly proportional to its input, that is, a plot of the output versus input is a straight line. (This means that in an ideal linear amplifier the output signal is an amplified replica of the input signal—except for phase.) The extent to which this ideal is approached in an actual amplifier is specified by the linearity of the amplifier. Linearity is usually described either in terms of a percentage of the output or a percentage of the full-scale value. Nonlinearities in an amplifier lead to *distortions* of the output signal.

Amplifier Output—Describes the maximum voltage, current, or power output of an amplifier.

Input/Output Impedance—An amplifier should have a high input impedance to avoid *loading* the signal source (or, if applicable, the previous amplifier stage.) When the amplifier output is to be connected to another amplifier, the output impedance of the first one should be low (for *efficient* power transfer). If the output is to be connected to a recording or indicating device, the output impedance should match the input impedance of the recording device (for *maximum* power transfer).

Noise—One source of distortion in amplifiers, besides nonlinearity, is *noise*. This quantity can be divided into two classes: (a) unwanted signals which are generated externally and enter the amplifier from the outside and (b) unwanted signals which are generated by the components of the amplifier itself. The greater the magnitude of the noise, the more the desired signal will be distorted and obscured. The noise signal created by the amplifier itself is usually listed in the specifications of the amplifier and is expressed in terms of an rms voltage value.

Differential Amplifiers

There is a special type of amplifier, called the *differential amplifier*, which is designed to amplify the difference between the voltage values of two input signals (V_1 and V_2). In such amplifiers, each of the two signals is applied to one of the input terminals of the amplifier (as shown in Fig. 16-3). This means that the term a_{in} of Eq. (16-1) becomes

$$a_{\text{in}} = (V_1 - V_2) \tag{16-2}$$

From this relation, it follows that for a differential amplifier Eq. (16-1) is written as:

$$a_{\text{out}} = V_{\text{out}} = K_D(V_1 - V_2) \tag{16-3}$$

In this equation, K_D is called the *differential gain* of the amplifier. (The reason for using this nomenclature will become clearer as we discuss the operation of actual differential amplifiers.)

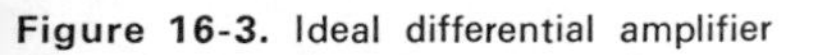

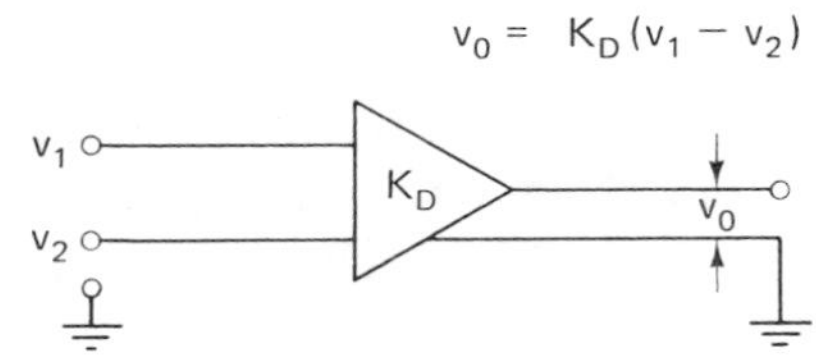

Figure 16-3. Ideal differential amplifier

We also note from Fig. 16-3 that in a differential amplifier neither of the input terminals is grounded. As a result, the differential amplifier can be used to amplify the difference between the voltages of two ungrounded points in a circuit. This makes differential amplifiers suitable for amplifying the signal output of such devices as the strain-gauge bridge shown in Fig. 16-4.

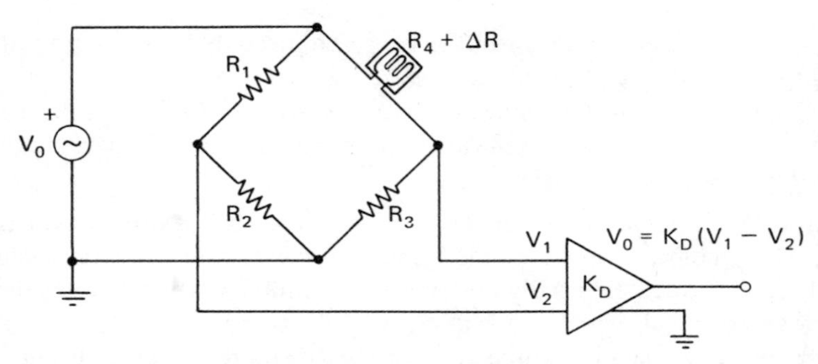

Figure 16-4. Use of a differential amplifier to measure the two ungrounded voltages of a Wheatstone bridge.

In Chapter 9 we also saw how oscilloscopes equipped with differential amplifiers are able to make use of this feature to measure nongrounded voltages. However, the fact that differential amplifiers can amplify nongrounded voltages is not the only characteristic which makes them so useful for certain applications. Some of these other unique characteristics will be examined as we continue with the discussion.

Since an ideal differential amplifier is designed to amplify only the difference between the voltage values applied to its inputs, two input signals of equal magnitudes should yield an output voltage of zero. As a result, the component of a signal which is *common* to both signals applied to the input of a differential amplifier should not be amplified. This component is called the *common-mode component* (V_C) and is defined as being equal to

$$V_C = \frac{1}{2}(V_1 + V_2) \tag{16-4}$$

The *difference component*, ΔV, which is the component we desire to be amplified, is defined as

$$\Delta V = (V_1 - V_2) \tag{16-5}$$

Example 16-1

If (a) $V_1 = 5$ and $V_2 = 3$, and if (b) $V_1 = 4 + 3 \sin \omega t$ and $V_2 = -5 + 5 \sin \omega t$, find V_C and ΔV for each case.

Solution: (a) If $V_1 = 5$ and $V_2 = 3$, then

$$V_C = \tfrac{1}{2}(V_1 + V_2) = 4$$

and

$$\Delta V = (V_1 - V_2) = 2$$

(b) If $V_1 = 4 + 3 \sin \omega t$ and $V_2 = -5 + 5 \sin \omega t$, then

$$V_C = -\tfrac{1}{2} + 4 \sin \omega t$$
$$\Delta V = 9 - 2 \sin \omega t$$

Actual differential amplifiers only approach the ideal characteristic of solely amplifying the ΔV component. In reality, instead of producing an output voltage according to Eq. (16-3), they also produce an output signal component that is proportional to V_C. That is, the the output signal of an actual differential amplifier is given by the equation

$$V_{\text{out}} = K_D(V_1 - V_2) + K_C \frac{V_1 + V_2}{2} = K_D \Delta V + K_C V_C \qquad (16\text{-}6)$$

where K_C is called the *common-mode gain.* This common-mode gain is unwanted, and amplifier-designers seek to minimize its value. As a method of specifying how closely an actual differential amplifier approaches the characteristics of its ideal counterpart, a quantity called the *common-mode rejection ratio* is used. The common-mode rejection ratio (CMRR) is defined to be the ratio of the differential gain (K_D) to the common-mode gain (K_C) of the amplifier:

$$\text{CMRR} = \frac{K_D}{K_C} \qquad (16\text{-}7)$$

An ideal differential amplifier would have an infinite CMRR. In practice, a well-designed commercially available differential amplifier is one with a CMRR of 1000 or more. CMRR is often expressed in terms of voltage decibels (dB) as well. In that case, the relation

$$\text{CMR (dB)} = 20 \log_{10} \text{CMRR} \qquad (16\text{-}8)$$

is used to convert the ratio (CMRR) to a voltage decibel value (CMR). Thus a differential amplifier which is specified as having a CMR of 80 dB has a CMRR of 10,000.

EXAMPLE 16-2

A differential amplifier receives two signals, V_1 and V_2, both of whose magnitudes are 10 V. The CMR of the amplifier is 90 dB and its differential gain is $K_D = 100$. Find the output voltage of the amplifier.

Solution: Using Eq. (16-8) we determine the CMRR and K_C:

$$\text{CMR} = 20 \log_{10} \text{CMRR}$$

$$90 = 20 \log_{10} \text{CMRR}$$

$$4.5 = \log_{10} \text{CMRR}$$

$$\text{CMRR} = 31{,}000$$

Then

$$\text{CMRR} = \frac{K_D}{K_C}$$

or

$$K_C = \frac{10^2}{3.1 \times 10^4} = 0.0032$$

Then using Eq. (16-6) we can find V_O

$$V_O = K_D(V_1 - V_2) + K_C\frac{(V_1 + V_2)}{2} = K_D(0) + 0.0032(10)$$

$$= 0.032 \text{ V} = 32 \text{ mV}$$

If the CMRR of a differential amplifier is high, the amplifier is able to reject hum (and other noise or interference) that appears simultaneously and in phase at the amplifier inputs. For example, if

$$V_1 = V_{\text{signal}} + V_{\text{noise}} = V_S + V_{\text{noise}}$$

and

$$V_2 = V_{\text{noise}}$$

then from Eq. (16-6) we can write

$$V_{\text{out}} = K_D\left(\Delta V + \frac{K_C}{K_D} V_C\right) = K_D V_S + \frac{K_D V_{\text{noise}}}{\text{CMRR}} + \frac{K_D V_S}{2\text{ CMRR}} \tag{16-9}$$

Equation (16-9) shows that if the CMRR of a differential amplifier is large, the component of voltage due to noise is largely rejected. This feature of differential amplifiers makes them extremely valuable for observing low-level signals in the presence of random noise. Electrical signals produced in biological systems are often detected along with a considerable amount of noise. As a result, differential amplifiers find wide use in measuring systems used to amplify bioelectrical signals.

Operational Amplifiers

Operational amplifiers are basically high-gain, dc amplifiers. Some are also designed to be differential amplifiers (although even these types are sometimes used with one input terminal grounded). The general symbol for operational amplifiers is shown in Fig. 16-5(a). The single-ended input types (i.e.,

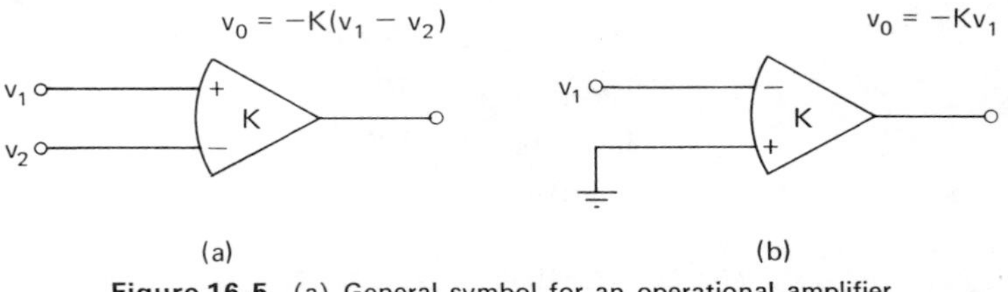

Figure 16-5. (a) General symbol for an operational amplifier (b) Operational amplifier with one input grounded (i.e. single-ended input)

one input terminal grounded) can be treated as a special case where the + input terminal of the general type is grounded (Fig. 16-5(b)). The major difference between operational amplifiers and ordinary dc, differential amplifiers is that operational amplifiers are designed to be used with external feedback networks. That is, a portion of the output signal of the operational amplifier is fed back to its input through various feedback paths, depending on the specific function of the operational amplifier. It is because of this feature that the operational amplifier can be used to perform such an extraordinary number of different functions.

The term *operational amplifier* originated because high-gain, dc feedback amplifiers were first developed to perform the mathematical *operations* of addition, subtraction, and integration in analog computers. This name has clung to them even though their use has spread to a vast number of measurement and control applications outside of analog computing.

The operational amplifier is being manufactured in complete units from integrated circuits as well as from discrete components. In either case, the designer of instrument systems can incorporate complete operational amplifiers into his system. Because of the ease of incorporating such complete building-blocks into instrumentation systems, the design and construction of new measurement systems has become a much simpler task.

CHARACTERISTICS OF OPERATIONAL AMPLIFIERS

The operational amplifier, without any feedback paths connected to it, is described as being operated in an *open-loop* mode (i.e., the feedback loop is not closed). The ideal characteristics of the operational amplifier in this open-loop mode are

1. Gain $= \infty$
2. Bandwidth $= \infty$
3. Input impedance $= \infty$
4. Output impedance $= 0$
5. Output signal, $V_o = 0$ when $V_2 = V_1$

Actual operational amplifiers, of course, cannot meet these ideal open-loop specifications. Instead they are designed to approximate these requirements as closely as possible. Actual operational amplifiers have open-loop gains ranging from 10^3 to 10^9 (typically 10^5) and a flat response (i.e., constant gain) from zero frequency out to several kHz. The input impedance is commonly in the range of 10 kΩ, while the output impedance is about 25–50 Ω.

The operational amplifier is invaluable as a general purpose amplifier primarily because of its high gain. This characteristic makes it possible to utilize external feedback connections with it in such a way that the *overall gain or the characteristics of the output signal from the amplifier depend primarily on the values of the elements in the feedback path.* The passive electrical components which are usually used to make up the elements of the feedback path have accurately known and stable values. Therefore, the overall output characteristics of the operational amplifier can be accurately controlled, *independently* of the properties of the amplifying element itself.

Once a feedback path is connected to the basic operational amplifier shown in Fig. 16-5, the amplifier is said to be operating in a *closed-loop mode*. The general representation of an operational amplifier with feedback is shown in Fig. 16-6(a). In this diagram, the impedances (Z_f and Z_{in}) can be any combination of resistive or reactive elements, and K represents the open-loop gain of the amplifier.

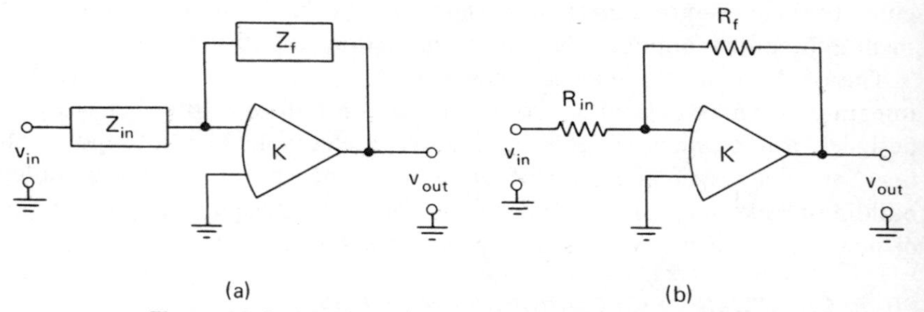

Figure 16-6. (a) Operational amplifier with feedback loop connected to it (b) Operational amplifier with resistive feedback path connected to it

It can be shown that when an amplifier with a large open-loop gain (i.e., $K \gg (1 + Z_f/Z_{in})$) is used with a feedback path like the one shown in Fig. 16-6(a), the output V_{out} is related to the input V_{in} by

$$V_{out} = -\frac{Z_{in}}{Z_f} V_{in} \tag{16-10}$$

If the impedances Z_f and Z_{in} are resistances R_f and R_{in}, we can rewrite Eq. (16-10) as

$$V_{out} = -\frac{R_f}{R_{in}} V_{in} \tag{16-11}$$

This equation shows that if we want to accurately multiply the value of some voltage signal by a (negative) constant, this can be done by connecting a feedback path of two accurately known resistors (whose ratio R_f/R_{in} is equal to the value of the multiplicative constant) to an op amp. Figure 16-7 shows how multiplication by 10 takes place. (Note that an additional unity-gain operational amplifier is included in the circuit to invert the minus

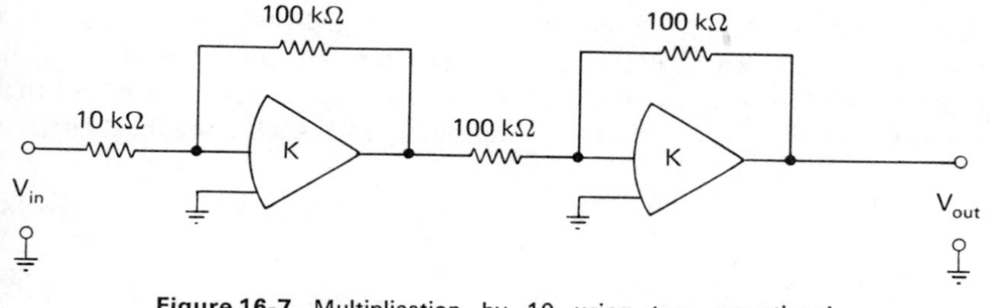

Figure 16-7. Multiplication by 10 using two operational amplifiers and appropriate feedback paths

sign.) However, the ratio of Z_f/Z_{in} must be kept sufficiently low that the condition of $K \gg (1 + Z_f/Z_{in})$ is maintained. If this condition is not well-satisfied, an error in the expected value of the output signal will occur.

If additional input branches containing resistances are connected to point A of Fig. 16-8, the output signal can be shown to be equal to

$$V_{out} = -\left(V_1\frac{R_f}{R_1} + V_2\frac{R_f}{R_2} + V_3\frac{R_f}{R_3}\right) \tag{16-12}$$

If $R_1 = R_2 = R_3 = R_f$ in this expression, then

$$V_{out} = -(V_1 + V_2 + V_3) \tag{16-13}$$

Thus the negative of the sum of the signals applied to the inputs of this circuit is provided at the output. Again we note that, to get a positive output value, a unity-gain inverting amplifier would have to be added in series to the summing amplifier.

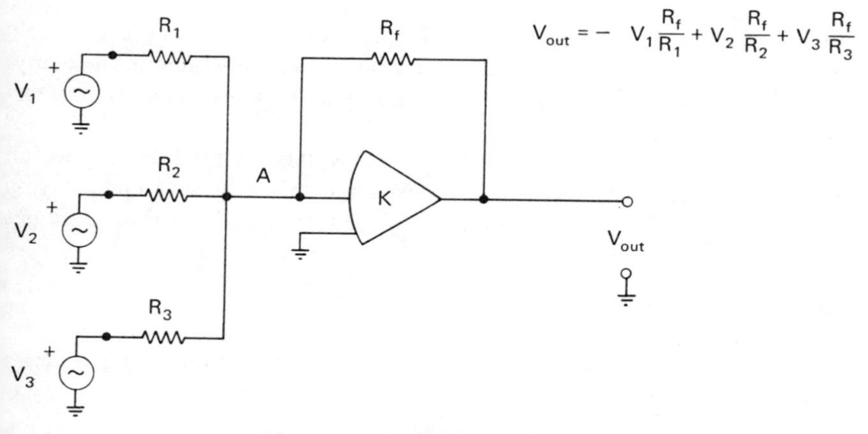

Figure 16-8. Summing network

If the resistor, R_f, in the feedback loop of Fig. 16-6(b) is replaced by a capacitor, C_f, the operational amplifier becomes an *integrator* (Fig. 16-9). From the general operational-amplifier equation [Eq. (16-10)], we find that the voltage output of an operational amplifier with such a feedback path is

$$v_{\text{out}} = -\frac{1}{RC}\int v_{\text{in}}\,dt \tag{16-14}$$

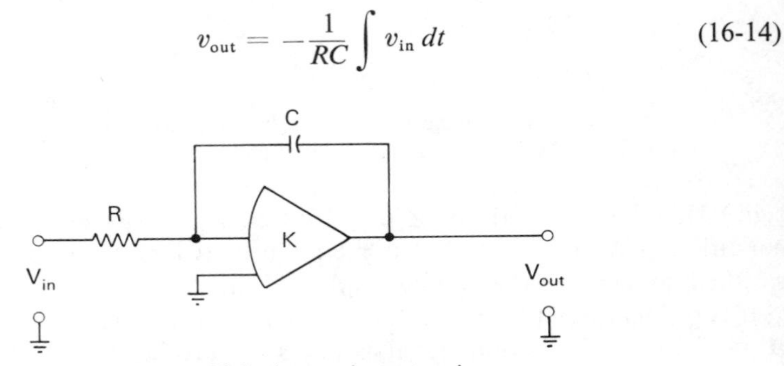

Figure 16-9. Integrating network

The capabilities of summation, multiplication, inversion, and integration allow operational amplifiers to solve linear differential equations as well as simultaneous linear algebraic equations. To perform these functions, the various amplifiers must be connected together in the proper configurations required by the equation or set of equations. The procedure of connecting operational amplifiers in an analog computer in the proper manner to solve such equations, is known as *programming* the analog computer. Further information on how to program analog computers to solve such equations can be found in texts devoted specifically to analog computers.[2]

ADDITIONAL USES OF OPERATIONAL AMPLIFIERS

To indicate some of the reasons why operational amplifiers are so widely used, we will describe a few more of the common circuits (among the many) in which operational amplifiers are being used. Figure 16-10 shows the circuits which we will consider.

In Fig. 16-10(a), the zener diode reference voltage is used to produce an accurately known output voltage, V_Z. If we apply this voltage to a multiplying operational amplifier circuit, the output of the operational amplifier becomes

$$V_{\text{out}} = -\frac{R_f}{R_{\text{in}}}V_z \tag{16-15}$$

[2]Joseph J. Blum, *Introduction to Analog Computation* (New York, N.Y.: Harcourt Brace Jovanovich, Inc., 1969).

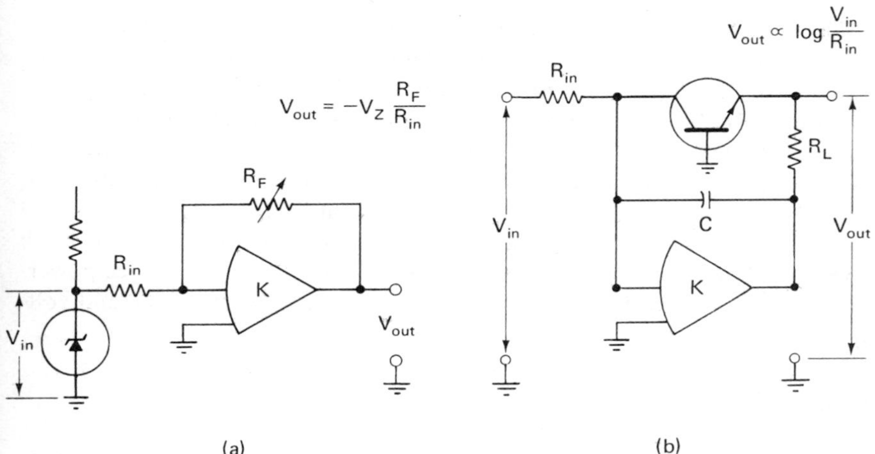

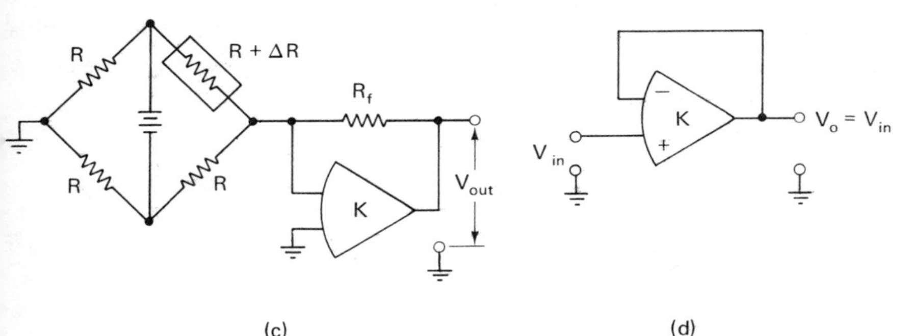

Figure 16-10. Use of the operation amplifier as a building block
(a) Precision variable-voltage source (b) Logarithmic amplifier
(c) Bridge amplifier (d) Voltage follower

Since R_f is a variable resistor, we can adjust the output voltage of this circuit quite accurately.

The circuit of Fig. 16-10(b) shows an amplifier whose output is proportional to the logarithm of the input. The logarithmic output variation is obtained with the help of a transistor connected in the feedback path.

An amplifier used to detect bridge imbalances is shown in Fig. 16-10(c). This circuit has an output voltage V_o:

$$V_o = E_B\left(\frac{\Delta R}{4R}\right)\left(\frac{R_f}{R}\right) \tag{16-16}$$

The *voltage follower* circuit shown in Fig. 16-10(d) has this name because the output, V_o, is equal in sign and magnitude to the input voltage. The function of this circuit is to isolate the output from the input of the device. Consequently, the input impedance is very high (100 MΩ) and the output impedance very low (less than 1 Ω). With this capability, less than 1 nA can be drawn from a circuit whose voltage is being measured by an instrument with such a voltage follower as its input element.

Some of the many other circuits (besides the ones listed above) in which operational amplifiers are used as building blocks, include dc voltage power-supply regulators, voltage-to-current converters, charge amplifiers, and voltage limiters.

Problems

1. Explain what the terms a) voltage gain (A_v) and b) power gain (A_p) signify.
2. Explain the difference between dc and ac amplifiers.
3. Demonstrate that the two definitions of bandwidth of an amplifier (that is, as defined in terms of power gain and in terms of voltage gain) are equivalent.
4. If the magnitudes of the input and output voltages of an amplifier are given as listed below, calculate the voltage gain of the amplifier:
   a) $v_i = 2$ V, $v_o = 50$ V
   b) $v_i = 75$ mV, $v_o = 3$ V
   c) $v_i = 50\ \mu$V, $v_o = 20$ mV
   d) $v_i = 25$ mV, $v_o = 10$ mV
5. If a 50 mV signal appears at the output of a dc amplifier when no signal is applied to its input, is this an unexpected event? Why or why not? How can the magnitude of this output signal be reduced?
6. A differential amplifier has a gain of $A_v = 120$. When the common-mode gain A_c is being measured, we get $v_i = 2$ V and $v_o = 20$ mV. Calculate the CMRR and the CMR of this amplifier.
7. Determine the output voltages for the "op amp" circuits shown in Fig. P16-1.

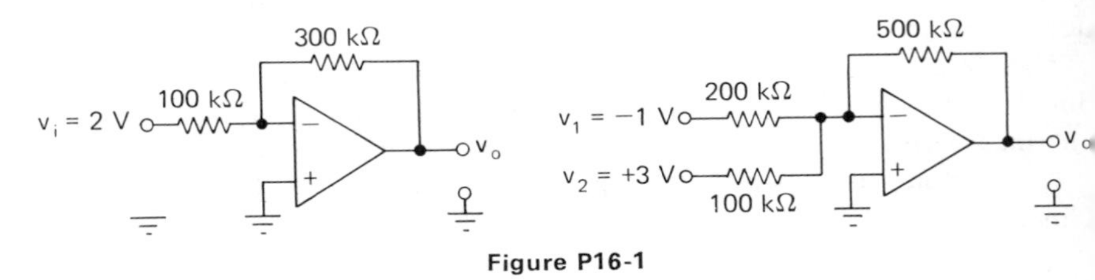

Figure P16-1

8. Since an amplifier is basically designed to multiply the magnitude of an input signal by some constant amount, what is the advantage of using an op amp with a feedback path to accomplish essentially the same purpose?

9. Explain why the property of having a high input impedance is often a desirable characteristic in amplifiers.
10. List some of the applications that make the op amp such a versatile building block in the design of measurement instrumentation.

References

1. Mandl, M., *Fundamentals of Electronics*, Chaps. 11, 12, 13, 15, and 16. Englewood Cliffs, N. J., Prentice-Hall, Inc., 1960.
2. Graeme, J. G., Tobey, G. E., and Huelsman, L. P., eds. *Operational Amplifiers*. New York, McGraw-Hill Book Company, 1971.
3. Malmstadt, H. V., Enke, C. G., and Toren, E. C., *Electronics for Scientists*, Chaps. 3, 4, and 8. New York, Benjamin, 1962.
4. Kahn, M., *The Versatile Op Amp*. New York, Holt, Rinehart and Winston, Inc., 1970.
5. Wedlock, B. D., and Roberge, J. K., *Electronic Components and Measurements*, Chap. 16. Englewood Cliffs, N. J., Prentice-Hall, Inc., 1969.

17

Uses of Electronic Instrumentation in Scientific Measurements

One of the objectives of this text was to describe electronic measuring instruments and techniques in such a way that they could be understood and used by workers in many different fields. The bulk of the material has thus been devoted to examining the operation of specific electrical and electronic devices and instruments. It was useful to present the information in this manner so that a rather complete summary of the basic principles of electrical instrumentation would be provided.

In this final chapter we will take a brief look at various scientific fields to see some of the ways in which the instruments we have described are actually put to use. The discussions will be short and we will not attempt to fully define all the new terms that will crop up in the surveys of each of the fields we will cover. What we hope to achieve by presenting such an overview is the following. First, it should give the specialist a definite idea of the type of electronic instruments utilized in his field. Being armed with this knowledge, he may wish to review those sections of the book which are devoted to the particular instruments that he is most likely to encounter. It may also spur him to examine some of the specialized texts which describe the instruments in greater detail. Second, it should give the general reader a better appreciation of the broad scope of measurements that can be undertaken with the aid of electronic instrumentation.

The areas into which we will divide our discussion are:

1. Physics
2. Chemistry and chemical engineering
3. Geology

4. Biology
5. Physiology and medicine
6. Pollution detection

A list of references describing the electronic instrumentation used in these areas is presented at the end of the chapter.

PHYSICS

The electronic measurements of such classical physical quantities as force, temperature, pressure, and light intensity are performed by a variety of electrical transducers. Many of the more common transducers used for such tasks were discussed in Chapter 15.

Current physics research makes use of some of these same transducers as well as a wide variety of additional electronic measuring equipment. In order to list some of the more widely used devices and techniques, we will consider three prominent areas that are presently of interest to physicists: high-energy and nuclear physics, solid-state physics, and low-temperature physics. We will also mention a few of the miscellaneous electronic devices being used in a cross section of physics research.

In high-energy and nuclear physics, the detection of radiation and interactions between fundamental particles is an important task. Such data yield information about the outcome of high-energy and nuclear experiments. The detection of the events taking place during these experiments is most often carried out with radiation detectors such as those also mentioned in Chapter 15. One class of detectors used in high-energy and nuclear work are *scintillators* connected to photomultiplier tubes. These devices produce light flashes when struck by radiation. The light flashes are then converted to electrical pulses by the instrument. The pulses can be fed to coincidence circuits or counters for recording. Ionization chambers, geiger counters, and solid-state radiation detectors are also heavily utilized in this line of research.

The investigations of solid-state physics pursue questions concerning the electrical, thermal, and magnetic properties of solids, as well as their structural makeup. The structural aspects of solids are studied with the help of piezoelectric crystals and X-ray diffraction techniques. Piezoelectric crystals are stimulated with electric signals to create acoustic waves in the solids to which they are attached. The propagation of these waves reveals some of the structural features of the solid lattices being studied. Oscilloscopes are used to detect the velocities and attenuations of these acoustic waves in solids. Likewise, X-ray beams incident on lattices are diffracted in patterns which depend on the arrangement of the atoms in the lattice. One popular X-ray-diffraction sensing device converts the diffracted X-ray beams to visible light and then detects the light with an electronic scintillation counter. The electrical properties of solids (such as conductivity and charge-carrier mobilities) are measured with Hall-effect transducers, together with voltmeters, power supplies, and pulse generators. Some additional electronic properties of solids, as revealed by cyclotron resonance measurements, are studied with the aid of rf oscillators and rf wattmeters. Absorbtion spectrophotometers and other spectroscopic techniques are utilized for the analysis of the composition of certain solids. Many such spectroscopic systems use photoconductive cells, photomultipliers, or thermistors to detect the radiation being emitted from the spectroscope, as well as null-balance recorders to indicate their responses.

Studies in *low-temperature* physics (20°K and below) make use of carbon or special semiconductor materials as temperature-sensing devices. The resistance of both of these materials changes sufficiently per degree at low temperatures to allow accurate temperature determinations to be made.

Some of the miscellaneous electronic devices being used in a wide variety of contemporary physical measurement systems include the *ion gauge*, the *mass spectrometer*, and instruments used to *extract weak signals from noise*. The ion gauge is used to measure the very low pressures associated with high vacuums. The mass spectrometer is an analytical tool which uses simultaneous applied electric and magnetic fields to separate ions according to their charge-to-mass ratio.

The instruments used to extract weak signals from surrounding noise are advanced electronic devices. The three common signal-extraction methods which these devices employ are signal-averaging, phase-locking, and correlation techniques.[1] *Signal-averaging* devices make use of the fact that random noise tends to have an average value of zero. Thus, by averaging the value of a signal over a long time period, the noise contribution diminishes in importance relative to the desired signal. *Phase-lock* devices (sometimes called *lock-in* amplifiers) extract signals by favoring the frequency of the signal being detected and rejecting the random frequencies of noise.

CHEMISTRY AND CHEMICAL ENGINEERING

Chemists and chemical engineers use a wide variety of electronic methods for analyzing chemical compounds. Some of these methods include ionic concentration determinations, gas chromatography, spectroscopic methods, polarography, and X-ray diffraction methods.

The ion concentration of a solution is often a measure of its composition or purity. The hydrogen ion concentration (or pH) is measured by the electronic pH meters described in Chapter 15. These meters consist of two special electrodes (which are immersed in the test solution) and a very-high-

[1]For more detailed but introductory information on these topics, see "Noise in Amplifiers," an article by Letzer and Webster, in *IEEE Spectrum*, Aug., 1970.

impedance dc voltmeter (called an *electrometer*). The voltage reading of the electrometer is calibrated to read pH directly. Another method for determining ionic concentrations in liquids involves making conductivity measurements of the solution. Such conductivity measurements are commonly used to measure the purity of distilled water. Two special electrodes and a Wheatstone bridge are utilized in this method to measure the resistance of the water.

The gas chromatograph and the absorption spectrophotometer are chemical analysis methods which use some type of electronic detecting device and recorder in their final stages. The gas chromatograph works best on gas samples, while the absorption spectrophotometer analyzes sample materials dissolved in a liquid solvent.

The atomic absorption method is another spectroscopic analysis method. It is used to detect the presence of trace metals (e.g., mercury or cadmium) in a variety of materials. In this method, the test sample in solution is atomized (as in a perfume atomizer) and its vapor heated by an acetylene flame. The energy absorbed by the atoms from the flame is reemitted at specific wavelengths according to the particular materials present in the sample. The various wavelengths of the emitted light are refracted at different angles and detected by photomultiplier tubes.

Polarography is another electrical method used for analyzing ionic solutions. The polarograph is a plot of the current versus voltage of a solution which is subjected to a particular electrical and chemical environment. In setting up a polarography measurement, the sample of interest is diluted in another special buffering solution. The ions of interest diffuse toward a mercury-drop electrode immersed in the total liquid solution. The rate of diffusion (and thus the concentration) of various ionic species toward the electrode can be determined by the polarograph. Each ionic type present in the solution can also be identified from the polarograph.

Nuclear magnetic resonance (NMR) techniques and the mass spectrometer are two additional measurement methods used by chemists and physicists for analyzing materials. NMR measurements are most commonly used to study the structure of complex organic and biological molecules. The mass-spectrometer (as mentioned in the section on physics measurements) uses simultaneously applied electric and magnetic fields to separate and identify atoms and molecules of various substances.

Chemical engineers are involved in monitoring and controlling such physical processes as temperature, fluid flow, and liquid level (although they are primarily concerned with questions of chemical composition). The various electrical transducers which were described in Chapter 15 are utilized almost exclusively for this type of monitoring. The data obtained by such devices are often recorded by either graphic recorders or digital computers.

GEOLOGY

Geological measurements which utilize electronic instrumentation can be divided into two broad categories:

1. Analysis of the composition of ores and minerals.
2. Geophysical measurements (including geophysical research and prospecting, and oceanographic research).

The analysis of ore and mineral compositions entails the use of many of the same instruments and methods used by chemists and physicists for analyzing chemical compounds. Those analytical instruments which rely (partially or fully) on electronic devices for their operation include mass spectrometers, X-ray diffraction devices (with photoelectric detectors), X-ray fluorescence techniques, electron microprobes, and atomic absorbtion spectrophotometers.

Geophysical measurements involve the studies of such phenomena as seismic events (earthquakes), the motion and properties of the Earth's crust, and the magnetic and gravitational fields of the Earth.

Seismic events and motions of the Earth's crust are monitored in many ways. Earthquakes are detected and measured with accelerometers such as those discussed in Chapter 15. The slippage of geological faults is checked by stringing strain gauges across the faults and by the use of laser interferometer methods. The laser interferometer determines changes in distance between two points on the earth's surface with use of a laser beam. The laser beam in this measurement serves as a yardstick whose length is very accurately known.

The *tiltmeter* is a relatively new device being used in connection with earthquake prediction. A long, mercury-filled tube is delicately balanced at its center. A metal plate is placed below one end of the tube. Any motion of the Earth's crust below the tube causes the distance (and thus the capacitance) between the metal and the mercury to change. These capacitance changes (and hence the motions of the Earth) are detectable as electric signals.

The Earth's magnetic field is measured with magnetometers. Changes in the uniformity of the magnetic field (called anomalies) yield information about large ore deposits, the composition of the crust, and the geological history of the Earth.

Geophysical prospecting (chiefly for oil) involves the propagation of low-frequency (5–100 Hz) acoustic waves through the Earth's crust. The waves are generated by explosions or the dropping of heavy weights onto the ground. The waves are then detected by *geophones* (similar to capacitive or electromagnetic microphones in their operation) and recorded on galvanometer recorders. The recorded data yield information about the densities and depths of layers existing in the Earth's crust.

Measurements of the ocean's depth and of sediments on the ocean floor are also carried out by this technique. The oxygen, salinity, and temperature of sea water are also measured by electrical transducers.

MEDICINE AND PHYSIOLOGY

Physiological and medical measurements made with the help of electronic instruments can be divided into three general categories:

1. Diagnostic measurements
2. Bodily function monitoring
3. Research measurements

When a measurement in any of these areas involves *electrical* phenomena taking place in the human body (e.g., electrical nerve impulses), *electrodes* are used to detect the resulting signals. (Such electrodes can be placed internally in the body as well as externally.) If the bodily phenomenon being measured is not an electrical quantity (such as blood pressure or respiration rate), a transducer must first be used to convert the quantity into an electrical signal.

Diagnostic measurements involving electronic measuring techniques include electrocardiograms (EKG), electroencephalograms (EEG), fetal EKGs, nerve-conduction measurements, and probing the internal organs and structure of the body using sound waves. The EKG is basically the waveform associated with the electrical activity of the heart. It is measured by placing electrodes at special locations on the body. The detected waveshape can help characterize a healthy or sick heart. The EEG is the waveform pattern produced by the electrical activity of the brain. It is detected by electrodes attached to special points on the head. The pattern of the EEG can also be interpreted to provide an indication of abnormal brain activity. To assist in the interpretation of the EEG, a *wave analyzer* is used to break down the complex EEG waveform into its various basic components.

Ultrasonography is a technique where ultrasonic energy is used to detect internal organs of the body. It operates by bouncing ultrasonic waves (produced by an electrical transducer attached to the body) off of the interfaces between internal organs. It is suitable for some applications where X rays cannot be used (such as detecting cysts in a woman's breast, locating possible brain tumors, and examining the fetus of a pregnant woman). Ultrasonic holography is also being studied as another method for exploring the internal structure of the body.

Patient monitoring involves the measurement of such bodily functions as heart rate, blood pressure, respiration rate, body temperature, and the EKG. If a patient-monitoring system is connected to a hospital patient, these bodily functions can be simultaneously monitored along with the bodily functions of many other patients. If any one of the functions goes awry, an alarm rings and a nurse is alerted to the patient's bedside. Closed-circuit televisions are also used as part of the patient-monitoring systems.

Heart catheters are plastic tubes which are inserted into a vein and slid along until the tip of the tube reaches the heart. Once emplaced, heart catheters yield information about blood pressure and flow taking place at the heart. This information can be made available during such surgical operations as valve replacement and open-heart surgery, as well as during the surgical recovery.

Physiological *research measurements* use many of the devices mentioned above and also some additional ones. The electroretinogram (used to study the electrical response of the retina to light), electrodes placed in the inner ear to detect the ear's response to sound, and electroanesthesic methods are all being currently utilized. Conscious control of the EEG and other bodily functions is being learned by individuals through the use of *biofeedback* methods. These methods monitor the bodily function of interest and alert the individual trying to exert control of them.

BIOLOGY

Biological measurements which make use of electronic instruments overlap with some of the measurements discussed in the *Physiology and Medicine* section. However, there are also many other electronic techniques and instruments used by biological workers besides the ones listed there. We shall list a few of them in this section.

The *cell* is the smallest unit of life. As part of the functioning of cells, a bioelectric potential is developed between points inside and outside of the cell. This potential difference arises because of the difference in concentrations of various ions within and external to the cell. Changes in this potential difference (voltage) are measures of the functioning and activity of the cell. In addition, if electric stimuli are applied to cells, an electrical response known as the *cell action potential* is produced. These bioelectric aspects of cells (and others) are studied with special voltmeters and amplifiers. In particular, the stimulation of nerve cells and their responses are measured with pulse and function generators and with differential amplifiers.

To study the internal conditions of living animals, ingestible oscillators made of transistors and integrated circuits are being used. These tiny oscillators are yielding information about temperature and pressure changes within living animals. The signals they generate belong in the rf portion of the spectrum and thus can propagate through the animal to be received in the space outside.

Other miscellaneous uses of electronics in biology include automatic bacteria counters, detectors of radioactive tracers, and light-level meters for studying plant growth and photosynthesis processes. Automatic bacteria counters use a scanner, photoconductive cell, and counter to determine the number of bacterial colonies growing in petri-dish cultures.

In general, the electrical signals created in biological systems are weak and obscured by noise; hence sensitive differential amplifiers are most commonly used to measure them. In addition, the procedures for extracting weak signals from noise discussed in the earlier section on physics instrumentation are also frequently used. Because of the weak signals encountered, proper grounding and shielding techniques also assume a position of paramount importance in both biological and physiological measurements.

POLLUTION MONITORING

Electronic methods currently available for the detection of pollutants involve many of the analytical techniques used by chemists and physicists. However, there is much research being done in this field and many more methods are being studied at present. The near future should see the development of a host of new detection devices.

In dealing with the pollution-monitoring devices being currently used, we can divide pollution detection into three broad areas:

1. Gas
2. Liquids
3. Solids

Gas pollutant detectors are being used to measure the quality of air. Absorption spectrophotometers are used to detect such gases as carbon monoxide (CO), carbon dioxide (CO_2), sulphur dioxide (SO_2), nitrogen oxides, ozone, and unburned hydrocarbons. Unfortunately, absorption spectrophotometric methods are quite slow; thus, investigations of methods which provide instantaneous detection of gas impurities over long distances (such as Raman spectroscopy using lasers) are being performed. Infrared semiconductor lasers are also being investigated as devices for monitoring the exhaust emissions of automobile engines.

Pollutants in liquids can be detected with the pH meters and conductivity meters described in Chapter 15. Other devices for sensing liquid pollutants include bacterial analyzers, dissolved-salt analyzers, and temperature meters (for measuring thermal pollution of water from electric power plants).

Pollution detection in solids is carried out primarily by soil and food analyzers. Gas chromatographs and atomic-absorption spectrophotometers (for sensing the presence of trace metals such as mercury) are used in both of these purposes.

References

1. BAIR, E. J., *Introduction to Chemical Instrumentation.* New York, McGraw-Hill Book Co., 1962.
2. DOBRIN, M. B., *Introduction to Geophysical Prospecting*, 2nd ed. New York, McGraw-Hill Book Co., 1960.
3. GRAEME, J. G., TOBEY, G. E., and HUELSMAN, L. P., eds., *Operational Amplifiers.* New York, McGraw-Hill Book Company, 1971.
4. TEKTRONIX, *Biophysical Measurements.* Beaverton, Ore., Tektronix, Inc., 1971.
5. TEKTRONIX, *Use of Transducers*, Beaverton, Ore., Tektronix, Inc., 1971.
6. KOWALSKI, E., *Nuclear Electronics.* New York, Springer-Verlag, 1969.
7. PRENSKY, S. D., *Electronic Instrumentation*, 2nd ed., Chap. 19. Englewood Cliffs, N.J., Prentice-Hall, Inc., 1970.
8. EWING, G. W., *Analytical Instrumentation, A Guide for Chemical Analysis.* New York, Plenum Press, 1966.
9. CHAPMAN, R. L. ed., *Environmental Pollution Instrumentation.* Pittsburgh, Pa., Instrument Society of America (ISA) Publ., 1969.

Appendix **A**

*Electrical Units and Conversion Factors**

Table A-1 Fundamental, Supplementary, and Derived Units

Quantity	Symbol	Dimension	Unit	Unit Symbol
Fundamental				
Length	l	L	meter	m
Mass	m	M	kilogram	kg
Time	t	T	second	s
Electric current	I	I	ampere	A
Thermodynamic temperature	T	Θ	degree Kelvin	°K
Luminous intensity			candela	cd
Supplementary				
Plane angle	α, β, γ	[L]°	radian	rad
Solid angle	Ω	$[L^2]$°	steradian	sr
Derived				
Area	A	L^2	square meter	m^2
Volume	V	L^3	cubic meter	m^3
Frequency	f	T^{-1}	hertz	Hz (l/s)
Density	ρ	$L^{-3}M$	kilogram per cubic meter	kg/m^3
Velocity	v	LT^{-1}	meter per second	m/s
Angular velocity	ω	[L]°T	radian per second	rad/s
Acceleration	a	LT^{-2}	meter per second squared	m/s^2
Angular acceleration	α	$[L]°T^{-2}$	radian per second squared	rad/s^2
Force	F	LMT^{-2}	newton	N(kg m/s^2)
Pressure, stress	p	$L^{-1}MT^{-2}$	newton per square meter	N/m^2
Work, energy	W	L^2MT^{-2}	joule	J (N m)

*Courtesy of W. D. Cooper, *Electronic Instrumentation and Measurement Techniques* (Englewood Cliffs, N.J.: Prentice-Hall, Inc., 1970), pp. 27–29.

Table A-1 Fundamental, Supplementary, and Derived Units (Continued)

Quantity	Symbol	Dimension	Unit	Unit Symbol
Power	P	L^2MT^{-3}	watt	W (J/s)
Quantity of electricity	Q	TI	coulomb	C (A s)
Potential difference, electromotive force	V	$L^2MT^{-3}I^{-1}$	volt	V (W/A)
Electric fieldstrength	E, ϵ	$LMT^{-3}I^{-1}$	volt per meter	V/m
Electric resistance	R	$L^2MT^{-3}I^{-2}$	ohm	Ω (V/A)
Electric capacitance	C	$L^{-2}M^{-1}T^4I^2$	farad	F (A s/V)
Magnetic flux	Φ	$L^2MT^{-2}I^{-1}$	weber	Wb (v s)
Magnetic fieldstrength	H	$L^{-1}I$	ampere per meter	A/m
Magnetic flux density	B	$MT^{-2}I^{-1}$	tesla	T (Wb/m²)
Inductance	L	$L^2MT^{-2}I^2$	henry	H (V s/A)
Magnetomotive force	U	I	ampere	A
Luminous flux			lumen	Im (cd sr)
Luminance			candela per square meter	cd/m²
Illumination			lux	Ix (Im/m²)

Table A-2 Electric and Magnetic Units

Quantity and Symbol	SI Unit			Conversion Factors	
	Name and Symbol		Defining Equation	CGSm	CGSe§
Electric current, I	ampere	A	$F_z = 10^{-7}I^2\frac{dN^*}{dz}$	10	10/c
Electromotive force, E	volt	V	p† = IE	10^{-8}	10^{-8}c
Potential, V	volt	V	p† = IV	10^{-8}	10^{-8}c
Resistance, R	ohm	Ω	R = V/I	10^{-9}	10^{-9}c
Electric charge, Q	coulomb	C	Q = It	10	10/c
Capacitance, C	farad	F	C = Q/V	10^9	$10^9/c^2$
Electric field strength, E	—	V/m	E = V/l	10^{-6}	10^{-6}c
Electric flux density, D	—	C/m²	$D = Q/l^2$	10^5	10^5/c
Permittivity, ϵ	—	F/m	ϵ = D/E	—	$10^{11}/4\pi c^2$
Magnetic fieldstrength, H	—	A/m	$\oint H\,dl = nI$	$10^{3/4}$	—
Magnetic flux, Φ	weber	Wb	E = dΦ/dt	10^{-8}	—
Magnetic flux density, B	tesla	T	B = Φ/l²‡	10^{-4}	—
Inductance, L, M	henry	H	M = Φ/I	10^{-9}	—
Permeability, μ	—	H/m	μ = B/H	$4\pi \times 10^{-7}$	—

**N* denotes Neumann's integral for two linear circuits each carrying the current *I*. F_z is the force between the two circuits in the direction defined by coordinate *z*, the circuits being in a vacuum.

†*p* denotes power.

‡l^2 denotes area.

§*c* = velocity of light in free space in cm/s = 2.997925×10^{10}.

Table A-3 British into Metric Conversion

	English Unit	Symbol	Metric Equivalent	Reciprocal
Length	1 foot	ft	30.48 cm	0.0328084
	1 inch	in.	25.4 mm	0.0393701
Area	1 square foot	ft^2	9.29030×10^2 cm^2	0.0107639×10^{-2}
	1 square inch	$in.^2$	6.4516×10^2 mm^2	0.155000×10^{-2}
Volume	1 cubic foot	ft^3	0.0283168 m^3	35.3147
Mass	1 pound (avdp)	lb	0.45359237 kg	2.20462
Density	1 pound per cubic foot	lb/ft^3	16.0185 kg/m^3	0.062428
Velocity	1 foot per second	ft/s	0.3048 m/s	3.28084
Force	1 poundal	pdl	0.138255 N	7.23301
Work, energy	1 foot-poundal	ft pdl	0.0421401 J	23.7304
Power	1 horsepower	hp	745.7 W	0.00134102
Temperature	degree F	t°F	$5(t - 32)/9$°C	—

Appendix **B**

Identification of Discrete Solid State Components

Semiconductor Diodes—Some of the more common packaging configurations in which semiconductor diodes are manufactured are shown in Fig. B-1. The cathode and anode of the diode are identified in the figures. A number with the prefix 1N is usually stamped on the body of the diode. This prefix

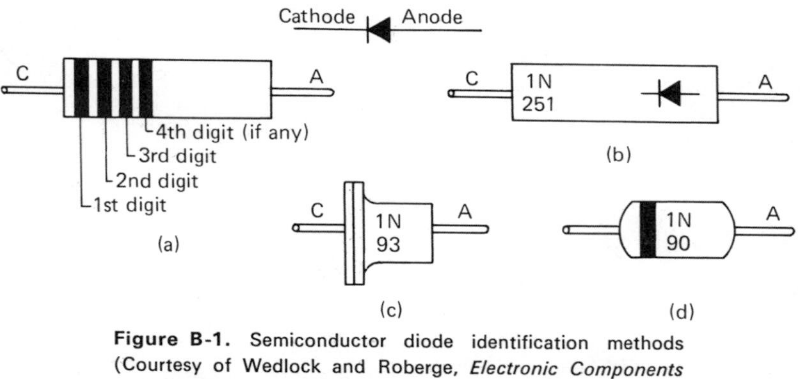

Figure B-1. Semiconductor diode identification methods (Courtesy of Wedlock and Roberge, *Electronic Components and Measurements,* Englewood Cliffs, N.J.: Prentice-Hall, Inc., 1969, p. 16.)

indicates that the device is a diode. If there are colored stripes on the body of the diode, the colors represent numbers (using the same color code as used in resistors). These numbers identify the diode type.

Discrete Transistors—Discrete transistors are manufactured in a wide variety of package configurations. Low-power transistors are usually packaged in metal containers or embedded in a plastic. High-power transistors are generally packaged in cases having large areas designed for being intimately attached to a heat sink. The more common packaging configurations are shown in Fig. B-2.

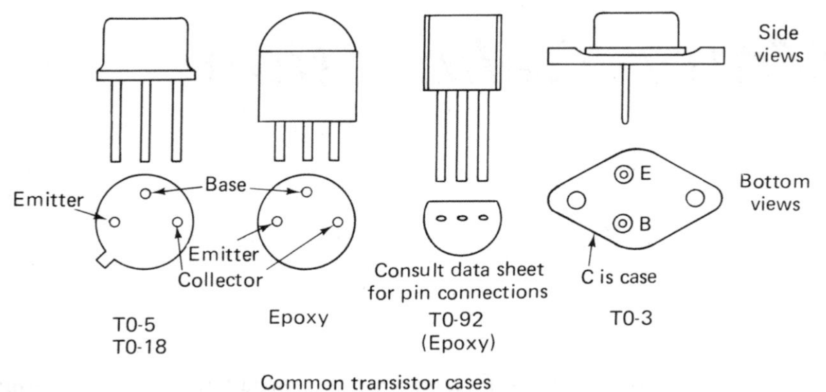

Figure B-2. Common transistor cases (Courtesy of Wedlock and Roberge, *Electronic Components and Measurements,* Englewood Cliffs, N.J.: Prentice-Hall, Inc., 1969, p. 16.)

The body of the transistor is usually stamped with a number. The prefix of the number is 2N and this indicates that the device is a transistor. The remaining digits of the number specify the transistor type.

Silicon Controlled Rectifiers (SCR)—Some configurations of the most common commercially available SCRs are shown in Fig. B-3.

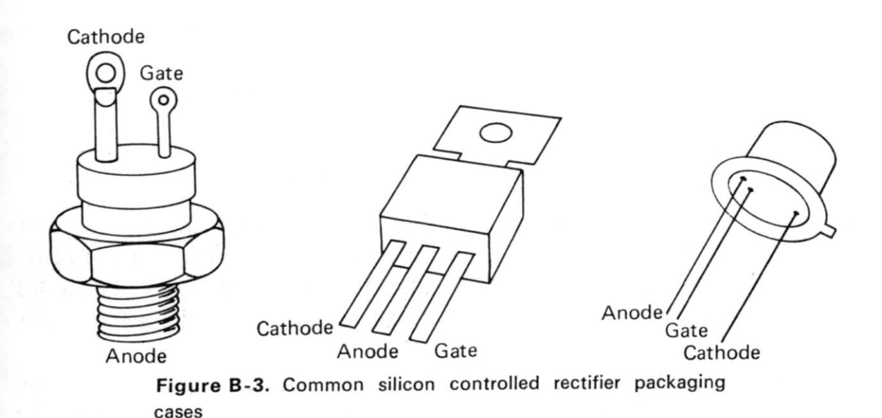

Figure B-3. Common silicon controlled rectifier packaging cases

Meter Calibration

The discussion in this appendix is only a brief introduction to the procedure of instrument calibration. However, the general procedures which are presented are quite useful as a means of increasing the accuracy of meters. For more details on instrument calibrations, see such references as E. Frank, *Electrical Measurement Analysis* (New York: McGraw-Hill Book Co., 1959).

In general, a meter is calibrated by comparing its readings to those of a highly accurate meter. Ammeters can be calibrated by using a test setup such as the one shown in Fig. C-1. In this setup, a source of constant current

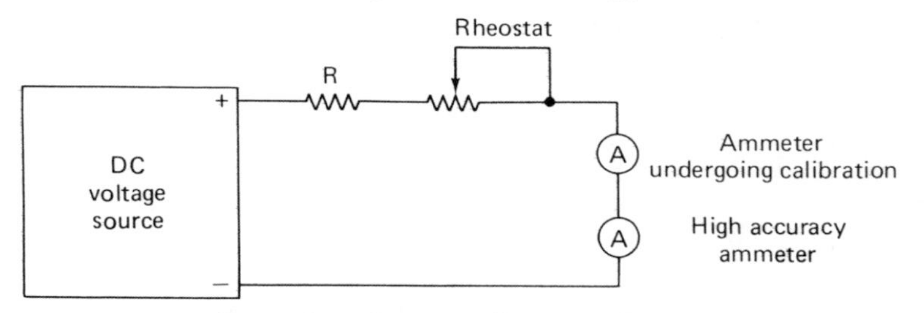

Figure C-1. Ammeter calibration test setup

should be provided by a precision power supply and rheostat. The value of the resistances, R, in the circuit should be at least 10 times the value of resistances of ammeters. The maximum current in the test circuit (V/R) should be about 10 percent greater than the full-scale deflection of the meter under test.

Voltmeters can be calibrated against other highly accurate voltmeters by using the test setup shown in Fig. C-2. In this calibration procedure, the

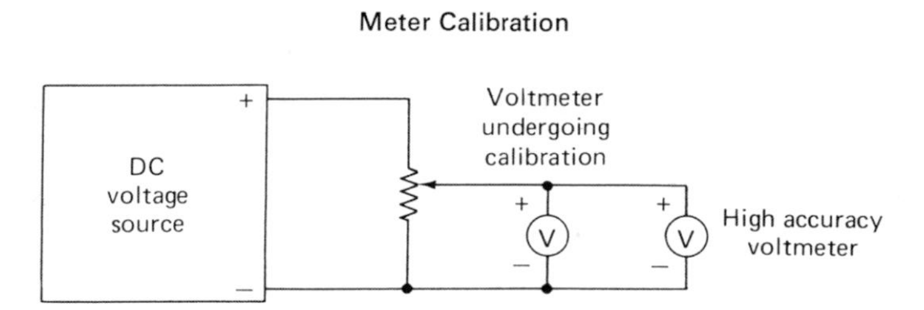

Figure C-2. Voltmeter calibration test setup

power supply voltage should be about 10 percent greater than the maximum full-scale voltage indicated by the voltmeter. The resistance of the variable resistance should be 10 percent or less than the combined resistance of the voltmeters.

A potentiometer can also be used to calibrate dc voltmeters and ammeters. The calibration accuracy derived from comparisons with the potentiometer can be very high. This is due to the fact that the potentiometer has an accuracy of 0.1 percent or better. The test setup for calibrating dc ammeters with a potentiometer is shown in Fig. C-3. The test setup for calibrating dc volt-

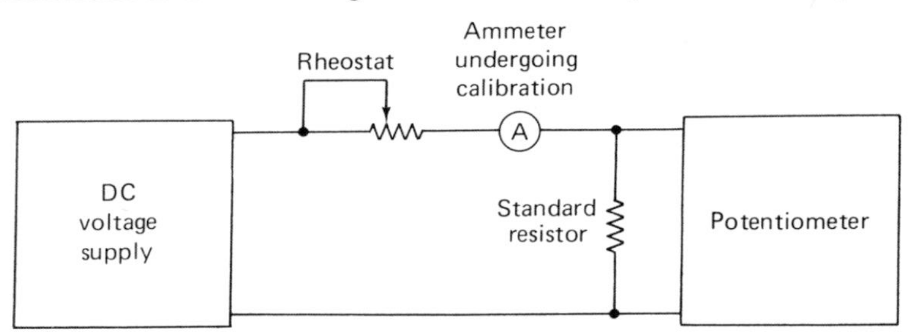

Figure C-3. Calibration of a dc ammeter using a potentiometer

meters with a potentiometer is shown in Fig. C-4. The voltbox in this figure is a device used to extend the measurement range of the potentiometer and has an accuracy on the order of 0.05 percent.

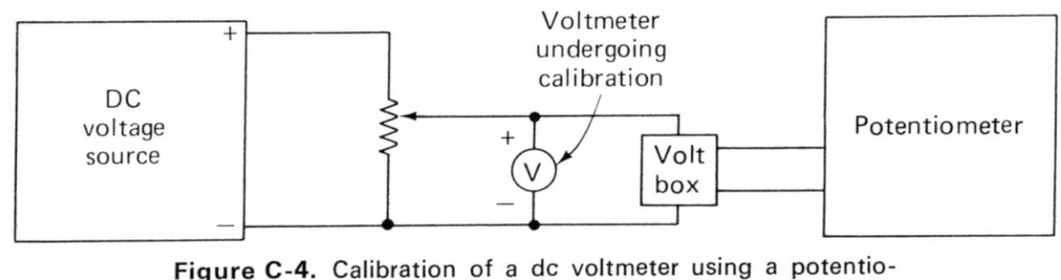

Figure C-4. Calibration of a dc voltmeter using a potentiometer

A calibration curve is the end result of the calibration procedure. This curve shows how much the reading of the meter must be corrected at each point along its scale so that it provides a correct reading at that point. (Fig. C-5).

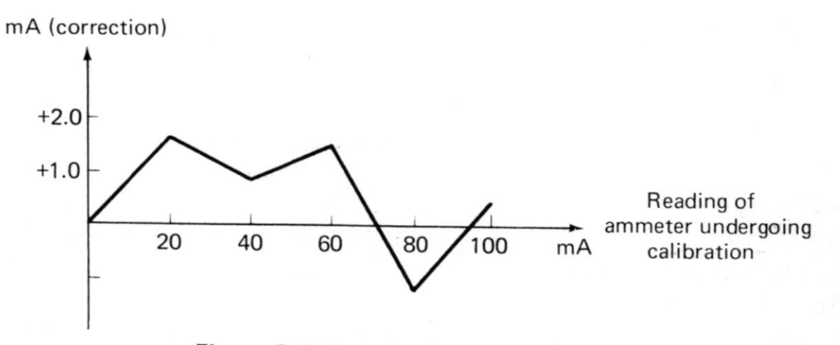

Figure C-5 Example of a calibration curve

Index